Turbulent Shear Flows 4

Selected Papers from the Fourth International
Symposium on Turbulent Shear Flows,
University of Karlsruhe, Karlsruhe, FRG,
September 12–14, 1983

Editors:

L. J. S. Bradbury F. Durst B. E. Launder
F. W. Schmidt J. H. Whitelaw

With 286 Figures

Springer-Verlag
Berlin Heidelberg New York Tokyo

Leslie J. S. Bradbury
Department of Mechanical Engineering, University of Surrey,
Guildford, Surrey GU2 5XH, England

Franz Durst
Lehrstuhl für Strömungsmechanik, Technische Fakultät
Friedrich-Alexander-Universität, Egerlandstraße 13,
D-8520 Erlangen, Fed. Rep. of Germany

Brian E. Launder
Department of Mechanical Engineering, University of Manchester,
Institute of Science and Technology, PO Box 88,
Manchester M60 1QD, England

Frank W. Schmidt
Mechanical Engineering Department, The Pennsylvania State University,
University Park, PA 16802, USA

James H. Whitelaw
Department of Mechanical Engineering, Imperial College of Science and
Technology, Exhibition Road, London SW7 2BX, England

ISBN-13:978-3-642-69998-6 e-ISBN-13:978-3-642-69996-2
DOI: 10.1007/978-3-642-69996-2

Library of Congress Cataloging in Publication Data. International Symposium on Turbulent Shear Flows (4th: 1983: University of Karlsruhe) Turbulent shear flows 4. Bibliography: p. Includes Index. 1. Shear flow–Congresses. 2. Turbulence–Congresses. I. Bradbury, L.J.S. (Leslie John Stanley), 1936–. II. Title. TA357.I59 1983 620.1'064 84-22143.

2153/3020-543210

Preface

The Fourth International Symposium on Turbulent Shear Flows took place at Karlsruhe University in Germany. The papers presented at this Symposium encompassed a similar range to that of the previous meetings, with greater emphasis placed on experimental work, and continued a trend towards the examination of complex flows. Once again, three-dimensional, recirculating and reacting flows featured strongly in the programme and were complemented by consideration of two-phase flows and discussions of both numerical and experimental techniques. The Symposium brought together some 300 participants from all over the world, and it was evident that there is a need for Turbulent Shear Flows Symposia, in order to obtain and communicate new information useful to researchers in the field of turbulent flows and of interest to engineers who design flow equipment.

This volume contains 27 papers selected from more than 100 presentations at the Symposium which have been reviewed and edited before publication. Together they provide an indication of the status of current knowledge on the subjects represented at the Symposium. They are grouped into four sections, namely:

- Fundamentals
- Free Flows
- Boundary Layers
- Reacting Flows

As in previous volumes in this series, each section begins with an introductory article considering the papers which follow in the broader context of available literature and current research. The editors believe that the volume communicates new material of interest to the scientific community and, at the same time, shows the advances which have taken place in recent years. Together with the other volumes on Turbulent Shear Flows Symposia, it provides a good indication of the present-day knowledge on turbulent flows.

It is noteworthy that, while new calculations of turbulent flow properties are being reported for increasingly complex flows, there is a growing awareness of the limitations imposed by numerical methods and the solution procedures of time-averaged equations. The recent Stanford Conference contributed to this awareness, for recirculating flows in particular, and current efforts to implement and evaluate new numerical assumptions and algorithms are growing, albeit in a diffuse way. Similarly, increased knowledge of the limitations of turbulence models associated with Reynolds-averaged equations appears to be leading to new interest in experiments and in the use of calculation methods more as a means to interpolate the measured results. In some cases, two-phase flows, in particular the ability to generate models, appears to have advanced beyond known physical information and further emphasis on improved understanding through carefully formulated experiments is required. This is reflected in the papers appearing in the present volume.

The Fourth Turbulent Shear Flows Symposium would not have been possible without the financial support of the research offices of the United States Army, Navy and Air Force and the National Science Foundation. The editors would like to thank these organizations for supporting the meeting at the University of Karlsruhe.

It is also the desire of the editors to thank the many individuals at the University of Karlsruhe who helped with both the organization and the running of the Symposium as well as the preparation of the social events. Thanks are also due to the many journals which helped to make the meeting successful by publishing announcements, calls for papers, etc. As with the earlier Symposia, the task of reviewing and selecting papers was carried out by a Papers Committee and an Advisory Committee whose members later acted as chairmen of technical sessions at the Symposium and who have also been a valuable source of helpful criticism. Their work was gratefully appreciated at all phases of the preparation. The committees consisted of:

- **Papers Committee:**

  L. J. S. Bradbury (University of Surrey, U.K.)
  F. Durst (University of Erlangen, F.R.G.)
  W. Leuckel (University of Karlsruhe, F.R.G.)
  H. McDonald (Scientific Research Assoc., Inc., USA)

- **Advisory Committee:**

R. J. Adrian	M. M. Gibson	P. A. Libby
J. C. André	V. W. Goldschmidt	J. L. Lumley
L. H. Back	K. Hanjalic	O. Martynenko
H. A. Becker	T. J. Hanratty	J. Mathieu
R. Borghi	J. R. Herring	W. C. Reynolds
S. Corrsin	M. Hino	W. Rodi
J. J. D. Domingos	M. Hirata	A. K. Runchal
R. Dumas	A. K. M. F. Hussain	K. Suzuki
H. Fiedler	P. N. Joubert	I. Wygnanski
I. Gartshore	E. Krause	J. Wyngaard

Finally, the editors would like to express their thanks to the authors of this book for meeting the various deadlines that were set. Thanks are also due to Springer Verlag for their help in producing this fourth volume of Post-Conference Proceedings of the Fourth Turbulent Shear Flow Symposium.

Erlangen, May 1984 The Editors

Contents

Part I

Fundamentals

Fundamental Aspects of Turbulent Shear Flows – Introductory Remarks

Jean-Claude André

Centre National de Recherches Météorologiques (DMN/EERM), Toulouse, France

The various papers in this section deal with different fundamental questions raised by the study of homogeneous or sheared turbulence. These questions concern firstly theoretical and experimental aspects of atmospheric (Schertzer and Lovejoy) and laboratory (Viets, Bethke and Bougine) flows, and in particular the nature of internal dynamics and intermittency. The other questions raised relate to modelling studies in the framework of either the probability density function (p.d.f.) (Anand and Pope), or two-point closure theories (Chollet), or, finally, one-point closure techniques (Janicka and Kollmann, Ettestad and Lumley, and Dekeyser and Launder). It is usually considered that modelling of the p.d.f. is most suitable for problems with highly non-quadratic mechanisms, like combustion, and that two-point closures can only be used for problems with oversimplified configurations allowing the assumption of homogeneity and isotropy. We shall see here that these two techniques are now beginning to be applied outside their initial restricted domain and are presently dealing with problems which a few years ago could only have been tackled by one-point closure models.

The first paper, "The dimension and intermittency of atmospheric dynamics" by Schertzer and Lovejoy, considers from a purely kinematical point of view the spectral properties of atmospheric turbulence over the whole range of energy-containing wave-numbers. Two main questions are raised:

(i): Is there any transitional regime in atmospheric motions between small-scale shear and/or buoyancy driven turbulence and large-scale turbulence where the effects of earth rotation are important? It is indeed known that rotation inhibits motions parallel to the rotation axis, so that large scales exhibit two-dimensional, or more properly quasi-two-dimensional, features (see e.g. André [1] and for more details Lesieur [2]). From a consideration of scale relations and probability distributions the authors argue that a dimensional transition between small-scale and large-scale turbulence is very uncertain. It should be noticed on the one hand that each set of data shown here is relative to at most 3 decades in vertical wave-number space, while the atmospheric spectrum extends over 9 decades in the horizontal (from approximately 1 cm to more than 10^4 km), so that the argument may not be definitive. On the other hand, there are indeed many instability mechanisms in the atmosphere (e.g. Charney [3]) which could account for a more-or-less evenly distributed energy injection over the scale range, and consequently for a rather continuous energy spectrum, thereby making it difficult to clearly display possibly different dynamical properties. It would, in this respect, be of particular importance to consider carefully the direct experimental measurements of Morel and Desbois [4], not referred to in this study, which support the idea of large-scale two-dimensional dynamics;

(ii): Does internal, or intrinsic, intermittency of atmospheric turbulence behave like a fractal of the type proposed by Mandelbrot [5]? The answer appears to be positive. The authors indeed show data which are fairly convincing and support the idea of an hyperbolic distribu-

In the following, we examine the possibility that the atmosphere is in fact scaling in the vertical as well as in the horizontal direction. To do so, we examine empirically the wind and temperature fields, attempting to capture two basic and conceptually distinct properties of these fields.

Scaling: The scaling provides the relationship between fluctuations of a field X ($\Delta X(\Delta z) = X(z_0 + \Delta z) - X(z_0)$, is the fluctuation between z_0 and $z_0 + \Delta z$), for large scales $\lambda \Delta z$ ($\lambda \gg 1$) to the small scales (Δz) by:

$$\Delta X(\lambda \Delta z) \stackrel{d}{=} \lambda^H \Delta X(\Delta z)$$

where H is the scaling parameter and "$\stackrel{d}{=}$" means equality in probability distributions ($X \stackrel{d}{=} Y$ means $\Pr(X > q) = \Pr(Y > q)$ for all q and Pr denotes probability). Note that the exponent of the corresponding power spectrum ($-\beta$) is related to H by $\beta = 2H + 1$, in the case of finite variance.

Hyperbolic Intermittency

Intermittency is directly connected with the probability law. One is particularly interested in the tail of this law, since it controls the relative frequency of extreme (intermittent) behaviour. For example, if the distribution has an algebraic fall-off at large fluctuations (ΔX), then the degree of intermittency can be characterised by the exponent α (the hyperbolicity) defined as follows:

$$\Pr(\Delta X' > \Delta X) \propto \Delta X^{-\alpha}$$

Random variables are distinguished from non-random ones by primes or by the use of upper case symbols for random, and lower case for non-random variables. The above equation therefore reads "the probability of a random fluctuation $\Delta X'$ exceeding a fixed ΔX is proportional to $\Delta X^{-\alpha}$".

Behaviour of this sort was first predicted by Mandelbrot (1974 a), who also showed that it was related to the problem of the dimension of the support of turbulence D_s, a question, we pursue in Section 8 in the context of anisotropy. Hyperbolic distributions have been invoked in other fields of physics (the "Holtsmark distribution", Feller (1971), or see Mandelbrot (1982) for other examples). Note that all moments of order α or higher diverge, a fact that has important consequences[2]. Levy (1937) and Feller (1971) are standard texts, in the case $\alpha < 2$, (the Levy-stable laws which form a convolution semi-group). We term this kind of intermittency "hyperbolic intermittency". In Section 9 we will discuss the connection with previous experimental results of this kind obtained for the rain field (Lovejoy 1981).

We shall primarily be interested in the vertical fluctuations of the horizontal velocity field (Δv) and in the buoyancy force per unit mass acting across a layer of thickness dz: $df = g\,d \ln \theta$, where θ is the potential temperature, and g the acceleration of gravity. These quantities are related to two fundamental frequencies: that of the vertical shear (s) and the Brunt-Vaisala frequency (n):

$$s = dv/dz \quad \text{and} \quad n^2 = g\,d \ln \theta/dz = df/dz.$$

The ratio of their squares defines the dimensionless Richardson number:

$$\mathrm{Ri} = n^2/s^2.$$

2 E.g. the experimental estimation increases without limit with the sample size (cf. Schertzer and Lovejoy 1983 b)

The shear frequency characterises the dynamical processes, and the Brunt-Vaisala frequency, the stability (and gravity waves). The dominant process has the highest frequency. Note that the above are defined locally (in time and space) and hence are often called "turbulent" quantities. Unfortunately, one is used to considering them in averaged forms.

To determine their scaling regime, Fourier analysis could be used. Here, we analyse directly the scaling of the probability law by measuring quantities across atmospheric layers of thickness Δz. This has the advantage that it enables the scaling parameters (H's) and the hyperbolicities (α's) to be obtained simultaneously. We therefore define:

$$s^2(\Delta z) = \Delta v^2(\Delta z)/\Delta z^2; \quad n^2(\Delta z) = g\Delta \ln|\theta|/\Delta z$$
$$\mathrm{Ri}(\Delta z) = n^2(\Delta z)/s^2(\Delta z).$$

4. Data Analysis

4.1 Radiosonde Analysis

s, n, Ri were evaluated from the high resolution radiosonde data obtained in the 1975 experiment in Landes, France. The quantities θ, $\underline{v}$ and the humidity were obtained every second ($\sim$ 3 m in the vertical) and processed to yield low noise data every 5 s ($\sim$ 15–20 m) (see Tardieu (1979) for more details). The velocity values were somewhat noisier than the θ values and were therefore smoothed, yielding low noise estimates every 50 m. All estimates of n, s, Ri were thus made over layers at least 50 m thick, and from the ground up to a somewhat arbitrary height of 6 km. Since the number of points in each layer varied somewhat (averaging $\sim$ 3 for each 50 m layer), mean θ, $\underline{v}$ were computed and $\Delta \ln \theta$, $\Delta \underline{v}^2$ were then determined. These are required to calculate s and n. The data examined are from 80 consecutive soundings taken at 3-hour intervals at Landes.

From the log-log plots shown in Figs. 1, 2, 3 it can easily be verified that the probability distributions of Δv, Δf and Ri exhibit both scaling and hyperbolic behaviour. Perhaps the easiest way to see this is to recall that for hyperbolic distributions:

$$\Pr(\Delta X' > \Delta X) = (\Delta X/\Delta X^*)^{-\alpha} \quad \text{for large } \Delta X$$

ΔX^* is the "width" of the distribution, or the amplitude of the fluctuations. Scaling implies then that the width grows with the separation as:

$$\Delta X^*(\Delta z) = (\Delta z/\Delta z_0)^H \, \Delta X^*(\Delta z_0) \quad \Delta z_0 \text{ is a constant.}$$

This is shown on the graph by the constant shift $H \log 2$ for each doubling of the separation Δz. The value $(-\alpha)$ is the slope of the straight line asymptote.

We obtain:

$$H_v = 3/5 \qquad \alpha_v = 5$$
$$H_{\ln\theta} = 9/10 \qquad \alpha_{\ln\theta} = 10/3$$
$$H_{\mathrm{Ri}} = 1 \qquad \alpha_{\mathrm{Ri}} = 1.$$

The H's and α's are given rational expressions, since, as explained below, they can often be deduced by dimensional considerations.

In a humid atmosphere, the buoyancy is determined by the wet bulb potential temperature (yielding n_w). We also plotted the distribution of n_w and found it to be virtually

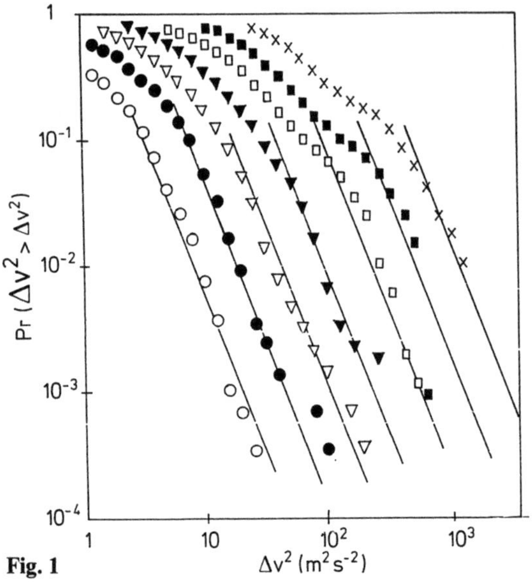

Fig. 1

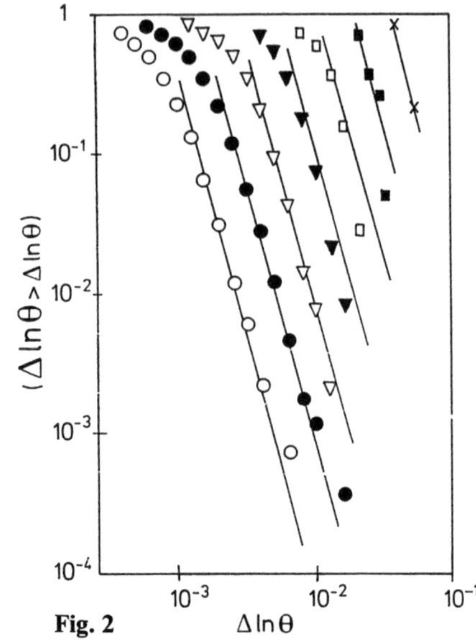

Fig. 2

Fig. 1. The probability distribution of fluctuations in the quantity $\Delta v^2(\Delta z)$ where $\underline{v}$ is the horizontal velocity for different layers as follows:

○: $\Delta z =$ 50 m, ●: $\Delta z =$ 100 m,
▽: $\Delta z =$ 200 m, ▼: $\Delta z =$ 400 m,
□: $\Delta z =$ 800 m, ■: $\Delta z =$ 1600 m,
×: $\Delta z =$ 3200 m.

The straight lines follow the equations in the text

Fig. 2. The probability distribution of fluctuations in the quantity $\Delta \ln \theta$, θ being the potential temperature. The symbols are the same as in Fig. 1. The straight lines follow the equation given in the text

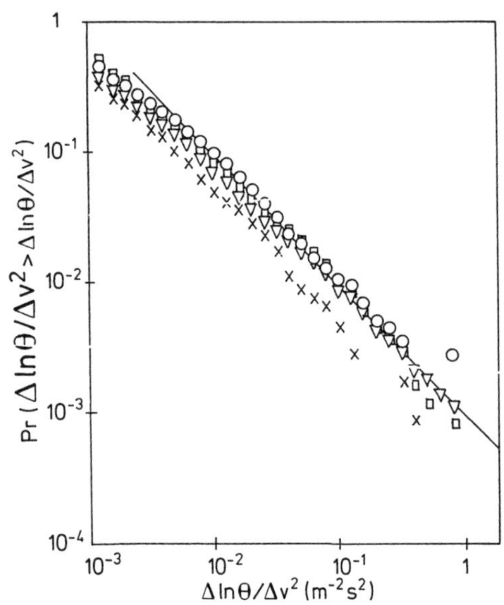

Fig. 3. The probability distribution of fluctuations in the quantity $\Delta \ln \theta/\Delta \underline{v}^2 = \mathrm{Ri} \cdot \Delta z^{-1}$, Ri being the Richardson number. The symbols are the same as in the previous figures. The curves for 100, 400, 1600 m have been supressed for clarity. (The 50 m curve probably falls below the others because of the effect of noise in the Δv^2 measurement in this value). The straight lines follow the equations given in the text.

indistiguishable from that of n. This is probably because Landes was quite dry – the relative humidity rarely exceeded 50%.

These exponents are quite robust i.e. the difference between the individual sounding is the width of the distributions, not the exponents, this is supported by the comparison of four subsamples of 20 soundings each. Although the most unstable samples had a width of the Ri distribution twice as small as the most stable sample, the exponents were unaffected (this was also true of s and n). These results are fully consistent with the pooled results shown in Figs. 1, 2, 3 if it is remembered that pooling data from hyperbolic distributions with similar exponents only changes the width of the distribution. Furthermore, comparisons of data from above and below 2 km show that the hyperbolic form is not due to the pooling of data from different altitudes.

4.2 Merceret's (1976) Aircraft Data

Another quantity of interest is the rate of turbulent energy transfer ($\bar{\varepsilon}$). This is the fundamental dynamical quantity in the turbulent cascade of energy from large to small scales. In this case, we obtained only a probability distribution by replotting Merceret's aircraft data. The result, which is shown in Fig. 4 can be represented by the following formula:

$$\Pr(\bar{\varepsilon}' > \bar{\varepsilon}) \sim \bar{\varepsilon}^{-\alpha_\varepsilon}; \quad \alpha_\varepsilon \sim 5/3.$$

Merceret obtained his value of $\bar{\varepsilon}$ by calculating the spectrum of horizontal wind fluctuations every second ($\sim$ 100 m – actually the 1 s spectrum is the average of literally thousands of spectra taken on millisecond time scales), from a research aircraft. Note that this 5/3 exponent is conceptually quite different from the famous Kolmogorov 5/3 exponent. Mandelbrot (1974a) points out that the value of α_ε depends on the dimension of the region over which it is averaged. We assume this average to have been taken over a straight line of dimension 1. In the following, we use the bar notation to indicate a spatial average and angular brackets $\langle . \rangle$ for ensemble averages.

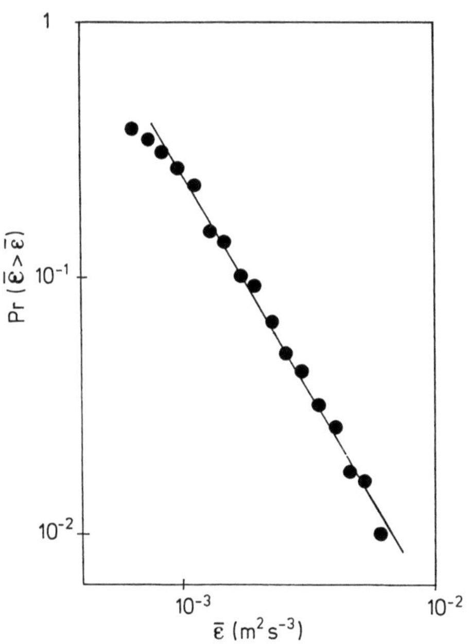

Fig. 4. The probability distribution of $\bar{\varepsilon}$ as replotted from Merceret (1976) Fig. 14. (These values are averages over 100 m in the horizontal, which are taken to be made over a line, thus $D_A = 1$. Linear regression yielded a correlation coefficient of 0.998, and a slope of -1.691, which implies $\alpha = 1.691 \sim 5/3$)

The spatially averaged flux $\bar{\varepsilon}_{l_n}$, which is the flux of eddies of typical scales l_n, averaged over a subspace A of dimension D_A (e.g. a line or a plane), can be defined as follows:

$$\bar{\varepsilon}_{l_n} = \sum_{(N_A(l_n))^{-1}} \varepsilon_{l_{n-1}}$$

where $N_A(l_n)$ is the number of eddies (empty or not) which intersect A:

$$N_A(l_n) = (l_n/l_0)^{-D_A} = \lambda^{D_A}.$$

A straightforward calculation, (Mandelbrot 1974a), shows that to the leading term:

$$\begin{aligned}\langle \bar{\varepsilon}_{l_n}^h \rangle &\sim \sum_{N_A(l_n)} \langle W^h \rangle \langle \bar{\varepsilon}_{l_{n-1}}^h \rangle / N_A(l_n)^h \\ &\sim N_A(l_n)^{1-h} \langle W^h \rangle \langle \bar{\varepsilon}_{l_{n-1}}^h \rangle.\end{aligned}$$

Thus, the of divergence of the hth moment of $\bar{\varepsilon}$ occurs when:

$$\langle W^h \rangle > \lambda^{(h-1)D_A}$$

because, if this inequality holds, then the moments increase at each level of the cascade process, and therefore diverge in the limit $n \to \infty$. The two functions appearing in this inequality are convex functions of h. This is also true of their logarithms (a stronger property), so that the equality has at most one root other than the trivial $h = 1$ (see Fig. 10).

Note that at least locally, around $h = 1$, the behaviour of $\langle W^h \rangle$ may be written as: $\lambda^{(h-1)(D-D_s)}$ which defines a new dimension D_s, which as shown below, is the dimension of the support of turbulence. D is the dimension of the space in which the cascade occurs, which Mandelbrot took to be 3. In the case of anisotropic cascades, D should be replaced by D_{el}.

This leads to:

$$(D - D_s) = d(\log_\lambda \langle W^h \rangle)/dh\,|_{h=1}$$

or

$$(D - D_s) = \langle W \log_\lambda W \rangle > \langle W \rangle \log_\lambda \langle W \rangle = 0.$$

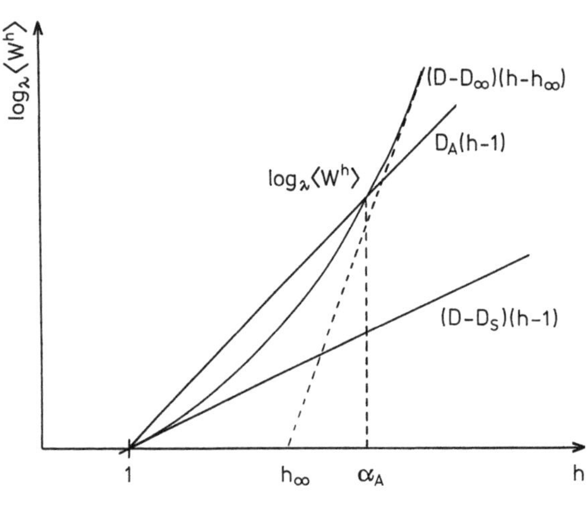

Fig. 10. An illustration of the Mandelbrot model of turbulent intermittency showing the functions discussed in the text

If $D_A > D - D_s$ then we have a non-divergent behaviour of $\bar{\varepsilon}^h$ for $h \to 1$. We therefore must average over a space of at least dimension $D_A > D - D_s$ in order for $\langle\bar{\varepsilon}\rangle$ to be finite. This gives a simple interpretation of D_s as the dimension of the support of turbulence.

8.3 The Simplest Case of the Mandelbrot Model

The simplest case is obtained when the local behaviour of $\langle W^h\rangle$ near $h = 1$, holds for all h (the slope in Fig. 10 is constant). In this case, we have:

$$\langle W^h\rangle = \lambda^{(h-1)(D-D_s)}$$

which corresponds to the Bernouilli process: $\Pr(W = 0) = 1 - \lambda^{-(D-D_s)}$ and $\Pr(W = \lambda^{D-D_s}) = \lambda^{-(D-D_s)}$. Clearly, the only parameter is D_s which uniquely determines the number of "active" eddies at each generation: $N_a(l_n) = \lambda^{D_s}$ all other eddies are considered "dead". This model gives convergence of all the moments of $\bar{\varepsilon}$ since $\langle W^h\rangle < \lambda^{(h-1)D_A}$ for all h if $D_A > D - D_s$. This model has been studied in detail in Frisch et al. (1978), who refer to it as the "β model".

8.4 The Next Simplest Case and the Divergence of Moments of $\bar{\varepsilon}$

It suffices to make a slight change in the above so as to yield divergent moments of $\bar{\varepsilon}$. Indeed, if rather than classifying eddies as "active" or "dead", we classify them "very active" and "weak", then, in general, as we show below, high order moments of $\bar{\varepsilon}$ no longer converge. Consider the following slightly modified Bernouilli model defined by the dimension D_∞ of the "very active regions":

$$\Pr(W = \lambda^{(D-D_\infty)}) = \lambda^{-(D-D_\infty)/h_\infty}, \qquad h_\infty > 1$$
$$\Pr(W = a_1) = 1 - \lambda^{-(D-D_\infty)h_\infty}$$

(note this reduces to the preceding case for $h_\infty = 1$).

Due to the linear asymptotic behaviour of $\log_\lambda\langle W^h\rangle$ for large h, higher moments of $\bar{\varepsilon}$, averaged over A, diverge when:

$$D - D_s < D_A < (D - D_\infty).$$

In concluding of this section, we point out that, in the general case, as h is increased the dimension of the support of ε^h, $D_s(h)$, decreases. In the case of the "α model" studied above, these dimensions were bounded below by D_∞ which is the dimension of the more extreme active regions ($D_\infty = D_s(\infty)$). The fact that intermittency is characterised by many fractal dimensions has been already discussed in Schertzer and Lovejoy (1983a)[3]. Recently, Parisi and Frisch (1984) have developed the idea of many fractal dimensions in the context of a somewhat different model that they call "multifractals".

9. The Phenomenology of Scaling, Hyperbolically Intermittent Fields

9.1 Introduction

If the scaling and intermittency described here hold under quite general conditions, then they must be compatible with the rich diversity of meteorological fields. Below we examine two

3 The exact formula is: $D_s(h) = D - \log_\lambda\langle W^h\rangle/(h-1)$

aspects of phenomenology and explain why we believe they can be better understood by using the scaling and intermittency described here.

9.2 Algebraic Formulae Relating Various Fields

One goal of phenomenology is to relate different meteorological fields. If scaling laws hold, and intermittency is expressed via hyperbolic distributions, then we may expect the form of the probability distributions (in particular the hyperbolic exponents) of different fields to be related by fairly simple algebraic equations. For example, if we define an "eddy" as a region surrounded by lines of constant horizontal velocity, then for a given ε and ϕ, the horizontal and vertical extents $(\Delta x, \Delta z)$ of the eddy may be obtained by equating the Δv's in Eqs. (1), (2):

$$\bar{\varepsilon}\Delta x \overset{d}{=} \bar{\phi}^{3/5}\, \Delta z^{9/5}$$

hence $\alpha_\varepsilon = 5\alpha_\phi/3$ as required by dimensional analysis (from Fig. 4 $\alpha_\varepsilon \sim 5/3$, $\alpha_\phi \sim 1$). Note that although $\bar{\varepsilon}$ and $\bar{\phi}$ are related in this way, they may or may not be dependent on each other. The sphero-scale, which can be defined by $\Delta x = \Delta z$ will therefore be a random variable unless $\bar{\phi}$ and $\bar{\varepsilon}$ are completely dependent.

One area where the investigation of algebraic relations is perhaps particularly urgent is in the case of the rainfield. This is a case where numerical modeling is notoriously difficult, but in which simple algebraic relations may exist. In particular, Lovejoy (1981) showed that $\alpha_R \sim 5/3$ in diverse areas of the world. (Note that this result applies to changes in the flux of rain (R) from isolated storms which is a 2-D lagrangian statistic. It is therefore unlikely to be trivially connected to the ε field whose value $\alpha_\varepsilon \sim 5/3$ is a 1-D eulerian statistic.)

9.3 Shapes, Structure and "Animals"

Another goal of phenomenology is the classification and understanding of the shapes and structures of individual meteorological fields. In fluid mechanics, the shapes and structures are so diverse that popular jargon refers to the different morphologies as fluid "animals", with the phenomenologist playing the role of a zoologist. In percolation theory the shapes encountered are so bizarre that the expression "lattice animals" has gained official acceptance. Below, we argue that the use of scaling and hyperbolic distributions may be important in taming the different meteorological animals including "fronts", "cells" and others. In this case, the animals are clearly fractals because of the scaling.

In appreciation of the significance of hyperbolic distributions (particularly for $\alpha < 2$), Mandelbrot and Wallis (1968) pointed out that they would have the effect of causing the largest of a set of such fluctuations to dominate over the others. In reference to the extreme meteorological fluctuation responsible for The Flood, they coined the expression "Noah effect" to describe this phenomenon. It is ironic that it has only recently been applied to actual meteorology (Lovejoy 1981).

Figure 11 shows a "hyperbolic fractal animal" obtained by Monte Carlo simulations in an anisotropic space. In contrast to the example of stochastic stratification with gaussian fluctuations (Fig. 6), here, the elementary fluctuations are hyperbolically distributed, hence the field is dominated by one main "animal" although other smaller ones are visible. In the extreme gaussian case, the morphology is entirely determined by the scaling because no matter what elementary fluctuations are used, they all have approximately the same intensity and thus their details tend to cancel.

In the case of hyperbolic fluctuations, on the contrary, the largest dominates and imposes its geometry, although it does so in an indirect fashion. Here, specifying H, α is insufficient

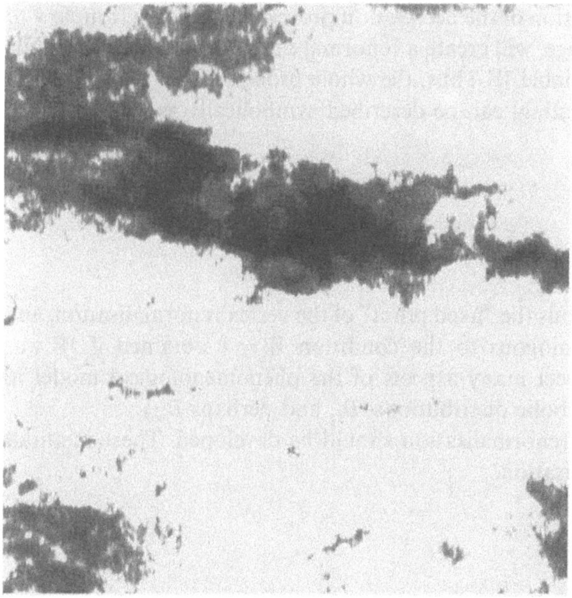

Fig. 11. An example of a hyperbolic fractal animal on an anisotropic space, with $D_{el} = 1.80$. The log intensities are indicated by the intensity of the grey. Note that in distinction to the gaussian case (Fig. 6b), one large "animal" dominates image, although others are visible, a manifestation of the "Noah" effect. (This model is on an 800×800 point grid, and the sphero-scale has the value 30 pixels. The fine structure is therefore oriented perpendicularly to the overall shape)

to determine the morphology – changes in the underlying metric (Section 6), or of the geometry of the elementary fluctuations strongly influence the observed shapes and structures.

Figure 11 is an anisotropic version of the fractal rain model discussed in (Lovejoy and Schertzer (1984); Lovejoy and Mandelbrot (1984)) which uses annuli (with intensities distributed with $\alpha \sim 5/3$) as a basic shapes. There it is argued that the existence of hyperbolic distributions is directly related to a hierarchy of straight-line structures, and hence to rain "fronts".

9.4 Hyperbolic Renormalisation

The phenomenological model of intermittency discussed in Section 8.2 may be placed in a dynamical context since it results from the fact that the characteristic time τ of transfer of energy from one scale l_n is:

$$\tau(l_n)^{-1} = W \Delta v(l_n)/l_n.$$

Hence, due to the presence of the random variable W, τ is far from having a characteristic value as in usual phenomenological models ($W = 1$). This can be understood as a phenomenological "renormalisation" of the vertex for the following reasons.

Renormalisation of the Navier-Stokes-Equations – as described for instance in Forster et al. (1978) – consists of systematically taking into account the direct interactions in a "decimation" process from high to low wave numbers. The basic problem encountered is the calculation of the effect of the indirect interactions of the low wavenumbers via the eliminated ones, which modify the interaction of the conserved wavenumbers. Renormalisation has therefore only been applied successfully in the case of low wavenumbers, where indirect interactions do not need to be taken into account (due to Gallilean invariance De Dominicis and Martin 1979). The indirect interactions seem however to be extremely important in the inertial range, as evidenced for example in the failure of the Direct Interaction Approximation in the inertial range (Kraichnan 1958). Until now, this failure has only been curred by *ad hoc* procedures (e.g. Herring et al. 1983).

We propose that repeated application of the decimation process, proceeding from low to high wave numbers in the inertial range, will create a renormalisation of the intensity of the vertex converging to the random variable W. Thus, the whole process of the application of the renormalisation group (with rescaling) can be described symbolically as:

$$l \to \lambda l$$
$$v \to \lambda^{H_v}$$
$$P \to W\lambda^{-1} P \quad \text{with } \langle W \rangle = 1$$

where P is the vertex intensity. W is thus the "fixed point" of the vertex renormalisation, and the flux conserving condition is analogous to the condition $W = 1$ obtained if W was deterministic. We may therefore expect many aspects of the phenomenological model to follow from the dynamics (e.g. hyperbolic distributions, D_s, and perhaps D_∞).

In our view, specific methods of renormalisation should be developed. These methods could be called hyperbolic renormalisation.

10. Conclusions

10.1 Summary of Principle Results

The object of this paper has been to elucidate the relationship between small and large scale structures in the atmosphere. The vertical shear and buoyancy force play a key role in the genesis of inhomogeneities, and were therefore investigated in detail using high resolution radiosonde and aircraft data. Two aspects of the data were investigated: the scaling exponents (H), and the hyperbolic exponents which characterise the intermittency (α). These parameters were analysed over atmospheric layers between 50 m and 3.2 km thick.

The principal results are as follows:

a) $H_v = 3/5$: This empirical value was also deduced by a dimensional argument similar to that first proposed by Obukhov (1959) and Bogliano (1959). Our derivation is analogous to the dimensional argument (based on ε) used to establish the horizontal (Kolmogorov) scaling ($H_h = 1/3$). It supposes that the fundamental dimensional quantity is ϕ, the flux of buoyancy force variance. Our result agrees with Adelfang (1971).

b) $\alpha_v = 5$: This empirical value was corroborated by a re-analysis of Merceret's (1976) measurements of $\bar{\varepsilon}$ from which we obtained $\alpha_\varepsilon = 5/3$. Simple physical arguments confirm $\alpha_v = 3\alpha_\varepsilon$. An immediate consequence is the divergence of the fifth and higher order moments of the velocity field, and the second and higher moments of $\bar{\varepsilon}$. The divergence of high moments of $\bar{\varepsilon}$ was first proposed by Mandelbrot (1974a).

c) $\alpha_{\mathrm{Ri}} = 1$: This shows that the Richardson number has such extreme fluctuations that even the mean may not exist. Monte Carlo methods showed that this result could be explained as the result of dividing two weakly correlated random variables. This points to the need to re-examine the validity of theories based on limited expansions of the equations in Ri.

d) $H_{\ln\theta} = 9/10$, $\alpha_{\ln\theta} = 10/3$: The potential temperature was also found to be scaling in the vertical and to be extremely intermittent-fourth and higher moments diverge.

e) $D_{\mathrm{el}} = 23/9$: To account for the different horizontal and vertical scaling, in the wind field, we introduced a special kind of fractal dimension (D_{el}) called an elliptical dimension because it expresses the fact that small eddies with circular cross-sections in the vertical appear as flattened ellipses at larger scales. Our definition of D_{el} leads to the equation

$D_{el} = 2 + H_h/H_v = 23/9 \sim 2.56$ (a convenient mnemonic to remember this result is "two plus Kolmogorov over Obukhov").

f) *The sphero-scale*: The velocity scales differently in the horizontal and vertical directions, and therefore there will be a distance at which the magnitude of the fluctuations will be equal. We call this distance the sphero-scale because at this scale, eddies are isotropic, and therefore, averaged eddies are spherical.

The sphero-scale is quite unlike an ordinary length scale – for example, it may vary in a very erratic manner.

g) *Stochastic Stratification*: A consequence of our anisotropic scaling space is that any structure, no matter how vertically oriented at small scales, will appear quite flat (i.e. stratified) at large scales. Several numerical examples were given. This concept may be sufficient to explain the observed large-scale stratification of meteorological fields, including the pressure field.

h) *hyperbolic intermittency*: The outlines of Mandelbrot's (1974a) model of fractally homogeneous turbulence were given, including a discussion of the relation between intermittency, the dimension of the support of turbulence (D_s), and D_{el}, and the various observed hyperbolic distributions. We conclude: atmospheric turbulence is fractally homogeneous and fractally anisotropic[4].

10.2 A New Direction for Meteorology?

The hypothesis that atmospheric motions are qualitatively quite different at small and large scales can be traced back to at least the 19th century. At first, this idea could only have had an *ad hoc*, phenomenological basis, since the equations of fluid mechanics and thermodynamics contain no length scales, and our knowledge of motions on scales less than hundreds of kilometers was inadequate. In recent years, theorists such as Charney (1971) have attempted to use turbulence theory to give this distinction a solid basis by arguing that large scale motions are "quasi-two dimensional", and are separated by a dimensional transition from small scale "quasi-three dimensional" motions. Experimentalists, from Pinus (1968) onwards, using high resolution satellite, aircraft, radiosonde and other data have searched for empirical evidence of a transition, apparently, without success in the range 10 cm to 4000 km.

We believe that the hypothesis of a dimensional transition is no longer tenable either theoretically, or empirically. The observed structure of the atmosphere can be explained by a simpler hypothesis: that it is anisotropic and scaling throughout, a fact that can be characterised by the intermediate "elliptical" dimension $D_{el} = 23/9 = 2.56$. If this hypothesis is true, then the atmospheric motions are never "flat" (two-dimensional), nor isotropic (three-dimensional), but always display aspects of both, according to a well defined fractal geometry.

In spite of the existence of an entire programme of theoretical and empirical research, the hypothesis of a dimensional transition is on more tenous ground than ever. The alternative, anisotropic scaling hypothesis proposed here will undoubtedly also require a large research program to substantiate, and cannot be said to be definitively established. For instance, the effects of the Coriolis force have not been considered. It is quite possible that this could be

4 In a celebrated poem, Richardson summarised atmospheric dynamics: "Big whorls have little whorls that feed on their velocity, and little whorls have smaller whorls and so on to viscosity – in the molecular sense." Anisotropic scaling adds a second stanza: "Flatter whorls have rounder whorls that feed on instability, and roundish whorls have rounder whorls, and so on to spherocity – in the statistical sense."

accounted for by a non-diagonal $\underset{\sim}{T}$, but *a priori*, there is no reason to suppose this to break the scaling. It is therefore possible that baroclinic instabilities will also be scaling.

This alternative, does offer several advantages, not the least of which is simplicity. If it is correct, then, small scale structures differ from large scale ones only in a continuous transformation in which they are progressively "flattened" or stratified at larger and larger scales. This transformation depends mainly on the action of two groups, the scaling group $\underset{\sim}{T}$, and the convolution group of stable Levy distribution (which characterise both the scaling and the intermittency respectively). Thus, atmospheric motions can be seen to be anisotropic, but scaling and intermittent at all scales.

One immediate practical consequence of this work is its impact on what we call the "stochastic coherence" of numerical weather forecasting procedures (see Schertzer et al. (1983)). This means that we verify that the scaling of the meteorological fields are not artificially modified through the different stages of the forecast procedure e.g. during the data assimilation, initialisation and numerical weather-prediction integrations model. This may prove an invaluable and general way of testing the validity of numerical forecasts because the capacity of modern computers is sufficient to allow scaling regimes to be obtained.

10.3 Theoretical Implications

The results of this paper may be expected to have an impact on at least two different areas of theoretical work.

The first is turbulence in anisotropic media, which should be approached in the fractally anisotropic framework proposed here (i.e. without separating the mean and fluctuating fields). Indeed, this framework may be essential in overcoming difficulties where so far only formal manipulations of renormalisation schemes have suceeded (e.g. Kraichnan (1964 a, b), Carnevale and Martin (1982)). The second way in which this work may have significant theoretical implications is in resolving difficulties of isotropic turbulence. In particular, this paper shows the need to develop methods appropriate for renormalising situations far from the usual one of "quasi-gaussian" distributions – methods that should perhaps be called "hyperbolic renormalisation".

Acknowledgements. We thank J. P. Gourtobe, and J. J. Tardieu for helpful discussions and comments on the data analysis. We are indebted to J. Peyriere and J. P. Kahane who helped sharpen the notion of elliptical dimension and sphero-scale. Special thanks are given to B. Mandelbrot who read and commented upon earlier drafts of this paper, including summaries which appeared in the preprint volume of the 4th Conference on Atmospheric and Oceanic Waves and Stability and the 6th Symposium on Turbulence and Diffusion (March 22–25 1983, AMS Boston). We thank G. Austin, K. Gage, J. R. Herring, D. K. Lilly, G. Therry, T. E. Van Zandt, T. Warn for helpful discussions and comments. By their interest, the following have stimulated this work: P. Bougeault, D. Cadet, M. Farge, M. Ghil, E. Mollo-Christensen, R. Peshansky, M. Rafy, D. Rousseau, G. Toulouse. Lovejoy would like to acknowledge the National Sciences and Engineering Council (Canada) for financial support. Schertzer acknowledges the Aspen Centre for Physics for hospitality during July 1982. This work was partially supported by the Centre National de Recherches Scientifiques (ATP RA (83) 28).

References

Adelfang, S. I. (1971): On the relation between wind shears over various intervals. J. Atmos. Sci. **10**, 138

Anselmet, F., Gagne, Y., Hopfinger, E. J., Antonia, R. A. (1984): High order velocity structure functions in turbulent shear flows. J. Fluid Mech. **140,** 63

Basdevant, C., Sadourny, R. (1975): Ergodic properties of inviscid truncated models of two-dimensional incompressible flows. J. Fluid Mech. **69**, 673

Batchelor, G. I., Townsend, A. A. (1949): The nature of turbulent motion at large wave numbers. Proc. Roy. Soc. A **199**, 238
Batchelor, G. K.: *The Theory of Homogeneous Turbulence* (University Press, Cambridge 1953) p. 198
Batchelor, G. K. (1969): Computation of energy spectra in homogeneous two-dimensional turbulence. Phys. Fluids **12**, 233
Boer, G. J., Shepherd, T. G. (1983): Large scale two-dimensional turbulence in the atmosphere. J. Atmos. Sci. **40**, 164
Bogliano, R. (1959): Turbulent spectra in a stably stratified atmosphere. J. Geophys. Res. **64**, 2226
Brachet, M. E., Meiron, D. I., Orsag, S. A., Nickel, B. G., Morf, R. H., Frisch, U. (1983): Small-scale structure of the Taylor-Green vortex. J. Fluid Mech. **130,** 411
Brown, P. S., Robinson, G. D. (1979): The variance spectrum of tropospheric winds over Eastern Europe. J. Atmos. Sci. **36**, 270
Carnevale, G. F., Frisch, U., Salmon, R. (1981): *H*-theorems in fluid dynamics. J. Phys. Math. Gen. A **14**, 1701
Carnevale, G. F., Martin, P. C. (1982): Field theoretical techniques in statistical mechanics: with application to non-linear wave dynamics. Geophys. Astrophys. Fluid Dynamics **20**, 131
Charney, J. (1971): Geostophic turbulence. J. Atmos. Sci. **28**, 6
Chollet, J. P., Lesieur, M. (1981): Parameterisation of small scales of three-dimensional isotropic turbulence utilising spectral closures. J. Atmos. Sci. **38**, 2747
Chorin, A. J. (1981): Estimates of intermittency, spectra, and blow-up in developed turbulence. Commun. Pure Appl. Math. **34**, 853
Curry, J. H., Herring, J. R., Loncqric, J., Orsag, S. A. (1982): Order and disorder in two- and three-dimensional Bernard convection. NCAR Preprint
De Dominicis, C., Martin, P. C. (1979): Energy spectra of certain randomly stirred fluids. Phys. Rev. A **19**, 419
Endlich, R. M., Singleton, R. C., Kaufman, J. W. (1969): Spectral analysis of detailed vertical wind speed profiles. J. Atmos. Sci. **26**, 1030
Feller, W.: *An Introduction to Probability Theory and Its Applications*, Vol. II (Wiley, New York 1971) p. 567
Fjortoft, R. (1953): On the changes in the spectral distribution of kinetic energy for two-dimensional, non-divergent flow. Tellus **5**, 225
Frisch, U., Sulen, P. L., Nelkin, M. (1978): A simple dynamical model of intermittent fully-developed turbulence. J. Fluid Mech. **87**, 719
Fox, D. G., Orszag, S. A. (1973): Inviscid dynamics of two-dimensional turbulence. Phys. Fluids **16**, 169
Gage, K. S. (1979): Evidence for a $k^{-5/3}$ law inertial range in meso-scale two-dimensional turbulence. J. Atmos. Sci. **36**, 1950
Gilet, M., Nicoloff, S., Klaus, V.: *Dual Doppler Radar Measurements of Winds and Turbulence a Head of a Cold Front*, in 19th Conf. on Radar Met. (AMS, Boston 1980) p. 30
Herring, J. R. (1977): On the statistical theory of two-dimensional topographic turbulence. J. Atmos. Sci. **34**, 1731
Herring, J. R. (1980): On the statistical theory of quasi-geostrophic turbulence. J. Atmos. Sci. **37**, 969
Herring, J. R., Schertzer, D., Lesieur, M., Newman, G. R., Chollet, J. P., Larchevêque, M. (1982): A comparative assessment of spectral closures as applied to passive scalar diffusion. J. Fluid Mech. **124**, 411
Holloway, G., Hendershot, M. C. (1974): The effects of topography on barotropic eddy fields. Mode Hot Line News **75**
Kahane, J. P. (1974): Sur le modèle de turbulence de Benoit Mandelbrot. C.R. Acad. Sci. Paris A, **278**, 621
Kolmogorov, A. N. (1962): A refinement of a previous hypothesis concerning the local structure of turbulence in a viscous incompressible fluid at high Reynolds number. J. Fluid Mech. **13**, 82
Kraichnan, R. H. (1958): The structure of isotropic turbulence at very high Reynolds numbers. J. Fluid Mech. **5**, 497
Kraichnan, R. H. (1964a): Direct-interaction approximation for shear and thermally driven turbulence. Phys. Fluids **7**, 1048
Kraichnan, R. H. (1964b): Diagonalising approximation for inhomogeneous turbulence. Phys. Fluids **7**, 1169
Kraichnan, R. H. (1967): Inertial ranges in two-dimensional turbulence. Phys. Fluids **10**, 1417
Kraichnan, R. H. (1974): On Kolmogorov's inertial-range theories. J. Fluid Mech. **62**, 305
Landau, L. D., Lifshitz, E. M.: *Fluid Mechanics* (Pergamon, New York 1959) p. 340
Leray, J. (1933): Etudes de diverses équations intégrales non-linéaires et de quelques problèmes que pose l'hydrodynamique. J. Math. Pures Appl. **12**, 1

Lesieur M., Herring, J. R. (1984): Diffusion of a passive scalar in two-dimensional turbulence. J. Fluid Mech. (submitted)
Lesieur, M., Sommeria, J., Holloway, G. (1981): Zones inertielles du spectre d'un contaminant passif en turbulence bidimensionalle. C.R. Acad. Sci. Paris **292**, Ser. II, 271
Levy, P.: *Théorie de l'addition des variables aléatoires* (Gauthier Villars, Paris 1937) p. 384
Lilly, D. K. (1983): Stratified turbulence and the meso-scale of the atmosphere. J. Atmos. Sci. **40,** 3, 749
Lorentz, E. N. (1963): Deterministic non-periodic flow. J. Atmos. Sci. **30**, 131
Lovejoy, S.: *A Statistical Analysis of Rain Areas in Terms of Fractals*, in 20th Conf. on Radar Met. (AMS, Boston 1981) p. 476
Lovejoy, S. (1982): The area-perimeter relation for rain and cloud areas. Science **216**, 185
Lovejoy, S., Mandelbrot, B. (1984): Fractal properties of rain and a fractal model. Tellus (in press).
Lovejoy, S., Schertzer, D. (1984): Generalised scale invariance in the atmosphere and a fractal model of rain. Wat. Resarces Res. (submitted).
MacPherson, J. I., Isaac, G. A. (1977): Turbulent characteristics of some Canadian cumulus clouds. J. Appl. Met. **16**, 81
Mandelbrot, B., Wallis, J. R. (1968): Noah, Joseph and operational hydrology. Wat. Resour. Res. **4**, 909
Mandelbrot, B. (1974a): Intermittent turbulence in self-similar cascades: divergence of high moments and dimension of the carrier. J. Fluid Mech. **62**, 331
Mandelbrot, B. (1974b): Multiplication aléatoires itérées, et distributions invariantes par moyenne pondéré. C.R. Acad. Sci. Paris A **278**, 289
Mandelbrot, B.: *Les Objects Fractals: Forme, Hasard et Dimension* (Flammarion, Paris 1975) p. 187
Mandelbrot, B.: *Fractals: Form, Chance and Dimension* (Freeman, San Francisco 1977) p. 365
Mandelbrot, B.: *The Fractal Geometry of Nature* (Freeman, San Francisco 1982) p. 461
Martin, P. C., De Dominicis, C. (1978): The long distance behaviour of randomly stirred fluids. Prog. Theor. Phys. Supp. **64**
Merceret, F. J. (1976): The turbulent microstructure of hurricane Caroline (1975). Mon. Wea. Rev. **104**, 1297
Monin, A. S.: *Weather Forecasting as a Problem in Physics* (MIT Press, Cambridge, MA 1972)
Monin, A. S., Yaglom, A. M.: *Statistical Fluid Mechanics*, Vol. II (MIT Press, Cambridge, MA 1975) p. 566
Morel, P., Larchevêque, M. (1974): Relative dispersion of constant level balloons in the 200 mb general circulation. J. Atmos. Sci. **31**, 2189
Nastrom, G. D., Gage, K. S. (1983): A first loom at wave-number spectra from GASP data. Tellus **35** 383.
Novikov, E. A., Stewart, R. W. (1964): Intermittency of turbulence and spectrum of fluctuations of energy dissipation. Izv. Akad. Nauk. SSSR Ser. Geofiz. **3**, 408
Obukhov, A. N. (1959): Effect of Archimedean forces on the structure of the temperature field in a turbulent flow. Dokl. Akad. Nauk SSSR, **125**, 1246
Orszag, S. A. (1970): Indeterminancy of the moment problem for intermittent turbulence. Phys. Fluids **13**, 2210
Parisi, O., Frisch, U.: *A Multifractal Model of Intermittency*, in Proc. of the Varenna School, ed. by M. Ghil (Italian Physical Soc. 1984) (in press)
Peyriere, J. (1974): Turbulence et dimension de Hausdorff. C.R. Acad. Sci. Paris A **278**, 567
Pinus, N. Z. (1968): The energy of atmospheric macro-turbulence. Izv. Atmos. Ocean. Phys. **4**, 461
Richardson, L. F. (1926): Atmospheric diffusion shown on a distance neighbour graph. Proc. Roy. Soc. London A **110**, 709
Riley, J. J., Metcalf, R. W., Weissman, M. A.: *Direct Simulations of Homogeneous Turbulence in Density-Stratified Fluids*, in Conference on Non-Linear Properties of Internal Waves, ed. by J. West (Am. Inst. Phys. 1981) p. 79
Schertzer, D., Simonin, S.: *A Theoretical Study of Radiative Cooling in Homogeneous and Isotropic Turbulence*, in Turbulent Shear Flows 3, ed. by L. J. S. Bradbury, F. Durst, B. E. Launder, F. W. Schmidt, J. H. Whitelaw (Springer, Berlin, Heidelberg, New York 1982) p. 262
Schertzer, D., Lovejoy, S., Thervy, G., Coiffier, J., Ernie, Y., Clochard, J. 1983: Are current NWP systems stochastically coherent? Preprint Vol. IAMAP/WMO Paris 29/8–4/9 1983, p. 325.
Schertzer, D., Lovejoy, S.: *On the Dimension of Atmospheric Motions (Extended Abstract)*, in Turbulence and Chaotic Phenomena in Fluids (Preprint Vol., IUTAM Symposium, Kyoto Univ. 1983a) p. 141
Schertzer, D., Lovejoy, S.: *On the Dimension of Atmospheric Motions*, in Turbulence and Chaotic Phenomena in Fluids, ed. by S. Tatsumi (North-Holland 1983b)
Schertzer, D., Lovejoy, S., Peyriere, J.: *Fractales elliptiques*. Preprint, Mét. Nat. EERM/CRMD (Paris, France 1983)
Simonin, O.: *Interaction turbulence, rayonnement*. Ph. D. Thesis (IMST, Marseille 1982)

Stewart, R. W., Townsend, A. A. (1951): Similarity and self-preservation in isotropic turbulence. Philos. Trans. Roy. Soc. A **243**, 359
Tardieu, J. J. (1979): Tech. Note 23, EERM, Mét. Nat., Paris
Van der Hoven, J. (1957): Power spectrum of horizontal wind speed in the frequency range from 0.0007 to 900 cycles per hour. J. Meteor. **14**, 160
Van Zandt, T. E. (1982): A universal spectrum of buoyancy waves in the atmosphere. Geophys. Res. Lett. **9**, 575
Vinnichencko, N. K. (1969): The kinetic energy spectrum in the free atmosphere from 1 second to 5 years. Tellus **22**, 158
Von Neumann, J. (1949): *Recent Theories of Turbulence*, in Collect. Works **6**, 437
Wilson, K. G. (1975): The renormalisation group: critical phenomenon and the Kondo problem. Rev. Mod. Phys. **47**, 773
Yaglom, A. M. (1966): The influence of the fluctuation in energy dissipation on the shape of turbulent characteristics in the inertial interval. Sov. Phys. Dokl. **2**, 26

Three-Dimensional Vortex Dynamics near a Wall

Hermann Viets *

College of Engineering, University of Rhode Island, Kingston, Rhode Island 02879, USA

Richard J. Bethke ** and David Bougine ***

Department of Engineering, Wright State University, Dayton, Ohio 45435 USA

Abstract

The three-dimensional deformation of an artificially generated line vortex near a flat wall is examined by means of flow visualization and quantitative conditioned sampling of a hot-wire anemometer. The vortex is observed to undergo a very rapid deformation from the rectilinear form to a hairpinlike shape. Similar shapes have been observed in the case of driven flows in laminar boundary layers and during the transition to turbulent flow. In the present case the vortices are driven by a rotor configuration of larger scale than the boundary layer thickness.

Introduction

Recent analysis by Feigenbaum (1980) and Kadanoff (1983) and others, indicate that an apparently chaotic behavior can be the result of rational analysis. Specifically, simple orderly mathematical processes can be shown to transition to complex or chaotic processes. As these analyses are extended to fluid mechanics, they offer the potential to analyze the transition from laminar to turbulent flow.

From the turbulent flow perspective, a feature which appears in laminar flow as well as during the transition to turbulence is the vortex, especially the three-dimensional vortex. The objective of the present paper is to examine the three-dimensional dynamics of a vortex which can be viewed as a building block of the transition process.

The dynamics of vortex structures has been of interest to many researches for a variety of other reasons. Much of the clearly interest was analytical and motivated by the ability to model the structure in terms of inviscid hydrodynamics. The ideal vortex structures were even employed at one time as a model for atomic structure and successfully predicted some of the earlier observations of atomic physics. Multiple vortex interactions were also computed and in some cases stability analyses were performed to reveal potential naturally occurring vortex configurations. Of course, the most successful application of inviscid vortex structures to predict an observed phenomenon is the well-known Karman vortex street analysis. Karman (1911) was able to predict, by actually computing the effect of a number of vortices on each other, that only a very specific spacing of vortices would yield a stable configuration.

Most of the early vortex analyses and experiments concerned free vortices (i.e. those removed from wall boundaries). This allowed the structure to be treated inviscidly. Most of these vortex structures were ring vortices and have been reviewed by Zaroodny (1973).

The dynamics of vortex ring motion has been analyzed inviscidly by Arms and Hama (1965). An experimental and analytical treatment was performed by Viets and Sforza (1972)

---

* Professor and Dean
** Associate Professor
*** Research Assistant

employing a viscous model due to Rankine (1858). Most recent fully viscous analyses are due to Widnall (1975) and Saffman (1978) and have been reviewed by Saffman and Baker (1979).

The present paper examines the three-dimensional vortex dynamics near a wall. The methods employed for the prediction of vortex dynamics far from a wall will be directly applicable if one is willing to model the wall in an inviscid fashion. Then the inviscid wall would amount to a plane of symmetry in the analysis.

Vortex Near a Wall

The initial studies of vortex structure near a wall were driven by the interest in boundary layer transition. The initial stability predictions of Tollmien (1929) and later Schlichting (1935) were verified by the classical experiment of Schubauer and Skramstad (1947) employing an electromagnetically oscillating ribbon in a laminar flow. The magnitude and frequency of the oscillation were controlled as well as the Reynolds number of the flow. The results, in terms of a stability diagram, are very well predicted by Tollmien's (1929) analysis.

Perhaps more interesting, from the point of view of vortex dynamics in the vicinity of a wall, is the structure of the unsteady flow produced near the surface. As calculated by Schlichting (1968) for a neutral (i.e. neither growing nor decaying) disturbance, the closed streamlines exhibit a vortex-like structure which is convected in the streamwise direction. This structure has been named Tollmien-Schlichting waves. Some excellent photographic evidence of this structure behind a driven oscillating ribbon of the type employed by Schubauer and Skramstad (1947) has been obtained by Wortmann (1977).

The deformation process of a straight vortex line of the Tollmien-Schlichting type has been conceptualized by Hinze (1975). The vortex distorts slightly and this small distortion is amplified by the self-induced velocity distribution set up by the vortex. The region of highest curvature of the vortex has the highest self-induced velocity. Thus the loop in the vortex is propelled fastest from the wall and finds itself in a higher streamwise velocity region. The loop is then convected faster in the streamwise directon than the straight portion deeper within the boundary layer and the vortex filiment is stretched until, by a mechanism which is not fully understood, it is said to "burst". This mechanism is currently being called into question as the actual process leading to boundary layer transition. Some recent evidence indicates that the initially turbulent region (or "spot") is actually composed of numerous vortex filiments instead of a single deformed vortex. However, from the present perspective of vortex dynamics near a wall, the model still offers a rational deformation process in a boundary layer.

Various investigators have attempted to model the transition process both analytically and experimentally. Coles and Barker (1975) produced a turbulent spot by causing a disturbance in a laminar boundary layer with an intermittent jet perpendicular to the wall. The two-dimensional slice through the spot looked very much like a convected vortex. Another investigation of a synthetic turbulent spot was performed by Haritonidis, Kaplan and Wygnanski (1978). A spark trip was employed and vortex-like structures were found in the boundary layer.

A viscous vortex model which results in the ongoing production of new vorticity, was developed by Doligalski, Smith and Walker (1980). In this model, the existence of the vortex caused an eruption of fluid from the wall region which then curls up and produces a vortex structure of the same sign.

Although it appears probable that the turbulent spot is not composed of a single vortex but rather a number of vortices, the dynamics of single vortices are required to treat the multiple vortex problem. In addition, the question of vortex deformation near a wall is of interest in relation to the energization of wall layers and the characteristics of unsteady flows. As will be discussed, the rotor device employed in the current investigation has been successfully applied to various situations where unsteady flow was found to have a beneficial effect.

Thus the objective of the study is to identify the deformation of an initially rectilinear vortex structure near a wall. The two techniques employed in the study are flow visualization and conditioned sampling of a hot-wire anemometer signal. The scale of the experiment is greater than the scale of the boundary layer thickness.

The Flow-Field Geometry

Various techniques are possible to generate the initially rectilinear vortex. The vibrating ribbon employed by Schubauer and Skramstad (1947) is one possibility as has been discussed. However, in the present experiment, the objective is to operate at a larger scale than the previous vibrating ribbon experiments. Another possibility is a technique employed by Francis et al. (1979) in their study of vorticity generated on the upper surface of an airfoil. In that case, the basic mechanism is a plate, positioned perpendicularly to the surface. The plate alternately extends from and retracts into the surface. Each time the plate is extended it creates a recirculation region behind itself. This region is shed and convected downstream each time the plate is retracted into the surface.

From the present point of view, it is desirable to have control over the strength of the vorticity produced and thus the flat plate is declined in favor of the rotor geometry shown in Fig. 1. The rotor has a discontinuity in its radius at the cusp shape. The mean flow is from

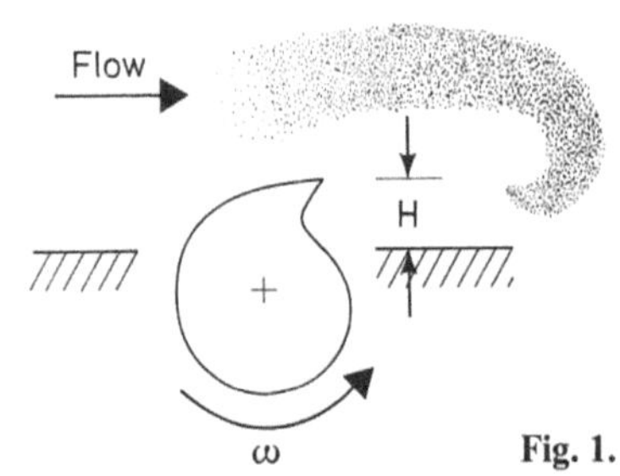

Fig. 1. Schematic of the vortex flow generated by a cam shaped rotor

left to right. If the rotor turns in a counterclockwise direction, the rotor tip appears periodically above the surface. Each time the cusp tip is exposed to the mean flow, a transverse vortex (turning in the clockwise direction) is produced, lying parallel to the surface and perpendicular to the mean flow. The strength of the vortex so produced is proportional to the relative velocity between the rotor tip and the mean flow. Thus control over the vortex strength can be achieved by simply changing the rotational speed of the rotor, including the possibility of turning the rotor in a clockwise direction.

The rotor geometry in Fig. 1 has been employed to study the vorticity produced as a quasi-two-dimensional structure in Viets et al. (1981 b). Only center line data was taken in that experiment.

In addition to a detailed examination of the vortex structure produced by the rotor, the geometry has been applied to various problems including the control of the flow over an airfoil, avoiding separation over a rearward facing ramp, driving the flow in a dump combustor and studying the properties of convected structures. These problems are discussed in the References. Each of the potential applications will be strongly affected by the three-dimensional nature of the vortex flow and thus the dynamics of this structure must be understood and predictable.

Flow Visualization Results

a. The Choice of Rotor Speed

The visualization of the flow-field is accomplished by the use of entrained kerosene vapor. The kerosene is vaporized by slowly dripping it on an inclined heated flat plate where it is vaporized and forced through a series of tubes which end at the inlet to the wind tunnel. Here the vapor is entrained into the tunnel, yielding the streaklines of the flow-field.

The initial study of the centerline plane was conducted at a rotor speed of 3000 rpm (in the counterclockwise direction) resulting in a velocity ratio of 2.0 relative to the freestream

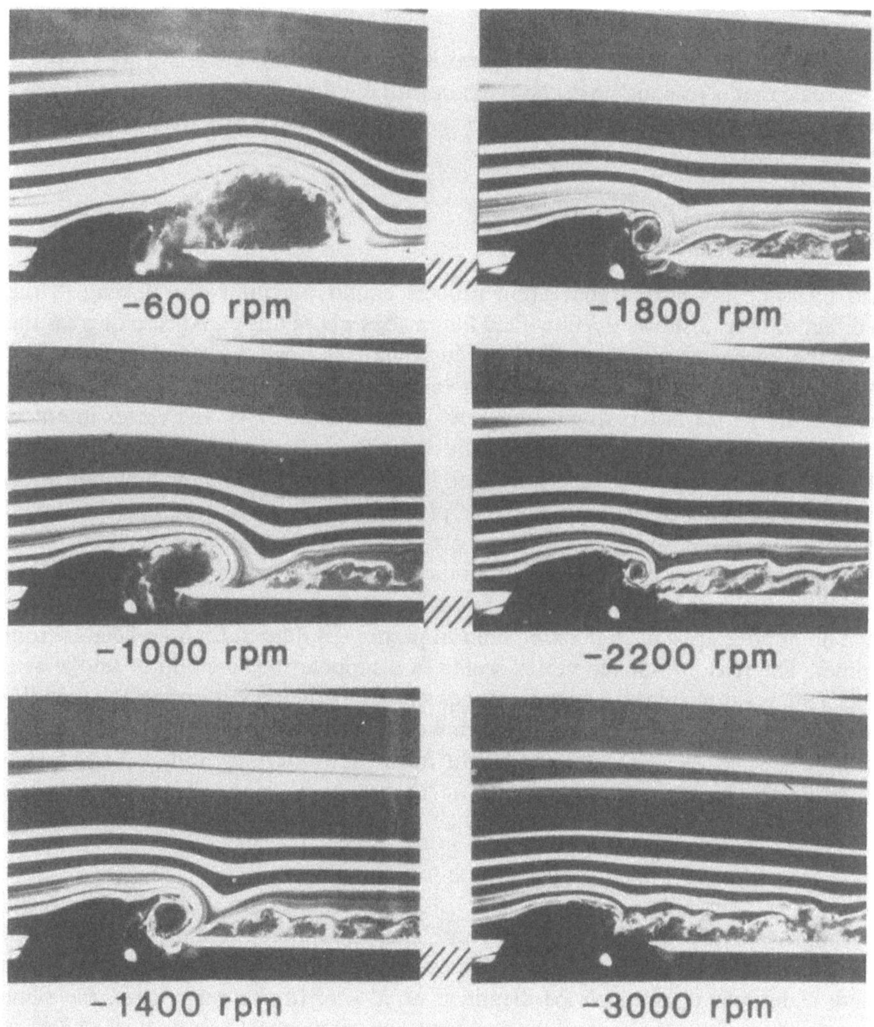

Fig. 2. Formation (at $\theta = 30°$) of the vortices formed at various rotor speeds, all in the clockwise direction

velocity. There are two problems with the 3000 rpm results from the point of view of studying the three-dimensional vortex structure. One is that the vortex scale is large, making it difficult to follow the details of the rather diffuse vortex. The second is that the vortices are relatively closely spaced, being only one transverse scale apart. This might create some problems in detecting the influence of one vortex on another. For these reasons it was determined that a lower frequency would be preferable. In addition, a well formed vortex was desired which led to the study illustrated in Fig. 2. The rotor is turning in a clockwise direction, as denoted by the negative sign. The angle θ is measured clockwise from the rotor position when the cusp is directly above the axis of rotation.

At a phase angle of $\theta = 30°$ the rotor position is such that the cusp point is rotated 30° from the upright position. For various rotational speeds between -600 and -3000 rpm, the flow-fields are shown in Fig. 2. In each case it is clear that a vortex rotating in the clockwise direction is produced, indicating that the freestream velocity, nominally 12 m/sec, is greater than the tip speed of the cusp shape. A possible exception is the case at -3000 rpm where the vortex is very weak or non-existent since the tip speed is approximately equal to the nominal freestream velocity.

From Fig. 2, the choice of a rotational speed is made. The objective is a well formed vortex with a scale that is reasonable in comparison with the tunnel scale. A rotational speed of -1600 rpm is chosen on this basis, resulting in a velocity ratio of 0.5 between the tip speed and freestream velocity.

b. Vortex Generation

In order to illustrate the vortex generation process employing the rotor turning in the clockwise direction, Fig. 3 shows the flow-field for various phase angles. At zero degrees the curling up of the smoke streaklines behind the cusp shape is clear. A secondary frequency is evident on the smokeline and may be related to an instability of the flow. As the phase angle increases, the vortex moves downstream. At a phase angle of 45° the vortex interacts with the corner of the downstream wall behind the rotor. By a phase angle of 67.5° the vortex is more clearly defined and by 90° the vortex is no longer evident. This is the first indication of the three-dimensionality of the flow, since the plane in which the photographs are taken is not the centerline of the tunnel but rather a plane closer to the far wall.

The data planes considered in the present investigation are shown in Fig. 4. In addition to the centerline plane, there are planes at five and ten centimeters to either side of the centerline. The photographs of Fig. 3 are taken in plane -4 (four rotor step heights from the centerline). The reason that the vortex seems to disappear into a cloud of smoke at a phase angle of 90° is simply that the vortex axis has shifted to an orientation more perpendicular (rather than parallel) to the wall. Farther downstream additional planes are examined including those labeled ± 1 ($\pm$ one step height from the centerline) and ± 3.5 (3.5 step heights from the centerline).

c. Visualization in Various Planes

The phase angle $\theta = 120°$ is chosen to illustrate the three-dimensionality of the vortex structure in Figs. 5 and 6. The flow-field in the centerline plane of the tunnel is shown in Fig. 5 where a discrete vortex may be identified at $X = 6$. In the next figure, the same flow-field is shown in the off-axis planes. At the outermost planes, ($Z = \pm 4$), the effect of the vortex is much reduced in scale and even the smaller structures are located much closer to the wall than is the case in the centerline plane.

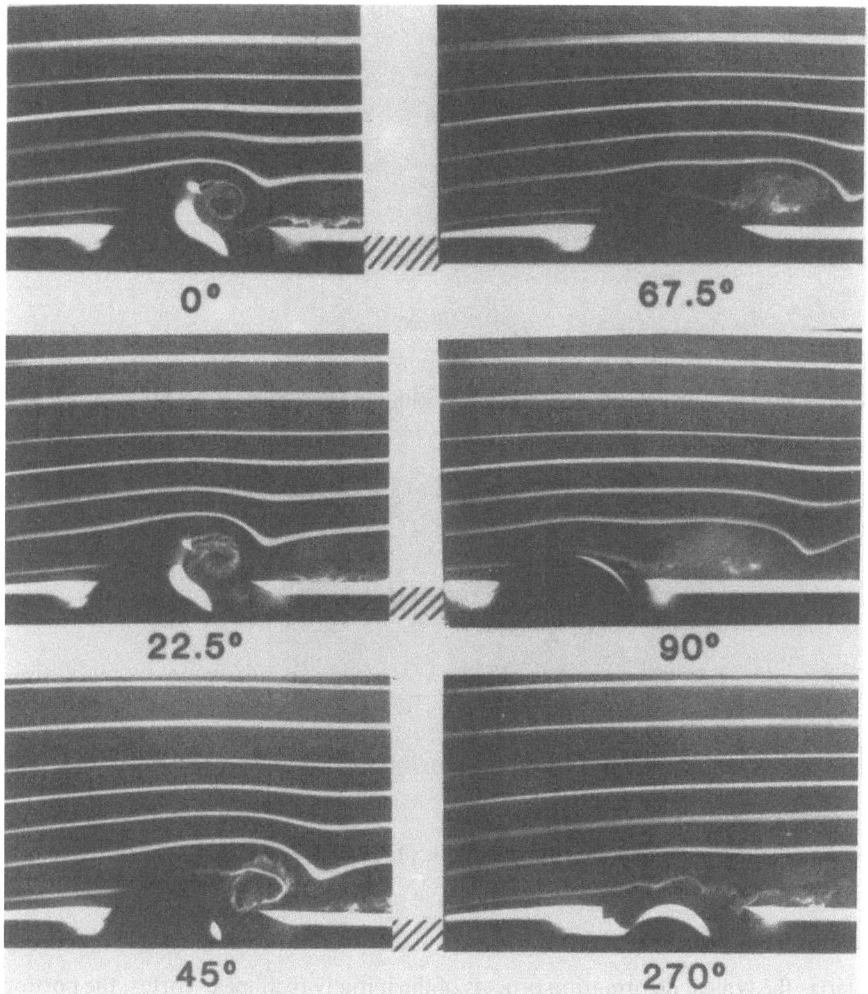

Fig. 3. Vortex formation at various phase angles

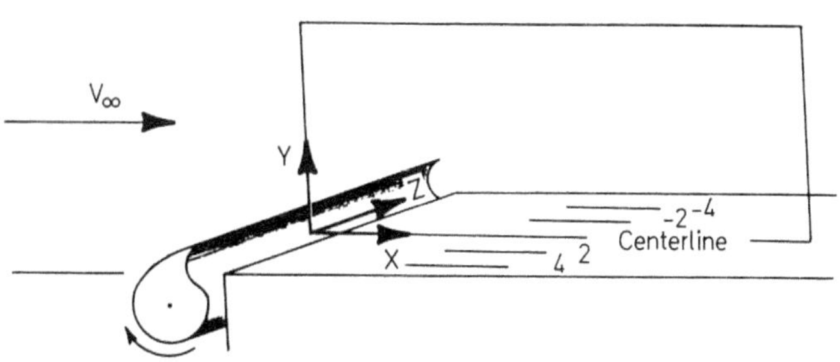

Fig. 4. Experimental schematic

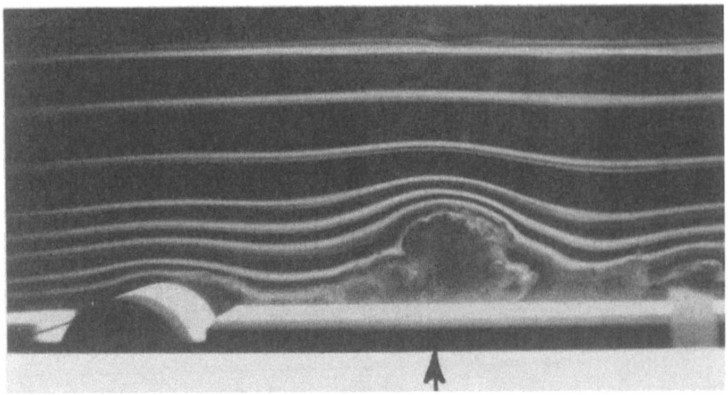

Fig. 5. Flow visualization in the $\bar{Z} = 0$ centerline plane with the phase angle $\theta = 120°$

Fig. 6. Flow visualization in the $\bar{Z} = +4$ plane with the phase angle $\theta = 120°$

d. Three-dimensional Structure

To characterize the typical deformation process of the initially rectilinear vortex, the portion near the centerline rises more rapidly from the wall than those elements away from the centerline. The effect is to produce a horseshoe-like shape of vortex with the ends of the horseshoe on the wall or attached to vortex elements that still have some rectilinear character and remain parallel to the wall as shown in Fig. 7. As the following data will show, these vortex elements may be found by flow visualization, but they become progressively more difficult to locate in a quantitative fashion.

The general deformation shown in Fig. 7 based on flow visualization is similar to that in a transitioning laminar boundary layer, but the scales are entirely different. It is interesting to note, however, that the dynamic character of the vortices is apparently not dependent upon the different scales.

When the flow visualization illustrated in Figs. 5 and 6 is repeated for various phase angles of the vortex generator, the vortex structures can be identified and the position of the vortices plotted in space at various times. In this way the trajectories of the vortices can be identified and compared to the schematic of Fig. 7.

The vortex trajectories identified by flow visualization in the centerline plane and the $Z = \pm 4$ planes is shown in Fig. 8. The circular data points indicate the centerline trajectory while the square symbols refer to the trajectories in the outer planes. The phase angles corresponding to each data point are so indicated. The centerline portion of the vortex rises

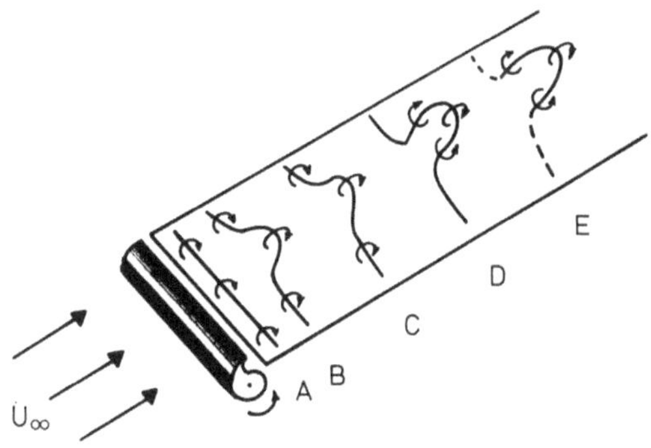

Fig. 7. Schematic of the vortex deformation based on the flow visualization results

very rapidly while the outer portions of the vortex rise very little. The resulting shape is very similar to that of Fig. 7. The only difference is that the vortex loops in Fig. 7 are inclined in the streamwise direction while the implication of Fig. 8 is that the loops are relatively perpendicular to the wall. This difference is simply due to the scale of the boundary layer relative to the scale of the vortex height above the plate. If the scales are comparable, the vortex rises into a higher velocity region and the uppermost portions of the vortex are convected downstream at a higher velocity than the lower portion. Thus the resulting structure will be inclined in the downstream direction. In the present case, the boundary layer thickness is small compared to the vortex scale, being less than a fourth of the height of the vortex generator. Thus the vortex deforms but is convected as a unit. Additional experiments currently underway will include the effect of a varying boundary layer thickness, including the case where the rotor generator is entirely submerged in the boundary layer.

Quantitative Results

The quantitative flow-field data was obtained by a Flow Corporation (now Datametrics) Model 900 constant temperature hot-wire anemometer in conjunction with a pair of Thermo Systems, Inc. linearizers. In order to obtain the instantaneous velocity field, the hot-wire data is conditioned by the circuitry shown schematically in Viets et al. (1981 b). The position of the rotor is determined by a magnetic pickup and this information is used to allow the circuitry to accept or reject the hot-wire signal. Only those signals generated at a given phase position are accepted. Thus the entire flow-field at any phase angle can be obtained and specific examples are analyzed in this section to verify and quantify the results of the flow visualization.

In order to identify a convected vortex structure, a coordinate frame moving with that structure may first be found. In that new coordinate frame, the structure will be very apparent and can be found with little difficulty. However the difficulty is to find the correct coordinate frame.

Bethke and Viets (1980, 1983) have identified two alternate methods to identify convected vortex structures. The first is based on discrete Fourier transformations while the second involves the decomposition of the flow into fundamental structures. An example of the application of the latter technique is shown in an upcoming section.

a. Quantitative Flow Structure, $\theta = 120°$

According to the flow visualization results of Fig. 8, the centerline plane shows evidence of a vortex structure at an X location of approximately 6. Searching the conditioned sampling

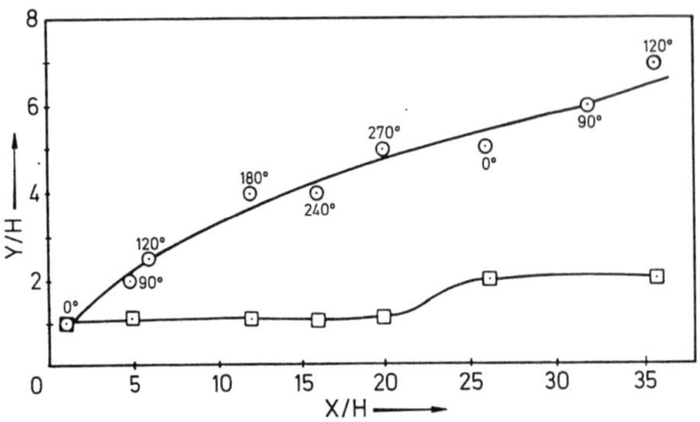

Fig. 8. Vortex trajectories in the centerline plane (○) and in the $\bar{Z} = \pm 4$ plane (□)

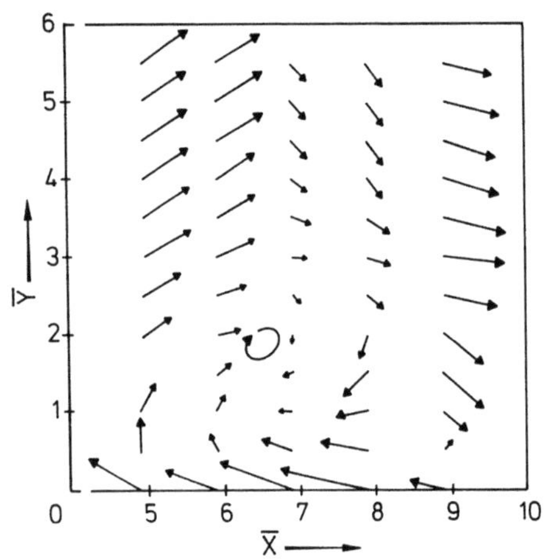

Fig. 9. Velocity field $\theta = 120°$, centerline plane

data at this streamwise location results in the structure shown in Fig. 9. This structure is, of course, only evident in a coordinate frame moving with the vortex. This frame was identified by employing the decomposition technique of Bethke and Viets (1983) and then subtracting the streamwise velocity from the flow-field to reveal the form of the structure. Thus the flow-field of Fig. 9 reveals a vortex structure approximately located at $X = 5.5$, $Y = 2$. That this position agrees well with the results of the flow visualization is seen in the composite of both hot-wire and flow visualization results in Fig. 10.

b. Quantitative Flow Structure, $\theta = 240°$

Moving farther downstream, the flow visualization results have identified a vortex at approximately $X = 16$. In the composite plot of Fig. 11, the local vortex structures identified in the three planes are shown. The vortex shape again verifies the results of flow visualization. In particular, the vortex is shown to be practically perpendicular to the plane of the wall. It is anticipated that the presence of a relatively thick boundary layer would cause the vortex structure to be inclined in the streamwise direction. This aspect is currently under investigation.

Data is available in the intermediate plane between planes 1 and 3.5 ($\theta = 240°$) in Fig. 12. Examination of that velocity field (at a distance of $Z = 2$ from the centerline) reveals no vortex structure. Of course, in order for the vortex to be continuous from point D to point

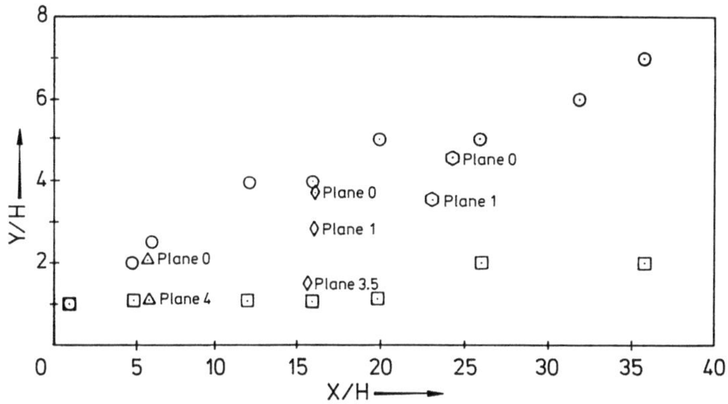

Fig. 10. Vortex trajectories with quantitative data at $\theta = 0°$ (⬡); $\theta = 120°$ (△) and $\theta = 240°$ (◇). Data (○) and (□) are flow visualization from Fig. 19

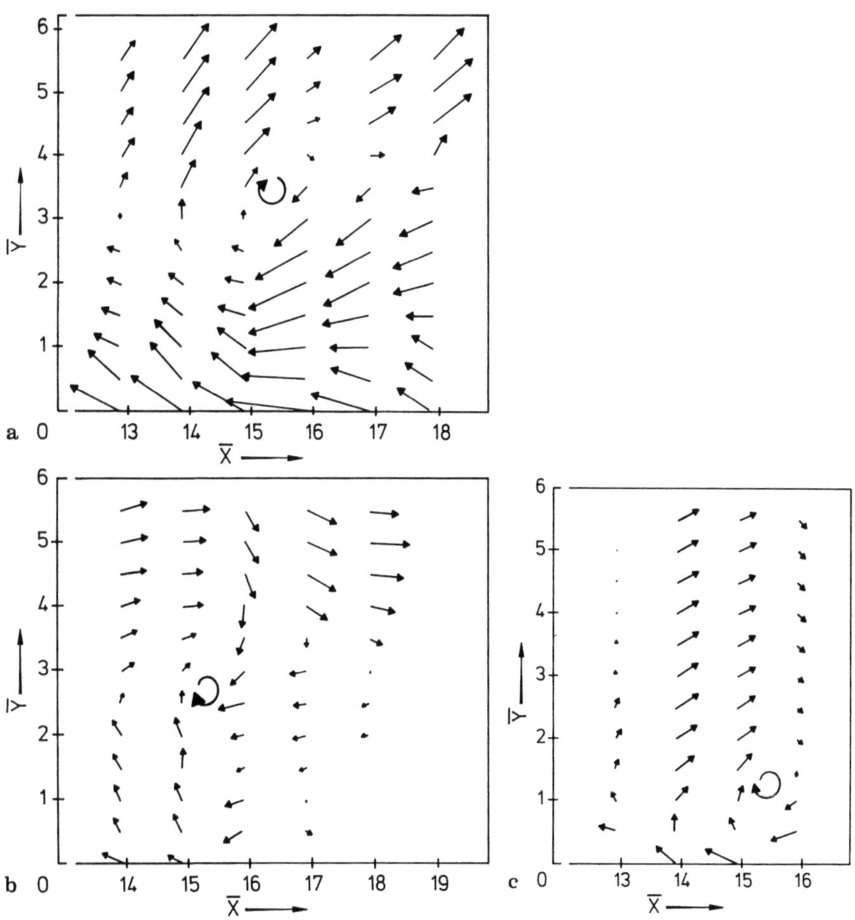

Fig. 11. Vortex structures found in three planes, $\bar{Z} = 0; 1; 3.5$

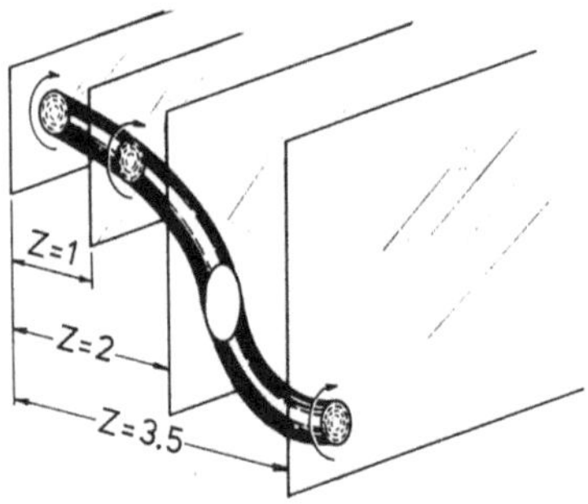

Fig. 12. Schematic of vortex intersection with planes

E in Fig. 10, it must also pass through this plane. The discrepancy is resolved in Fig. 12 where the vortex is shown to penetrate the intermediate $Z = 2$ plane at such an acute angle, that the vortex signature is lost. The inclined vortex does not yield the typical pattern and as the angle between the vortex and the plane decreases, the scale of the signature goes to zero.

c. Quantitative Flow Structure, $\theta = 0°$

At the 0° phase angle, the vortex is in the creation process and is therefore too small and close to the rotor to be found by this technique. The next streamwise appearance of the vortex at 0° phase is at $X = 25$, according to the flow visualization results. From the quantitative point of view, vortex structures are found in planes $Z = 0$ and $Z = 1$ as shown in Fig. 10. At larger distances from the centerline plane, the signature of the vortex is so weak as to be indistinguishable with this technique.

Thus the vortex structure has formed a configuration twice as high above the flat plate as its width.

Conclusions

The three-dimensional flow-field structure revealed by the flow visualization results and the conditioned sampling hot-wire results are summarized in Fig. 7. The initially rectilinear vortex deforms very rapidly as it is convected downstream. Initially there is some evidence that the vortex lines extend out to the walls of the tunnel. However, farther downstream the connection is tenuous and it is possible that the vorticity lines end on the surface of the flatplate instead of the wind tunnel walls.

It is probable that the single deformed vortex is due to the end conditions on the wind tunnel walls. That is, if the rotor were infinitely long, the initially rectilinear vortex would still deform into a pattern similar to that in Fig. 7 and the pattern would be repeated with a scale length determined by the stability of the rectilinear vortex to an applied disturbance. In the present experimental case, the applied disturbance is the end walls which supply boundary layers which retard the vortex ends relative to the vortex center. This initial deformation then progresses into the shape observed. The aspect ratio of the deformed vortex (i.e. its height above the plane-distance between the legs perpendicular to the plate) is approximately two at a position 30 generator heights downstream.

Acknowledgment. The effort reported here was funded by the Air Force Office of Scientific Research under Grant No. AFOSR-81-0025 monitored by Capt. (Dr.) M. S. Francis. The contribution of Mont Ball in the construction of the apparatus and acquiring the data is much appreciated.

References

Arms, R. J., Hama, F. R. (1965): Localized induction concept on a curved vortex and motion of an elliptical vortex rings. Phys. Fluids **8**, 553

Bethke, R. J., Viets, H. (1980): Data analysis to identify coherent flow structures. AIAA Paper No. 80-1561

Bethke, R. J., Viets, H. (1983): Identification of convected flow structures by decomposition techniques. AIAA Paper No. 83-0048

Coles, D., Barker, S. J. (1975): Some remarks on a synthetic turbulent boundary layer, in *Turbulent Mixing in Non-reactive and Reactive Flows* (Plenum, New York 1975)

Doligalski, T. L., Smith, C. R., Walker, J. D. A. (1980): Production mechanism for turbulent boundary layer flows. Progr. Aeronaut. **72**, 47

Feigenbaum, M. J. (1983): Universal behavior in non-linear systems. Los Alamos Science, Summer, 4

Francis, M. S., Keese, J. E., Lang, J. D., Sparks, G. W., Jr., Sisson, G. E. (1979): Aerodynamic characteristics of an unsteady separated flow. AIAA **17**, 1332

Haritonidis, J. H., Kaplan, R. E., Wygnanski, I. (1980): Interaction of a turbulent spot with a turbulent boundary layer flows. Progr. Astro. Aero. **72**, 47

Hinze, J. O.: *Turbulence*, 2nd edition (McGraw-Hill, New York 1975)

Karman, Th. Von (1911): Über den Mechanismus des Widerstandes, den ein bewegter Körper in einer Flüssigkeit erfährt. Göttinger Nachrichten, Mathematisch-Physikalische Klasse, 509

Kadanoff, L. P. (1983): Roads to chaos. Physics Today **46**

Rankine, W. J. M.: *Manual of Applied Mechanics*, 1st edition (Griffen, London 1858)

Saffman, P. G., Baker, G. R. (1978): The number of waves on unstable vortex rings. J. Fluid Mech. **84**, 625

Saffman, P. G., Baker, G. R. (1979): Vortex interactions. Fluid Mech. **11**, 96

Schlichting, H.: *Boundary Layer Theory*, 6th edition (McGraw-Hill, New York 1968)

Schubauer, G. B., Skramstad, H. K. (1947): Laminar boundary layer oscillations and stability of laminar flow. J. Aerosp. Sci. **14**, 69

Tollmien, W. (1929): Über die Entstehung der Turbulenz. 1. Mitteilung, Nachr. Ges. Wiss. Göttingen, Math. Phys. 21–44 (also NACA TM 609, 1931)

Viets, H. (1979): Coherent structures in time dependent flow. NATO/AGARD CP-271

Viets, H., Piatt, M. (1981 a): Induced unsteady flow in a dump combustor. AIAA Progr. Astro. Aero. **76**, 611

Viets, H., Piatt, M., Ball, M. (1981 b): Boundary layer control by unsteady vortex generation. Wind Eng. Ind. Aero. **7**, 135

Viets, H., Piatt, M., Ball, M. (1981): Forced vortices near wall. AIAA Paper 81-02565

Viets, H., Piatt, M., Ball, M. (1979): Unsteady wing boundary layer energization. AIAA Paper 79-1631

Viets, H., Sforza, P. M. (1972): Dynamics of bilaterally symmetric vortx rings. Phys. Fluids **15**, 230

Widnall, S. E. (1975): The structure and dynamics of vortex filiments. Ann. Rev. Fluid Mech. **7**, 141

Wortmann, F. X. (1977): The incompressible fluid motion downstream of two-dimensional Tollmien-Schlichting waves. AGARD/NATO CP-224 on Laminar-Turbulent Transition

Zaroodny, S. J. (1973): Revised theory of vortex rings – a simplified review of the state of the art. U.S. Ballistic Research Laboratory, BRL MR 2305

Diffusion Behind a Line Source in Grid Turbulence

M. S. Anand and S. B. Pope

Sibley School of Mechanical and Aerospace Engineering,
Upson Hall, Cornell University, Ithaca, NY 14853

Abstract

The flow considered is the thermal wake downstream of a fine heated wire in grid turbulence. The inability of generally-applicable turbulent diffusion models and second-order closures to calculate the mean and variance of the temperature is reviewed. Calculations based on the transport equation for the joint probability density function (pdf) of velocity and temperature show good agreement with measurements of mean temperature but not with those of the variance. Better calculations of the variance are obtained with a method based on the joint pdf conditional on the lateral velocity at the source.

Introduction

The thermal wake downstream of a heated wire in grid turbulence has been the subject of several experimental [1–4] and theoretical studies ([5–10], for example). The flow is of fundamental theoretical importance and is also relevant to the practical problem of the dispersion of heat and pollutants in the environment [11].

In comparison to free shear flows – jets, wakes, mixing layers – the thermal wake appears simple: the mean velocity is uniform and the turbulence is isotropic. But turbulence models that are applicable to free shear flows (e.g. $k - \varepsilon$ and Reynolds-stress models) produce qualitatively incorrect results for the thermal wake [9]. This is because the thermal wake thickness σ is, initially, much smaller than the integral length scale, and because we are interested in the wake at convection times that are much smaller than the integral time scale of the turbulence.

In this work, the thermal wake is studied using velocity-temperature joint pdf transport equations [12, 13]. It is found that the standard (unconditional) pdf method is successful in calculating the mean temperature field, but not the temperature variance. A second method – based on the joint pdf conditional on the lateral velocity at the heated wire – is successful in calculating the normalized variance profiles and calculates the absolute magnitude of the variance to within a factor of two. This level of accuracy is comparable with that obtained using two-particle dispersion models [4].

Thermal Wake

The thermal wake is sketched on Fig. 1. A fine heated wire of diameter d is placed normal to the flow, a distance x_0 downstream of a turbulence-generating grid of mesh size M. The mean velocity U is uniform and the turbulence intensity normal to the flow and the wire is found to vary according to the power law:

$$\langle w^2 \rangle / U^2 = A(x/M)^{-m}. \tag{1}$$

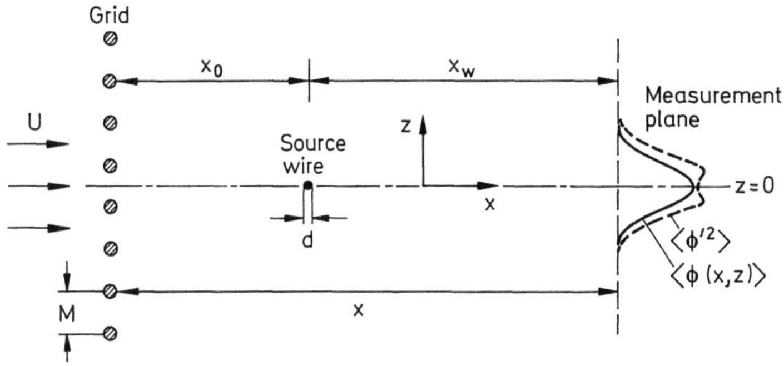

Fig. 1. Sketch of a heated wire downstream of a turbulence generating grid

In the measurements of Warhaft [3] – with which the calculations are compared – the conditions are: $U = 7.0$ m/s, $M = 25$ mm, $x_0/M = 52$, $A = 0.0714$, $m = 1.32$, $d = 0.025$ mm (for measurements up to $x_w/M = 8.1$), and $d = 0.127$ mm (for measurements beyond $x_w/M = 8.1$).

The wire is intended to be sufficiently fine that it does not significantly affect the mean or fluctuating velocity field. Although the wire is quite hot, the excess temperature rapidly falls to a few degrees Celsius. Thus, except in the immediate vicinity of the wire, the excess temperature is a conserved, passive scalar.

The principal quantities of interest are the mean $\langle\phi\rangle$ and variance $\langle\phi'^2\rangle$ of the normalized excess temperature $\phi(\underset{\sim}{x}, t)$. (Angled brackets and primes denote means and fluctuations about the mean, respectively.) At any distance x_w downstream of the source (i.e. the heated wire), the profile of $\langle\phi\rangle$ is found to be Gaussian [1–4],

$$\langle\phi(x_w, z)\rangle = (\sigma\sqrt{2\pi})^{-1} \exp(-\tfrac{1}{2}z^2/\sigma^2). \tag{2}$$

The integral $\int_{-\infty}^{\infty} \langle\phi(x_w, z)\rangle\, dz$ is a conserved quantity, and ϕ is normalized so that this integral is unity. Profiles of $\langle\phi^2\rangle$ (but not of $\langle\phi'^2\rangle$) also appear to be Gaussian [4].

There is an analogy [1] between the thermal wake and the temperature field resulting from the instantaneous production (at time $t = 0$) of an excess temperature distribution along the plane $z = 0$. Thermal wake statistics at (x_w, z) are similar to those at (t, z) resulting from the plane source, where $t = x_w/U$. The analogy depends upon the turbulence intensity $\langle u^2\rangle^{1/2}/U$ being small and mean axial gradients being small in comparison with lateral gradients. In common with previous work, our calculations are of the statistically one-dimensional, time dependent evolution of the temperature field from the plane source in decaying homogeneous turbulence. The results at time t are compared with the thermal wake data of Warhaft at $x_w = Ut$. (Henceforth t is used interchangeably with x_w/U.)

Previous Work

Mean Temperature

A theory should predict a Gaussian mean temperature profile (Eq. 2) and should determine the evolution of the width of the profile $\sigma(t)$. In high Reynolds-number, non-decaying, homogeneous turbulence, there are three regimes:

Molecular diffusion,	$t \ll \Gamma/\langle w^2\rangle$:	$\sigma^2 \approx 2\Gamma t$.
Turbulent convection,	$\Gamma/\langle w^2\rangle \ll t \ll T$:	$\sigma^2 \approx \langle w^2\rangle t^2$.
Turbulent diffusion,	$t \gg T$:	$\sigma^2 \approx 2\Gamma_t t$.

Here Γ is the thermal diffusivity, T is the Lagrangian integral time scale, and Γ_t is the turbulent diffusion coefficient. Taylor [5] neglected molecular diffusion and obtained an expression for σ in terms of the Lagrangian velocity autocorrelation function. This expression (which can readily be generalized to the case of decaying turbulence [4]) correctly accounts for the turbulent convection and diffusion regimes. The effect of molecular diffusion is simply to increase σ^2 by $2\Gamma t$ [2].

The exact equation for the mean of ϕ is

$$\frac{\partial\langle\phi\rangle}{\partial t} = \Gamma\frac{\partial^2\langle\phi\rangle}{\partial z^2} - \frac{\partial\langle w\phi\rangle}{\partial z}. \tag{3}$$

Turbulence models attempt to approximate the scalar flux $\langle w\phi\rangle$ so that Eq. 3 can be solved for $\langle\phi\rangle$.

Turbulent diffusion models (mixing length, $k-\varepsilon$ etc.) approximate the scalar flux by

$$\langle w\phi\rangle = -\Gamma_t\frac{\partial\langle\phi\rangle}{\partial z}. \tag{4}$$

As their name implies, these models are applicable only to the turbulent diffusion regime: early on ($t \ll T$) they predict far too rapid spreading. For example, according to the $k-\varepsilon$ model (see e.g. [14]) the turbulent diffusion coefficient is

$$\Gamma_t = \frac{C_\mu}{\sigma_f}k^2/\varepsilon, \tag{5}$$

where k is the turbulent kinetic energy ($k = 3/2\,\langle w^2\rangle$ in isotropic turbulence) and ε is the rate of dissipation

$$\varepsilon = -\,dk/dt. \tag{6}$$

The model constants are taken to be $C_\mu = 0.09$ and $\sigma_f = 0.7$. With the turbulence decaying according to Eq. 1, the $k-\varepsilon$ model yields (neglecting molecular diffusion)

$$\sigma^2 = \int_0^t 2\Gamma_t(t')\,dt' = \frac{3AM^2C_\mu}{m(2-m)\,\sigma_f}\left\{\left(\frac{x}{M}\right)^{(2-m)} - \left(\frac{x_0}{M}\right)^{(2-m)}\right\}. \tag{7}$$

On Fig. 2, the broken line shows this predicted evolution of σ compared to Warhaft's data (triangles). It may be seen that Eq. 7 is qualitatively and quantitatively incorrect. At the first measurement station, the calculated width is over three times that measured.

Turbulent diffusion models can be made to work, by making Γ_t an explicit function of σ or t [9, 10]. But such ad hoc modifications are inconsistent with the principles of invariant modelling [15, 16] and have little general utility.

Second-order closures experience the same problem [7, 9]. The correct evolution of σ can be obtained only by modifying the transport coefficient for the third moment $\langle w^2\phi\rangle$ to be a function of the temperature field.

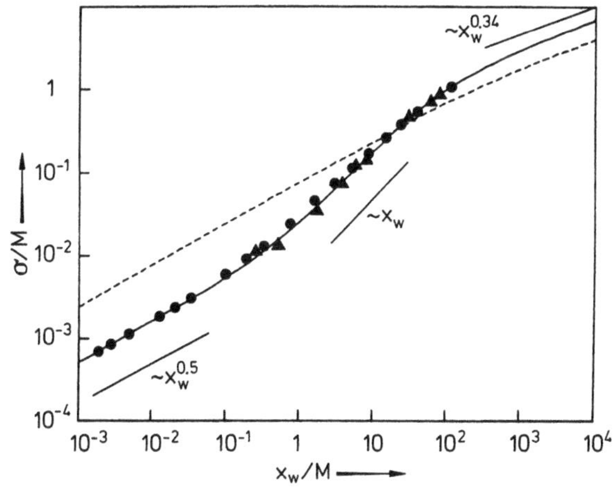

Fig. 2. Thermal wake thickness against distance from the wire. ---- $k-\varepsilon$ model, ● unconditional pdf method, ——— conditional pdf method, ▲ experimental data [3]

Temperature Variance

The exact equation for the variance of ϕ is

$$\frac{\partial \langle \phi'^2 \rangle}{\partial t} + \frac{\partial \langle w \phi'^2 \rangle}{\partial z} = \Gamma \frac{\partial^2 \langle \phi'^2 \rangle}{\partial z^2} - 2 \langle w \phi' \rangle \frac{\partial \langle \phi \rangle}{\partial z} - 2 \varepsilon_\phi, \tag{8}$$

where the scalar dissipation is

$$\varepsilon_\phi \equiv \Gamma \left\langle \frac{\partial \phi'}{\partial x_i} \frac{\partial \phi'}{\partial x_i} \right\rangle. \tag{9}$$

The simplest modelling assumption for ε_ϕ is that the scalar-dissipation time scale $1/2 \langle \phi'^2 \rangle / \varepsilon_\phi$ is proportional to the mechanical-dissipation time scale k/ε: then ε_ϕ is given by

$$\varepsilon_\phi = C_\phi \tfrac{1}{2} \langle \phi'^2 \rangle \, \varepsilon / k. \tag{10}$$

The data of Warhaft and Lumley [17] suggest that C_ϕ is not a universal constant, but a value of $C_\phi = 2.0$ is commonly employed (e.g. [12]).

The validity of Eq. 10 for the thermal wake can be tested rather directly. Let $I(t)$ be the integral of the scalar variance:

$$I \equiv \int_{-\infty}^{\infty} \langle \phi'^2 \rangle \, dz. \tag{11}$$

Then, from Eqs. 3 and 8 and the knowledge that the profile of $\langle \phi \rangle$ is Gaussian, we obtain

$$\frac{dI}{dt} = \bar{P} - C_\phi \frac{\varepsilon}{k} I, \tag{12}$$

where

$$\bar{P} \equiv \left(\frac{1}{2} \frac{d\sigma^2}{dt} - \Gamma \right) \Big/ (2 \sigma^3 \sqrt{\pi}). \tag{13}$$

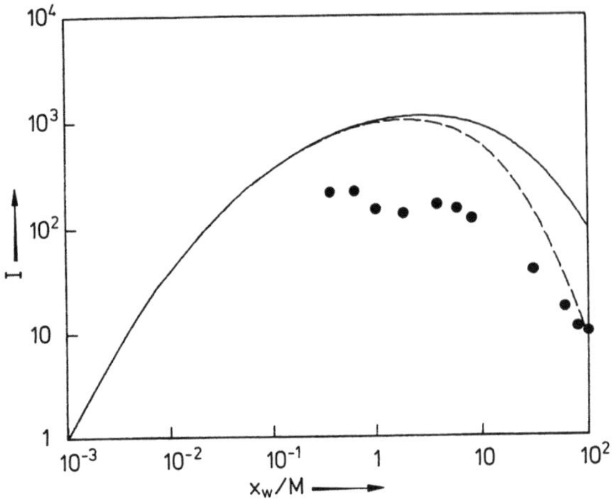

Fig. 3. Integral variance against distance from the wire. ——— unconditional model (Eqs. 10–13) $C_\phi = 2.0$, – – – unconditional model $C_\phi = 4.0$, ● from experimental data [3]

It may be noted that a knowledge of $\sigma(t)$ is sufficient to determine the production term $\bar{P}$. In the section after next, a model that accurately determines σ is described (see the solid line on Fig. 2). Using this result (Eq. A 20) $\bar{P}(t)$ was evaluated and then Eq. 12 was integrated numerically to yield $I(t)$. The result is compared with Warhaft's data on Fig. 3.

It may be seen from Fig. 3 that with the conventional value $C_\phi = 2.0$, the calculated integral variance I exceeds the measured value by, typically, a factor of 7 to 12. With the higher value $C_\phi = 4.0$, this range is reduced to 4 to 7. Since $\sigma(t)$ given by Eq. A 20 agrees extremely well with the data (see Fig. 2), the only significant assumption in the analysis is the modelling of the scalar dissipation, Eq. 10. The results shown on Fig. 3 clearly indicate, therefore, that this model is grossly in error for the thermal wake. If ε_ϕ were determined instead from a consistently-modelled transport equation [16] the result would be little different.

In light of these observations, Sykes, Lewellen and Parker [7] replaced Eq. 10 with a model for ε_ϕ in terms of a scalar length scale, for which a modelled ordinary differential equation was solved. They were then able to obtain accurate calculations of the scalar variance, including the effect of the source size. While this method succeeded for the case considered, because of the way in which both $\langle w^2 \phi \rangle$ and ε_ϕ are modelled, the model does not appear to be applicable to the general case (i.e. more than one transported scalar [16], or more than one source [7]).

A completely different method of calculating the temperature variance is based on the relative dispersion of two fluid particles or, better, of two molecules. There has been considerable work on this method recently (see [8] for references) culminating in the comparison of theory and experiment by Stapountzis, Sawford, Hunt and Britter [4]. The mean thermal wake width $\sigma(t)$ is calculated to within experiment error; the variance is calculated to within a factor of two; and normalized variance profiles agree well with the data.

Unconditional pdf Method

Turbulent diffusion models are unsuccessful in calculating the mean wake thickness $\sigma(t)$ because the second moment $\langle w\phi' \rangle$ is approximated by a gradient diffusion model. Similarly, in second-order closures a gradient-diffusion model is used for the third moment $\langle w^2 \phi \rangle$ and consequently σ is not calculated correctly. In this instance gradient-diffusion models fail because the length scale of variation of the temperature field is much smaller than the turbulence integral scale. But gradient-diffusion models can be entirely avoided by making use of the velocity-scalar joint pdf equation.

For a general flow, $f(\boldsymbol{V}, \psi; \boldsymbol{x}, t)$ is the probability density of the joint events $\boldsymbol{U}(\boldsymbol{x}, t) = \boldsymbol{V}$ and $\phi(\boldsymbol{x}, t) = \psi$. The four new independent variables V_1, V_2, V_3, ψ form the velocity-temperature space. An exact equation for the joint pdf can be derived [12]: for the flow under consideration, the equation for $f(\boldsymbol{V}, \psi; z, t)$ is

$$\frac{\partial f}{\partial t} + V_3 \frac{\partial f}{\partial z} = -\frac{\partial}{\partial \psi}\{f\langle \Gamma \nabla^2 \phi \,|\, \boldsymbol{V}, \psi\rangle\} - \frac{\partial}{\partial V_i}\{f\langle a_i \,|\, \boldsymbol{V}, \psi\rangle\}, \tag{14}$$

where, for any function Q, $\langle Q \,|\, \boldsymbol{V}, \psi\rangle$ is the expectation of Q conditional upon the joint events $\boldsymbol{U} = \boldsymbol{V}$ and $\phi = \psi$. The force (per unit mass) on a fluid particle is

$$a_i = \nu \frac{\partial^2 U_i}{\partial x_j \, \partial x_j} - \frac{1}{\varrho} \frac{\partial p}{\partial x_i}, \tag{15}$$

where ν is the kinematic viscosity, p the pressure, and ϱ the density.

In the joint pdf equation only the terms on the right-hand side need to be modelled. It may be seen that the convective term $V_3 \, \partial f / \partial z$ is in closed form and hence no transport model is required. Stochastic models for the terms in $\nabla^2 \phi$ and a_i have been described previously [12, 13]. The joint pdf equation incorporating these models can be integrated to form transport equations for the scalar flux and scalar variance:

$$\begin{aligned}\frac{\partial}{\partial t}\langle w\phi\rangle + \frac{\partial}{\partial z}\langle w^2\phi'\rangle + \langle w^2\rangle \frac{\partial}{\partial z}\langle\phi\rangle - \Gamma \frac{\partial^2}{\partial z^2}\langle w\phi\rangle \\ = -C_m \frac{\varepsilon}{k}\langle w\phi\rangle,\end{aligned} \tag{16}$$

$$\begin{aligned}\frac{\partial}{\partial t}\langle\phi'^2\rangle + \frac{\partial}{\partial z}\langle w\phi'^2\rangle + 2\langle w\phi\rangle \frac{\partial}{\partial z}\langle\phi\rangle - \Gamma \frac{\partial^2}{\partial z^2}\langle\phi'^2\rangle \\ = -C_\phi \frac{\varepsilon}{k}\langle\phi'^2\rangle.\end{aligned} \tag{17}$$

These equations illustrate the effect of the model constants which were chosen to be $C_m = 2.9$ and $C_\phi = 2.0$.

The modelled joint pdf equation was solved by a Monte Carlo method [13] for the conditions of Warhaft's experiment. The profiles of $\langle\phi\rangle$ were found to be Gaussian and the development of $\sigma(t)$ (circles on Fig. 2) is in excellent agreement with the data. In high-Reynolds number decaying turbulence ($\langle w^2\rangle \propto x_w^{-m}$), σ can be expected to vary as $\sqrt{x_w}$, x_w, and $x_w^{(1-m/2)}$, respectively, in the three regimes – molecular diffusion, turbulent convection, turbulent diffusion. The calculations exhibit the correct variations in the first two regimes, and are consistent with the third regime. Neither the calculations nor the measurements extend far enough to show the third regime distinctly.

It may be seen that the equation for the variance derived from the modelled pdf equation (Eq. 17) is just the exact equation (Eq. 8) with the scalar dissipation modelled according to Eq. 10. Consequently, the integrated variance $I(t)$ obtained from the pdf equation is the same as that obtained previously (Eqs. 12–13, Fig. 3), and is greater than the measured value by, typically, a factor of 7–12.

Thus this unconditional pdf method is successful in calculating the mean spreading, but the calculated variance is grossly in error.

Conditional pdf Method

In order to obtain more accurate calculations of the temperature variance, we seek a more complete statistical description of the thermal wake. In the early stages, the random variable that has most influence upon the wake's development is w_0 – the lateral velocity at the source at time zero:

$$w_0 \equiv w(\boldsymbol{x}=0,\ t=0). \tag{18}$$

Indeed, at very early times, the temperature profile is completely determined by w_0. It is a laminar wake of thickness $\sqrt{2\Gamma t}$ that has been convected laterally a distance $w_0 t$.

We shall therefore consider statistical quantities conditioned on $w_0 = \tilde{w}$ – primarily $\tilde{f}(V,\psi\,|\,\tilde{w};\boldsymbol{x},t)$, the joint probability of $\boldsymbol{U}=\boldsymbol{V}$ and $\phi=\psi$, conditional upon $w_0 \equiv w(\boldsymbol{x}=0,\ t=0)=\tilde{w}$. The unconditional joint pdf (at any $\boldsymbol{x},t$) can readily be recovered by

$$f(V,\psi) = \int_{-\infty}^{\infty} \tilde{f}(V,\psi\,|\,\tilde{w})\, g(\tilde{w})\, d\tilde{w}, \tag{19}$$

where $g(\tilde{w})$ is the pdf of w_0 which (in grid turbulence) is known to be Gaussian [18]:

$$g(\tilde{w}) = \frac{1}{\sqrt{2\pi\,\langle w_0^2\rangle}} \exp(-\tfrac{1}{2}\tilde{w}^2/\langle w_0^2\rangle). \tag{20}$$

The conditional mean profile is

$$\tilde{\phi} \equiv \langle \phi\,|\,\tilde{w}\rangle = \iint \psi\, \tilde{f}(V,\psi\,|\,\tilde{w})\, dV\, d\psi, \tag{21}$$

and, again the unconditional mean is recovered by

$$\langle\phi\rangle = \int \langle\phi\,|\,\tilde{w}\rangle\, g(\tilde{w})\, d\tilde{w}. \tag{22}$$

(Here, and henceforth, the limits of integration $\pm\infty$ are not explicitly indicated. Note also that $\tilde{\phi}$ and ϕ'' (defined below) depend on $\tilde{w}$ as well as on $\boldsymbol{x},t$.)

The principal virtue of this approach is that fluctuations in ϕ can be decomposed into two parts: the difference between conditional and unconditional means $(\tilde{\phi}-\langle\phi\rangle)$, and fluctuations about the conditional mean $\phi'' \equiv \phi - \tilde{\phi}$. The unconditional variance can then be expressed as

$$\langle\phi'^2\rangle = \int g(\tilde{w})\,(\tilde{\phi}-\langle\phi\rangle)^2\, d\tilde{w} + \int g(\tilde{w})\, \widetilde{\phi''^2}\, d\tilde{w}, \tag{23}$$

where the conditional variance is

$$\widetilde{\phi''^2} \equiv \langle \phi''^2\,|\,\tilde{w}\rangle = \langle(\phi - \langle\phi\,|\,\tilde{w}\rangle)^2\,|\,\tilde{w}\rangle. \tag{24}$$

The first contribution to $\langle\phi'^2\rangle$ is due to the meandering or flapping of the wake [6, 4]. The second contribution is due to fluctuations about the conditional mean.

The exact transport equation for $\tilde{f}$ is

$$\frac{\partial \tilde{f}}{\partial t} + V_i \frac{\partial \tilde{f}}{\partial x_i} = -\frac{\partial}{\partial \psi}\{\tilde{f}\langle \Gamma \nabla^2 \phi \mid V, \psi, \tilde{w}\rangle\}$$
$$-\frac{\partial}{\partial V_i}\{\tilde{f}\langle a_i \mid V, \psi, \tilde{w}\rangle\}. \tag{25}$$

Since the conditional joint pdf contains two-time Lagrangian information, the modelling of the conditional expectations should be consistent with our knowledge of such statistics [19]. The following model, which is based on Langevin's equation [20], is consistent with Kolmogorov's inertial range scaling laws [22]:

$$\langle a_i \mid V, \psi, \tilde{w}\rangle = -\frac{1}{\varrho}\frac{\partial}{\partial x_i}\langle p \mid V, \psi, \tilde{w}\rangle$$
$$-\left(\tfrac{1}{2} + \tfrac{3}{4}C_0\right)\frac{\varepsilon}{k}\{V_i - \langle U_i \mid \tilde{w}\rangle\} - \tfrac{1}{2}C_0\,\varepsilon\frac{\partial \ln \tilde{f}}{\partial V_i}. \tag{26}$$

And, according to Kolmogorov's second hypothesis, C_0 is a universal constant.

For the scalar dissipation term, the model

$$-\frac{\partial}{\partial \psi}\{\tilde{f}\langle \Gamma \nabla^2 \phi \mid V, \psi, \tilde{w}\rangle\} = \Gamma \nabla^2 \tilde{f} + \tilde{\varepsilon}_\phi \frac{\partial}{\partial \psi}\{\tilde{f}(\psi - \tilde{\phi})\}/\widetilde{\phi''^2}, \tag{27}$$

is exact for the mean and variance (although it may be inaccurate for higher moments). And, although it proved inapplicable in the unconditional case, we hypothesize that the conditional scalar dissipation can be modelled by:

$$\tilde{\varepsilon}_\phi \equiv \left\langle \frac{\partial \phi''}{\partial x_i}\frac{\partial \phi''}{\partial x_i} \middle| \tilde{w} \right\rangle = C_\phi \tfrac{1}{2}\widetilde{\phi''^2}\,\varepsilon/k. \tag{28}$$

In this conditional case, $\tilde{\varepsilon}_\phi$ is due solely to turbulent straining (and not to flapping) and so may be expected to scale according to Eq. 28.

Equations 25–28 could be solved (for given $\tilde{w}$) by a Monte Carlo method to yield $\tilde{f}$. Then the unconditional pdf could be obtained from Eq. 19. Instead of following this approach, we obtain (in the Appendix) approximate analytical solutions.

The conditional mean profile of ϕ is Gaussian

$$\langle \phi \mid \tilde{w}\rangle = \frac{1}{\sqrt{2\pi}\,\tilde{\sigma}}\exp\left(-\tfrac{1}{2}(z - \tilde{z})^2/\tilde{\sigma}^2\right). \tag{29}$$

Expressions are given in the Appendix for the conditionally-expected location of the center of the wake $\tilde{z}(t)$ and its width $\tilde{\sigma}$. The unconditional wake thickness σ, obtained from Eqs. 22 and 29, is shown as the full line on Fig. 2. Even though the constant value $C_0 = 2.083$ was chosen with respect to this data, the agreement over the whole range of x_w/M lends support to the modelling. (C_0 was not determined to four figures; rather the quantity $(3\,C_0)^{1/2} = 2.5$ was determined to two figures.)

The experimental data shown on Fig. 2 are for one set of conditions in Warhaft's experiment (i.e. $x_0/M = 52$, $U = 7.0$ m/s). It is shown in the Appendix that the theory

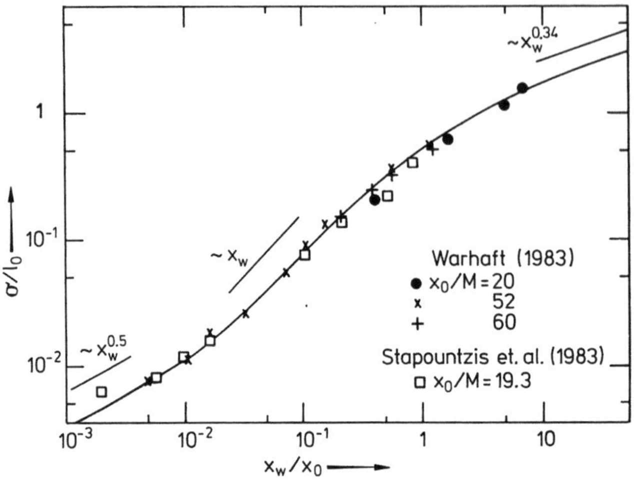

Fig. 4. Thermal wake thickness (σ) normalized by the turbulent length scale at the wire (l_0) against downstream distance from the wire (x_w) normalized by the distance of the wire from the grid (x_0). The solid line is from theory (Eq. A 20) and the symbols represent various experimental data

predicts that (except in the molecular diffusion regime) σ/l_0 is a unique function of x_w/x_0, where l_0 is the length scale $k^{3/2}/\varepsilon$ at the wire. Figure 4 shows σ/l_0 plotted against x_w/x_0 for the three values of x_0/M considered by Warhaft, and the single value of x_0/M considered by Stapountzis et al. It may be seen that the theory is in excellent agreement with the data which do, indeed, collapse in this plot.

We now turn our attention to the determination of the unconditional variance $\langle\phi'^2\rangle$. As previously mentioned, Eq. 23, $\langle\phi'^2\rangle$ can be decomposed into a contribution from the wake flapping

$$\langle\phi'^2\rangle_{\text{flap}} \equiv \int g(\tilde{w})\,(\tilde{\phi} - \langle\phi\rangle)^2\, d\tilde{w}, \tag{30}$$

and a contribution from fluctuations about the conditional mean

$$\langle\phi'^2\rangle_c \equiv \int g(\tilde{w})\, \widetilde{\phi''^2}\, d\tilde{w}, \tag{31}$$

Similarly, the integral variance I, Eq. 11, can be decomposed as

$$I = I_{\text{flap}} + I_c, \tag{32}$$

where I_{flap} and I_c are the integrals over z, of $\langle\phi'^2\rangle_{\text{flap}}$ and $\langle\phi'^2\rangle_c$.

Since expressions for both $\langle\phi\rangle$ and $\tilde{\phi}$ have been obtained, $\langle\phi'^2\rangle_{\text{flap}}$ can be determined from Eq. 30, and I_{flap} by integrating the result: the results are given in the Appendix. The conditional integral variance I_c can be determined in the same way as I was determined for the unconditional case. It satisfies Eq. 12 (with I_c replacing I), with the production $\bar{P}$ given by Eq. 13 with $\tilde{\sigma}$ replacing σ (see Appendix). This equation was integrated numerically to determine I_c. I was then determined from Eq. 32.

The calculated integral variance I and the flapping contribution I_{flap} are shown on Fig. 5 for two values of the model constant C_ϕ. It may be seen that up to $x_w/M \approx 1.0$, I_{flap} contributes more than 90% of the variance, whereas beyond $x_w/M \approx 10$ it contributes less than 25%. At the first two measurement locations ($x_w/M = 0.36$ and 0.62) the calculations are in excellent agreement with the data. By $x_w/M = 8.1$, where the influence of C_ϕ begins to appear, the calculated variance is about twice the measured value. With $C_\phi = 2.0$, the calculated variance is four times the measured value at $x_w/M = 100$, whereas with $C_\phi = 4.0$,

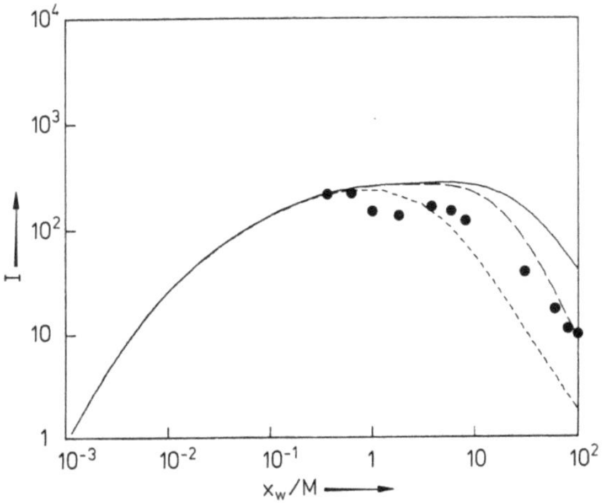

Fig. 5. Integral variance against distance from the wire. ——— conditional pdf method $C_\phi = 2.0$, — — — conditional pdf method $C_\phi = 4.0$, - - - - flapping contribution I_{flap}, ● from experimental data [3]

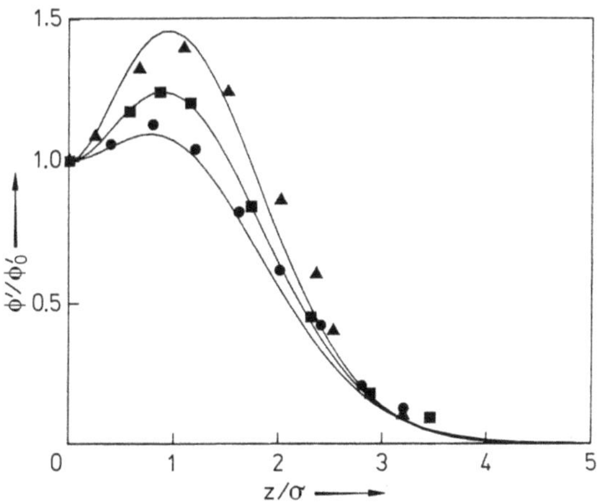

Fig. 6. Normalized standard deviation against cross-stream distance. ——— conditional pdf method. ▲, ■, ● data [3] at $x_w/M = 0.36$, 0.62, 1.00, respectively

the calculations never exceed the measurements by more than a factor of two. This level of agreement (at least with $C_\phi = 4.0$) is comparable with that obtained by Stapountzis et al. [4]. (It may be noted that these authors' calculations extend to $x_w/M = 20$ even though their measurements extend to $x_w/M = 85$. As with our calculation, it appears that the agreement between their theory and experiment deteriorates beyond $x_w/M = 10$.)

In order to calculate profiles of $\langle \phi'^2 \rangle$ we need one more piece of information. Assuming $\langle \phi^2 \rangle$ to have a Gaussian profile [4] with variance $X\sigma^2$, then the profiles of $\langle \phi'^2 \rangle$ can be determined from σ, I and X. But from our analysis X is not known. If there were only flapping (i.e. $I_c = 0$), then X could be determined, since $\langle \phi'^2 \rangle_{\text{flap}}$ is known: the value of X obtained thus we denote by X_{flap}. In the early stages of the wake, flapping is the dominant contribution to I, and X_{flap} can be expected to be a good approximation to X. Figure 6 shows normalized rms profiles

$$\phi'/\phi'_0 \equiv (\langle \phi'^2 \rangle / \langle \phi'^2 \rangle_{z=0})^{1/2}, \tag{33}$$

at the first three measuring stations based on the calculated values of σ, I and X_{flap}. It may be seen that there is good agreement with the data. The location and magnitude of the peak away from the plane of symmetry is calculated reasonably well.

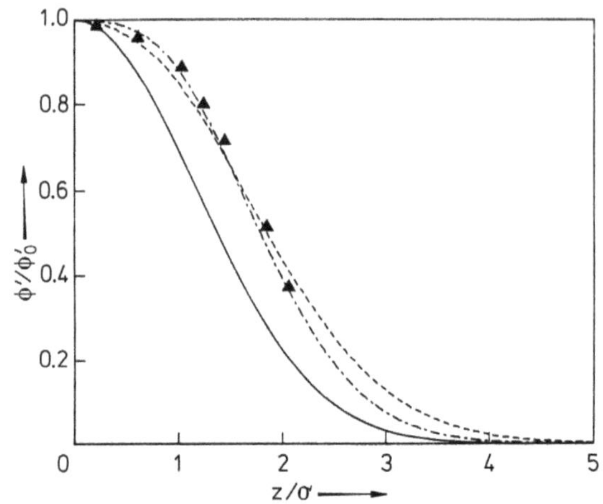

Fig. 7. Normalized standard deviation against cross-stream distance at $x_w/M = 100$: conditional pdf method. ——— $C_\phi = 2.0$, $X = X_{\mathrm{flap}} = 1.59$, ---- $C_\phi = 2.0$, $X = 1.00$, —·—·— $C_\phi = 4.0$, $X = 1.39$, ▲ data [3]

At the measurement stations further downstream, the peak value of $\langle \phi'^2 \rangle$ is on the plane of symmetry and the profiles are approximately self similar. Figure 7 shows a comparison of the measured and calculated profiles at the last measurement station $x_w/M = 100$. Here, I_c is the dominant contribution to I, and it would be fortuitous if X_{flap} approximated X. It can be seen that the calculated profile (with $C_\phi = 2.0$) based on σ, I and $X_{\mathrm{flap}} = 1.59$ is significantly narrower than the experimental profile: but with the value $X = 1.0$, the agreement is improved. Better agreement still is obtained with $C_\phi = 4.0$, $X = 1.39$.

Discussion

Both the conditional and unconditional pdf methods are capable of describing the mean spreading of the thermal wake through all three stages of its development. The ability is due to the fact that convection appears in closed form in the pdf equations, and gradient-diffusion models are avoided.

The conditional pdf method succeeds in calculating the temperature variance (to within a factor of 2) whereas the unconditional method does not. The crucial difference between the two methods is revealed in the (unconditional) integral variance equation:

$$\frac{dI}{dt} = \bar{P} - 2\bar{\varepsilon}_\phi, \tag{34}$$

where

$$\bar{\varepsilon}_\phi \equiv \int_{-\infty}^{\infty} \varepsilon_\phi \, dz. \tag{35}$$

The integral production $\bar{P}$ is the same for both models (see Eq. 13). For the unconditional model, the integral dissipation is

$$\bar{\varepsilon}_\phi = 1/2\, C_\phi \frac{\varepsilon}{k} I, \tag{36}$$

whereas for the conditional model it is

$$\tilde{\varepsilon}_\phi = 1/2\, C_\phi \frac{\varepsilon}{k}(I - I_{\text{flap}}) + \frac{\Gamma}{4\sqrt{\pi}}(\tilde{\sigma}^{-3} - \sigma^{-3}). \tag{37}$$

Figures 3 and 5 show that the modelled terms (that contains C_ϕ) have no effect until beyond the first measurement station. The success of the conditional model in calculating the variance at $x_w/M = 0.36$ is due, therefore, to the last term in Eq. 37. (At this location, the unconditional model yields a variance larger by a factor of 3.) The term – which is exact (given σ and $\tilde{\sigma}$) – is due to the steep gradients in the conditional mean profile. From a broader perspective, the success of the conditional model is due to its ability to separate the flapping of the wake (largely determined by w_0) from the more random spreading.

Several approximations are made in the Appendix in order to obtain analytic solutions. The errors thus introduced can be avoided by solving the modelled conditional pdf transport equation by a Monte Carlo method.

Acknowledgement. We are grateful to Professor Z. Warhaft for making available his unpublished data. This work was supported in part by Grant No. DAA929-82-K0017 from the U.S. Army Research Office.

Appendix

An analytic solution for the conditional mean $\langle \phi \,|\, \tilde{w} \rangle$ can be obtained from the modelled joint pdf equation (Eqs. 25–28) with three approximations. First, the conditional pressure gradient in Eq. 26 is neglected. This term is small and its neglect is an excellent approximation. Second, the scalar dissipation term (Eq. 27) is omitted. The principal justification here is that the scalar dissipation does not affect the mean directly. Third, the condition $w(\boldsymbol{x} = 0, t = 0) = \tilde{w}$ is replaced by the (unlikely) condition $w(x, y, z = 0, t = 0) = \tilde{w}$ (all x and y). This is identical to the conventional assumption that dispersion can be approximated as a one-dimensional phenomenon.

With these approximations the modelled conditional joint pdf equation becomes

$$\frac{\partial \tilde{f}}{\partial t} + V_i \frac{\partial \tilde{f}}{\partial x_i} = \Gamma \nabla^2 \tilde{f} + \left(\frac{1}{2} + \frac{3}{4} C_0\right) \frac{\varepsilon}{k} \frac{\partial}{\partial V_i} \{\tilde{f}(V_i - \langle U_i \,|\, \tilde{w} \rangle)\}$$
$$+ \frac{1}{2} C_0 \varepsilon \frac{\partial^2 \tilde{f}}{\partial V_i \,\partial V_i}. \tag{A 1}$$

This is also the joint pdf equation [13] corresponding to the following random walk of a particle with position $\hat{\boldsymbol{x}}(t)$ and velocity $\hat{\boldsymbol{U}}(t)$

$$\hat{\boldsymbol{x}}(t + \Delta t) = \hat{\boldsymbol{x}}(t) + \hat{\boldsymbol{U}}(t)\, \Delta t + (2\Gamma \Delta t)^{1/2}\, \boldsymbol{\xi}, \tag{A 2}$$

$$\hat{\boldsymbol{U}}(t + \Delta t) = \hat{\boldsymbol{U}}(t) \left[1 - \left(\frac{1}{2} + \frac{3}{4} C_0\right) \frac{\varepsilon}{k} \Delta t\right] + (C_0 \varepsilon \Delta t)^{1/2}\, \boldsymbol{\xi}', \tag{A 3}$$

(in the limit of $\Delta t \to 0$), where $\boldsymbol{\xi}$ and $\boldsymbol{\xi}'$ are independent joint-normal random vectors with zero mean and unit diagonal covariance matrix. For the case considered, motion in all but the $z = x_3$ direction is irrelevant. The appropriate initial conditions are $\hat{z}(t = 0) = 0$, $\hat{w}(t = 0) = \tilde{w}$.

A standard analysis of the random walk [20] shows that the mean profile of a diffusing substance (the excess temperature in the present case) is represented by the pdf of the position of particles leaving the source (at $z = 0$) at time $t = 0$. Further, it is shown that this pdf is Gaussian [20]. We can therefore write

$$\tilde{\phi} \equiv \langle \phi(z,t) \mid \tilde{w} \rangle = \frac{1}{\sqrt{2\pi}\,\tilde{\sigma}} \exp(-(z-\tilde{z})^2/2\tilde{\sigma}^2), \tag{A4}$$

where $\tilde{z}(\tilde{w},t)$ and $\tilde{\sigma}^2(\tilde{w},t)$ are the conditional mean and variance. These quantities can be expressed in terms of the Lagrangian position $\hat{z}(t)$ and velocity $\hat{w}(t)$ of fluid particles originating from $z = 0$ at time $t = 0$ with velocity $\hat{w}(0) = \tilde{w}$:

$$\tilde{z}(\tilde{w},t) = \langle \hat{z}(t) \mid \tilde{w} \rangle = \int_0^t \langle \hat{w}(t') \mid \tilde{w} \rangle \, dt', \tag{A5}$$

and

$$\begin{aligned} \tilde{\sigma}^2(\tilde{w},t) &= \langle \hat{z}(t)^2 \mid \tilde{w} \rangle - \tilde{z}(\tilde{w},t)^2 \\ &= 2\int_0^t \int_0^{t''} \langle \hat{w}(t')\, \hat{w}(t'') \mid \tilde{w} \rangle \, dt'\, dt'' + 2\Gamma t - \tilde{z}(\tilde{w},t)^2. \end{aligned} \tag{A6}$$

In the following analysis, $\tilde{z}$ and $\tilde{\sigma}^2$ are determined.

For the decaying turbulent flow under consideration, ε and k are functions of t; $\varepsilon \propto (t_0 + t)^{-(m+1)}$ and $\varepsilon/k = m/(t_0 + t)$ where $t_0 \equiv x_0/U$. From the random walk in Eq. A 3 we then obtain

$$\begin{aligned} \langle \hat{w}(t) \rangle &= \hat{w}(0) \exp\left(-\beta \int_0^t (\varepsilon/k)\, dt \right) \\ &= \hat{w}(0)\,(1 + t/t_0)^{-\beta m}, \end{aligned} \tag{A7}$$

where

$$\beta = (\tfrac{1}{2} + \tfrac{3}{4} C_0). \tag{A8}$$

Thus

$$\langle \hat{w}(t) \mid \tilde{w} \rangle = \tilde{w}(1 + t/t_0)^{-\beta m}. \tag{A9}$$

Substitution of Eq. A 9 in Eq. A 5 yields

$$\tilde{z}(\tilde{w},t) = \tilde{w}\, t_0 [1 - (1 + t/t_0)^{-s}]/s, \tag{A10}$$

where $s \equiv \dfrac{m}{2}(\tfrac{3}{2} C_0 + 1) - 1$.

Now, in order to determine $\tilde{\sigma}^2$, let

$$\tilde{\sigma}^2 = 2\Gamma t + \Delta_0^2, \tag{A11}$$

where, from Eq. A 6, Δ_0^2 is defined by

$$\Delta_0^2 = 2\int_0^t \int_0^{t''} \langle \hat{w}(t')\, \hat{w}(t'') \mid \tilde{w} \rangle \, dt'\, dt'' - \tilde{z}^2. \tag{A12}$$

We write

$$\langle \hat{w}(t')\, \hat{w}(t'') \,|\, \tilde{w} \rangle = \int_{-\infty}^{\infty} g'(w', t' \,|\, \tilde{w})\, \langle \hat{w}(t')\, \hat{w}(t'') \,|\, \hat{w}(t') = w' \rangle \, dw', \tag{A 13}$$

where $g'(w', t' \,|\, \tilde{w})$ is the pdf of $\hat{w}$ at time t' conditional upon $\hat{w}(0) = \tilde{w}$ which is also Gaussian [21]:

$$g'(w', t' \,|\, \tilde{w}) = \frac{1}{\sqrt{2\pi}\, W} \exp\left(-\left(w' - \langle \hat{w} \,|\, \tilde{w} \rangle\right)^2 / 2 W^2\right), \tag{A 14}$$

where

$$W^2(\tilde{w}, t') = \langle \hat{w}(t')^2 \,|\, \tilde{w} \rangle - \langle \hat{w}(t') \,|\, \tilde{w} \rangle^2. \tag{A 15}$$

$W^2(\tilde{w}, t)$ is the solution of the following differential equation derived from the random walk in Eq. A 3:

$$\partial W^2 / \partial t = C_0 \varepsilon - 2\beta(\varepsilon/k)\, W^2, \tag{A 16}$$

with the initial condition $W^2(\tilde{w}, 0) = 0$. It may be seen from Eq. A 16 that, with the specified initial condition, W^2 is independent of $\tilde{w}$.

Further, the two-time Lagrangian velocity correlation for the random walk in Eq. A 3 can be shown to be

$$\langle \hat{w}(t')\, \hat{w}(t'') \,|\, w(t') = w' \rangle = w'^2 [(t_0 + t')/(t_0 + t'')]^{s+1} \qquad \text{for } t' \leqq t''. \tag{A 17}$$

Substitution of Eq. A 17 and the solution of Eq. A 16 into Eq. A 13 and subsequently into Eq. A 12 yields

$$\Delta_0^2 = 2 \langle w_0^2 \rangle\, t_0^2 \left\{ \frac{-(r+s)}{2 s^2 (r-s)} + \frac{(1 + t/t_0)^{r-s}}{r(r-s)} + \frac{(r+s)(1 + t/t_0)^{-s}}{r s^2} - \frac{(1 + t/t_0)^{-2s}}{2 s^2} \right\}, \tag{A 18}$$

where $r \equiv 1/2\, m(3/2\, C_0 - 1) + 1$. Thus the conditional profile $\langle \phi \,|\, \tilde{w} \rangle$ is completely determined from Eqs. A 4, A 10, A 11 and A 18.

From the conditional profile and Eq. 22 it can be seen that the unconditional profile is indeed Gaussian and the unconditional wake width σ is determined to be

$$\sigma^2 = \tilde{\sigma}^2 + [\tilde{z}(\langle w_0^2 \rangle^{1/2}, t)]^2 \tag{A 19}$$

$$= 2\Gamma t + \Delta^2 \tag{A 20}$$

where

$$\Delta^2 = \Delta_0^2 + [\tilde{z}(\langle w_0^2 \rangle^{1/2}, t)]^2 = 2 \langle w_0^2 \rangle\, t_0^2 \left\{ \frac{(1 + t/t_0)^{r-s}}{r(r-s)} + \frac{(1 + t/t_0)^{-s}}{r s} - \frac{1}{s(r-s)} \right\}. \tag{A 21}$$

Equation A 20 shows that σ^2 is composed of a contribution $(2\Gamma t)$ due to molecular diffusion, and a contribution (Δ^2) due to turbulent dispersion. The latter contribution scales entirely with turbulent quantities. It is readily deduced from the power law decay Eq. 1 that $(\langle w_0^2 \rangle\, t_0^2)^{1/2}$ is proportional to the length scale $l_0 \equiv k^{3/2}/\varepsilon$ at the wire:

$$\langle w_0^2 \rangle^{1/2}\, t_0 = (2/3)^{1/2}\, m\, l_0. \tag{A 22}$$

Thus Eq. A 8 can be rewritten

$$(\Delta/l_0)^2 = \frac{4m^2}{3}\left\{\frac{(1 + x_w/x_0)^{r-s}}{r(r-s)} + \frac{(1 + x_w/x_0)^{-s}}{rs} - \frac{1}{s(r-s)}\right\}. \tag{A 23}$$

This equation predicts that, for different conditions (i.e. different U, x_0/M, and power-law constant A), the quantity Δ/l_0 depends only on x_w/x_0 and m. (From flow to flow, m may vary by 15%.)

From the conditional and unconditional mean profiles, the flapping contribution to $\langle\phi'^2\rangle$ can be determined. Performing the integration in Eq. 30 yields

$$\langle\phi'^2\rangle_{\text{flap}} = \frac{1}{2\pi}\left\{\frac{\exp(-\frac{1}{2}z^2/\sigma_*^2)}{\sqrt{2}\,\tilde{\sigma}\,\sigma_*} - \frac{\exp(-z^2/\sigma^2)}{\sigma^2}\right\}, \tag{A 24}$$

where

$$\sigma_*^2 = \tfrac{1}{2}\tilde{\sigma}^2 + [\tilde{z}(\langle w_0^2\rangle^{1/2}, t)]^2. \tag{A 25}$$

And, integrating over z, the integral variance due to flapping is found to be

$$I_{\text{flap}} = [1/\tilde{\sigma} - 1/\sigma]/(2\sqrt{\pi}). \tag{A 26}$$

Noting that the shape of the profile of $\tilde{\phi}$ (though not its position) is independent of $\tilde{w}$ Eq. A 18, we assume, similarly, that the shape of the profile of $\widetilde{\phi''^2}$ is also independent of $\tilde{w}$. Then, the integral variance due to conditional fluctuations is, from Eqs. 31 and 32,

$$I_c = \int \widetilde{\phi''^2}\, dz. \tag{A 27}$$

From the modelled conditional pdf equation, a modelled equation for ϕ''^2 can be obtained. This can be integrated over z to obtain

$$\frac{dI_c}{dt} = \bar{P}_c - C_\phi \frac{\varepsilon}{k} I_c, \tag{A 28}$$

where, from the fact that $\tilde{\phi}$ is Gaussian,

$$\bar{P}_c \equiv \left(\frac{1}{2}\frac{d\tilde{\sigma}^2}{dt} - \Gamma\right)\Big/(2\tilde{\sigma}^3\sqrt{\pi}). \tag{A 29}$$

(It may be noted that these equations are the same (mutadis mutandis) as Eqs. 12 and 13 for the unconditional model.)

The solution described above applies to a line source. In the experiment, the heated wire is a source of finite size which Stapountzis et al. [4] suggest modelling as a Gaussian temperature profile of size $\sigma_0 = 1.25\,d$. Such an initial profile can be accommodated in the analysis simply by changing $2\Gamma t$ to $\sigma_0^2 + 2\Gamma t$ in Eqs. A 11 and A 20. The reported variance calculations were performed with $\sigma_0 = 1.25\,d$, but this produced only a small change in I compared to calculations with $\sigma_0 = 0$.

References

1. Townsend, A. A. (1954): The diffusion behind a line source in homogeneous turbulence. Proc. Roy. Soc. **A 224**, 487
2. Uberoi, M. S., Corrsin, S. (1953): Diffusion of heat from a line source in isotropic turbulence. NACA Report 1142
3. Warhaft, Z. (1984): Interference of line sources in grid turbulence. J. Fluid Mech. **44,** 363
4. Stapountzis, H., Sawford, B. L., Hunt, J. C. R., Britter, R. E. (1983): Structure of the temperature field downwind of a line source in grid turbulence. Submitted to J. Fluid Mech.
5. Taylor, G. I. (1921): Diffusion by continuous movements. Proc. Lond. Math. Soc. **20**, 196
6. Gifford, F. (1959): Statistical properties of a fluctuating plume dispersal model. Adv. Geophys. **6**, 117
7. Sykes, R. I., Lewellen, W. S., Parker, S. F. (1984): A turbulent transport model for concentration fluctuations and fluxes. J. Fluid Mech. **139,** 193
8. Sawford, B. L., Hunt, J. C. R. (1983): Effects of turbulence structure, molecular diffusion and source size on fluctuations of concentration in homogeneous turbulence. Submitted to J. Fluid Mech.
9. Deardorff, J. W. (1978): Closure of second- and third-order rate equations for diffusion in homogeneous turbulence. Phys. Fluids **21,** 525
10. Libby, P. A., Scragg, C. A. (1973): Diffusion of heat from a line source downstream of a turbulence grid. AIAA J. **11**, 562
11. Fischer, H. B., List, E. J., Koh, R. C. Y., Imberger, J., Brooks, N. H.: *Mixing in Inland and Coastal Waters*, (Academic, New York 1979)
12. Pope, S. B. (1981): Transport equation for the joint probability density function of velocity and scalars in turbulent flow. Phys. Fluids **24**, 588
13. Pope, S. B. (1985): PDF methods for turbulent reactive flows. To be published in Prog. Energy Combust. Sci.
14. Launder, B. E., Spalding, D. B.: *Mathematical Models of Turbulence*, (Academic, New York 1972)
15. Lumley, J. L. (1978): Computational modelling of turbulent flows. Adv. Appl. Mech. **18**, 123
16. Pope, S. B. (1983): Consistent modelling of scalars in turbulent flows. Phys. Fluids **26**, 404
17. Warhaft, Z., Lumley, J. L. (1978): An experimental study of the decay of temperature fluctuations in grid generated turbulence. J. Fluid Mech. **99**, 545
18. Tavoularis, S., Corrsin, S. (1981): Experiments in nearly homogeneous turbulent shear flow with a uniform mean temperature gradient. Part I. J. Fluid Mech. **104**, 311
19. Monin, A. S., Yaglom, A. M.: *Statistical Fluid Mechanics*, Vol. 2, pp. 358–361 (MIT Press, Cambridge 1975)
20. Chandrasekhar, S.: In *Noise and Stochastic Processes*, ed. by N. Wax (Dover, New York 1954)
21. Uhlenbeck, G. E., Ornstein, L. S.: In *Noise and Stochastic Processes*, ed. by N. Wax (Dover, New York 1954)
22. Pope, S. B. (1983): A Lagrangian two-time probability density function equation for inhomogeneous turbulent flows. Phys. Fluids **26**, 3448

Two-Point Closure Used for a Sub-Grid Scale Model in Large Eddy Simulations

J. P. Chollet

Institut de Mécanique, Université de Grenoble, France/
National Center for Atmospheric Research* Boulder, Colorado 80307, USA

Abstract

A statistical two-point closure – the Eddy Damped Quasi Normal Markovian (EDQNM) – approximation is used to model the sub-grid scales (SGS) in a large-eddy simulation (LES) of three-dimensional homogeneous turbulence. First, eddy viscosities and diffusivities are derived from the hypothesis of a $k^{-5/3}$ energy spectrum to parameterize the small scales of velocity and passive scalar fields: in that case, the LES yields fairly good $k^{-5/3}$ inertial spectra up to the cutoff wavenumber. Then, a complete coupling between the LES in the large scales and the EDQNM calculation of the small scales is developed. Finally, the numerical simulation of large eddies is assessed from the standpoint of predictability studies.

1. Introduction and Equations

The velocity field of a three-dimensional high Reynolds number flow cannot be computed over the whole range of scales of interest, even when using the most powerful computers currently available. Therefore, the velocity field is computed explicitly in the large scales only, taking account of the small scales statistically, in order to ensure at least some dissipation of energy. We have to couple a large-eddy simulation (LES) – numerical solution of the equations of the flow field – to a statistical model of the small scales. Two-point closures are able to represent in a statistical way the distribution of energy among the various scales and the interactions between the large scales (completely calculated through the LES) and the small scales (Sub-grid scales, modeled statistically). We use the eddy damped quasi-normal Markovian (EDQNM) approximation which has been extensively described with references to satisfactory comparisons with numerical results and experiments, in Larchevêque et al. (1980) and Herring et al. (1982). For the purpose of tractable EDQNM computations, we have to assume homogeneity and isotropy.

We focus our attention on freely-evolving turbulence whose kinetic energy decays with time. When stationary flows are considered, some empirical ways of modeling can be adjusted (as illustrated for instance by Siggia and Patterson, 1978), even if it takes some computing time and lacks strong theoretical justification. Here, we consider initial velocity fields confined in the large scales, and study the establishment of a Kolmogorov $k^{-5/3}$ range.

The equations to be considered for a numerical LES of the velocity field are, in Fourier space:

For the large scales, $|\boldsymbol{k}| < k_c$, the time evolution of the velocity field $u_i(\boldsymbol{k},t)$:

$$\left[\frac{\partial}{\partial t} + \nu k^2\right] u_i(\boldsymbol{k},t) = -ikm\, P_{ij} \int_{\substack{\boldsymbol{p}+\boldsymbol{q}=\boldsymbol{k} \\ |\boldsymbol{p}|,\, |\boldsymbol{q}| < k_c}} u_j(\boldsymbol{p},t)\, u_m(\boldsymbol{q},t)\, d\boldsymbol{p} + t^{>}(k\,|\,k_c,t) \tag{1.1}$$

with the usual notations and $P_{ij} = \delta_{ij} - k_i k_j/k^2$.

* The National Center for Atmospheric Research is sponsored by the National Science Foundation

For the small scales $|\boldsymbol{k}| > k_c$: the time-evolution of the energy spectrum $E(k,t)$:

$$\left[\frac{\partial}{\partial t} + 2\nu k^2\right] E(k,t) = \iint\limits_{\substack{\boldsymbol{p}+\boldsymbol{q}=\boldsymbol{k} \\ p,q>k_c}} S(k,p,q,t)\,dp\,dq + T^<(k\,|\,k_c,t) \tag{1.2}$$

$$S(k,p,q,t) = \theta_{kpq}\frac{xy+z^3}{q}\,E(q,t)\,\{k^2 E(p,t) - p^2 E(k,t)\} \tag{1.3}$$

x, y, z are the cosines of the interior angles of the triangle k, p, q. θ_{kpq} is a relaxation time for the triple correlations $\langle uuu\rangle$; its value depends on the energy spectrum and on a numerical constant λ.

The interactions between large and small scales are represented by $t^>$ and $T^<$. The large scales are affected by the small scales through $t^>$ (i.e., a global effect of energy dissipation). $t^>$ has to be modeled since the velocity field is unknown in the small scales ($p, q > k_c$)

$$t^>(k\,|\,k_c,t) = -\,ikm\,P_{ij}\int\limits_{\boldsymbol{p}+\boldsymbol{q}=\boldsymbol{k}} u_j(\boldsymbol{p},t)\,u_m(\boldsymbol{q},t)\,d\boldsymbol{p}, \quad |\boldsymbol{p}| \text{ or } |\boldsymbol{q}| > k_c,\ |\boldsymbol{k}| < k_c. \tag{1.4}$$

The sub-grid scale modeling we define is based upon the distribution of energy $E(k,t)$. An eddy-viscosity formulation in spectral space, proposed by Kraichnan (1976), is derived from a part $T^>$ of the energy transfer.

$$\nu(k\,|\,k_c,t) = -\,T^>(k\,|\,k_c,t)/2k^2\,E(k,t) \tag{1.5}$$

with

$$T^>(k\,|\,k_c,t) = \iint\limits_{\boldsymbol{p}+\boldsymbol{q}=\boldsymbol{k}} S(k,p,q,t)\,dp\,dq, \quad p \text{ or } q > k_c,\ k < k_c. \tag{1.6}$$

Then, this eddy-viscosity is introduced into the equation (1.1) for the velocity field, with the term:

$$t^>(k\,|\,k_c,t) = -\,\nu(k\,|\,k_c,t)\,k^2\,u_i(\boldsymbol{k},t). \tag{1.7}$$

The large scales affect the small scales through $T^<$ (i.e., a global effect of energy production).

$$T^<(k\,|\,k_c,t) = \iint\limits_{\boldsymbol{p}+\boldsymbol{q}=\boldsymbol{k}} S(k,p,q,t)\,dp\,dq, \quad p \text{ or } q < k_c,\ k > k_c. \tag{1.8}$$

We notice that the coupling terms of the transfer $T^<$ and $T^>$ in (1.6) and (1.8) depend on the distribution of energy $E(k,t)$ over the whole range of wave numbers.

We also consider the advection of a passive scalar Ψ, that is any fluctuation of temperature or any tracer concentration provided that it does not react on the velocity field. The equations for the scalar field in the large scales and for the scalar variance spectrum in the small scales are very similar to (1.1) to (1.7), (see Chollet and Lesieur, 1982). We consider an eddy-diffusivity:

$$D(k\,|\,k_c,t) = -\,T_\theta^>(k\,|\,k_c,t)/2k^2\,E_\theta(k,t) \tag{1.9}$$

with the sub-grid transfer:

$$T_\theta^{>}(k \mid k_c, t) = \iint_{p+q=k} S_\theta(k, p, q, t)\, dp\, dq, \quad p \text{ or } q > k_c,\ k < k_c \tag{1.10}$$

and

$$S_\theta(k, p, q, t) = \theta'_{kpq} \frac{kp}{q^3} (1 - z^2)\, E(q, t)\, \{k^2\, E_\theta(p, t) - p^2\, E_\theta(k, t)\} \tag{1.11}$$

θ'_{kpq} is a relaxation time for the triple correlations $\langle \Psi \Psi u \rangle$; its value depends on the energy spectrum and on two numerical constants, λ' and λ''.

The sub-grid-scale models used here will be based upon the eddy viscosity and diffusivity defined by (1.5) and (1.9). In Section 2, a simplified model is defined, assuming that the cut-off wavenumber belongs to an extended inertial subrange. In Section 3, we present a more complete coupling of statistical modeling with large eddy computations. We consider high Reynolds number turbulence; also, we cannot compute a "reference solution" including dissipative scales, and we only check the existence of inertial and inertial-convective spectra up to the cut-off wave-number.

Aupoix and Cousteix (1982) used the same method at lower Reynolds number and obtained results which compared satisfactorily with the experiments by Comte-Bellot and Corrsin.

2. Simplified Sub-Grid-Scale Modeling

To a first approximation, the small scales Eq. (1.2) reduces to: $E(k, t) \sim k^{-5/3}$ for $k \in [k_c, \infty[$, which makes sense for high Reynolds number flows with $k_D \gg k_c$ (k_D is the Kolmogorov dissipation wave number).

As $E(k, t)$ is known for $k > k_c$, $t^{>}$ is the only coupling term to be computed. The large scales are still calculated with (1.1), but we consider a large-scale spectrum $E(k) \sim k^{-5/3}$ to compute $t^{>}$; this is realistic if k_c is not located at the very beginning of the inertial range. $\nu(k \mid k_c, t)$ is formulated in a non-dimensionalized form:

$$\nu(k \mid k_c, t) = \nu_t^{+}(k/k_c)\,(E(k_c, t)/k_c)^{1/2} \tag{2.1}$$

as extensively described in Chollet and Lesieur (1981).

We notice, from (2.1), that the eddy viscosity is proportional to a "turbulent velocity" $(k_c\, E(k_{c,t}))^{1/2}$ and a "mixing length" $1/k_c$.

A non-dimensionalized eddy diffusivity is defined, using $k^{-5/3}$ energy and scalar variance spectra, by:

$$D(k \mid k_c, t) = D_t^{+}(k/k_c)\,(E(k_c, t)/k_c)^{1/2}. \tag{2.2}$$

The eddy Prandtl number $\mathrm{Pr}_t = \nu_t^{+}/D_t^{+}$ is the ratio of the eddy viscosity (2.1) to the eddy diffusivity (2.2). Eddy viscosity, diffusivity and Prandtl number are plotted as a function of k/k_c, in Fig. 1. The numerical constants used to calculate these quantities are only those adjusted for the relaxation times of the EDQNM two-point closure: $\lambda = 0.361$, for the kinetic energy (André and Lesieur, 1977), $\lambda' = 0$, $\lambda'' = 3.61\,\lambda$ for the passive scalar variance (Herring et al., 1982).

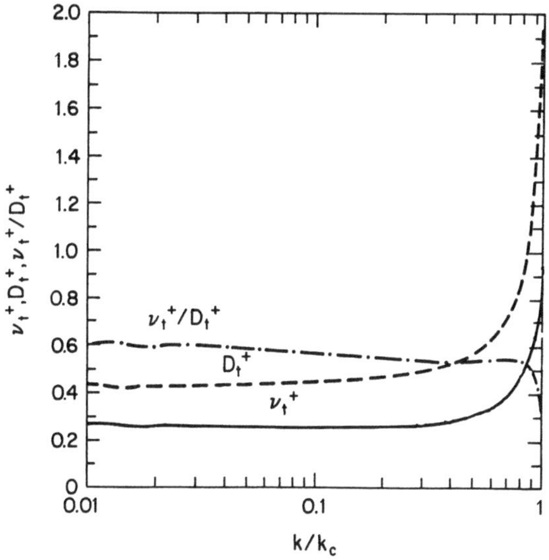

Fig. 1. Non-dimensionalized eddy viscosity, eddy diffusivity and eddy Prandtl number.

The SGS modeling has been tested first in an EDQNM computation over the whole range of scales since it allows the calculation of a reference solution which includes dissipative scales. Results of such parameterized statistical computations are presented for energy and scalar spectra in Chollet and Lesieur (1982), especially to study long-term decays of energy and scalar variance for infinite values of the Reynolds number. The sub-grid-scale modeling has been used in a spectral code of large eddy simulation of turbulence in a cubical box with periodical boundary conditions; the resolution is 32^3, both for the velocity and the passive scalar fields. The eddy Prandtl number Pr_t was assumed to be constant and equal to 0.6 as suggested by Fig. 1. In Figs. 2a) and b), we notice the decay of spectra and their tendency to obey the inertial and inertial-convective $k^{-5/3}$ laws.

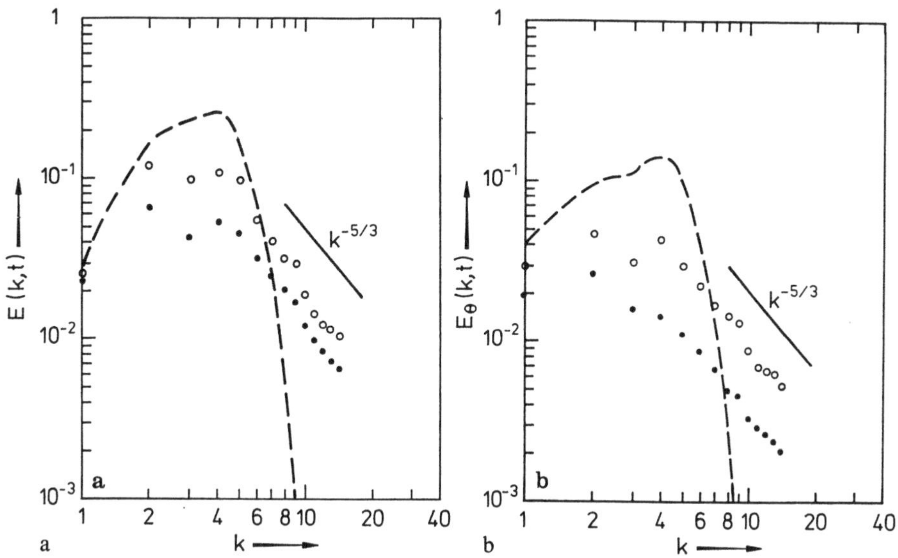

Fig. 2. LES with simplified sub-grid scale modeling (ν_t^+ of Fig. 1 and $Pr_t = \nu_t^+/D_t^+ = 0.6$). — — — $t = 0.0$, ○ ○ ○ $t = 1.0$ sec, ● ● ● $t = 2.0$ sec; a) kinetic energy: $E(k, t)$; b) scalar variance $E_\theta(k, t)$

Upon close examination of the cusp of $v_t^+(k/k_c)$ (Fig. 1), it has been noticed by Chollet (1983) that this cusp can be approximated by $v_t^+ \sim (k/k_c)^4$; consequently, the eddy viscosity (2.1) can be viewed as the addition of an eddy viscosity operator $(-\nabla^2)$ and a higher order dissipativity $(-\nabla^2)^3$, the latter is effective for the local interactions close to k_c. These dissipation operators can be compared to the very high dissipativity used by Basdevant et al. (1981) in large eddy simulations of two-dimensional turbulence whose dynamics is fairly different from the three-dimensional turbulence considered here. As suggested by Aupoix and Cousteix (1982) and Yoshizawa (1982), the usual Smagorinsky formulations in physical space can be recovered from the spectral viscosities; the equations for the velcocity field $\boldsymbol{u}(\boldsymbol{x}, t)$ would take account of the small scales through the term:

$$\langle S_{ij} S_{ij} \rangle^{1/2} \left(\frac{\Delta x}{2\pi}\right)^2 \{c_1(-\nabla^2) + c_2(-\nabla^2)^3\} \tag{2.3}$$

where:

$\Delta x \sim 2\pi/k_c$, is the mesh size
S_{ij} is the rate of strain
c_1 and c_2 are two numerical constants whose value depend on the closure constants, the filter and the numerical scheme in physical space

The fact that the spectra exhibit quite satisfactory $k^{-5/3}$ ranges can be interpreted as an *a posteriori* justification for the values of the closure-constants, especially for the more debatable values of λ' and λ''. The cusps of v_t^+ and D_t^+ can be significantly modified when k_c is close to the dissipative range or close to the integral scale, (Chollet, 1983).

3. Full Statistical Modeling of the Small Scales

The use of a $k^{-5/3}$ spectrum to calculate the eddy viscosity is not realistic:

(1) for freely evolving turbulence, in the initial stage of decay, before the establishment of a $k^{-5/3}$ law
(2) when k_c is close to the Kolmogorov wavenumber k_D
(3) when k_c is in the neighbourhood of the energy containing range.

To take full account of the distribution of energy over the whole range of wave numbers, we have to solve completely (1.1) and (1.2) and formulate the coupling terms $T^<$ and $t^>$.

The large scales are computed with the 32^3 spectral code mentioned in Section 2 and the small scales are computed with the EDQNM code used earlier in statistical computations (Chollet and Lesieur, 1981).

The time evolutions of the large scales ($k < k_c$) and the small scales ($k > k_c$) are linked only by the coupling terms (1.7) and (1.8). At every time-step, the evolution of the energy spectrum of the small scales is calculated with (1.2) and the evolution of the velocity field in the large scales is calculated with (1.1). The coupling terms $T^>$ and $T^<$ are derived from an energy spectrum $E(k)$ whose:

(1) small-scale contributions come from the EDQNM evolution (1.2)
(2) large-scale contributions come from averaging spectra on the computed velocity field.

The results pertaining to a time evolution of a realization of the velocity field are given in Fig. 3 (the initial energy is still zero for $k > k_c$). We observe the tendency toward the inertial range and the continuity of the energy spectrum when crossing k_c.

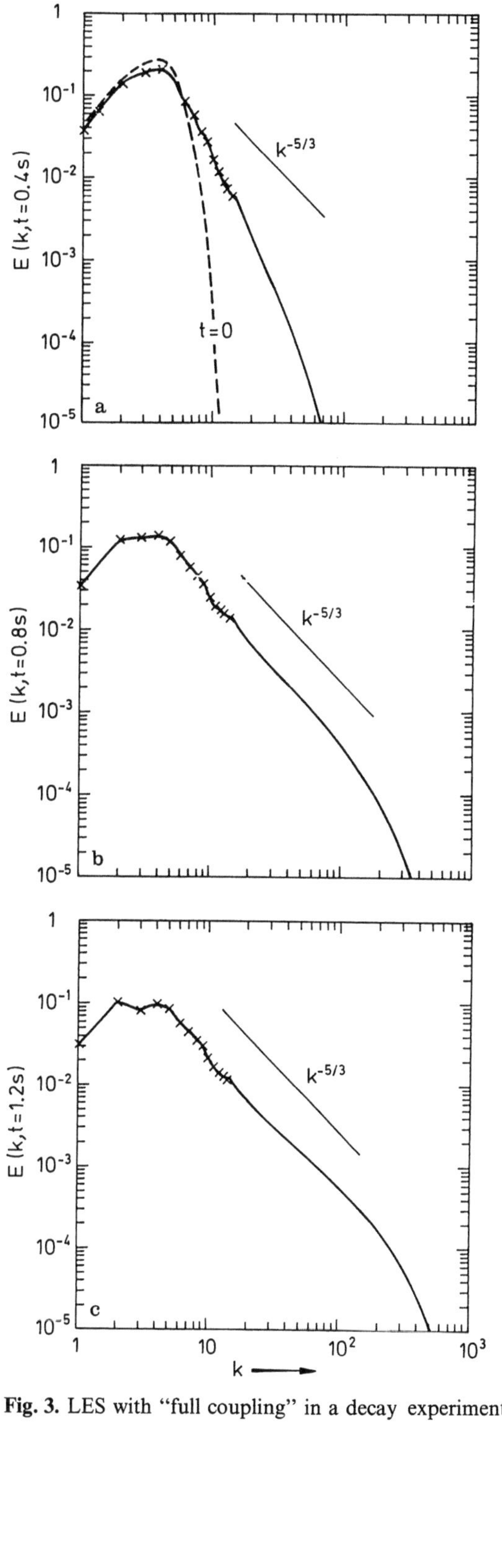

Fig. 3. LES with "full coupling" in a decay experiment of velocity field (× × × LES)

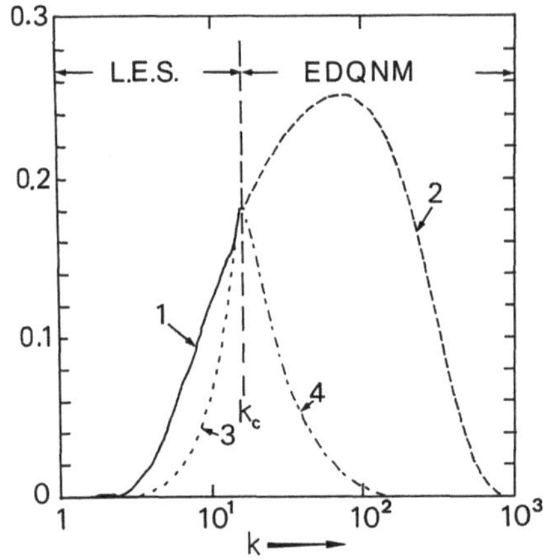

Fig. 4. Energy flux in a LES with full coupling (at $t = 1.2$ sec). (1) total flux from LES: $-\int_0^k T(k)\,dk$, (2) total flux from EDQNM: $\int_k^\infty T(k)\,dk$, (3) from SGS modeling: $-\int_\infty^k T^>(k)\,dk$, (4) from large scales modeling: $\int_k^0 T^<(k)\,dk$

This continuity is also observed on Fig. 4, which shows the total energy flux and the flux associated with the "coupling terms" $T^<$ and $t^>$. We have not attempted to study in detail the "coupling" procedure since it is much more expensive that the simplified model of Section 2, especially with our numerical code whose EDQNM part could not be vectorized. We only aim at proving the feasibility of the procedure. From what we know about the EDQNM and LES computations of the scalar field, we can infer that such a coupling procedure is also valid for any numerical simulation of a scalar field.

4. Predictability of the Large Scales from LES Results

In a large eddy simulation, the sub-grid scales are not calculated in detail; by comparison with the large scales, they can be considered as computed with some unspecified (in nature and amount) inaccuracies and errors. These errors contaminate the large scales in a finite time, which impairs the accuracy of the numerical simulation; this is a problem of "predictability" of the large scales.

The initial fields we consider correspond to evolved solutions of simulations with their energy distributed over a wide range of wave-numbers including scales beyond the cut-off k_c. The corresponding initial energy spectrum exhibits a $k^{-5/3}$ inertial range. We compare two realizations $u^{(1)}$ and $u^{(2)}$ of the velocity field beginning with the same initial field except in the smallest scales. We consider here two methods of creating the initial decorrelation in the small scales (characteristics are given in Table 1):

(1) two different cut-off wave-numbers $k_c^{(2)} < k_c^{(1)}$ but the fields are identical for $k < k_c^{(2)}$; this straightforward method corresponds to a study of the dependence of results on the amount of small scales used by the numerical simulation.

(2) $k_c^{(1)} = k_c^{(2)}$, a range of high wave numbers is altered in a proper way – a rotation of each velocity vector whose axis is in the direction of the wave vector. Unfortunately the intensity of this alteration has to be quite strong in order not to be swept out by the eddy viscosity.

In this section, all the computations use the SGS simplified model described in Section 2.

Instead of the cross-spectrum $E_w(k)$ previously used in Chollet and Lesieur (1981) for preliminary studies of decorrelation, we prefer the more significant "error spectrum" $E_\Delta(k) = E(k) - E_w(k)$. The initial distribution of error-energy $E_\Delta(k)$ is not controlled since

Table 1

Run	$k_c^{(1)}$	$k_c^{(2)}$	Alteration of $u^{(2)}$	Symbols in Fig. 7
3031	15.0	14.0	(1)	+ + +
3047	15.0	10.0	(1)	■ ■ ■
3041	15.0	15.0	(2) at $k = 10.0$	○ ○ ○
3046	15.0	15.0	(2) for $k \in [10.0, k_c^{(2)}]$	▲ ▲ ▲
1920	15.0	13.0	(1)	● ● ●

it is just a consequence of the initial alteration of the field in the smallest scales; this is a difference between these computations and the predictability calculations done by Metais et al. (1983) with a statistical (EDQNM) approach in the same case of freely decaying turbulence. The schematic spectra of Fig. 5 show differences and similarities between a closure and a LES approach for predictability studies, especially because we simulate only a short range of wave-numbers in the explicitly computed scales of the LES.

A time-evolution of energy spectra and error spectra for LES calculations is shown in Fig. 6. We note that in spite of the different methods used to create the initial decorrelation in the small scales, the evolutions are quite similar. Moreover, the increasing range of the error spectra seems to behave like k^4, in agreement with the analysis by Metais et al. (1983), using properties of the EDQNM closure. The same kind of wavenumber interactions give rise to the $E(k) \sim k^4$ energy spectra (for $k \to 0$), calculated by Rogallo (1981) with a LES.

To characterize the contamination of the scales by the error, we define a wave-number k_E:

$$f = E_\Delta(k_E, t)/E(k_E, t).$$

This "local" parameter is more suitable than the relative amount of the error:

$$r(t) = \int_0^\infty E_\Delta(k,t) \Big/ \int_0^\infty E(k,t)$$

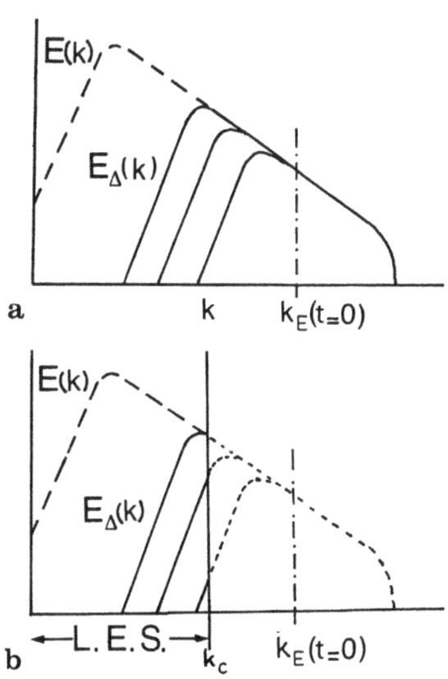

Fig. 5. Contamination of energy spectra by small scale errors. a) in statistical computations, b) in LES computations

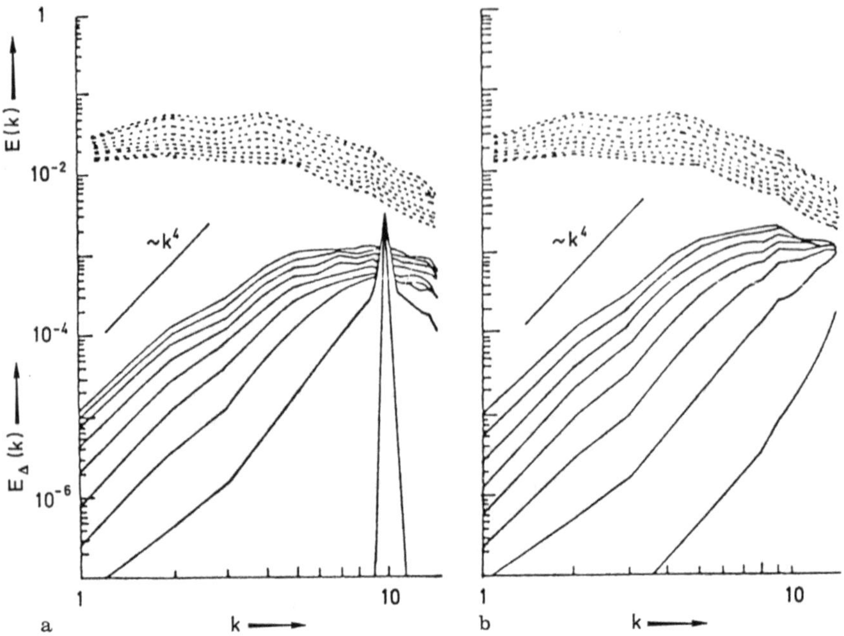

Fig. 6. Energy spectra $E(k)$ and error spectra $E_{\Delta}(k)$, every 0.4 sec (the initial large eddy turnover time is 0.18 sec). a) run 3031, b) run 3041

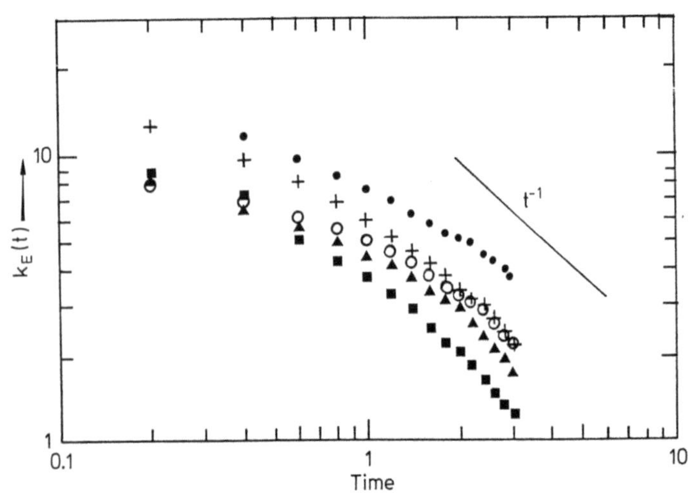

Fig. 7. Evolution with time of the error wave number k_E (Symbols in Table 1)

in order to use quantities which do not depend explicitly on small scales ($> k_c^{(2)}$), or on large scales which cannot be calculated very accurately. In order to include k_E in the range of calculated wave-numbers, we choose $f = 0.01$. The evolution of k_E is drawn in Fig. 7; we note that it does not depend on the way the initial decorrelation is generated, at least after some time. The t^{-1} law is in agreement with the EDQNM results of Metais et al. (1983). This t^{-1} law is a transient, from a $t^{-3/2}$ generalized Richardson law to an ultimate stage of complete decorrelation.

From these results, it appears that the decorrelation evolves with time independently of the way it is generated. There is no major difference between dropping a range of small scales or keeping them after some alteration. The contamination of the large scales of two-dimensional turbulence would be quite different (Metais et al. 1983) with an error spreading first over the whole range of scales; a paradoxical conclusion could be that three-dimensional LES are more accurate than two-dimensional LES, even with a lower resolution.

Conclusions

The statistical theories (here, the EDQNM approximation) provide a consistent model of the sub-grid scales in three-dimensional turbulence. Large Eddy Simulations of velocity and scalar fields exhibit good $k^{-5/3}$ inertio-convective spectra up to the cut-off wavenumber even when there is no energy in the initial field for $k > k_c$.

As the LES computation is independent of the EDQNM code except for the SGS model, these results can be interpreted as a validation of the values attributed to the closure constants; on the other hand, it justifies the concepts of eddy viscosity and diffusivity.

To drop any assumption of an inertial range, we have considered the coupling of a full statistical description of the small scales and the numerical computation of the velocity field for large scales. This method can be used to compute turbulence at any value of Reynolds number, with any shape of energy spectrum, and with no constraints on the relative location of the cut-off with respect to the integral scales. Even with this improvement, the SGS model is only statistically relevant and any information about phase correlations on either side of k_c is lost. Predictability concepts can be used to estimate the relevance of results from LES, giving rise to other tests for the validation of SGS modeling.

We have developed a model for the sub-grid scales in the frame of homogeneous isotropic turbulence. Bertoglio and Mathieu (1983) have extended the use of the EDQNM closure to homogeneous shear flows in order to model the small scales; they point out the limits of isotropic eddy-viscosities for real flows. Nevertheless, there is experimental evidence that many non-isotropic flows exhibit fairly good $k^{-5/3}$ inertial spectra. It follows that, if the small scales remain almost isotropic, our method could be extended to flows whose large scales are determined by specific non-isotropic mechanisms. Of course, it would presumably be more valid for free layers than for flows close to rigid boundaries.

Acknowledgements. We are grateful to M. Lesieur and J. R. Herring for comments and suggestions, to N. Dalfes for help in using efficiently the computer at NCAR, to M. Meneguzzi for providing the numerical code used as the basis for developing the present simulation. The simulations were performed on the CRAY 1 at NCAR.

References

André, J. C., Lesieur, M. (1977): Influence of helicity on the evolution of isotropic turbulence at high Reynolds number. J. Fluid Mech. **81**, 181

Aupoix, B., Cousteix, J. (1982): Modèles simples de tensions de sous-maille en turbulence homogène isotrope. Rech. Aerosp. **4**, 273

Basdevant, C., Legras, B., Sadourny, R., Beland, M. (1981): A study of barotropic model flows: Intermittency, waves and predictability. J. Atmos. Sci. **38**, 2305

Bertoglio, J. P., Mathieu, J.: Study of sub-grid models for sheared turbulence. 4th Symposium presented at the *Turbulent Shear Flows 4* (Karlsruhe 1983)

Chollet, J. P. (1983): Statistical closure to derive a sub-grid-scale modeling for large eddy simulations of three-dimensional turbulence. NCAR Technical Note, TN-206 + STR

Chollet, J. P., Lesieur, M. (1981): Parametrization of the small scales of three-dimensional isotropic turbulence utilizing spectral closures. J. Atmos. Sci. **38**, 2747
Chollet, J. P., Lesieur, M. (1982): Modélisation sous-maille des flux de quantité de mouvement et de chaleur en turbulence tridimensionnelle isotrope. La Météorologié, VIe Série **29–30**, 183
Herring, J. R., Schertzer, D., Lesieur, M., Newman, G. R., Chollet, J. P., Larchevêque, M. (1982): A comparative assessment of spectral closures as applied to passive scalar diffusion. J. Fluid Mech. **124**, 411
Kraichnan, R. H. (1976): Eddy viscosity in two- and three-dimensions. J. Atmos. Sci. **33**, 1521
Larchevêque, M., Chollet, J. P., Herring, J. R., Lesieur, M., Newman, G. R., Schertzer, D.: Two-point closure applied to a passive scalar in decaying isotropic turbulence, in *Turbulent Shear Flow 2*, ed. by L. J. S. Bradbury, F. Durst, B. E. Launder, F. W. Schmidt, J. H. Whitelaw (Springer, Berlin, Heidelberg, New York 1980) p. 50
Metais, O., Chollet, J. P., Lesieur, M.: Predictability of the large scales of freely evolving three- and two-dimensional turbulence, AIP Conference Proceedings No. 106, in *Predictability of Fluid Motions*, La Jolla Institute, ed. by G. Holloway and B. J. West
Rogallo, R. S. (1981): Numerical experiments in homogeneous turbulence. NASA Technical Memorandum 81315, NASA Ames Research Center, USA
Siggia, E. D., Patterson, G. S. (1978): Intermittency effects in a numerical simulation of stationary three-dimensional turbulence. J. Fluid Mech. **86**, 567
Yoshizawa, A. (1982): A statistically derived sub-grid model for the large-eddy simulation of turbulence. Phys. Fluids **25**, 1532

Reynolds-Stress Closure Model for Conditional Variables

J. Janicka

Lehrstuhl für Techn. Thermodynamik, RWTH Aachen, Fed. Rep. of Germany

W. Kollmann

Department of Mechanical Engineering, University of California, Davis, California 95616, USA

1. Introduction

Free boundaries of turbulent shear flows show an intermittent character with a sharp interface separating instantaneously turbulent and non-turbulent zones shown by Corrsin and Kistler (1954). Prediction models for turbulent shear flows did not take this intermittent character into account so far. In this paper a second order closure model for turbulent shear flows based on conditional variables (Dopazo 1977) is presented, that allows prediction of intermittency factor, zone-conditioned mean velocities and Reynolds-stresses. It is based on an earlier closure for intermittency factor, mean velocities and kinetic energy and dissipation rate (Byggstøyl and Kollmann 1981) which used the concept of turbulent pseudo-viscosity for the turbulent zone. The equations for conditional moments contain explicitly terms accounting for the relative motion of turbulent and non-turbulent zones and the transport of various quantities through it. Some of those terms are non-closed on the second-order level and their closure will be discussed. The resulting conditional Reynolds-stress model will be evaluated by comparison of calculated properties of thin shear flows with experimental results.

2. Closure Model

The definiton of conditional moments requires the notion of a non-negative discriminator variable $\phi(\boldsymbol{x}, t)$ such that $\phi \geqq \varepsilon > 0$ corresponds to turbulent flow at the point $(\boldsymbol{x}, t)$ considered and $\phi < \varepsilon$ to non-turbulent. If randomness and nonzero vorticity are accepted as properties defining the turbulent state of the flow, the scalar ϕ is given by

$$\phi(\boldsymbol{x}, t) = \omega'_\alpha \omega'_\alpha$$

where ω'_α is the vorticity fluctuation. The threshold ε for discrimination must be larger than zero for finite Reynolds numbers, because of the elliptic nature of the differential equations governing the motion. Then the indicator function $I(\boldsymbol{x}, t)$ can be defined

$$I(\boldsymbol{x}, t) = \begin{cases} 0 & \text{for } \phi(\boldsymbol{x}, t) < \varepsilon \\ 1 & \text{for } \phi(\boldsymbol{x}, t) \geqq \varepsilon. \end{cases}$$

The equation

$$S(\boldsymbol{x}, t) \equiv \phi(\boldsymbol{x}, t) - \varepsilon = 0$$

defines implicitly a surface within the instantaneous interface separating turbulent and non-turbulent zones. If the interface velocity is denoted by v^s_α, the speed V with which the interface progresses along its normal n_α (positive into turbulent zone) relative to the fluid is given by

$$n_\alpha V = v_\alpha - v_\alpha^s$$

then follows that the indicator function is transported with points on the interface, which are not material points. Thus

$$\frac{\partial I}{\partial t} + v_\alpha^s \frac{\partial I}{\partial x_\alpha} = 0 \tag{1}$$

and the derivatives of the indicator function can be expressed in terms of the interface properties

$$\frac{\partial I}{\partial t} = - v_\alpha^s n_\alpha \delta(s) \tag{2}$$

and

$$\frac{\partial I}{\partial x_\alpha} = n_\alpha \delta(s). \tag{3}$$

Turbulent-zone moments are defined by

$$\bar{\bar{\varphi}} \equiv \frac{\langle I \varphi \rangle}{\langle I \rangle}$$

$$\varphi^* \equiv \varphi - \varphi$$

where $\gamma \equiv \langle I \rangle$ is the intermittency factor and non-turbulent zone moments by

$$\tilde{\bar{\varphi}} \equiv \frac{\langle (1 - I) \varphi \rangle}{1 - \langle I \rangle}$$

$$\varphi^0 \equiv \varphi - \tilde{\bar{\varphi}}.$$

The equations for intermittency factor γ and zone-conditioned moments follow then using standard methods (see Dopazo (1977) and Byggstøyl and Kollmann (1981) Appendix). The equations for first and second order conditional moments contain correlations analogous to the unconditional case and new correlations representing the interface movement and transport of mass and momentum through it. The closure of these new correlations will be discussed subsequently in some detail.

2.1 Intermittency Factor

The exact transport equation for γ follows from (1) by averaging $(\bar{\ } = \langle \cdot \rangle)$

$$\frac{\partial \gamma}{\partial t} + \bar{v}_\alpha \frac{\partial \gamma}{\partial x_\alpha} = - \frac{\partial}{\partial x_\alpha} [\gamma (1 - \gamma) (\bar{\bar{v}}_\alpha - \tilde{\bar{v}}_\alpha)] + \langle V \delta(s) \rangle. \tag{4}$$

The intermittency source $\langle V\delta(s)\rangle$ which is the rate of entrainment of non-turbulent fluid per unit volume requires closure. The model of Byggstøyl and Kollmann (1981) can be applied in the form

$$\langle V\delta(s)\rangle \cong -C_{g1}\gamma(1-\gamma)\frac{\widetilde{v_\alpha^* v_\beta^*}}{\bar{\bar{k}}}\left(\frac{\partial\bar{\bar{v}}_\alpha}{\partial x_\beta}+\frac{\partial\bar{\bar{v}}_\beta}{\partial x_\alpha}\right)+C_{g2}\frac{\bar{\bar{k}}^2}{\bar{\bar{\varepsilon}}}\frac{\partial\gamma}{\partial x_\alpha}\frac{\partial\gamma}{\partial x_\alpha}-C_{g3}\gamma(1-\gamma)\frac{\bar{\bar{\varepsilon}}}{\bar{\bar{k}}} \tag{5}$$

with $C_{g1}=1.8$, $C_{g2}=0.14$, $C_{g3}=0.05$. The first term on the right hand side represents the growth of the turbulent zone due to the mean strain, the second the enhanced transport due to spatial inhomogeneity and the last term the shrinking of the turbulent zone due to viscous effects. The last term in (5) reflects the fact that for a finite Re number flows a finite threshold level for the variable ϕ discriminating between turbulent and non-turbulent zones is required. Consequently C_{g3} should be a function of this level such that C_{g3} vanishes if the threshold value for ϕ vanishes. For the present calculations however C_{g3} was simply kept at the given value.

The difference of the conditional mean velocities, which acts as a turbulent flux for the intermittency factor, is included in the closure model. For parabolic flows, however, only the longitudinal components are calculated and it is advantageous to model the difference of the cross-flow components. Thus

$$\bar{\bar{v}}_\beta-\tilde{\bar{v}}_\beta \cong C_8\frac{\bar{\bar{k}}}{\bar{\bar{\varepsilon}}}\widetilde{v_\alpha^* v_\beta^*}\frac{1}{\gamma}(1-\gamma)^{z_1-1}\frac{\partial\gamma}{\partial x_\alpha} \tag{6}$$

is applied for $\beta=2$ only ($C_8=0.16$, $z_1=1.0$).

2.2 Conditional Momentum Sources

Conditional averaging of the momentum equations leads to new correlations representing momentum sources. The turbulent zone mean velocity $\bar{\bar{v}}_\alpha$ satisfies

$$\frac{\bar{D}}{Dt}\bar{\bar{v}}_\alpha=-\frac{1}{\gamma}\frac{\partial}{\partial x_\beta}(\gamma\,\widetilde{v_\alpha^* v_\beta^*})-\frac{1}{\varrho}\frac{\partial\bar{\bar{p}}}{\partial x_\alpha}+\nu\Delta\bar{\bar{v}}_\alpha+\bar{\bar{S}}_\alpha+(1-\gamma)(\bar{\bar{v}}_\beta-\tilde{\bar{v}}_\beta)\frac{\partial\bar{\bar{v}}_\alpha}{\partial x_\beta} \tag{7}$$

where

$$\frac{\bar{D}}{Dt}\cdot\equiv\frac{\partial}{\partial t}\cdot+\bar{v}_\alpha\frac{\partial}{\partial x_\alpha}\cdot$$

The source term $\bar{\bar{S}}_\alpha$ can be cast in the form $\bar{\bar{S}}_\alpha=\frac{1}{\gamma}F_\alpha^*$

$$F_\alpha^i\equiv\left\langle\left(v_\alpha^i V+p^i n_\alpha-\nu n_\beta\frac{\partial v_\alpha^i}{\partial x_\beta}\right)\delta(s)\right\rangle-\nu\frac{\partial}{\partial x_\beta}\langle v_\alpha^i n_\beta\,\delta(s)\rangle;\qquad i=*,0.$$

The non-turbulent zone mean velocity $\tilde{\bar{v}}_\alpha$ is governed by

$$\frac{\bar{D}}{Dt}\tilde{\bar{v}}_\alpha=-\frac{1}{1-\gamma}\frac{\partial}{\partial x_\beta}[(1-\gamma)\,\widetilde{v_\alpha^0 v_\beta^0}]-\frac{1}{\varrho}\frac{\partial\tilde{\bar{p}}}{\partial x_\alpha}+\nu\Delta\tilde{\bar{v}}_\alpha+\tilde{\bar{S}}_\alpha-\gamma(\bar{\bar{v}}_\beta-\tilde{\bar{v}}_\beta)\frac{\partial\tilde{\bar{v}}_\alpha}{\partial x_\beta} \tag{8}$$

with source term

$$\tilde{\tilde{S}}_\alpha = -\frac{1}{1-\gamma} F^0_\alpha .$$

Since the equations for $\bar{\bar{v}}_\alpha$ and $\tilde{\bar{v}}_\alpha$ can be combined according to (Byggstøyl and Kollmann (1981), Appendix)

$$\bar{v}_\alpha = \gamma \bar{\bar{v}}_\alpha + (1-\gamma) \tilde{\bar{v}}_\alpha$$

and

$$\overline{v'_\alpha v'_\beta} = \gamma \overline{v^*_\alpha v^*_\beta} + (1-\gamma) \widetilde{v^0_\alpha v^0_\beta} + \gamma (1-\gamma) (\tilde{\bar{v}}_\alpha - \bar{\bar{v}}_\alpha)(\tilde{\bar{v}}_\beta - \bar{\bar{v}}_\beta)$$

to the equation for the unconditional mean $\bar{v}_\alpha$, the following relation for the conditional momentum sources is obtained

$$\gamma \bar{\bar{S}}_\alpha + (1-\gamma) \tilde{\bar{S}}_\alpha = (\tilde{\bar{v}}_\alpha - \bar{\bar{v}}_\alpha) S_\gamma + \frac{1}{\varrho} (\tilde{\bar{p}} - \bar{\bar{p}}) \frac{\partial \gamma}{\partial x_\alpha} \tag{9}$$

where $S_\gamma \equiv \langle v \delta(s) \rangle$ is the source term of the intermittency Eq. (4). An order of magnitude analysis shows that both terms on the right hand side of (9) are of the same order for the longitudinal component in thin shear layers. For the cross-flow component the pressure-difference term is dominant and balances the left hand side of (9). Relation (9) is important for the closure of Eq. (8) for the non-turbulent zone mean velocity.

2.3 Non-Turbulent Zone Stress Tensor

If the discriminating scalar ϕ is defined as enstrophy $\omega'_\alpha \omega'_\alpha$, then it follows that the vorticity fluctuations in the non-turbulent zone are below the threshold and therefore negligible. This condition can be expressed analytically in the Corrsin-Kistler (1954) equation (see Dopazo 1977) for the non-turbulent zone stress tensor

$$\frac{\partial}{\partial x_\beta} [(1-\gamma) \widetilde{v^0_\alpha v^0_\beta}] = \frac{\partial}{\partial x_\alpha} [(1-\gamma) \tilde{\bar{k}}] - \langle (v^0_\alpha v^0_\beta n_\beta - k^0 n_\alpha) \delta(s) \rangle \tag{10}$$

where $k^0 \equiv \frac{1}{2} v^0_\alpha v^0_\alpha$ and $\tilde{\bar{k}} \equiv \tilde{\bar{k}}^0$ denote kinetic energy in the non-turbulent zone. This relation has interesting properties: The diffusive term

$$\frac{\partial}{\partial x_\beta} [(1-\gamma) \widetilde{v^0_\alpha v^0_\beta}]$$

which appears in (8), can be split into diffusion due to normal stresses analogous to the mean pressure and diffusion due to shear and normal stresses which are created at the interface. Furthermore, it follows that the interface term can be represented as the divergence of a flux

$$- \langle (v^0_\alpha v^0_\beta n_\beta - \delta_{\alpha\beta} k^0 n_\beta) \delta(s) \rangle = \frac{\partial}{\partial x_\beta} [(1-\gamma) F_{\alpha\beta}]. \tag{11}$$

The equation for the non-turbulent zone mean velocity appears now in a form consistent with (9) and (11) as

$$\begin{aligned}\frac{\bar{D}}{Dt}\tilde{\bar{v}}_\alpha = & -\frac{1}{1-\gamma}\frac{\partial}{\partial x_\beta}[(1-\gamma)F_{\alpha\beta}] - \frac{1}{\varrho}\frac{\partial \tilde{\bar{p}}}{\partial x_\alpha} + \nu\Delta\tilde{\bar{v}}_\alpha \\ & -\frac{1}{1-\gamma}\frac{\partial}{\partial x_\alpha}[(1-\gamma)\tilde{\bar{k}}] + (\tilde{\bar{v}}_\alpha - \bar{\bar{v}}_\alpha)\frac{S_\gamma}{1-\gamma} - \frac{\gamma}{1-\gamma}\bar{\bar{S}}_\alpha \\ & +\frac{1}{\varrho}(\tilde{\bar{p}} - \bar{\bar{p}})\frac{1}{1-\gamma}\frac{\partial \gamma}{\partial x_\alpha} - \gamma(\tilde{\bar{v}}_\beta - \bar{\bar{v}}_\beta)\frac{\partial \tilde{\bar{v}}_\alpha}{\partial x_\beta}. \end{aligned} \tag{12}$$

It should be noted that in the derivation of (12) only the fluctuations of velocity in the non-turbulent zone are assumed irrotational. The mean vorticity may be non-zero.

2.4 Turbulent-Zone Mean Velocity

All conditional moment equations can be cast in a form analogous to the equation for the unconditional moment plus a collection of interface terms (denoted by A with the appropriate indices). For $\bar{\bar{v}}_\alpha$ these terms follow from (7) as

$$\bar{\bar{A}}_\alpha = \bar{\bar{S}}_\alpha + (1-\gamma)(\tilde{\bar{v}}_\beta - \bar{\bar{v}}_\beta)\frac{\partial \bar{\bar{v}}_\alpha}{\partial x_\beta}. \tag{13}$$

From the closure of the intermittency-factor equation the expression for the second part follows without new assumptions as

$$(1-\gamma)(\tilde{\bar{v}}_\beta - \bar{\bar{v}}_\beta)\frac{\partial \bar{\bar{v}}_\alpha}{\partial x_\beta} \cong C_8\frac{\bar{\bar{k}}}{\bar{\bar{\varepsilon}}}\widetilde{v_\beta^* v_\delta^*}\frac{1}{\gamma}(1-\gamma)^{z_1}\frac{\partial \gamma}{\partial x_\gamma}\frac{\partial \bar{\bar{v}}_\alpha}{\partial x_\beta}. \tag{14}$$

For the closure of the first part, the exchange of momentum between the turbulent and non-turbulent zone is considered. If the turbulent zone is faster it will lose momentum to the non-turbulent zone proportional to the velocity difference of the zones and vice versa. The amount of momentum transferred depends on the turbulent time scale and the crossing frequency of the interface which is given by S_γ. Then the following expression emerges

$$\bar{\bar{S}}_\alpha \cong -(\tilde{\bar{v}}_\alpha - \bar{\bar{v}}_\alpha)\left[C_4(1-\gamma)^{z_2}\frac{\bar{\bar{\varepsilon}}}{\bar{\bar{k}}} + \frac{1}{\gamma}S_\gamma\right] \tag{15}$$

where the model for the intermittency source is given by (5) and $C_4 = 2.0$, $z_2 = 2.0$.

2.5 Non-Turbulent Zone Mean Velocity

For the closure of Eq. (12) two assumptions are required. First the flux $F_{\alpha\beta}$ must be expressed in terms of known moments and secondly the pressure difference must be dealt with. For the flux we adopt:

$$F_{\alpha\beta} \cong C_D^*\gamma\frac{\bar{\bar{k}}^2}{\bar{\bar{\varepsilon}}}\left(\frac{\partial \tilde{\bar{v}}_\alpha}{\partial x_\beta} + \frac{\partial \tilde{\bar{v}}_\beta}{\partial x_\alpha}\right) \tag{16}$$

which is a turbulent viscosity model with $C_D^* = 0.06$. The pressure difference terms is assumed to balance with the following group

$$\frac{1}{\varrho}(\tilde{\bar{\bar{p}}} - \bar{\tilde{p}}) \frac{\partial \gamma}{\partial x_\alpha} \cong \frac{1}{1-\gamma} \frac{\partial}{\partial x_\alpha} [(1-\gamma)\, \tilde{\bar{\bar{k}}}] - \frac{F_{\alpha\beta}}{1-\gamma} \frac{\partial \gamma}{\partial x_\beta}. \tag{17}$$

Thus the closed equation for $\tilde{\bar{\bar{v}}}_\alpha$ is obtained using (15). It is given in the appendix where the complete model equations are assembled.

2.6 Turbulent-Zone Stress Tensor

The terms analogous to the unconditional forms are modelled by carrying over the closure assumptions of the unconditional case (Lumley (1978) and Launder, Reece, and Rodi (1975)) to the turbulent zone variables. Closure assumptions like quasi-Gaussianity of the fourth moment for velocity fluctuations have certainly more justification for the conditional case than the unconditional because the center spike corresponding to the small fluctuations in the non-turbulent zone is removed in the conditional case, thus bringing the flatness factors closer to the Gaussian value. The interface group (exact forms see Byggstøyl and Kollmann 1981) requires new considerations. The turbulent zone propagates into the non-turbulent by increasing the vorticity of irrotational parcels of fluid. This propagation is only possible if the net effect on the turbulent zone stress is gain at the expense of non-turbulent-zone fluctuations. Thus

$$\bar{\bar{S}}_{\alpha\beta} \cong C_5\, \gamma(1-\gamma) \frac{\bar{\bar{\varepsilon}}}{\tilde{\bar{\bar{k}}}} \overline{\widetilde{v_\alpha^* v_\beta^*}}. \tag{18}$$

The contribution due to relative motion follows from model assumptions on entrainment and momentum transport introduced above. Then the interface group is given by

$$\bar{\bar{A}}_{\alpha\beta} \cong C_8 (1-\gamma)^{z_1} \frac{\tilde{\bar{\bar{k}}}}{\bar{\bar{\varepsilon}}} \widetilde{v_\gamma^* v_\delta^*} \frac{\partial}{\partial x_\gamma} \widetilde{v_\alpha^* v_\beta^*} \frac{1}{\gamma} \frac{\partial \gamma}{\partial x_\delta} + \left(2\nu \frac{\partial}{\partial x_\delta} \widetilde{v_\alpha^* v_\beta^*} - \widetilde{v_\alpha^* v_\beta^* v_\delta^*} \right) \frac{1}{\gamma} \frac{\partial \gamma}{\partial x_\delta}$$

$$+ \frac{1}{\gamma} (\nu \Delta \gamma - \langle V \delta(s) \rangle)\, \overline{v_\alpha^* v_\beta^*} + \bar{\bar{S}}_{\alpha\beta} \tag{19}$$

where $C_5 = 1.7$.

2.7 Non-Turbulent Zone Stress Tensor

The principal axes of turbulent-zone and non-turbulent zone stress tensors are not aligned. Experimental evidence (Wygnanski and Fiedler (1969); and Gutmark and Wygnanski 1976) suggests that the gradient of the intermittency factor is close to a principal axis of the non-turbulent stress tensor. In the present closure model these considerations were taken into account simply by setting $\widetilde{v_\alpha^0 v_\beta^0}$ proportional to $\gamma \cdot \widetilde{v_\alpha^* v_\beta^*}$ with a factor $C_{11} = 0.5$ for normal stresses and $C_{12} = 0.2$ for the shear stress. These *ad hoc* relations do not enter the calculations directly and are applied to determine unconditional stress components only.

2.8 Dissipation Rate

The dissipation rate in the non-turbulent zone is negligible, thus the propagation of the interface provides a gain of dissipation rate. The model for $\bar{\bar{S}}_\varepsilon$ is then

$$\bar{\bar{S}}_\varepsilon \cong C_6 \gamma (1-\gamma) \frac{\bar{\bar{\varepsilon}}^2}{\tilde{\bar{\bar{k}}}} \tag{20}$$

where $C_6 = 1.9$. The complete interface group is given by

$$\bar{\tilde{A}}_\varepsilon \cong C_8 \frac{\bar{\tilde{k}}}{\bar{\tilde{\varepsilon}}} \overline{\widetilde{v_\gamma^* v_\delta^*}} (1-\gamma)^{z_1} \frac{1}{\gamma} \frac{\partial \gamma}{\partial x_\gamma} \frac{\partial \bar{\tilde{\varepsilon}}}{\partial x_\delta} + \left(2\nu \frac{\partial \varepsilon}{\partial x_\delta} - \overline{\widetilde{v_\gamma^* \varepsilon^*}} \right) \frac{1}{\gamma} \frac{\partial \gamma}{\partial x_\delta}$$
$$+ \frac{1}{\gamma} (\nu \Delta \gamma - \langle V \delta(s) \rangle)\, \bar{\tilde{\varepsilon}} + \bar{\tilde{S}}_\varepsilon . \tag{21}$$

3. Applications

The conditional Reynolds-stress closure model was evaluated by calculating several parabolic shear flows and comparing the results with experimental data. Two flows will be discussed in detail: the plane and round jet.

The boundary conditions in both cases where Dirichlet conditions at the outer edge of the flow and homogeneous Neumann conditions at the symmetry axis with the exception of the shear stress. For the plane jet the transport equation for the shear stress was solved, which requires the stress to be zero at the symmetry axis. For the round jet the equation for $\frac{1}{r} \overline{\widetilde{u^* v^*}}$ was solved, which has to satisfy

$$\frac{\partial}{\partial r} \left(\frac{\overline{\widetilde{u^* v^*}}}{r} \right) = 0$$

at the symmetry axis. For initial conditions several possibilities exist:

1) Fully turbulent flow at the entrance: all profiles including intermittency factor and conditional stresses must be known.
2) Nearly laminar flow: the intermittency factor is set initially to a small (but non-zero) value and mean values in both zones are set equal. Turbulent zone stresses are set to a small initial value.
3) Nearly laminar flow: the intermittency factor is set initially to be proportional to the mean velocities varying from unity to a small external value. The turbulent zone stresses are set to such a small value that a corresponding turbulent viscosity does not exceed the laminar viscosity.

For the calculations presented here the third option was chosen. The second option gives results that are different in the developing region of the jet but approach the same profiles as for the third set of conditions as approximate self-similarity is reached.

For the numerical solution by means of a finite-difference procedure (Patankar and Spalding 1970) all first-order moments were defined at the nodes of the grid, whereas all second-order moments were defined at the midpoint. This leads to nearly second-order accurate representation of production terms in the stress equations and the diffusion term in the mean velocity equation (turbulent zone). No tendency to wiggle as described by Launder and Morse (1979) was encountered.

3.1 Results

The intermittency factor for the plane jet in Fig. 1 and the round jet in Fig. 8 agrees well with the experiments of Gutmark and Wygnanski (1970). In both figures the derivative $\frac{\partial \gamma}{\partial y}$ is shown as broken line indicating a shape similar to the crossing frequency in Fig. 2 of

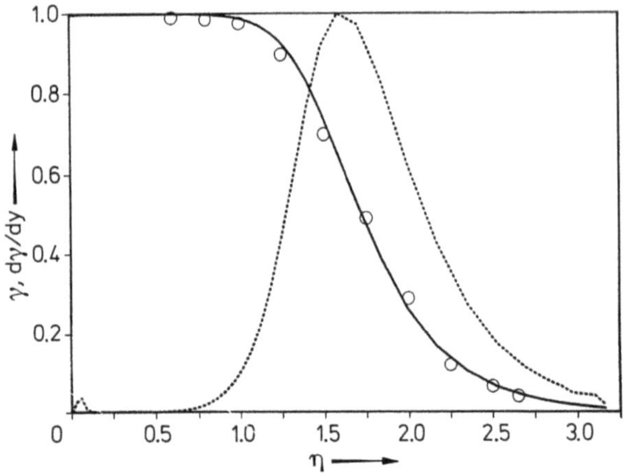

Fig. 1. Intermittency factor γ and $d\gamma/dy$ as function of $\eta = y/y_{0.5}$ ($y_{0.5}$: half-width of $\bar{u}$) for a plane jet at $x/D = 40$. Calculation: line, experiment [6]: symbols

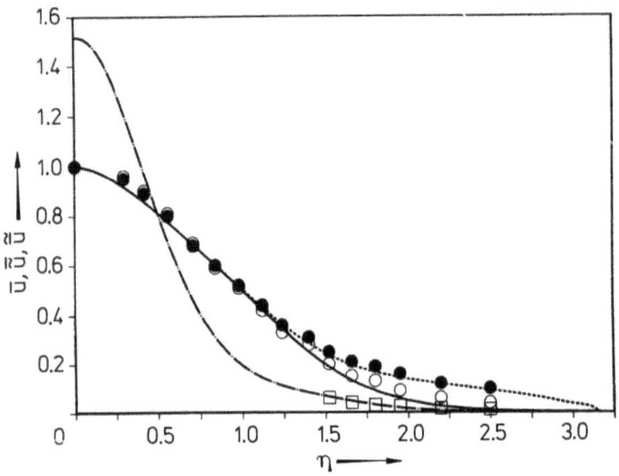

Fig. 2. Mean velocities for a plane jet at $x/D = 40$. Unconditional: ——— calculated, ○ experiment [6]. Turbulent zone: ········ calculated, ● experiment [6]. Non-turbulent zone: —·—·— calculated, □ experiment [6]

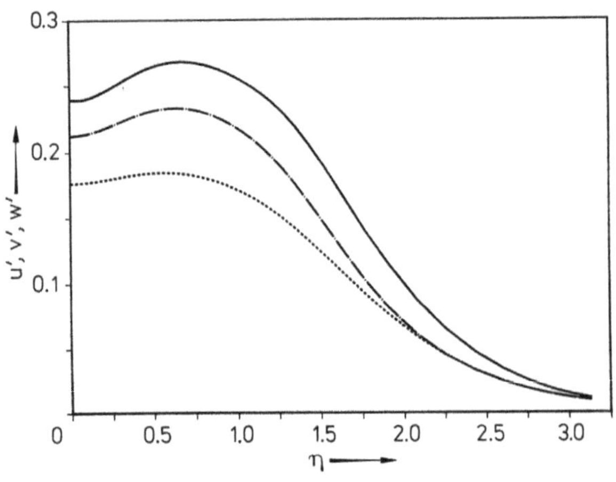

Fig. 3. Unconditional normal stress components for a plane jet at $x/D = 40$. $(\overline{u'^2})^{1/2}/(\bar{u}_0 - \bar{u}_\infty)$: ———, $(\overline{v'^2})^{1/2}/(\bar{u}_0 - \bar{u}_\infty)$: ······, $(\overline{w'^2})^{1/2}/(\bar{u}_0 - \bar{u}_\infty)$: —·—·—

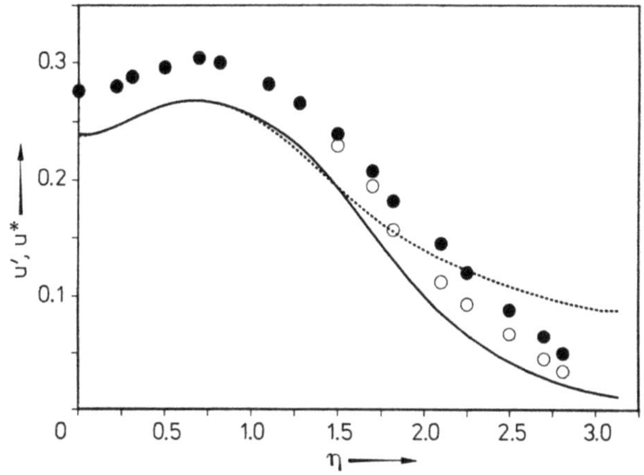

Fig. 4. Unconditional and turbulent zone normal stress component for a plane jet at $x/D = 40$. $(\overline{u'^2})^{1/2}/(\bar{u}_0 - \bar{u}_\infty)$: ——— calculated, ○ experiment [6], $(\widetilde{u^{*2}})^{1/2}/(\bar{u}_0 - \bar{u}_\infty)$: ········ calculated, ● experiment [6]

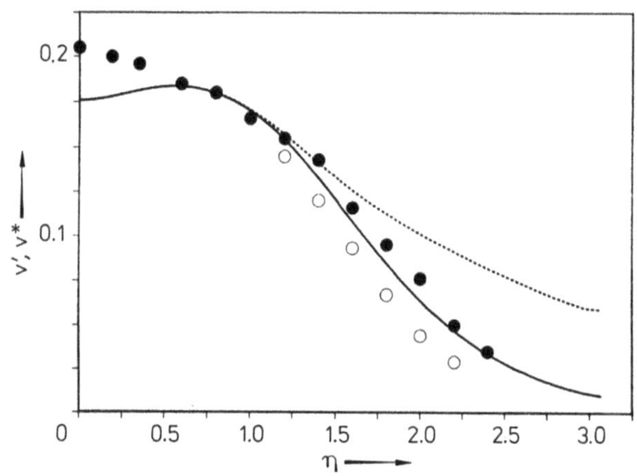

Fig. 5. Unconditional and turbulent zone normal stress component for a plane jet at $x/D = 40$. $(\overline{v'^2})^{1/2}/(\bar{u}_0 - \bar{u}_\infty)$: ——— calculated, ○ experiment [6], $(\widetilde{v^{*2}})^{1/2}/(\bar{u}_0 - \bar{u}_\infty)$: ········ calculated, ● experiment [6]

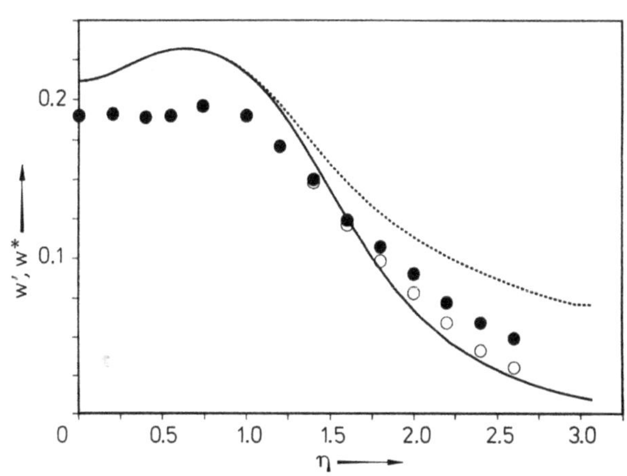

Fig. 6. Unconditional and turbulent zone normal stress component for a plane jet at $x/D = 40$. $(\overline{w'^2})^{1/2}/(\bar{u}_0 - \bar{u}_\infty)$: ——— calculated, ○ experiment [6], $(\widetilde{w^{*2}})^{1/2}/(\bar{u}_0 - \bar{u}_\infty)$: ········ calculated, ● experiment [6]

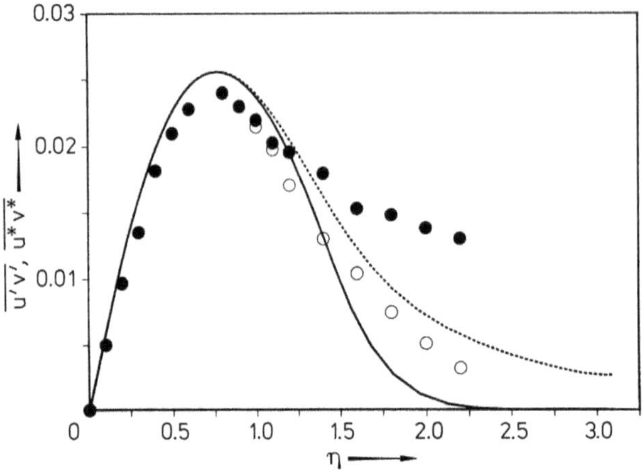

Fig. 7. Unconditional and turbulent zone normal stress component for a plane jet at $x/D = 40$. $(\overline{u'v'})/(\bar{u}_0 - \bar{u}_\infty)^2$: —— calculated, ○ experiment [6], $(\widetilde{u^* v^*})/(\bar{u}_0 - \bar{u}_\infty)^2$: ········ calculated, ● experiment [6]

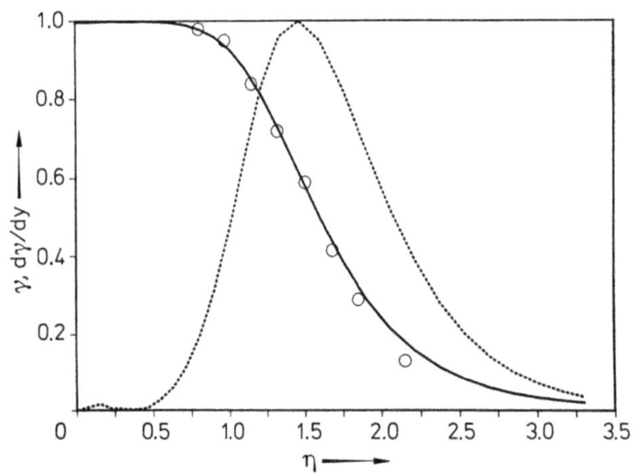

Fig. 8. Intermittency factor γ and $d\gamma/dy$ as function of $\eta = y/y_{0.5}$ ($y_{0.5}$: half-width of $\bar{u}$) for a round jet at $x/D = 60$.

Gutmark and Wygnanski (1976). The mean velocity for both zones and the unconditional mean in Fig. 2 (plane jet) and Fig. 9 (round jet) show the typical result produced by a conditional closure model: the non-turbulent zone mean decays much slower than the turbulent zone mean, because in the non-turbulent zone fluctuations are induced by the random movement of the interface only and decay rapidly with distance from it. If a turbulent pseudo-viscosity is established for the turbulent zone, then the corresponding pseudo-viscosity in the non-turbulent fluid is roughly proportional to the intermittency factor times the turbulent zone viscosity, because the deeper a non-turbulent packet of fluid penetrates the turbulent zone, the longer it must follow the movement of the surrounding turbulent fluid. Since γ can be interpreted as time spent in the turbulent zone divided by total time the proportionality indicated above follows. This is incorporated in the closure model (16) and is reflected in the results for $\tilde{u}$ and $\tilde{\tilde{u}}$. Only a few experimental points for the non-turbulent zone $\tilde{\tilde{u}}$ are available but they seem to support the calculated results in Fig. 2. The unconditional normal stress components in Fig. 3 and Fig. 10 show their relative magnitude and form. A detailed comparison of the turbulent zone and unconditional stress components with the experiments as far as available is shown in Fig. 4 to Fig. 7 for the plane jet and Fig. 11 to Fig. 14 for the round jet. In the core region of the flows (γ near unity) the calculations are determined by the same closure assumptions as the unconditional model

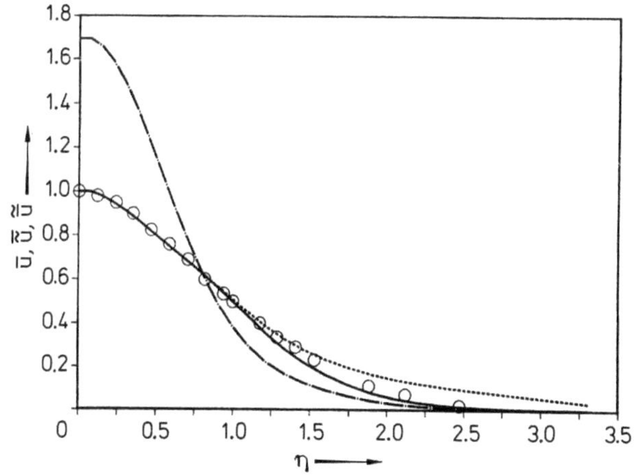

Fig. 9. Mean velocities for a round jet at $x/D = 60$. Unconditional: —— calculated, o experiment [7]. Turbulent zone: ······ calculated. Non-turbulent zone: —·—·— calculated

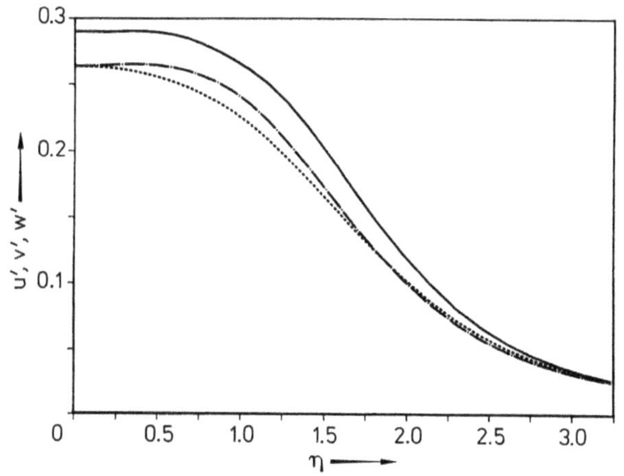

Fig. 10. Unconditional normal stress components for a round jet at $x/D = 60$. $(\overline{u'^2})^{1/2}/(\bar{u}_0 - \bar{u}_\infty)$: ——, $(\overline{v'^2})^{1/2}/(\bar{u}_0 - \bar{u}_\infty)$: ······, $(\overline{w'^2})^{1/2}/(\bar{u}_0 - \bar{u}_\infty)$: —·—·—

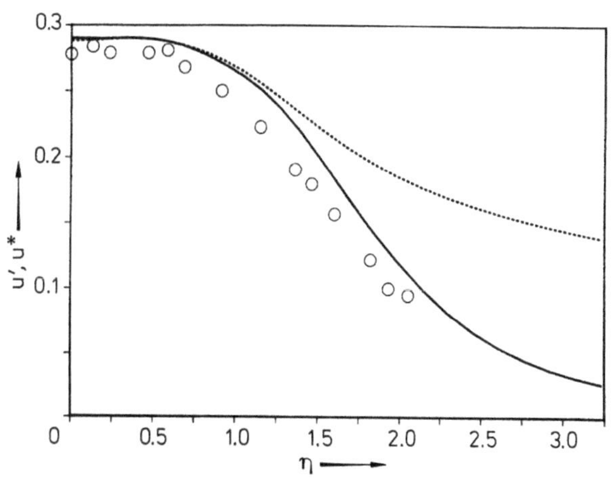

Fig. 11. Unconditional and turbulent zone normal stress component for a round jet at $x/D = 60$. $(\overline{u'^2})^{1/2}/(\bar{u}_0 - u_\infty)$: —— calculated, o experiment [7], $(\widetilde{u^{*2}})^{1/2}/(\bar{u}_0 - \bar{u}_\infty)$: ······ calculated

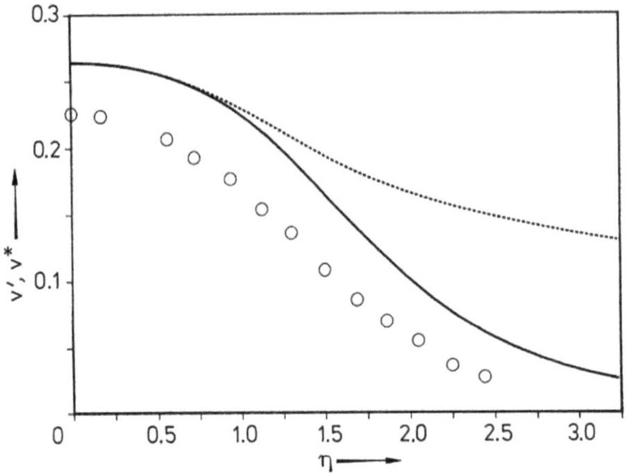

Fig. 12. Unconditional and turbulent zone normal stress component for a round jet at $x/D = 60$. $(\overline{v'^2})^{1/2}/(\bar{u}_0 - u_\infty)$: —— calculated, o experiment [7], $(\overline{v^{*2}})^{1/2}/(\bar{u}_0 - \bar{u}_\infty)$: ······ calculated

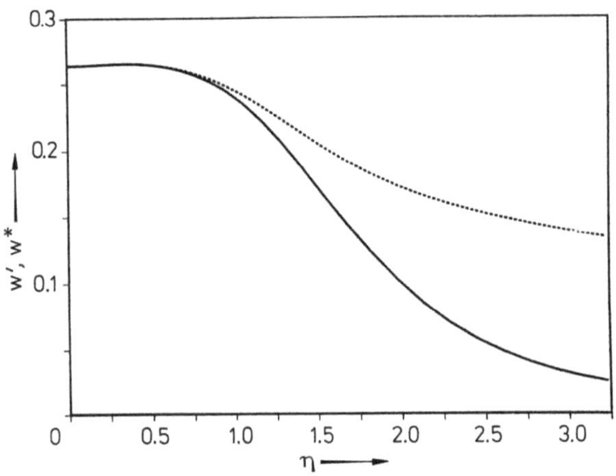

Fig. 13. Unconditional and turbulent zone normal stress component for a round jet at $x/D = 60$. $(\overline{w'^2})^{1/2}/(\bar{u}_0 - u_\infty)$: —— calculated, o experiment [7], $(\widetilde{w^{*2}})^{1/2}/(\bar{u}_0 - \bar{u}_\infty)$: ······ calculated

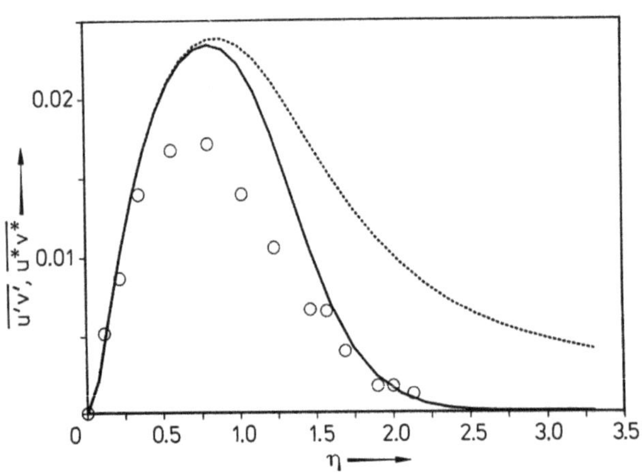

Fig. 14. Unconditional and turbulent zone normal stress component for a round jet at $x/D = 60$. $(\overline{u'v'})/(\bar{u}_0 - u_\infty)^2$: —— calculated, o experiment [7], $(\widetilde{u^*v^*})/(\bar{u}_0 - \bar{u}_\infty)^2$: ······ calculated

(Launder, Reece, and Rodi 1975), whereas in the outer part of the flows the interface terms become significant. It is noticeable that the calculated normal stresses (turbulent zone) are in both cases higher than the measurements in the outer part, whereas the shear stress is lower than the measured values in this part (plane jet Fig. 7). This is a consequence of the model (18) for the interface terms in the stress equations and (20) for the dissipation rate. This part of the closure is currently undergoing further development.

4. Conclusions

The closure of conditional moment equations leads to detailed information on the properties of the flow in turbulent and non-turbulent zones. The interaction of the zones requires some new model assumptions, which are however guided by the condition that conditional sources must combine appropriately to zero for the unconditional moment. This approach has potential and further extensions and applications can be considered.

Turbulent flows with combustion: the instantaneous flame front can be considered an interface separating burnt and unburnt regions (premixed flames) or fuel and oxidizer (diffusion flames) zones if the reactions are sufficiently fast (large Damköhler numbers).

The notion of an interface can be generalized to define a family of fluctuating surfaces enclosing parts of the flow with distinct properties in terms of the discriminating scalar. This allows the constructing of length and time scales appropriate to such parts.

Appendix

The complete set of model equations for Cartesian coordinates is given here in the form such that convection along unconditional mean streamlines is balanced by diffusion and sources.

Intermittency factor γ:

$$\frac{\bar{D}\gamma}{Dt}=\frac{\partial}{\partial x_\alpha}\left[C_8\frac{\tilde{\bar{\bar{k}}}}{\tilde{\bar{\bar{\varepsilon}}}}\,\overline{\widetilde{v^*_\alpha v^*_\beta}}\,(1-\gamma)^{z_1}\frac{\partial\gamma}{\partial x_\beta}\right]-C_{g1}\gamma(1-\gamma)\frac{\overline{\widetilde{v^*_\alpha v^*_\beta}}}{\tilde{\bar{\bar{k}}}}\left(\frac{\partial\bar{\bar{v}}_\alpha}{\partial x_\beta}+\frac{\partial\bar{\bar{v}}_\beta}{\partial x_\alpha}\right)$$
$$-C_{g2}\frac{\tilde{\bar{\bar{k}}}^2}{\tilde{\bar{\bar{\varepsilon}}}}\frac{\partial\gamma}{\partial x_\alpha}\frac{\partial\gamma}{\partial x_\alpha}-C_{g3}\gamma(1-\gamma)\frac{\tilde{\bar{\bar{\varepsilon}}}}{\tilde{\bar{\bar{k}}}}.\qquad\text{(A 1)}$$

Turbulent zone mean velocity v_α:

$$\frac{\bar{D}}{Dt}\bar{\bar{v}}_\alpha=-\frac{1}{\gamma}\frac{\partial}{\partial x_\beta}\left(\gamma\,\overline{\widetilde{v^*_\alpha v^*_\beta}}\right)-\frac{1}{\varrho}\frac{\partial\bar{\bar{p}}}{\partial x_\alpha}+\nu\Delta\bar{\bar{v}}_\alpha+(\tilde{\bar{v}}_\alpha-\bar{\bar{v}}_\alpha)\left[C_4(1-\gamma)^{z_2}\frac{\tilde{\bar{\bar{\varepsilon}}}}{\tilde{\bar{\bar{k}}}}+\frac{S_\gamma}{\gamma}\right]$$
$$+C_8\frac{\tilde{\bar{\bar{k}}}}{\tilde{\bar{\bar{\varepsilon}}}}\,\overline{\widetilde{v^*_\beta v^*_\gamma}}\,\frac{1}{\gamma}(1-\gamma)^{z_1}\frac{\partial\gamma}{\partial x_\gamma}\frac{\partial\bar{\bar{v}}_\alpha}{\partial x_\beta}.\qquad\text{(A 2)}$$

The intermittency source S_γ is given by (5).

Non-turbulent zone mean velocity $\tilde{\bar{v}}_\alpha$:

$$\frac{\bar{D}}{Dt}\tilde{\bar{v}}_\alpha=\frac{\partial}{\partial x_\beta}\left[C^*_D\frac{\tilde{\bar{\bar{k}}}^2}{\tilde{\bar{\bar{\varepsilon}}}}\left(\frac{\partial\tilde{\bar{v}}_\alpha}{\partial x_\beta}+\frac{\partial\tilde{\bar{v}}_\beta}{\partial x_\alpha}\right)\right]-\frac{1}{\varrho}\frac{\partial\tilde{\bar{p}}}{\partial x_\alpha}+\nu\Delta\tilde{\bar{v}}_\alpha-(\tilde{\bar{v}}_\alpha-\bar{\bar{v}}_\alpha)\,C_4\gamma(1-\gamma)^{z_2-1}\frac{\tilde{\bar{\bar{\varepsilon}}}}{\tilde{\bar{\bar{k}}}}$$
$$+C_8\frac{\tilde{\bar{\bar{k}}}}{\tilde{\bar{\bar{\varepsilon}}}}\,\overline{\widetilde{v^*_\beta v^*_\gamma}}\,(1-\gamma)^{z_1-1}\frac{\partial\gamma}{\partial x_\gamma}\frac{\partial\bar{\bar{v}}_\alpha}{\partial x_\beta}.\qquad\text{(A 3)}$$

Turbulent zone stress tensor $\widetilde{v_\alpha^* v_\beta^*}$:

$$\frac{\bar{D}}{Dt}\widetilde{v_\alpha^* v_\beta^*} = \frac{\partial}{\partial x_\gamma}\left(C_5 \frac{\bar{\bar{k}}}{\bar{\bar{\varepsilon}}}\widetilde{v_\delta^* v_\gamma^*}\frac{\partial}{\partial x_\delta}\widetilde{v_\alpha^* v_\beta^*}\right) - \widetilde{v_\alpha^* v_\gamma^*}\frac{\partial \bar{\tilde{v}}_\beta}{\partial x_\gamma} - \widetilde{v_\beta^* v_\gamma^*}\frac{\partial \bar{\tilde{v}}_\alpha}{\partial x_\gamma} - \frac{2}{3}\delta_{\alpha\beta}\bar{\bar{\varepsilon}}$$

$$- C_1 \frac{\bar{\bar{\varepsilon}}}{\bar{\bar{k}}}\left(\widetilde{v_\alpha^* v_\beta^*} - \frac{2}{3}\delta_{\alpha\beta}\bar{\bar{k}}\right) + \Phi_{\alpha\beta} + \bar{\bar{A}}_{\alpha\beta}. \qquad \text{(A 4)}$$

$\bar{\bar{A}}_{\alpha\beta}$ is given by (19) and the fast response part of the pressure-strain correlation by Launder, Reece, and Rodi (1975)

$$\Phi_{\alpha\beta} = -\frac{C_2 + 8}{11}\left(P_{\alpha\beta} - \frac{2}{3}\delta_{\alpha\beta}P\right) - \frac{30\,C_2 - 2}{55}\bar{\bar{k}}\left(\frac{\partial \bar{\tilde{v}}_\alpha}{\partial x_\beta} + \frac{\partial \bar{\tilde{v}}_\beta}{\partial x_\alpha}\right)$$

$$- \frac{8\,C_2 - 2}{11}\left(D_{\alpha\beta} - \frac{2}{3}\delta_{\alpha\beta}P\right) \qquad \text{(A 5)}$$

where

$$P_{\alpha\beta} \equiv -\widetilde{v_\alpha^* v_\gamma^*}\frac{\partial \bar{\tilde{v}}_\beta}{\partial x_\gamma} - \widetilde{v_\beta^* v_\gamma^*}\frac{\partial \bar{\tilde{v}}_\alpha}{\partial x_\gamma}$$

$$P \equiv \tfrac{1}{2}P_{\alpha\alpha}$$

$$D_{\alpha\beta} \equiv -\widetilde{v_\alpha^* v_\gamma^*}\frac{\partial \bar{\tilde{v}}_\gamma}{\partial x_\beta} - \widetilde{v_\beta^* v_\gamma^*}\frac{\partial \bar{\tilde{v}}_\gamma}{\partial x_\alpha}.$$

The constants are $C_1 = 1.5$ ($C_1 = 2.5$ for round jets) and $C_2 = 0.4$ of Launder, Reece, and Rodi (1975).

Dissipation rate $\bar{\bar{\varepsilon}}$:

$$\frac{\bar{D}}{Dt}\bar{\bar{\varepsilon}} = \frac{\partial}{\partial x_\alpha}\left(C_\varepsilon \frac{\bar{\bar{k}}}{\bar{\bar{\varepsilon}}}\widetilde{v_\alpha^* v_\beta^*}\frac{\partial \bar{\bar{\varepsilon}}}{\partial x_\beta}\right) - C_{\varepsilon 1}\frac{\bar{\bar{\varepsilon}}}{\bar{\bar{k}}}\widetilde{v_\alpha^* v_\beta^*}\frac{\partial \bar{\tilde{v}}_\alpha}{\partial x_\beta} - C_{\varepsilon 2}\frac{\bar{\bar{\varepsilon}}^2}{\bar{\bar{k}}} + \bar{\bar{A}}_\varepsilon \qquad \text{(A 6)}$$

$\bar{\bar{A}}_\varepsilon$ is given by (21) and the constants are $C_\varepsilon = 0.15$, $C_{\varepsilon 1} = 1.44$, $C_{\varepsilon 2} = 1.92$ of Launder, Reece, and Rodi (1975).

Acknowledgement. The second author (WK) was supported by DOE Grant AS03-76SF00034. Part of the work was done at the Combustion Research Facility of Sandia Livermore.

References

1. Corrsin, S., Kistler, A. L. (1954): NACA TN-3133
2. Dopazo, C. (1977): J. Fluid Mech. **81**, 433
3. Byggstøyl, S., Kollmann, W. (1981): Int. J. Heat Mass Transfer **24**, 1811
4. Lumley, J. L. (1978): Adv. Appl. Mech. **18**, 123
5. Launder, B. E., Reece, G. J., Rodi, W. (1975): J. Fluid Mech. **68**, 537
6. Gutmark, E., Wygnanski, I. (1976): J. Fluid Mech. **73**, 465
7. Wygnanski, I., Fiedler, H. (1969): J. Fluid Mech. **38**, 577
8. Patankar, S. V., Spalding, D. B.: *Heat and Mass Transfer in Boundary Layers* (Intertext, London 1970)
9. Launder, B. E., Morse, A.: In *Turbulent Shear Flows 1*, ed. by F. Durst, B. E. Launder, F. W. Schmidt, J. H. Whitelaw (Springer, Berlin, Heidelberg, New York 1979) p. 279

Parameterization of Turbulent Transport in Swirling Flows* – I: Theoretical Considerations

David Ettestad and John L. Lumley

Cornell University, Ithaca, New York 14853, USA

Abstract

A second order closure scheme is proposed for turbulent transport in swirling flows. Specifically, a scheme is constructed according to the same principles that led to the successful description of buoyancy-driven turbulent flows (Zeman, 1975; Zeman and Lumley, 1976a, b, c; Lumley et al. 1978). In the narrow gap approximation the scheme takes on a particularly simple form, which can be rationalized by a simple physical model.

1. Introduction

By swirling flows, we mean axisymmetric flows with a non-negligible component of azimuthal velocity, in which the mean axial velocity is everywhere positive (see Fig. 1). That is, we exclude from consideration here flows which are recirculating. Since the recirculation is caused by the adverse axial pressure gradient resulting from the swirl, we are considering here only weak swirl.

Flows of this type occur frequently in nature and technology. The flow in a tornado is of this type, as is that in an aircraft trailing vortex. Flow in a gas turbine combustor is generally recirculating and not axisymmetric; hence, our considerations will not be directly applicable to this flow. We hope, however, that a successful parameterization of swirl under our simplified conditions will lead to an understanding of the more complex flow.

Swirling flows present a challange to the modeler due to the influence on radial turbulent transport of the radial distribution of azimuthal velocity. This is weakly analogous to the effect of density stratification in a flow with gravity. That is, a small-disturbance stability analysis of a layer of fluid heated from below (Pellew and Southwell, 1940) and of isothermal circular Couette flow (Taylor, 1923) leads to the same equation, where the radial distribution of angular momentum in the Couette flow plays the role of the vertical distribution of density in the heated flow.

The analogy is not strict, however, since the mechanisms differ. In the heated flow (see Fig. 2), a parcel of fluid displaced upward (in the limit of no molecular transport) does not exchange heat with its surroundings, and hence moves at constant entropy, or constant potential temperature – the temperature adjusted by the adiabatic change in temperature due to the change in pressure (Lumley and Panofsky, 1964). The parcel will thus experience a density anomaly at its new height, if the local potential temperature is different. If potential temperature is increasing upward (a stable situation), the parcel will be cooler, producing a

* Supported in part by the U.S. Office of Naval Research under the following programs: Physical Oceanography (Code 422PO), Fluid Dynamics (Code 438), Power (Code 473); in part by the U.S. National Science Foundation under grant no. ATM 79-22006 and CME 79-19817; and in part by the U.S. Air Force Office of Scientific Research

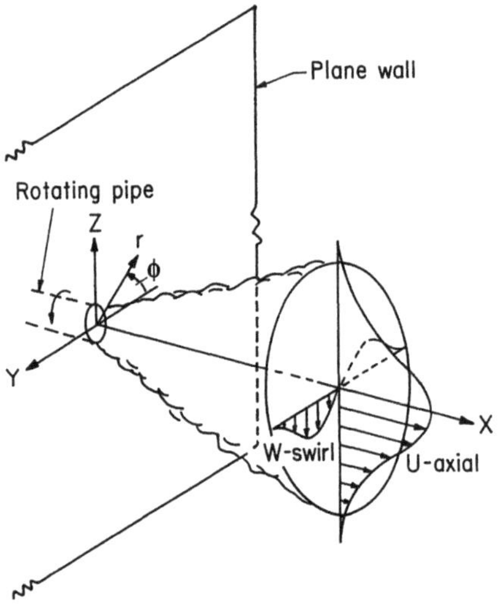

Fig. 1. Definition sketch (after Pratte and Keffer, 1972). Note that, in the tensor notation used in the mathematical sections, $u^2 = W/r$, $u_2 = r\,W$, $u^1 = V$, $u^3 = U$

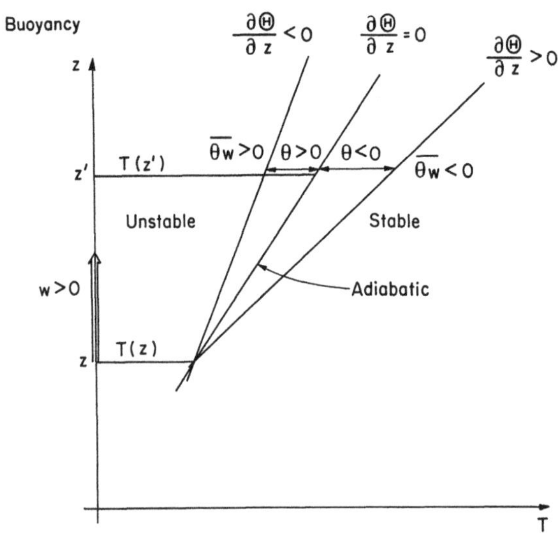

Fig. 2. Change of temperature of a wandering parcel of fluid in the atmosphere. The parcel rises along the adiabatic line, producing a positive temperature anomaly relative to the surroundings (and hence a positive heat flux) if the potential temperature is falling with altitude

negative heat flux, and vice versa. The analysis of Zeman (1975), Zeman and Lumley (1976a, b, c) and Lumley et al. (1978) indicates that the vertical diffusivity can be written as

$$\nu_T \simeq T(\overline{w^2} + c\,\overline{\theta w}\,Tg/\Theta_0) \tag{1.1}$$

where c is a constant, Θ_0 is the temperature of an adiabatic atmosphere, T is a Lagrangian integral time scale, θ is the temperature fluctuation and w is the vertical velocity fluctuation. In the absence of buoyancy, the vertical diffusivity would be simply the first term. The presence of the second term has the effect of reducing the vertical diffusivity when the heat flux is negative, in stable conditions, and of increasing it in unstable conditions, when the heat flux is positive. This, of course, is in accordance with observation: in unstable stratification, the buoyantly driven eddies increase the vertical transport, while stable stratification suppresses the vertical motions, reducing the vertical transport. Incidentally, the fact that the heat flux and the gradient of the potential temperature change sign together suggests something like

$$\overline{\theta w} \approx -\overline{w^2}\, T\, \partial\Theta/\partial z \tag{1.2}$$

and this is the standard mixing length or gradient transport approximation.

In the case of swirl, a particle displaced radially approximately conserves its angular momentum (see Fig. 3); at each radius, there is a radial pressure gradient that corresponds to the existing gradient of mean angular momentum. Whether the particle will return to its original location or continue depends on the radial pressure gradient it needs for equilibrium compared to that at its new location. If the mean angular momentum is increasing outward, the radial pressure gradient at the new location is stronger than needed for equilibrium, and the particle returns toward its original location, for a stable situation. The azimuthal velocity of the particle displaced to a larger radius is, under these circumstances, below the local value, so that the azimuthal velocity fluctuation is negative, producing a negative radial-azimuthal Reynolds stress, and vice versa. We will show below that, in analogy with the case of buoyancy, the radial diffusivity is given by

$$\nu_T \cong T(\overline{v^2} + c\,\overline{vw}\, TW/r). \tag{1.3}$$

In the absence of swirl, the radial turbulent diffusivity is given by the first term. The effect of the second term is to decrease the radial diffusivity when the mean angular momentum distribution is stable, and vice versa. Again, this corresponds to observation: growth of a boundary layer on a concave surface (unstable) is accelerated, while growth of a boundary layer on a convex surface (stable) – presuming separation is avoided – is inhibited. Again, the fact that the radial-azimuthal Reynolds stress and the radial gradient of angular momentum change sign at the same time suggests something like

$$\overline{vw} \cong -T\overline{v^2}\, r^{-1}\, \partial(rW)/\partial r \tag{1.4}$$

as a sort of mixing length or gradient transport approximation.

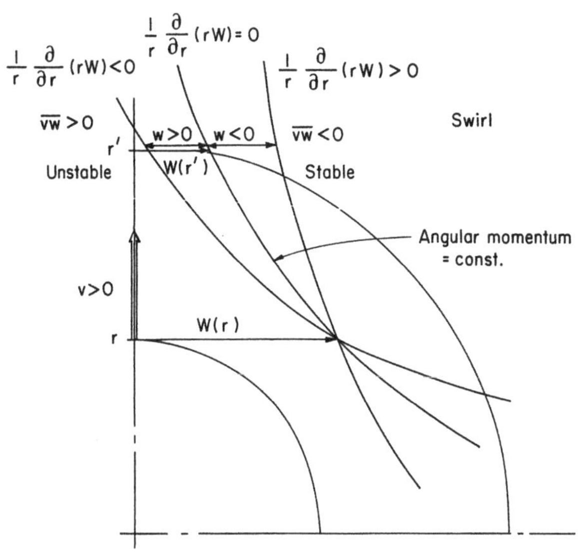

Fig. 3. Change of azimuthal velocity of a wandering parcel of fluid in a swirling flow. The parcel moves out along a line of constant angular momentum, producing a positive swirl velocity anomaly (and a positive radial-swirl Reynolds stress) if the angular momentum falls with radius

The essential difference is the necessary presence of an azimuthal fluctuation in the case of swirl, so that a Reynolds stress is induced in the azimuthal plane. In the heated flow case, no horizontal fluctuation is necessarily associated with a vertical fluctuation, in the absence of wind shear. Hence, in the swirl case we will have strong coupling between the radial and azimuthal components of all fluxes.

Thus we expect that a radial distribution of angular momentum increasing outward will be a stabilizing influence, and will suppress radial transport and reduce the rate of spread of a scalar marker, while a radial distribution of angular momentum decreasing outward will have a destabilizing influence, and will augment radial transport, and increase the rate of spread of a scalar marker. This is borne out by experiment (Eskinazi and Yeh, 1956; Wyngaard, 1968). We can construct a simple physical model to illustrate the mechanism, which is, of course, essentially Rayleigh's stability criterion. If we consider motion confined to planes perpendicular to the axis, and neglect viscosity and non-linear and unsteady terms, and assume that fluid displaced radially conserves angular momentum, we can write for the radial displacement ξ of a material point

$$\ddot{\xi} = -2(W/r)(W' + W/r)\,\xi \tag{1.5}$$

where $W(r)$ is the azimuthal velocity. Thus, if $W' + W/r \geqq 0$ (radial gradient of angular momentum positive) the displacement is stable. The induced azimuthal velocity fluctuation is

$$w = -(W' + W/r)\,\xi \tag{1.6}$$

so that, if $W' + W/r \geqq 0$, a point displaced outward will be moving slower than its surroundings. Following now the same procedure as was used in Lumley et al. (1978), we imagine a velocity field v^0, w^0 which evolves from $t = 0$ without the influence of rotation, so that we may write to first order

$$v = v^0 - 2(W/r)(W' + W/r)\int_0^t d\tau \int_0^\tau d\tau'\, v^0(\tau'). \tag{1.7}$$

That is, we are saying that, to first order, the radial velocity is composed of the non-rotating radial velocity plus the integral of the radial acceleration (1.5) produced by the rotation, the radial displacement being given by the integral of the (non-rotating) radial velocity.

Hence, to first order we have

$$\overline{v^2} = \overline{v^{02}}\,[1 - 4(W/r)(W' + W/r)\,\beta T^2] \tag{1.8}$$

(squaring and averaging (1.7)), where T is a Lagrangian integral time scale for the 0 field undisturbed by rotation, and β is a constant. We see that, if $W' + W/r \geqq 0$, $\overline{v^2}$ is indeed reduced below the value for a field without rotation. We may rewrite (1.4) in a form similar to that referred to above. Since, to first order

$$w = w^0 - (W' + W/r)\int_0^t v^0\, d\tau \tag{1.9}$$

(i.e. – using (1.6) for the induced additional azimuthal velocity, with the integral of the non-rotating radial velocity as an estimate for the radial displacement) we may write

$$\overline{vw} = -(W' + W/r)\,\alpha T\overline{v^{02}} \tag{1.10}$$

if we take the v^0, w^0 field as having no Reynolds stress – that is, all the stress is induced by the rotation. Hence, we may write (1.8) as

$$\overline{v^2} = \overline{v^{02}}\,[1 + 4\beta(W/r)\,T\,\overline{vw}/\alpha\,\overline{v^{02}}]. \tag{1.11}$$

It is evident that this is essentially identical to Eq. (1.3).

While it is tempting to try to parameterize swirl directly in a $k - \varepsilon$ model (Jones and Launder, 1973) using forms such as (1.8) and (1.10), we will give our attention here to the development of a more complete Reynolds stress model similar to that developed by Zeman (1975), Zeman and Lumley (1976a, b, c) and Lumley et al. (1978). From the experience gained from this model, we may hope to be in a better position to develop a simple parameterization.

In modeling of swirling flows, there are more problems than the parameterization of turbulent transport. In particular, the equation for the dissipation, or equivalently for a length scale, has not been satisfactorily extended to axisymmetric non-swirling flows; a beginning has been made on its extension to swirling flows (Launder et al. 1977), but it is expected that bringing this effort to a satisfactory conclusion will be non-trivial. There may also be difficulties with the model for the rapid part of the pressure-strain correlation. However, we will not treat these difficult problems here. Recent experimental evidence (Ramaprian and Shivaprasad, 1981) indicates that it is the differences in the turbulent transport that are largely responsible for the observed differences in structure in turbulent boundary layers; hence, by attacking the transport first we feel we are according appropriate priorities.

In the following several sections we develop mathematically the form of the third moments in a swirling flow. The reader uninterested in these mathematical details should turn directly to the Discussion.

2. Mathematical Preliminaries

We will work in cylindrical coordinates. If x^1, x^2, x^3 is a Cartesian system, we define the cylindrical system $\bar{x}^1$, $\bar{x}^2$, $\bar{x}^3$ by

$$\begin{aligned} x^1 &= \bar{x}^1 \cos \bar{x}^2 \\ x^2 &= \bar{x}^1 \sin \bar{x}^2 \\ x^3 &= \bar{x}^3. \end{aligned} \tag{2.1}$$

We use the notation of Lumley (1970), Appendix 2. The transformation matrices are given by

$$\underset{\sim}{X}{}^i_j = \begin{vmatrix} \cos \bar{x}^2 & -\bar{x}^1 \sin \bar{x}^2 & 0 \\ \sin \bar{x}^2 & \bar{x}^1 \cos \bar{x}^2 & 0 \\ 0 & 0 & 1 \end{vmatrix} \quad \bar{X}^i_j = \begin{vmatrix} \cos \bar{x}^2 & \sin \bar{x}^2 & 0 \\ -\sin \bar{x}^2/\bar{x}^1 & \cos \bar{x}^2/\bar{x}^1 & 0 \\ 0 & 0 & 1 \end{vmatrix}. \tag{2.2}$$

The only non-zero Christoffel symbols are $\Gamma^2_{12} = \Gamma^2_{21} = 1/\bar{x}^1$, $\Gamma^1_{22} = -\bar{x}^1$. The metric tensor is diagonal, with $g_{11} = g_{33} = 1$, $g_{22} = (\bar{x}^1)^2$. Rather than attempting to obtain equations for the fluctuating velocities, and to form from them the equations for the Reynolds stress and higher moments, we will write the equations for the moments directly in general tensor form and specialize them for our particular case.

Note that we will use the co- and contravariant velocities obtained from the Cartesian velocities by formal transformation, so that

$$\bar{u}_2 = \underline{X}_2^j u_j = \bar{x}^1 W$$
$$\bar{u}^2 = \bar{X}_j^2 u_j = W/\bar{x}^1 = \bar{g}^{22}\bar{u}_2 \tag{2.3}$$

where W is the physical azimuthal velocity. Hence, neither u^2 nor u_2 is physically a velocity; u^2 is an angular velocity, and u_2 an angular momentum. These are mathematically the most convenient coordinates; however, for comparison with experiment it is necessary to change to physical velocities. Only the azimuthal component need be changed (since $g_{11} = g_{33} = 1$); for the physical component we have

$$\bar{u}^2(\bar{g}_{22})^{1/2} = \bar{u}_2(\bar{g}^{22})^{1/2} = W. \tag{2.4}$$

To simplify the notation, where no confusion can arise, we will omit the bars.

3. The "Rapid" Pressure Terms

Working in Cartesian coordinates for the moment, we may split the pressure into two components in a well-known way (Lumley, 1978):

$$-p_{,ii}^{(1)}/\varrho = 2U_{i,j}u_{j,i}, \quad -p_{,ii}^{(2)}/\varrho = u_{i,j}u_{j,i} \tag{3.1}$$

where $p^{(1)}$ is referred to as the linear or "rapid" part of the pressure, while $p^{(2)}$ is the return-to-isotropy part.

In Lumley (1978) it was shown that, in the equation for the third moment, the convection and production terms, and the "rapid" pressure correlation terms, were of smaller order than the other terms, and could be neglected. Here, however, we will have to retain these terms when they are multiplied by the swirl component, since we are presuming that to be the dominant effect. Thus, we will need a form for the "rapid" pressure correlations.

We can write (still in Cartesian coordinates)

$$-\overline{p_{,l}^{(1)}u_r u_s}/\varrho = 2U_{i,j}\int \psi_{jrs}\varkappa_i\varkappa_l/\varkappa^2\, d\boldsymbol{\varkappa} \tag{3.2}$$

where

$$\int \psi_{jrs}\, d\boldsymbol{\varkappa} = \overline{u_j u_r u_s}. \tag{3.3}$$

If we write

$$I_{iljrs} = \int \psi_{jrs}\varkappa_i\varkappa_l/\varkappa^2\, d\boldsymbol{\varkappa} \tag{3.4}$$

then I_{iljrs} must satisfy the following conditions:

$$I_{iljrs} = I_{lijrs}; \quad I_{iijrs} = \overline{u_j u_r u_s};$$
$$I_{illrs} = 0; \quad I_{iljrs} = I_{iljsr}. \tag{3.5}$$

If we take a general linear form for I_{iljrs} in terms of $\overline{u_j u_r u_s}$, we find that the conditions are sufficient to determine all the coefficients, giving

$$I_{iljrs} = (2/5)\,\delta_{il}\,\overline{u_j u_r u_s} - (1/10)\,(\delta_{ij}\,\overline{u_l u_r u_s} + \delta_{lj}\,\overline{u_i u_r u_s}). \tag{3.6}$$

This is the same form which produces the pressure transport expression

$$-\,\overline{p^{(2)}\,u_k}/\varrho = I_{ijkij} = \overline{q^2\,u_k}/5 \tag{3.7}$$

(Lumley, 1978) which has been so successful (Janicka and Lumley, 1981 a, b; Taulbee and Lumley, 1981).

The full $p^{(1)}$ correlation in the equation for $\overline{u_i u_j u_k}$ consists of expression (3.2) plus the two permutations. On the left side of the equations we have the terms (still in Cartesian coordinates)

$$U_{i,l}\,\overline{u_l u_j u_k} + U_{j,l}\,\overline{u_l u_i u_k} + U_{k,l}\,\overline{u_l u_i u_j}. \tag{3.8}$$

When these are combined with the rapid pressure terms obtained from (3.2) by use of (3.6), we find that they combine to give

$$(2/5)\,S_{il}\,\overline{u_l u_j u_k} + (2/5)\,S_{jl}\,\overline{u_l u_i u_k} + (2/5)\,S_{kl}\,\overline{u_l u_i u_j} \tag{3.9}$$

where S_{ij} is the strain rate, $(1/2)\,(U_{i,j} + U_{j,i})$.

Note that the expression resulting from the combination of (3.2) and (3.6) satisfies realizability in the sense that it vanishs when the velocity components in question vanish.

Hence, the net effect of the "rapid" pressure is to remove the contribution of the vorticity to the production, and to reduce the contribution of the strain rate to the production.

4. The Viscous Terms

Appearing in each equation for a third moment, we have a viscous term. For example, in the equation for $\overline{u_i u_j u_k}$ (now in general coordinates) we have a term of the form

$$-\,(2/3)\,(\bar{\varepsilon}/\overline{q^2})\,(g_{ik}\,\overline{q^2\,u_j} + g_{ij}\,\overline{q^2\,u_k} + g_{jk}\,\overline{q^2\,u_i}) \tag{4.1}$$

(see Lumley, 1978). In a simple situation in which the equations are not cross-coupled (no buoyancy, no rotation) it is straightforward to calculate $\overline{q^2\,u_i}$ from the equation for $\overline{u_i u_j u_k}$ and substitute it.

Here, however, the equations are strongly cross-coupled, and the complexity introduced by these viscous terms is formidable. We propose a simplification which will retain some of the physics.

In the viscous terms only, in cylindrical coordinates, we propose to write

$$\overline{q^2\,u_k}/3 = \overline{u^1\,u_1\,u_k} = \overline{u^2\,u_2\,u_k} = \overline{u^3\,u_3\,u_k} \tag{4.2}$$

so that, for example, the viscous terms in the equation for $\overline{u_3^2\,u_1}$, can be replaced in the following way:

$$-\,(2/3)\,(\bar{\varepsilon}/\overline{q^2})\,g_{33}\,\overline{q^2\,u_1} = -\,2(\bar{\varepsilon}/\overline{q^2})\,g_{33}\,g^{33}\,\overline{u_3^2\,u_1} = -\,2(\bar{\varepsilon}/\overline{q^2})\,\overline{u_3^2\,u_1}. \tag{4.3}$$

The term is now of the same form as the return to isotropy term (Lumley, 1978), and can be combined with it, the net result being a change in the value of the time scale. We are saying, effectively, that

$$\overline{\varepsilon_{\alpha\alpha} u_i} = (\bar{\varepsilon}/\overline{q^2})\, \overline{u_\alpha^2 u_i} \tag{4.4}$$

where $\varepsilon_{\alpha\alpha}$ is the α-component of the dissipation. This is surely somewhat in error, because the dissipation should be more nearly isotropic than the velocity at large Reynolds numbers. The error, however, should be small relative to the errors introduced by our other approximations, and the simplification is considerable. Equation (4.4) is not as serious an approximation as writing $\overline{\varepsilon_{ij}} \propto \overline{u_i u_j}$ (see Corrsin, 1972), since we are dealing here only with the diagonal components.

5. The Equations for the Mechanical Third Moments

We follow the approach outlined in Lumley (1978). There, however, we showed that the convection and production terms were of higher order relative to the term retained, and could be neglected. Here, we suppose that at least some of the mean velocities and their gradients are large enough to keep the convection and production terms; in particular, we expect the azimuthal (physical) velocity to be large, and the radius small, so that

$$(V_\theta/r)\,(l_t/u) = 0(1) \tag{5.1}$$

even as $l_t/r \to 0$, where l_t is a turbulent length scale, defined, say, by $\varepsilon = u^3/l_t$, $3u^2 = q^2$, and l is a scale characteristic of cross-stream gradients of axial velocity,

$$\partial V_z/\partial r = |\Delta V_z|_{\max}/l \tag{5.2}$$

(where $|V_z|_{\max}$ is the maximum difference in V_z, the axial velocity). Reasoning in this way, we will keep all convection and production terms involving the azimuthal velocities, and neglect the rest. The order of magnitude analysis of Lumley (1978) can be carried over exactly, otherwise, since we may work in Cartesian coordinates and transform to generalized coordinates afterward.

In general coordinates, then, the equation for $\overline{u_i u_j u_k}$ becomes (incorporating the results of Section 3)

$$\begin{aligned}(\overline{u_i u_j u_k})_{,p}\, U^p &+ (2/5)\, S_i^p\, \overline{u_p u_j u_k} + (2/5)\, S_j^p\, \overline{u_p u_i u_k} \\ &+ (2/5)\, S_k^p\, \overline{u_p u_i u_j} + (\overline{u_i u_j})_{,p}\, \overline{u^p u_k} + (\overline{u_i u_k})_{,p}\, \overline{u^p u_j} \\ &+ (\overline{u_j u_k})_{,p}\, \overline{u^p u_i} = -\, 3\, C_1 (\bar{\varepsilon}/\overline{q^2})\, \overline{u_i u_j u_k} \\ &+ (\bar{\varepsilon}/\overline{q^2})\,(C_1 - 2)\,(g_{ij}\overline{q^2 u_k} + g_{ik}\overline{q^2 u_j} + g_{jk}\overline{q^2 u_i})/3\end{aligned} \tag{5.3}$$

where, for simplicity of notation in this section and below we are suppressing the overbars which would otherwise indicate that these velocities are written in a general non-Cartesian coordinate system. We will, of course, apply also the results of Sect. 4 to the last term when we write (5.3) in component form.

The only velocity we retain is the azimuthal, U^2; hence

$$(\overline{u_i u_j u_k})_{,p}\, U^p = (\overline{u_i u_j u_k})_{,2}\, U^2 = (-\, \Gamma_{i2}^p\, \overline{u_p u_j u_k} - \Gamma_{j2}^p\, \overline{u_i u_p u_k} - \Gamma_{k2}^p\, \overline{u_i u_j u_p})\, U^2 \tag{5.4}$$

since $\partial/\partial \bar{x}^2$ of all quantities is taken to be zero by axisymmetry.

The mean velocity gradients we retain are

$$U_{1,2} = -U_2/\bar{x}^1, \quad U_{2,1} = \partial U_2/\partial \bar{x}^1 - U_2/\bar{x}^1 \tag{5.5}$$

so that the non-zero components of the strain rate are

$$S_{12} = S_{21} = (1/2)\,\partial U_2/\partial \bar{x}^1 - U_2/\bar{x}^1. \tag{5.6}$$

If for simplicity we write

$$\begin{aligned} &3C_1\bar{\varepsilon}/\overline{q^2} = 1/T, && A = \Gamma^2_{12}U^2 - (2/5)S^2_1 \\ &(\bar{\varepsilon}/\overline{q^2})(C_1 - 2) = K/T, && B = (2/5)S^1_2 - \Gamma^1_{22}U^2 \\ &(1-3K)/T = 1/T'', \quad (1-K)/T = 1/T' && \end{aligned} \tag{5.7}$$

then our equation can be written as

$$-3A\overline{u_2u_1^2} + 3(\overline{u_1^2})_{,p}\overline{u^pu_1} = -\overline{u_1^3}/T'' \tag{5.8}$$

$$-2A\overline{u_1u_2^2} + B\overline{u_1^3} + (\overline{u_1^2})_{,p}\overline{u^pu_2} + 2(\overline{u_1u_2})_{,p}\overline{u^pu_1} = -\overline{u_1^2u_2}/T' \tag{5.9}$$

$$2B\overline{u_1^2u_2} - A\overline{u_2^3} + (\overline{u_2^2})_{,p}\overline{u^pu_1} + 2(\overline{u_1u_2})_{,p}\overline{u^pu_2} = -\overline{u_1u_2^2}/T' \tag{5.10}$$

$$-A\overline{u_2u_3^2} + 2(\overline{u_1u_3})_{,p}\overline{u^pu_3} + (\overline{u_3^2})_{,p}\overline{u^pu_1} = -\overline{u_1u_3^2}/T' \tag{5.11}$$

$$3B\overline{u_1u_2^2} + 3(\overline{u_2^2})_{,p}\overline{u^pu_2} = -\overline{u_2^3}/T'' \tag{5.12}$$

$$B\overline{u_1u_3^2} + 2(\overline{u_2u_3})_{,p}\overline{u^pu_3} + (\overline{u_3^2})_{,p}\overline{u^pu_2} = -\overline{u_2u_3^2}/T' \tag{5.13}$$

$$\begin{aligned} -A\overline{u_2^2u_3} + B\overline{u_1^2u_3} &+ (\overline{u_1u_2})_{,p}\overline{u^pu_3} \\ &+ (\overline{u_1u_3})_{,p}\overline{u^pu_2} + (\overline{u_2u_3})_{,p}\overline{u^pu_1} = -\overline{u_1u_2u_3}/T \end{aligned} \tag{5.14}$$

$$-2A\overline{u_1u_2u_3} + (\overline{u_1^2})_{,p}\overline{u^pu_3} + 2(\overline{u_1u_3})_{,p}\overline{u^pu_1} = -\overline{u_1^2u_3}/T' \tag{5.15}$$

$$2B\overline{u_1u_2u_3} + (\overline{u_2^2})_{,p}\overline{u^pu_3} + 2(\overline{u_2u_3})_{,p}\overline{u^pu_2} = -\overline{u_2^2u_3}/T'. \tag{5.16}$$

These equations may be separated into non-interacting groups. For example, (5.11) and (5.13) are dependent only on each other; (5.14–5.16) also form an autonomous group; and (5.8–5.10), (5.12) are also mutually dependent.

Let us take the largest group first. If we define

$$\Omega = 1 + 3ABT'T'', \quad \Omega' = 1 + 3ABT'T'' + 4ABT'^2 \tag{5.17}$$

then we may define a matrix

$$\beta_{ij} = \frac{-\Omega T'}{\Omega' + \Omega(\Omega - 1)} \left| \begin{array}{cccc} \Omega'T''/\Omega T' & 3AT'' & 6A^2T'T''/\Omega & 6A^3T'T''^2/\Omega \\ -BT'' & 1 & 2AT'/\Omega & 2A^2T'T''/\Omega \\ 2B^2T'T''/\Omega & -2BT'/\Omega & 1 & AT' \\ -6B^3T'T''^2/\Omega & 6B^2T'T''/\Omega & -3BT'' & \Omega'T''/\Omega T' \end{array} \right| \tag{5.18}$$

and we may write the solution in terms of β_{ij}:

$$\begin{vmatrix} \overline{u_1^3} \\ \overline{u_1^2 u_2} \\ \overline{u_1 u_2^2} \\ \overline{u_2^3} \end{vmatrix} = \beta \begin{vmatrix} 3(\overline{u_1^2})_{,p}\overline{u^p u_1} \\ (\overline{u_1^2})_{,p}\overline{u^p u_2} + 2(\overline{u_1 u_2})_{,p}\overline{u^p u_1} \\ 2(\overline{u_1 u_2})_{,p}\overline{u^p u_2} + (\overline{u_2^2})_{,p}\overline{u^p u_1} \\ 3(\overline{u_2^2})_{,p}\overline{u^p u_2} \end{vmatrix} . \tag{5.19}$$

We may now collect the terms and write in matrix form

$$\begin{aligned} &\begin{vmatrix} \overline{u_1^3} \\ \overline{u_1^2 u_2} \\ \overline{u_2^2 u_1} \end{vmatrix} \\ &= \begin{vmatrix} 3\beta_{11}\overline{u^p u_1} + \beta_{12}\overline{u^p u_2} & 2\beta_{12}\overline{u^p u_1} + 2\beta_{13}\overline{u^p u_2} & \beta_{13}\overline{u^p u_1} + 3\beta_{14}\overline{u^p u_2} \\ 3\beta_{21}\overline{u^p u_1} + \beta_{22}\overline{u^p u_2} & 2\beta_{22}\overline{u^p u_1} + 2\beta_{23}\overline{u^p u_2} & \beta_{23}\overline{u^p u_1} + 3\beta_{24}\overline{u^p u_2} \\ 3\beta_{31}\overline{u^p u_1} + \beta_{32}\overline{u^p u_2} & 2\beta_{32}\overline{u^p u_1} + 2\beta_{33}\overline{u^p u_2} & \beta_{33}\overline{u^p u_1} + 3\beta_{34}\overline{u^p u_2} \end{vmatrix} \\ &\times \begin{vmatrix} (\overline{u_1^2})_{,p} \\ (\overline{u_1 u_2})_{,p} \\ (\overline{u_2^2})_{,p} \end{vmatrix} . \end{aligned} \tag{5.20}$$

The fourth component can also be written in this form:

$$\begin{aligned} \overline{u_2^3} = {} & (\overline{u_1^2})_{,p}[3\beta_{41}\overline{u^p u_1} + \beta_{42}\overline{u^p u_2}] + (\overline{u_1 u_2})_{,p}[2\beta_{42}\overline{u^p u_1} + 2\beta_{43}\overline{u^p u_2} \\ & + (\overline{u_2^2})_{,p}[\beta_{43}\overline{u^p u_1} + 3\beta_{44}\overline{u^p u_2}]. \end{aligned} \tag{5.21}$$

We see the appearance here of the same matrix form found by Zeman (1975), Zeman and Lumley (1976 a, b, c), Lumley et al. (1978), and the coefficients are of the general type (1.7).

The group of three equations may be attacked in a similar way. If we define a matrix

$$\beta_{ij}^1 = \frac{-T'}{1 + 4ABTT'} \begin{vmatrix} 1 + 2ABTT' & 2AT & 2A^2TT' \\ -BT & T/T' & AT \\ 2B^2TT' & -2BT & 1 + 2ABTT' \end{vmatrix} \tag{5.22}$$

then the solution may be written as

$$\begin{vmatrix} \overline{u_1^2 u_3} \\ \overline{u_1 u_2 u_3} \\ \overline{u_2^2 u_3} \end{vmatrix} = \beta^1 \begin{vmatrix} (\overline{u_1^2})_{,p}\overline{u^p u_3} + 2(\overline{u_1 u_3})_{,p}\overline{u^p u_1} \\ (\overline{u_1 u_2})_{,p}\overline{u^p u_3} + (\overline{u_1 u_3})_{,p}\overline{u^p u_2} + (\overline{u_2 u_3})_{,p}\overline{u^p u_1} \\ (\overline{u_2^2})_{,p}\overline{u^p u_3} + 2(\overline{u_2 u_3})_{,p}\overline{u^p u_2} \end{vmatrix} . \tag{5.23}$$

Now, however, when we expand, we cannot directly write the result in the form (5.20) due to the presence of such terms as $\overline{u^p u_3}$.

It is tempting to suppose that u_3 is produced from the mean 3-velocity gradient by the u_1 fluctuation, and that the u_2 fluctuations are principally due to angular momentum conservation under 1-excursions. This is a kind of mixing length assumption, leading to

$u_3 \propto u_1 \propto u_2$. Then we can write

$$\begin{aligned} \overline{u^p u_3}\,\overline{u_1^2} &= \overline{u_1 u_3}\,\overline{u^p u_1} \\ \overline{u^p u_3}\,\overline{u_2^2} &= \overline{u_2 u_3}\,\overline{u^p u_2} \\ \overline{u_1 u_2}\,\overline{u^p u_3} &= (1/2)\,[\overline{u_1 u_3}\,\overline{u^p u_2} + \overline{u_2 u_3}\,\overline{u^p u_1}]. \end{aligned} \tag{5.24}$$

This, however, is not quite what we need. We need the derivatives $(\overline{u_1^2})_{,p}\,\overline{u^p u_3}$ etc. If we attempt to write

$$(\overline{u_1^2})_{,p}\,\overline{u^p u_3} = (\overline{u_1 u_3})_{,p}\,\overline{u^p u_1} \tag{5.25}$$

even assuming that the coefficients of proportionality are not functions of position, we find that (5.25) fails due to the different groups of Christoffel symbols on the two sides.

The only way the issue can be simply resolved is to assume that $l/r \ll 1$, where l is the scale of the radial gradient of the Reynolds stress. This assumption amounts to either very thin shear layers relative to the radius, or to large radius; essentially this is the small gap approximation familiar in stability analyses of circular Couette flow.

This is not inconsistent with our previous assumptions, which were simply that W/r is of the same order as u/l. If $l/r \ll 1$ and $l_t/r \ll 1$, but $W/r \sim u/l_t$, then $W/u \sim (r/l)\,(l/l_t) \gg 1$. Hence, we must have very large velocities relative to turbulent intensities.

If we make the narrow gap assumption, then for turbulent quantities the only important covariant derivative is radial, and that is essentially given by the ordinary radial derivative.

Now we can write

$$\begin{aligned} (\overline{u_1^2})_{,1}\,\overline{u^1 u_3} &= (\overline{u_1 u_3})_{,1}\,\overline{u^1 u_1} \\ (\overline{u_2^2})_{,1}\,\overline{u^1 u^3} &= (\overline{u_2 u_3})_{,1}\,\overline{u^1 u_2} \\ (\overline{u_1 u_2})_{,1}\,\overline{u^1 u_3} &= (1/2)\,[(\overline{u_1 u_3})_{,1}\,\overline{u^1 u_2} + (\overline{u_2 u_3})_{,1}\,\overline{u^1 u_1}]. \end{aligned} \tag{5.26}$$

With the approximation (5.26), we can write

$$\begin{vmatrix} \overline{u_1^2 u_3} \\ \overline{u_1 u_2 u_3} \end{vmatrix} = \begin{vmatrix} 3\beta_{11}^1\,\overline{u^1 u_1} + (3/2)\,\beta_{12}^1\,\overline{u^1 u_2} & (3/2)\,\beta_{12}^1\,\overline{u^1 u_1} + 3\beta_{13}^1\,\overline{u^1 u_2} \\ 3\beta_{21}^1\,\overline{u^1 u_1} + (3/2)\,\beta_{22}^1\,\overline{u^1 u_2} & (3/2)\,\beta_{22}^1\,\overline{u^1 u_1} + 3\beta_{23}^1\,\overline{u^1 u_2} \end{vmatrix} \begin{vmatrix} (\overline{u_1 u_3})_{,1} \\ (\overline{u_2 u_3})_{,1} \end{vmatrix} \tag{5.27}$$

and the third component also can be written in this form:

$$\begin{aligned} \overline{u_2^2 u_3} = (\overline{u_1 u_3})_{,1}\,&[3\beta_{31}^1\,\overline{u^1 u_1} + (3/2)\,\beta_{32}^1\,\overline{u^1 u_2}] \\ &+ (\overline{u_2 u_3})_{,1}\,[(3/2)\,\beta_{32}^1\,\overline{u^1 u_1} + 3\beta_{33}^1\,\overline{u^1 u_2}]. \end{aligned} \tag{5.28}$$

Finally, we may solve the group of two, defining

$$\beta_{ij}^2 = \frac{-T'}{1 + ABT'^2} \begin{vmatrix} 1 & +AT' \\ -BT' & 1 \end{vmatrix} \tag{5.29}$$

and

$$\begin{vmatrix} \overline{u_1 u_3^2} \\ \overline{u_2 u_3^2} \end{vmatrix} = \boldsymbol{\beta}^2 \begin{vmatrix} (\overline{u_3^2})_{,p}\,\overline{u^p u_1} + 2(\overline{u_1 u_3})_{,p}\,\overline{u^p u_3} \\ (\overline{u_3^2})_{,p}\,\overline{u^p u_2} + 2(\overline{u_2 u_3})_{,p}\,\overline{u^p u_3} \end{vmatrix}. \tag{5.30}$$

To place this in the same form as (5.20) and (5.27), we must take the narrow gap assumption, discard $(\ )_{,2}$, and assume

$$\begin{aligned}(\overline{u_1 u_3})_{,1}\, \overline{u^1 u_3} &= (\overline{u_3^2})_{,1}\, \overline{u^1 u_1} \\ (\overline{u_2 u_3})_{,1}\, \overline{u^1 u_3} &= (\overline{u_3^2})_{,1}\, \overline{u^1 u_2}\end{aligned} \tag{5.31}$$

which can be justified in the same way. Then

$$\begin{aligned}\overline{u_1 u_3^2} &= (\overline{u_3^2})_{,1}\, [3\,\beta_{11}^2\, \overline{u^1 u_1} + 3\,\beta_{12}^2\, \overline{u^1 u_2}] \\ \overline{u_2 u_3^2} &= (\overline{u_3^2})_{,1}\, [3\,\beta_{21}^2\, \overline{u^1 u_1} + 3\,\beta_{22}^2\, \overline{u^1 u_2}].\end{aligned} \tag{5.32}$$

6. Discussion

To summarize the strategy we have used in deriving the expressions for the third moments:

1) we first form an equation for the moment generating function, equivalent to the probability density function for the fluctuating velocities;
2) we require that the equation produce a density that relaxes to a Gaussian with a single time constant in the absence of inhomogeneity, swirl, buoyancy, chemical reactions or any other disturbing influence – this determines the form of the pressure correlation;
3) we require that the equation produce at second order the modeled equations for the fluxes and variances in the homogeneous situation – this determines all remaining unknown terms. This is similar to determining the form of a Gaussian distribution by specifying its second moment.
4) we assume that the inhomogeneity is weak – that is, that the turbulence is fine-grained relative to the scale of the inhomogeneity. This has the effect of eliminating many additional terms, and leaving algebraic expressions for the third moments which contain no new constants; there is nothing adjustable, the only constants appearing being the relaxation constants from the second moment equations, which are determined by comparison with experiment in homogeneous flows.

In carrying out Step 4), in the atmospheric situation we assume that the buoyant terms are of equal importance, so that the ratio

$$g\,\theta\, l_t / \Theta_0 u^2 = 0(1) \tag{6.1}$$

during the limit process $l_t/l \to 0$, where l_t is the length scale of the turbulence and l is the length scale of the inhomogeneity. That is, we assume that the turbulent velocities are of a magnitude that could be produced by the observed temperature differences accelerating in the gravitational field. Here we make a similar assumption for the swirl terms

$$W l_t / r u = 0(1) \tag{6.2}$$

during the same limiting process: we assume that the azimuthal velocity fluctuation induced by a turbulent radial excursion is of the order of observed turbulent velocities.

In the atmosphere we obtain third moments of the following form:

$$\begin{vmatrix} \overline{\theta w^2} \\ \overline{\theta^2 w} \\ \overline{w^3} \\ \vdots \end{vmatrix} = -\,\boldsymbol{F}\,\partial/\partial z \begin{vmatrix} \overline{\theta w} \\ \overline{\theta^2} \\ \overline{w^2} \\ \vdots \end{vmatrix} \tag{6.3}$$

where individual terms of the matrix $\boldsymbol{F}$ are of the form (1.1); the constants in the different terms are different. In the case of swirl, we obtain third moments of the form

$$\begin{vmatrix} \overline{wv^2} \\ \overline{w^2v} \\ \overline{v^3} \\ \vdots \end{vmatrix} = -\boldsymbol{F}\,\partial/\partial r \begin{vmatrix} \overline{wv} \\ \overline{w^2} \\ \overline{v^2} \\ \vdots \end{vmatrix} \tag{6.4}$$

where the individual terms of the matrix $\boldsymbol{F}$ are of the form (1.3), again with different constants in different terms.

The structure of the entire approach is reminiscent of non-equilibrium thermodynamics of a mixture, where the variances and fluxes play the role of the various species. In the kinetic theory case, one has, for example, a flux of heat produced not only by a gradient of temperature, but also by a gradient of salinity, while the flux of salt is produced not only by the salinity gradient, but also by the temperature gradient. Here, gradients of all the variances and fluxes contribute to all of the third moments.

Computations have been carried out using these forms for the third moments in the atmosphere and in other flows. In Zeman (1975); Zeman and Lumley (1976a, b, c; 1979); Lumley et al. (1978), early versions of this approach were used, which retained some adjustable constants. More recently, computations have been carried out by Mansfield at Cornell using the form for the third moments described here, and presented fully in Lumley (1978), which contains no adjustable constants. We show in Fig. 4 Mansfield's results for the vertical transport of the turbulent energy in a buoyancy-driven atmospheric surface mixed layer, compared with several experiments. It is evident that the agreement is within experimental error.

Now, of course, this formalism must be applied to the calculation of well documented swirling flows. It will probably be necessary first, of course, to model homogeneous swirling flows, like that of Traugott (1958) and of Wigeland and Nagib (1978), to calibrate the other aspects of the model (notably the dissipation equation) which we have ignored here.

We should then attempt modeling of such swirling flows as the swirling jets of Morse (1980); Pratte and Keffer (1972); Chigier and Chervinsky (1967) and Rose (1962), the curved

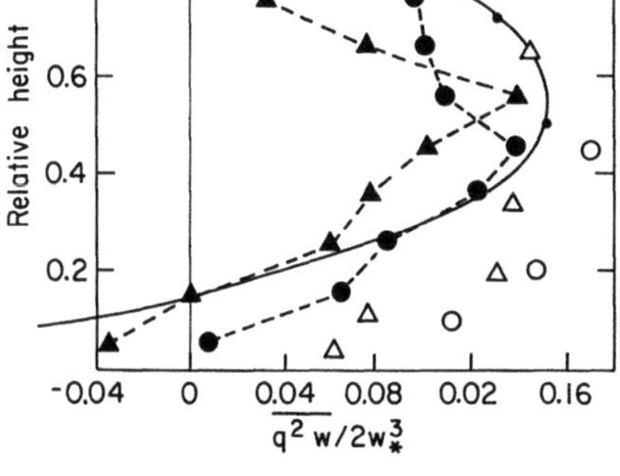

Fig. 4. Vertical flux of turbulent kinetic energy in the buoyancy-driven atmospheric surface mixed layer computed by Mansfield using the non-adjustable model of Lumley (1978). Solid circles and triangles, Willis and Deardorff (1974); open circles and triangles, Lenschow (1968, 1970, 1974). For references, see Zeman and Lumley (1976b)

channel flow of Eskinazi and Yeh (1956), the curved mixing layer of Wyngaard et al. (1968) and perhaps even the rotating channel flow of Lezius and Johnston (1972) (see also Schetz, 1980). In all of these, (but particularly the spinning jet), new modeling of the dissipation equation will doubtless be necessary.

It is at this point entirely unclear whether the difficulties experienced in Launder and Morse (1977) with the Reynolds stress closure (the wrong sign of the axial-azimuthal correlation) will be alleviated by the present approach; it appears quite possible that the problem there lies in the form of the so-called rapid terms of the pressure-strain correlation, which appears (based on the conclusions of the Stanford Conference; see Kline et al., 1981) not to respond properly to rotation. If this is the case, it will be necessary to develop a new model for this term before successful modeling can proceed.

It is interesting that the primary effect of swirl on the transport is not through the pressure terms, as one might have expected. From Section 3 we see that the effect of the pressure terms is to cancel the effect of rotation (at least as far as the mean velocity gradients are concerned) so that under solid body rotation ($S_{ij} = 0$) the terms in the mean velocity gradients contribute nothing. Rather, the effect of rotation comes about entirely through the rotation of the coordinate system, an essentially geometrical effect (felt through the convective derivative). Of course, it is felt indirectly through the modification of the value of $\overline{u_1 u_2}$, which will come about through the second moment equations, and which appears in the diffusion coefficient in the third moment equations due to the rotation of the frame.

References

Chigier, N. A., Chervinsky, A. (1967): Experimental investigation of swirling vortex motion in jets. J. Appl. Mech. **34**, 443

Corrsin, S. (1972): Comment on transport equations in turbulence. Phys. Fluids **16**, 157

Eskinazi, S., Yeh, H. (1956): An investigation on fully developed turbulent flows in a curved channel. J. Aersp. Sci. **23**, 23

Janicka, J., Lumley, J. L. (1981 a): A note concerning the pressure transport terms in second order modeling, (in preparation)

Janicka, J., Lumley, J. L.: Second order modeling in non-constant density flows. Sibley School of Mechanical and Aerospace Engineering Report No. FDA-81-01. Cornell Univ. Ithaca, NY 1981 b)

Jones, W. P., Launder, B. E. (1973): The calculation of low-Reynolds number phenomena with a two-equation model of turbulence. Int. J. Heat Mass Transf. **16**, 1119

Kline, S. J., Cantwell, B. J., Lilley, G. M. (eds.): *The 1980–81 AFOSR-HTTM-Stanford Conference on Complex Turbulent Shear Flows: Comparison of Computation and Experiment.* (Stanford University, Thermosciences Div. Stanford CA, 1981)

Launder, B. E., Morse, A.: Numerical Prediction of Axisymmetric Free Shear Flows with a Second Order Reynolds Stress Closure, in *Turbulent Shear Flows 1*, ed. by F. Durst, B. E. Launder, F. W. Schmidt, J. H. Whitelaw (Springer, Berlin, Heidelberg, New York 1979) p. 279

Launder, B. E., Priddin, C. H., Sharma, B. I. (1977): The calculation of turbulent boundary layers on spinning and curved surfaces. J. Fluids Engineering **99**, 231

Lezius, D., Johnston, J. P. (1972): The structure and stability of turbulent wall layers in rotating channel flow. J. Fluid Mech. **56**, 533

Lumley, J. L.: *Stochastic Tools in Turbulence.* (Academic, New York 1970)

Lumley, J. L.: Computational Modeling of Turbulent Flows, in *Advances in Applied Mechanics* Vol. 18, ed. by C.-S. Yih (Academic, New York 1978) p. 123

Lumley, J. L., Panofsky, H. A.: *The Structure of Atmospheric Turbulence.* (Interscience, New York 1964)

Lumley, J. L., Zeman, O., Siess, J. (1978): The influence of buoyancy on turbulent transport. J. Fluid Mech. **84**, 581

Morse, A. (1980): Axisymmetric turbulent shear flows with and without swirl. Ph. D. Thesis, University of London

Pratte, B. D., Keffer, J. F. (1972): The swirling turbulent jet. J. Basic Eng. 94, 739

Pellew, A., Southwell, R. V. (1940): On maintained convective motion in a fluid heated from below. Proc. Roy. Soc. **A 176**, 312
Ramaprian, B. R., Shivaprasad, B. G. (1981): The instantaneous structure of mildly curved turbulent boundary layers. Submitted for publication to J. Fluid Mech.
Rose, W. G. (1962): A swirling round turbulent jet. J. Appl. Mech. **29**, 615
Schetz, J. A.: *Injection and Mixing in Turbulent Flow*, Progress in Astronautics and Aeronautics, Vol. 68, ed. by M. Summerfield (American Institute of Aeronautics and Astronautics, New York 1980)
Taulbee, D. B., Lumley, J. L.: Prediction of the turbulent wake with a second order closure model. Sibley School of Mechanical and Aerospace Engineering Report No. FDA-81-04. (Cornell University Ithaca, NY 1981)
Taylor, G. I. (1923): Stability of a viscous fluid contained between rotating cylinders. Phil. Trans. **223**, 289
Traugott, S. C. (1958): Influence of solid body rotation on screen-produced turbulence. NACA TN 4135
Wigeland, R. A., Nagib, H. M.: Grid generated Turbulence with and without rotatin about the streamwise direction. Fluids and Heat Transfer Report R78-1. (Illinois Institute of Technology, Chicago 1978)
Wyngaard, J. C., Tennekes, H., Lumley, J. L., Margolis, D. P. (1968): Structure of turbulence in a curved mixing layer. Phys. Fluids **11**, 1251
Zeman, O. (1975): The dynamics of entrainment in the plantary boundary layer: a study in turbulence modeling and parameterization. Ph. D. Thesis, The Pennsylvania State University, University Park
Zeman, O., Lumley, J. L.: Turbulence and Diffusion Modeling in Buoyancy-Driven Mixed Layers, in *Proceedings of Third Symposium on Atmospheric Turbulence, Diffusion Air Quality*, Raleigh, NC (American Meterological Society, Boston, MA 1976a) p. 38
Zeman, O., Lumley, J. L. (1976b): Modeling buoyancy-driven mixed layers. J. Atmos. Sci. **33**, 1974
Zeman, O., Lumley, J. L.: A Second Order Model for Buoyancy-Driven Mixed Layers, in *Proceedings 9th ICHMT International Seminar Turbulent Buoyant Convection*, (Hemisphere Washington, DC 1976c) p. 65
Zeman, O., Lumley, J. L.: Buoyancy Effects in Entraining Turbulent Boundary Layers: A Second Order Closure Study, in *Turbulent Shear Flows 1*, ed. by F. Durst, B. E. Launder, F. W. Schmidt, J. H. Whitelaw (Springer, Berlin, Heidelberg, New York 1979) p. 295

A Comparison of Triple-Moment Temperature-Velocity Correlations in the Asymmetric Heated Jet with Alternative Closure Models

I. Dekeyser
IMST, Marseille, France

B. E. Launder
UMIST, Manchester, England

Abstract

Measurements are reported of triple moments of velocity and temperature in a heated asymmetric two-dimensional turbulent jet involving velocity fluctuations in the $x_1 - x_2$ plane. The data thus obtained have been compared with those given by alternative algebraic models of the triple moments using, in the model formulae, experimental values of the second-moment quantities and the dissipation rate of kinetic energy. The study supports the view that in strongly asymmetric flows the contribution of mean temperature gradients to the triple moments can be appreciable. The comparison also provides some support for the use of the generalized gradient transport hypothesis in approximating dissipation of the triple moments.

Nomenclature

- d — width of exit slot
- d_{ijk} — diffusion of $\overline{u_i u_j u_k}$ (likewise for velocity-temperature 3rd moments; subscript identifies quantity)
- k — turbulence kinetic energy
- l_{γ_u} — x_2 distance between the positions where $\gamma_u = 1/2$
- l_{γ_θ} — u_2 distance between the positions where $\gamma_\theta = 1/2$
- p — fluctuating pressure
- P_{ij} — generation rate of Reynolds stress
- P_{ijk} — generation rate of $\overline{u_i u_j u_k}$ due to Reynolds stress gradients (1) and mean strain (2)
- q' — $\sqrt{2k}$
- T — turbulent time scale (k/ε)
- u_i — fluctuating velocity
- U_i — mean velocity
- U — mean streamwise velocity (in x_1 direction)
- u', v', w' — rms velocity fluctuations in x_1, x_2 and x_3 directions
- x_1 — streamwise coordinate
- x_2 — cross stream Cartesian coordinate
- x_{2m} — x_2 position of velocity maximum
- $x_{2\theta}$ — x_2 position of temperature maximum
- γ_u — intermittency of turbulent velocity signal
- γ_θ — intermittency of turbulent temperature sequal
- ε — dissipation rate of turbulence energy
- ε_{ijk} — "dissipation" rate of $\overline{u_i u_j u_k}$
- η_u — $(x_2 - x_m)/l\gamma_u$
- η_θ — $(x_2 - x_{2\theta})/l\gamma_\theta$
- ν — kinematic viscosity
- ϱ — density of fluid
- θ — fluctuating temperature
- θ' — $\sqrt{\overline{\theta^2}}$
- Θ — mean temperature
- $\Delta\Theta$ — temperature above ambient
- $\Delta\Theta_m$ — maximum temperature difference across jet
- ϕ_{ijk} — non-dispersive pressure correlation in $\overline{u_i u_j u_k}$

1. Introduction

Most current efforts in second-moment closure adopt highly simplistic representations of the triple correlations appearing as unknowns in the transport equations for the second-moments. This practice may be regarded as both an invocation of the principle of receding

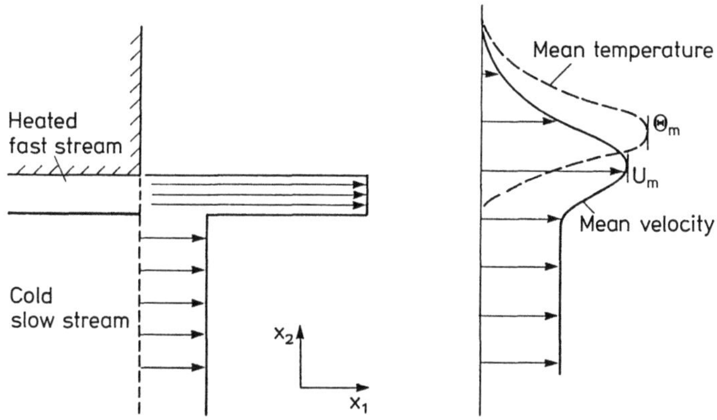

Fig. 1. The flow configuration

influence and an appreciation that, since the non-zero triple moments are more numerous than those of second-rank, to adopt other than a rudimentary approximation may let the computer budget get out of hand. While undoubtedly a cruder approximation suffices for third than for second-moments it is not by any means established that the currently popular algebraic models give sufficient accuracy or width of applicability. Indeed, Lumley, Zeman and Siess (1978) have argued persuasively that the diurnal growth of the mixed layer in the atmospheric boundary layer is crucially dependent on the triple-moment approximations and that an explicit account of gravitational contributions to the triple-moment balances must be included [1].

The present paper has arisen from an examination of another flow – one with more direct connection with engineering flow problems – in which diffusive transport is more than usually important. This is the heated asymmetric jet shown in Fig. 1. A heated plane jet mixes on one side with stagnant surroundings and on the other with a moving stream. The shear flow has two features of particular interest: firstly the scale and the intensity of turbulent motions on the side of the jet mixing with stagnant fluid is markedly higher than on the other; secondly the unequal entrainment rates that result produce, after some development distance, a flow in which the maximum temperature is shifted laterally with respect to that for velocity. Thus, unlike the symmetric jet or wake, the regions where the generation rates of turbulence energy k and mean-square temperature variance, $\overline{\theta^2}$ occur are not coincident. In these circumstances there is liable to be departure from the commonly prescribed simple connections between the scalar and dynamic fields – such as a uniform turbulent Prandtl number or time-scale ratio R (Launder, 1976). The experiment, which is one of several on asymmetric turbulent flows to have been performed at the IMST in recent years (for earlier studies see Beguier et al. 1978, Keffer et al. 1979) has given special attention to the measurement of the triple products of velocity and temperature which appear as the diffusive transport terms in the conservation equations for the second-moments. These measurements are directly compared with various simple models that express the triple-moments in terms of second-moments and their gradients. In addition to forms already in the literature the present work considers the addition of two further elements that extrapolate to the triple products ideas that have in the past been widely used in closing the second-moment equations.

1 This point of view is implicitly supported by the earlier third-moment closure study of the same problem reported by André et al. (1979)

2. The Experimental Programme

Apparatus

A full account of the complete experimental programme is provided by Dekeyser (1982, 1983). Accordingly we here provide only a brief summary with emphasis on the measurement of the triple-moments.

The flow configuration is shown in Fig. 1. The shear flow is formed by a heated high-velocity air jet (≅ 30 m/s) discharged through a slot 10 mm high and 200 wide. The air is heated in passage through a low-velocity settling chamber upstream of the jet contraction. The chamber consists of a dozen grills, each comprising 8 cylindrical electrical heaters. In the experiments reported here the air at discharge from the slot is warmed to approximately 32 °C above ambient temperature. Buoyancy effects are entirely negligible. On its lower side the heated jet mixes with what, for the purposes of the present study, may be considered an infinite uniform-density stream. It is in fact a low velocity (8 m/s) unheated air stream discharged through a square opening 200 mm high. The low-velocity stream is bounded on its lower side by a smooth wall. The hot jet is free to entrain stagnant ambient air on its upper edge. Side walls are provided to constrain the shear flow to an essentially two-dimensional development; measurements showed that the streamwise enthalpy flux was conserved within 5% over a development from 40–100 slot heights downstream.

Instrumentation

The mean velocity and temperature fields were mapped respectively by a pitot tube and chromel-constantan thermocouple, the sensitivity of the latter offering a precision of about 0.2 K. To improve stability the cold junction was in fact placed in the unheated uniform stream.

Temperature fluctuations were measured by a "cold" platinum wire 1 μ in diameter and 0.4 mm long with a resistance of about 100 Ω. The wire is placed on a Wheatstone bridge circuit and heated by a constant current of about 150 μA. There is a 3000:1 signal amplifi-

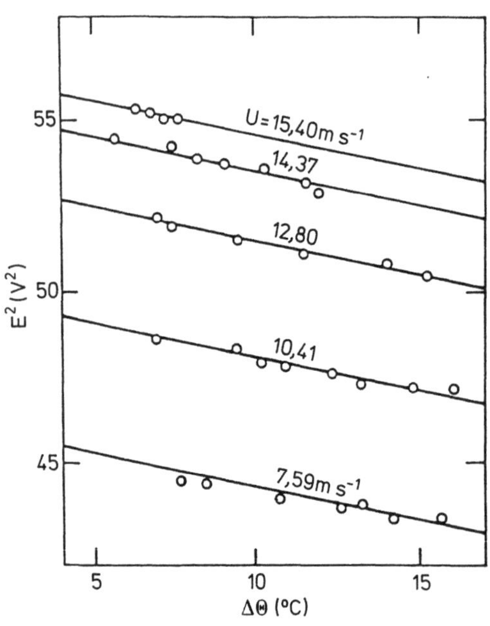

Fig. 2. Hot-wire sensitivity to velocity and temperature

cation and an electronic compensation circuit for thermal inertia. The overall sensitivity is approximately 50 μV/K. For the majority of the measurements the cold wire, which was aligned normal to the flow, formed part of a 3-wire probe. The other two wires, of 5 μ diameter platinum were arranged as cross-wires (the angle between the wires being approximately 90°); the planes formed by each of the three wires with the mean flow vector were mutually parallel. They were operated with an overheat coefficient of 0.8 giving a wire temperature of approximately 200 K above ambient. The sensitivities of the wires to velocity and temperature fluctuations were determined by constructing calibration curves for wire emf as a function of velocity for a series of different mean temperatures spanning the range encountered in the experiment. The cold wire sensibly responded only to temperature fluctuations; the hot wires were significantly sensitive to both velocity and temperature fluctuations though, as we see in Fig. 2, the rate of change of E^2 with voltage was independent of velocity. This encouraged the decision that the hot wires should be employed without linearizers. The separation of the instantaneous fluctuating velocity components and temperatures was achieved by means of an analogue circuit consisting of operational amplifiers similar to those used by Schon and Baille (1972). To obtain the triple correlations the separated velocity and temperature signals were passed through two successive amplifiers (the second a quadratic amplifier), a multiplier and a high-pass filter with a low-frequency cut-off set at 1 Hz to remove the continuous part of the quadratic signal. The same circuit allowed simultaneous measurement of the double and triple correlations with the help of a DISA correlator.

The Measurements

All the measurements reported hereunder were obtained 50 jet slot heights downstream of discharge. The mean velocity and temperature profiles in Fig. 3 indicate that the temperature maximum is displaced relative to the velocity maximum in the direction of the stagnant surroundings. Figures 4 and 5 show the profiles of rms velocity and temperature fluctuations. The data of Fig. 4 were taken without heating the jet. The different sets were taken with the

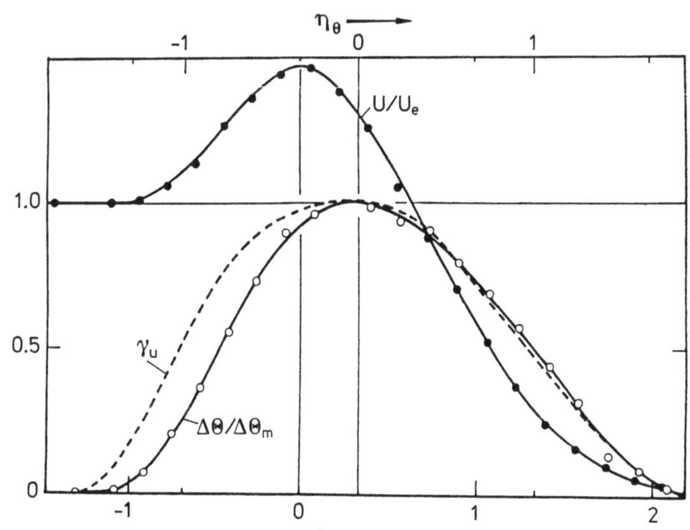

Fig. 3. Mean velocity and temperature profiles at $x/d = 60$

wires in the $x_1 - x_2$ plane (▵, ○, ▲, ◬, ●) or the $x_1 - x_3$ plane (▿, ◬, ▼); as noted above, unlinearized signals were used except for one case (▵). There is an encouraging uniformity obtained from the various runs. As would be expected, the peak turbulence intensities occur on the upper side of the jet mixing with stagnant surroundings. The temperature variance is more symmetric than the turbulence intensities, reflecting the fact that the mean temperature profile is also fairly symmetric. The three sets of data included in Fig. 5 were obtained

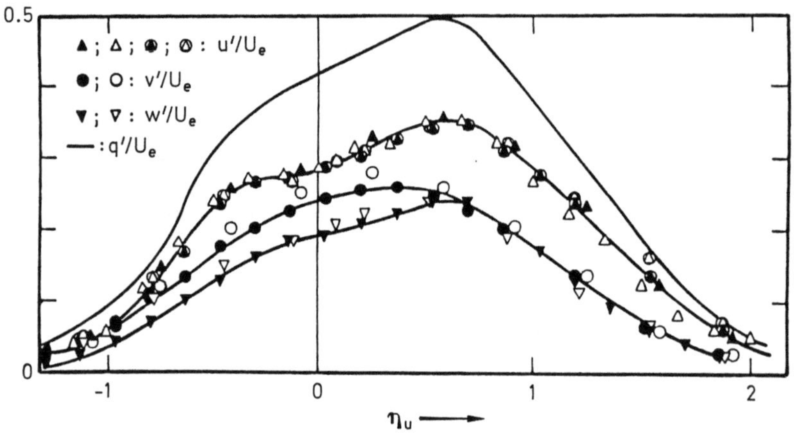

Fig. 4. Turbulence intensity profiles across jet

from a single "cold" wire alone (○), a cold wire incorporated in the 3-wire probe but with the two hot wires turned off (◓), and from the 3-wire probe in which the hot wires were functioning (◒). There is close agreement between the first two profiles but some contamination of the cold probe signal by the hot wires is evident on the side of the jet mixing with stagnant surroundings though this only becomes serious for values of η_u greater than unity. It is, of course, necessary to use this last operating mode for finding θ when forming correlations with the velocity field.

The experimental data for the triple-moments are compared with the various model proposals in Section 4.

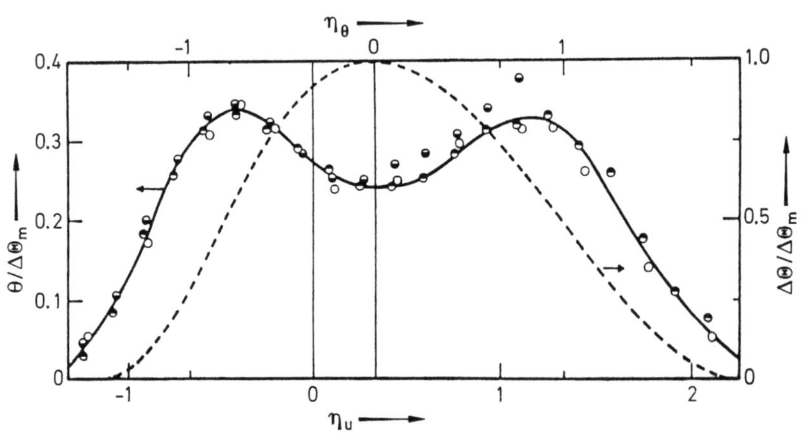

Fig. 5. Intensity of temperature fluctuations across jet

3. The Models Considered

The present section outlines the models of the triple moments to be compared with data of the asymmetric jet in the next section; approximations are provided for $\overline{u_i u_j u_k}$, $\overline{u_i \theta u_k}$ and $\overline{\theta^2 u_k}$ which appear in the second-moment equations for the Reynolds stress ($\overline{u_i u_j}$), scalar flux ($\overline{u_i \theta}$) and scalar variance ($\overline{\theta^2}$). It is convenient to present the various models as closures, albeit very crude ones, of the exact third-moment equations. It is reasonable to regard the formula resulting from any closure as a "package" whose usefulness does not necessarily reflect the correctness (or otherwise) of the individual closure assumptions made along the way. For example, all models considered here explicitly neglect convective transport of the triple-moments and it is likely that the more successful schemes are implicitly making some accounting for this in the choices made for the various empirical constants. Our attention is confined to relatively simple schemes that can be adopted without greatly adding to the computational labour involved in numerical simulations of turbulent shear flows with second-moment closures. More elaborate treatments have been developed by Lumley (1978) and André, Lacarrère and Traoré (1982), the latter within the framework of a third-moment closure.

The exact equation for the transport of triple velocity products for a uniform density shear flow unaffected by external force fields may be written:

$$\frac{D\overline{u_i u_j u_k}}{Dt} = \underbrace{P_{ijk1} + P_{ijk2}}_{\text{"Generation"}} + \underbrace{d_{ijk}}_{\text{Diffusion}} + \underbrace{\phi_{ijk}}_{\text{Non-dispersive pressure interactions}} - \underbrace{\varepsilon_{ijk}}_{\text{Dissipation}} \tag{1}$$

where

$$P_{ijk1} \equiv +\left(\overline{u_i u_j}\,\overline{u_k u_{m,m}} + \overline{u_k u_i}\,\overline{u_j u_{m,m}} + \overline{u_j u_k}\,\overline{u_i u_{m,m}}\right)$$

$$P_{ijk2} \equiv -\left(\overline{u_i u_j u_m}\,U_{k,m} + \overline{u_k u_i u_m}\,U_{j,m} + \overline{u_j u_k u_m}\,U_{i,m}\right)$$

$$d_{ijk} \equiv -\left(\overline{u_i u_j u_k u_m} + \overline{\frac{p}{\varrho} u_j u_k}\,\delta_{im} + \overline{\frac{p}{\varrho} u_k u_i}\,\delta_{jm} + \overline{\frac{p}{\varrho} u_i u_j}\,\delta_{km} - \overline{\nu u_i u_j u_{k,m}}\right)_{,m}$$

$$\phi_{ijk} \equiv \overline{(p/\varrho)\left(u_i u_{j,k} + u_k u_{i,j} + u_j u_{k,i}\right)}$$

$$\varepsilon_{ijk} \equiv 2\nu\left(\overline{u_k u_{i,m} u_{j,m}} + \overline{u_j u_{k,m} u_{i,m}} + \overline{u_i u_{j,m} u_{k,m}}\right).$$

The notation is given under Nomenclature but is, in any case, what is standardly adopted. The pressure-containing correlation ϕ_{ijk} may be decomposed into a non-linear part ϕ_{ijk1} composed purely of fluctuating quantities and a linear part ϕ_{ijk2} comprising products of triple-moments and mean velocity gradients. Cormack et al. (1978) in drawing comparisons which several simple triple-velocity closure proposals concluded that the scheme of Hanjalic and Launder (1972) performed the best. This model emerged from:

- neglecting convection and diffusive transport by pressure and molecular action
- approximating the pressure interactions by

$$\phi_{ijk1} \propto -\overline{u_i u_j u_k}/T \tag{2}$$

  and neglecting ϕ_{ijk2}
- making the Gaussian approximation for the quadruple velocity moment in d_{ijk}:

$$\overline{u_i u_j u_k u_m} = \overline{u_i u_j}\;\overline{u_k u_m} + \overline{u_j u_k}\;\overline{u_i u_m} + \overline{u_k u_i}\;\overline{u_j u_m} \tag{3}$$

- retaining P_{ikj1} exactly but neglecting P_{ijk2} and ε_{ijk}.

The resultant algebraic equation may be reorganized to give:

$$\overline{u_i u_j u_k} = c_s T(\overline{u_k u_l}\,\overline{u_i u_j}_{,l} + \overline{u_j u_l}\,\overline{u_k u_i}_{,l}\,\overline{u_i u_l}\,\overline{u_j u_k}_{,l})$$

or, for brevity,

$$-\overline{u_i u_j u_k} = c_s T G_{ijk}. \tag{4}$$

The time-scale T in Eqs. (2) and (4) is taken as (k/ε) and the recommended value of the empirical coefficient c_s is 0.11. A similar, though simpler, version of the above form was earlier proposed by Daly and Harlow (1970):

$$-\overline{u_i u_j u_k} = c_s' T\,\overline{u_k u_l}\,\overline{u_i u_j}_{,l}. \tag{5}$$

Equation (5) may be thought of either as a simplification of (4) (in which just the first term is retained) or, more directly, as invoking the general gradient transport hypothesis

$$-\overline{u_k \phi} \propto T\,\overline{u_k u_l}\,\phi_{,l} \tag{6}$$

a proposition extensively employed by the Los Alamos group in the late '60's and early '70's. In fact the value for c_s' of 2.0 suggested by Daly and Harlow (1970) has not been found to be usable by other groups. Wyngaard and Coté (1974) employed the same model but with $c_s' = 0.3$ while Launder et al. (1975) adopted 0.22.

Besides the established simple proposals represented by Eqs. (4) and (6), certain elaborations of them have been here examined. The first is purely the inclusion of mean-strain generation, P_{ijk2}. No approximations are needed and the triple-moment model that results may be expressed:

$$-\overline{u_i u_j u_k} = c_s'' T(G_{ijk} - P_{ijk2}). \tag{7}$$

This form has been compared with the present measurements of Dekeyser (1982) with some overall improvement on Eq. (6). However, a weakness (at least in concept) of Eq. (7) is that if one decides to include P_{ijk2}, one should also introduce a model for the mean-strain contribution to ϕ_{ijk} (the part referred to as ϕ_{ijk2}). At *second*-moment level a simple yet useful and very widely-used approximation for the effects of pressure fluctuations on mean-strain generation is that known as the "isotropization-of-production" model:

$$\phi_{ij2} = -c_2(P_{ij} - \tfrac{1}{3}\delta_{ij} P_{kk}) \tag{8}$$

where ϕ_{ij2} and P_{ij} are respectively the non-dispersive pressure interaction and stress-generation rates in the $\overline{u_i u_j}$ transport equations associated with mean-strain (for further details see, for example, Launder et al., 1975). This same idea has been applied to model pressure effects on stress generation rates by buoyant forces and to the corresponding non-dispersive pressure effects on scalar flux generation rates, Launder (1975). The coefficient c_2 and the corresponding constants that arise in approximating the other processes mentioned by this isotropization-of-production idea are usually chosen in the range 0.5 ± 0.1. In the case of the triple-moments, it is an obvious extension to suppose:

$$\phi_{ijk2} = -0.5\,P_{ijk2}. \tag{9}$$

Here, as in corresponding approximations later, we set the coefficient equal to 0.5 on the grounds that that value is roughly what has been used in models involving the double moments and that, in view of other uncertainties, it is unlikely that any more refined tuning of the individual coefficients can be contemplated at present in view of the relative imprecision of the experimental data. The only previous application of the "isotropization-of-production" idea to triple-moments has been by André et al. (1982) who, in their study of buoyant convection between parallel planes adopted a coefficient of 0.2 rather than 0.5. The difference between these values can be attributed at least partly to the fact that pressure reflections from the plane walls would have been important in the buoyant experiment (Gibson and Launder 1978), a feature absent from the present free shear flow. In fact, if we are to apply the isotropization-of-production idea consistently we should assume that pressure fluctuations remove 50% of P_{ijk1} too. Launder (1982) points out that the exact (but unclosed) expression for ϕ_{ijk1} obtained by eliminating the pressure (via solution of its Poisson equation) in favour of velocity fluctuations does indeed suggest the need for such a process. Accordingly we would take

$$\phi_{ijk1} = -c^{-1}\frac{\overline{u_i u_j u_k}}{T} - 0.5\,P_{ijk1}. \tag{10}$$

The neglect of the dissipative terms in Eq. (1) is usually justified on the grounds of local isotropy. Wyngaard and Sundararajan (1979) among others have shown from considering the θ^3 budget in the atmospheric boundary layer that the usual assumption of local isotropy cannot be too freely applied to the triple-moments. The viscous destruction terms in (1) consist of products of fluctuating velocities and velocity gradients. Local isotropy would lead us to discard these terms because fine-scale motions (wherein the velocity gradients are dominant) are generally held to be uninfluenced by large-scale activities. It is now accepted, however, that there is some measure of linkage between the two: the most energetic large eddies are also the ones carrying the most intense fine-scale activity. In view of the foregoing it might appear reasonable to apply the general gradient transport hypothesis to model ε_{ijk}. From Eq. (6) and the definition of ε_{ijk} (following Eq. (1)) we obtain:

$$\varepsilon_{ijk} \propto -T(\overline{u_k u_l}\,\varepsilon_{ij,l} + \overline{u_j u_l}\,\varepsilon_{ik,l} + \overline{u_i u_l}\,\varepsilon_{jk,l}) \tag{11}$$

where $\varepsilon_{ij} \equiv 2\nu\overline{u_{i,m}u_{j,m}}$. Moreover, if we now apply to the *second*-moment dissipation term the usual assumption of local isotropy $\varepsilon_{ij} = \frac{2}{3}\delta_{ij}\varepsilon$ we finally obtain:

$$\varepsilon_{ijk} = -c_\varepsilon T\varepsilon_{,l}(\overline{u_k u_l}\,\delta_{ij} + \overline{u_j u_l}\,\delta_{ki} + \overline{u_i u_l}\,\delta_{jk}). \tag{12}$$

Three models of $\overline{u_i u_j u_k}$ are compared with the measurements in the following section: These are: Eq. (4) (the Hanjalic-Launder model); Eq. (7) modified by the inclusion of dissipation, Eq. (12); and the model which emerges by the further inclusion of the isotropization-of-production model for ϕ_{ijk} (i.e. Eq. (9) together with Eq. (10) in place of Eq. (2)). The very simple gradient diffusion version given by Eq. (5) did somewhat less well overall than Eq. (4) (Dekeyser, 1982) and to avoid unnecessary clutter its variation is not shown on the curves in question.

Precisely parallel treatments to those discussed above at some length have been applied to the correlations giving rise to scalar flux diffusion $\overline{u_k u_i \theta}$ and to diffusion of the scalar variance $\overline{u_k \theta^2}$. For the former case, omitting molecular transport effects, the exact equation may be written

$$\frac{D\overline{u_k u_i \theta}}{Dt} = P_{ki\theta 1} + P_{ki\theta 2} + d_{ki\theta} + \phi_{ki\theta 1} + \phi_{ki\theta 2} - \varepsilon_{ki\theta} \tag{13}$$

where the symbols on the right side of (13) have an analogous significance to those in Eq. (1).

The most widely applied algebraic approximations to (13) neglect convective transport, molecular dissipation, generation due to mean field gradients $P_{ki\theta 2}$ and the contribution of pressure transport in $d_{ki\theta}$. Closure is effected by approximating the pressure interaction as:

$$\phi_{ki\theta 1} = -c_\theta^{-1}\,\overline{u_k u_i \theta}/T, \tag{14a}$$

$$\phi_{ki\theta 2} = 0, \tag{14b}$$

assuming a normal distribution for the quadruple correlation in $d_{ki\theta}$:

$$d_{ki\theta} \equiv -\overline{u_k u_i \theta u_m}_{,m} = (\overline{u_k u_i}\;\overline{u_m \theta} + \overline{u_m u_k}\;\overline{u_i \theta} + \overline{u_i u_m}\;\overline{u_k \theta})_{,m} \tag{15}$$

and by retaining $P_{ki\theta 1}$ without the need for approximation:

$$P_{ki\theta 1} \equiv \{\overline{u_i u_k}\;\overline{\theta u_m}_{,m} + \overline{\theta u_i}\;\overline{u_k u_m}_{,m} + \overline{u_k \theta}\;\overline{u_i u_m}_{,m}\}. \tag{16}$$

Introduction of Eq. (14)–(16) into (13) produces:

$$-\overline{u_i u_k \theta} = c_\theta T(\overline{u_k u_m}\;\overline{u_i \theta}_{,m} + \overline{u_i u_m}\;\overline{u_k \theta}_{,m} + \overline{\theta u_m}\;\overline{u_i u_k}_{,m}) \tag{17}$$

a form essentially the same as that proposed by Deardorff (1973). This is equivalent in its modelling concept to Eq. (4). Simpler derivatives have been used in which the last term on the right is discarded (Owen, 1973) or the last *two* terms (Wyngaard and Coté, 1974). These simpler versions are less successful than (17) in reproducing the distribution of $\overline{u_i u_k \theta}$ across the asymmetric jet, Dekeyser (1982), and will therefore not be included in the present comparisons.

The dissipative term $\varepsilon_{ik\theta}$ denotes the following group of terms:

$$\varepsilon_{ij\theta} \equiv (\lambda + \nu)\,(\overline{u_k u_{i,m} \theta_{,m}} + \overline{u_i u_{k,m} \theta_{,m}}) + 2\nu \overline{\theta u_{i,m} u_{k,m}}. \tag{18}$$

Application of the gradient transport idea to the first pair of items on the right of (18) produces a quantity proportional to

$$T(\overline{u_k u_l}\;\varepsilon_{i\theta,l} + \overline{u_i u_l}\;\varepsilon_{k\theta,l}).$$

However, the molecular dissipation of heat flux $\varepsilon_{i\theta}$ is zero if, as before, we assume local isotropy to apply to all the fine-scale, *second*-moment correlations; the terms are thus discarded. The third term, while strictly not in a form appropriate to the gradient transport idea may nevertheless be treated in this way if we regard the instantaneous temperature as simply a "marker" of velocity fluctuations. With that assumption the model for $\varepsilon_{ik\theta}$ emerges as:

$$\varepsilon_{ik\theta} = -c_{\varepsilon\theta} T \overline{\theta u_l}\,\varepsilon_{,l}\,\delta_{ik}. \tag{19}$$

Finally, if one adopts the "isotropization-of-production" (IP) concept Eq. (14a) is replaced by

$$\phi_{ki\theta 1} = \tilde{c}_\theta^{-1}\,\frac{\overline{u_k u_i \theta}}{T} - 0.5\,P^*_{ki\theta 1} \tag{20}$$

where $P^*_{ki\theta 1}$ is $P_{ki\theta 1}$ with the first term on the right of Eq. (16) dropped; the term is excluded because it originates from the temperature fluctuation equation from which pressure fluctuations are absent (an analogous principle is usually applied in closing the $\overline{u_i\theta}$ equation where $\phi_{i\theta 2}$ is approximated as $+\,0.5\,\overline{u_k\theta}\,U_{i,k}$). Strictly, for consistency, Eq. (20) should be taken in the form

$$- c_\theta^{-1}\left(\frac{\overline{u_k u_i \theta}}{T} - \frac{\delta_{ik}}{3}\frac{\overline{u_m u_m \theta}}{T}\right) - 0.5\left(P^*_{ki\theta 1} - \frac{\delta_{ik}}{3}P^*_{mm\theta 1}\right)$$

$$- c_{\theta 1}^{-1}\,\delta_{ik}\,\overline{u_m u_m \theta}/T - c_{\theta 2}\,\delta_{ik}\,P^*_{mm\theta 1}\,. \tag{20'}$$

The first line of Eq. (20′) was first suggested by André et al. (1979) and the second follows naturally from it. In their original study these workers took $c_{\theta 1}^{-1}$ equal to $c_\theta^{-1}/3$ a choice that simplifies their approximation to that of Eq. (20). (In the more recent study by André et al. (1982), however, a different choice of coefficients was made.) In the present enquiry we have retained Eq. (20) rather than Eq. (20′) principally to avoid the introduction of two further unknown coefficients to be optimized against only two experimental profiles (of $\overline{u_1^2\theta}$ and $\overline{u_2^2\theta}$).

The mean-field generation rate $P_{ki\theta 2}$ arises from both velocity and temperature gradients:

$$P_{ki\theta 2} \equiv -\left(\overline{u_i u_l \theta}\,U_{k,l} + \overline{u_k u_l \theta}\,U_{i,l} + \overline{u_i u_k u_l}\,\Theta_{,l}\right). \tag{21}$$

Dekeyser (1982) concludes, however, that the contribution of mean velocity gradients is barely significant compared with that due to temperature gradients. Moreover, $\phi_{ki\phi 2}$ is associated only with velocity gradients (not temperature gradients) and according to the IP model its effect will be to halve an already marginal contribution to $P_{ki\theta 2}$. Accordingly, $\phi_{ki\theta 2}$ and the first two terms on the right of Eq. (21) are omitted.

In Section 4 we compare the distributions of $\overline{u_i u_k \theta}$ resulting from Eq. (17); from Eq. (13) used in conjunction with Eqs. (14–16) and (19); and Eq. (13)[1] closed by means of Eqs. (15), (16), (19) and (20).

The exact equation for the transport of $\overline{u_k\theta^2}$ may be written:

$$\frac{D\overline{u_k\theta^2}}{Dt} = P_{k\theta^2 1} + P_{k\theta^2 2} + d_{k\theta^2} + \phi_{k\theta^2 1} + \phi_{k\theta^2 2} - \varepsilon_{k\theta^2}\,. \tag{22}$$

Again the symbols retain a physical significance consistent with Eqs. (1) and (13). The specific correlations denoted by the production and dissipation terms in (22) are

$$P_{k\theta^2 1} \equiv \overline{\theta^2}\;\overline{u_k u_l}_{,l} + 2\,\overline{u_k\theta}\;\overline{u_l\theta}_{,l}$$

$$P_{k\theta^2 2} \equiv \left(\overline{u_l\theta^2}\,U_{k,l} + 2\,\overline{u_k u_l \theta}\,\Theta_{,l}\right)$$

$$\varepsilon_{k\theta^2} \equiv 2(\nu + \lambda)\,\overline{\theta\theta_{,j}u_{k,j}} + 2\lambda\,\overline{u_k\theta_{,j}\theta_{,j}}\,.$$

The counterpart of Eqs. (4) and (17) is obtained from (22) by neglecting convective transport, molecular and pressure contributions to $d_{k\theta^2}$, and mean field gradients. On introducing the Gaussian approximation of quadruple moments and the usual return-to-isotropy representation for $\phi_{ki\theta 1}$, the following form, due to Deardorff (1973) emerges:

$$-\overline{u_k\theta^2} = c_\theta^*\,T\left(\overline{u_k u_l}\;\overline{\theta^2}_{,l} + 2\,\overline{u_k\theta}\;\overline{u_l\theta}_{,l}\right). \tag{23}$$

1 Convective transport of $u_i u_k \theta$ is discarded in each case

Some workers (e.g. Launder and Samaraweera, 1979) have retained just the first term in parentheses to facilitate numerical solution. However, this truncation is less successful than (23) in describing the measured distribution of $\overline{u_k \theta^2}$, Dekeyser (1982).

Two elaborations of Eq. (23) are considered here. One is the form which results from retaining the constituents of (23) but now adding mean temperature gradients from $P_{k\theta^2 2}$ and approximating $\varepsilon_{k\theta^2}$ by the generalized gradient transport hypothesis. Assuming, s in approximating $\varepsilon_{ki\theta}$, that $\varepsilon_{k\theta} \cong 0$ leads to

$$\varepsilon_{k\theta^2} = -\hat{c}_{\varepsilon\theta} T \overline{u_k u_l}\, \varepsilon_{\theta,l} \tag{24}$$

where ε_θ is the dissipation rate of $\frac{1}{2}\overline{\theta^2}$. The other form considered includes, in addition, the IP model for $\phi_{k\theta^2 1}$:

$$\phi_{k\theta^2 1} = -\hat{c}_\theta^{-1} \frac{\overline{u_k \theta^2}}{T} - 0.5\, P_{k\theta^2 1}. \tag{25}$$

The two forms precisely parallel the models considered above for $\overline{u_i u_j u_k}$ and $\overline{u_i u_k \theta}$.

4. Comparison Between Models and Experiments

The previous section has presented models for velocity and scalar triple moments in terms of second-moment quantities and their gradients. Here we compare the predicted distributions of the triple moments with directly measured values. In doing so we take from experiment the values of the second-moment quantities appearing in the various models. This goes equally for the dissipation rates ε and ε_θ though in this case the values used were those deduced as the closing term in the kinetic energy and $\overline{\theta^2}$ budgets (in the former case with pressure diffusion neglected).

In the comparisons the three coefficients of (4), (17) and (23) (i.e. c_2, c_θ and c_θ^*) are, for uniformity, all set equal to 0.11 (no significant improvement would result from optimizing them separately). The corresponding coefficients in the more elaborate models (i.e. c, $\tilde{c}_\theta$ and $\hat{c}_\theta$) are taken as 0.075 irrespective of whether or not the IP model is included. The dissipation coefficients c_ε, $c_{\varepsilon\theta}$ and $\hat{c}_{\varepsilon\theta}$ take the values 0.2, 0.1 and 0.3 which amounts to choosing the value of the coefficients in Eq. (6) to be 0.15, the value adopted by several workers in approximating the same correlations in the dissipation rate transport equations (wherein they appear as *diffusion* terms).

Comparisons between experiment and model prediction are provided in Figs. 6, 7 and 8 for respectively the triple velocities, the scalar flux transport and the scalar variance transport. The usual thin shear flow approximation has been made with only cross-stream gradients of mean or turbulence quantities being retained. Let us consider first the measurements of the triple-velocity correlation. In all figures the stagnant stream is on the right. The large scale mixing on this side is reflected in the large values of $\overline{u_1^3}$, $\overline{u_1^2 u_2}$, $\overline{u_1 u_2^2}$ compared with corresponding levels on the side of the shear flow entraining moving fluid. In contrast the $\overline{u_2^3}$ profile is almost antisymmetric with the peak magnitudes on each side being nearly the same. This probably reflects the fact that this is the only one of the triple velocities presented in which mean-strain generation is negligible. The two profiles involving odd powers of u_1 are fairly similar in shape with positive values on the edges and negative values near the centre. The most complicated behaviour is displayed by $\overline{u_1^2 u_2}$ with two minima and two maxima.

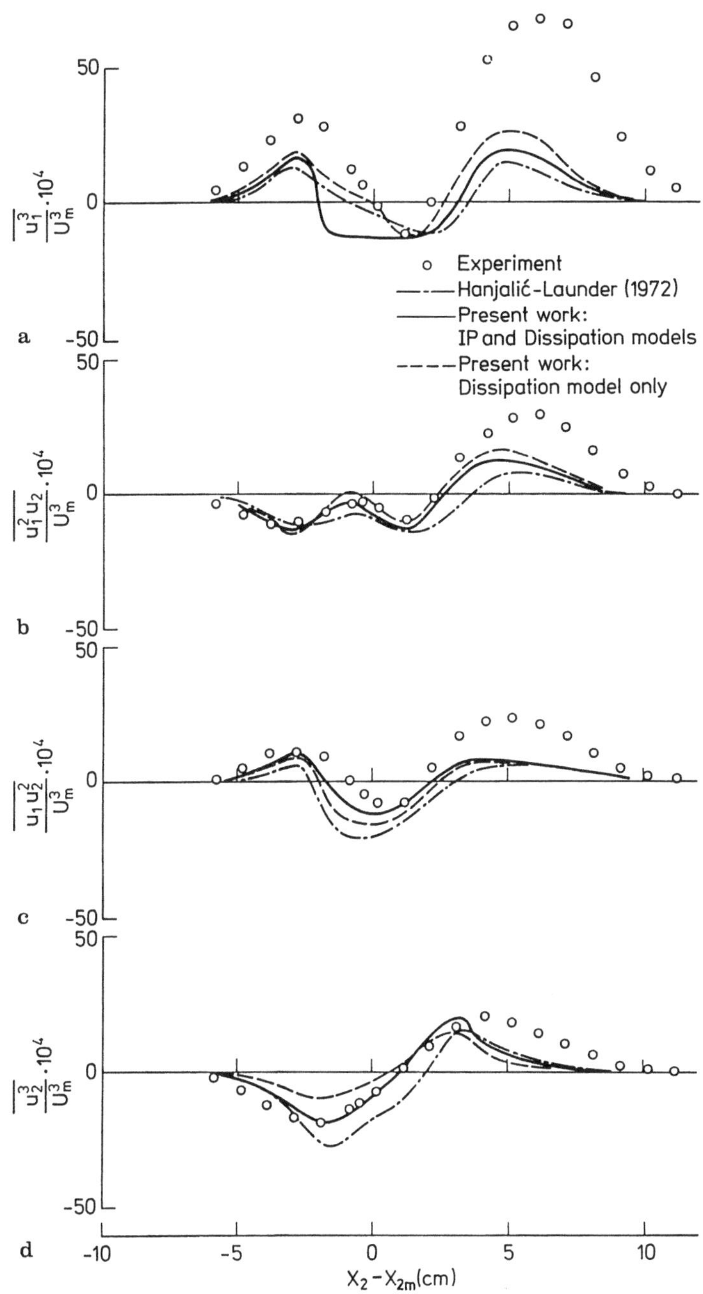

Fig. 6. Profiles of $\overline{u_k u_j u_k}$: comparisons between experiment and models

The three models for $\overline{u_i u_j u_k}$ all succeed in mimicking the basic features of the profile shape; for example, they unerringly provide the correct number of maxima and minima for each component. The simplest of the three schemes, Eq. (5), performs less well, however, than the others. Of the two new models tested, for all except $\overline{u_1^3}$, the version that includes the IP model for non-dispersive pressure effects does slightly better; it gives a particularly impressive account of the undulations in $\overline{u_1^2 u_2}$ for $(x_2 - x_{2m})$ less than 3 cm. For the exceptional

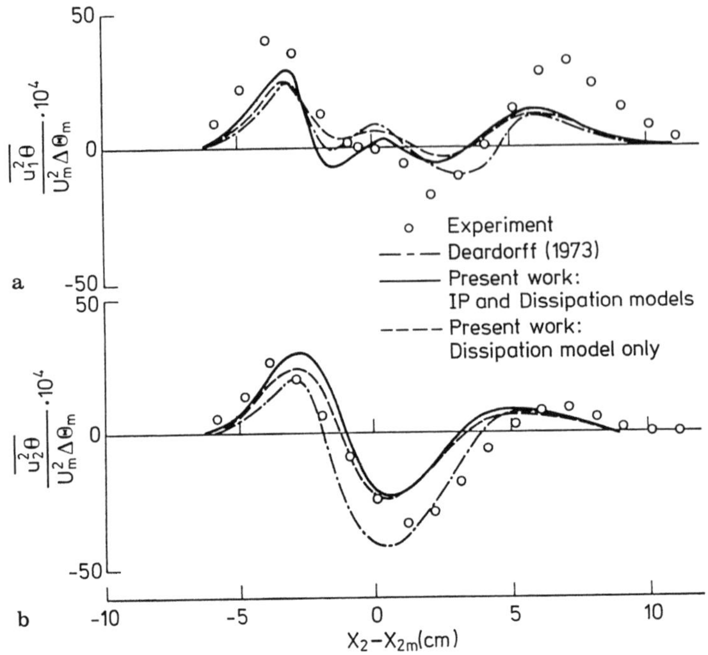

Fig. 7. Profiles of $\overline{u_k u_i \theta}$: comparison between experiment and models

case, however, it produces an abrupt change of sign of the correlation in the moving-stream side which is not present in the data. This turns out to be provoked by a term proportional to $\overline{u_1^2}\ \overline{u_1 u_{2,2}}$ which does not appear in either of the other models. All schemes produce markedly lower values than measured on the stagnant stream side. Hot-wire errors associated with intermittent, high-intensity turbulence will almost certainly produce measured values that are too high for $(x_2 - x_{2m}) > 5$. Nevertheless, it seems unlikely that experimental error in $\overline{u_i u_j u_k}$ is the principal cause of the difference. The poor agreement could point to a fundamental weakness in the models and/or it might reflect an estimate of ε (obtained as noted as the closing term in the kinetic energy equation) that was also too high near the low velocity boundary [2].

The scalar flux diffusion shown in Fig. 7 exhibits (like the temperature and scalar variance profiles) a closer degree of symmetry about the two sides of the jet. Interestingly, slightly higher peak values now occur on the moving-stream side. While $\overline{u_2^2 \theta}$ displays an uncomplicated variation that for $\overline{u_1^2 \theta}$ shows inflectional behaviour near x_{2m}. The three models also show quite different distributions for $\overline{u_1^2 \theta}$ and $\overline{u_2^2 \theta}$ in the central region broadly in line with measurements. In the case of $\overline{u_1^2 \theta}$ the measured levels near the shear flow edges are markedly higher than given by any of the models. It would have to be said that though the two models introduced in the present study give somewhat different distributions than Eq. (17) they cannot really be said to be in any *better* overall accord. The use of Eq. (20′) in place of Eq. (20) may, as commented earlier, offer scope for improvement.

The scalar variance diffusion in Fig. 8 exhibits a complex variation across the jet. The most striking feature is that $\overline{u_1 \theta^2}$ takes on large negative values over the interior of the shear flow. This reflects the fact that negative values of u_1 (brought by newly-entrained fluid) are

2 Since ε appears in the denominator of the time scale too large values depress all the triple moments

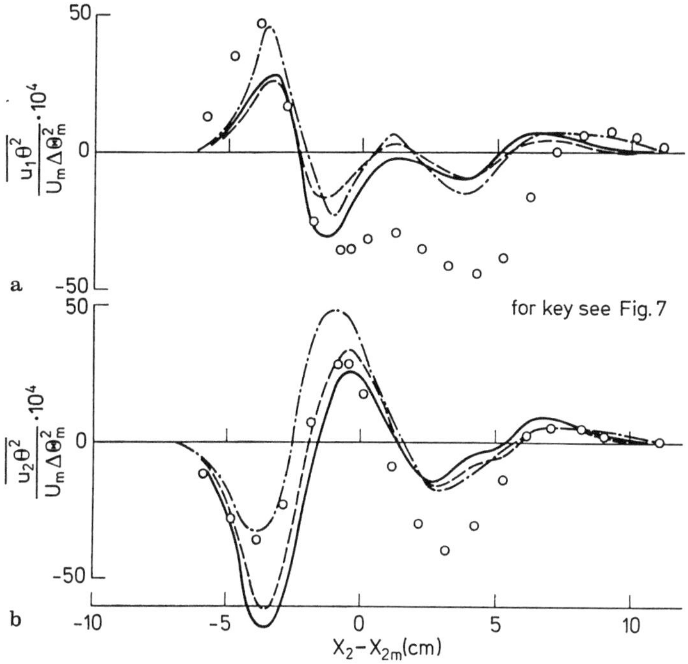

Fig. 8. Profiles of $\overline{u_k \theta^2}$: comparison between experiment and models

associated with larger fluctuations about the mean temperature than those of positive u_1. The same feature of the flow is responsible for the large, negative readings of $\overline{u_2 \theta^2}$ for $(x_2 - x_{2m})$ in the range 2 – 5 cm. None of the models succeeds in reproducing these features. Apart from this (serious) weakness, however, all three models capture broadly the trend of the distributions including, for example, the fact that the $\overline{u_1 \theta^2}$ profile should display three maxima and two minima. The two models here introduced give rather different peak levels in different regions, but as with the scalar flux, it is not possible on the basis of the two sets of data shown here to declare one form conclusively superior to another.

5. Concluding Remarks

The paper has presented some of the data of the triple moments of velocity and temperature fluctuations recently obtained in an asymmetric heated jet. The flow is one where production of these triple products due to mean velocity and temperature gradients is comparable with that due to inhomogeneities in the second-moment distributions. Because of the structural asymmetry, the triple moment data display in many cases a complicated pattern.

The three models considered here all achieve a better than qualitative agreement with the third-moment data but none can be said to be distinctly successful. For the triple velocity data the models based on a gradient approximation of the "dissipative" term ε_{ijk} with or without the isotropization of production model of the non-dispersive pressure correlations are more successful than the earlier Hanjalic-Launder (1972) formulation which neglected both agencies. The discrepancies between experiment and prediction near the stagnant boundary are still very serious however. This behaviour is also present for $\overline{u_1^2 \theta}$ but not, to any extent, for any of the other temperature-velocity triple products.

For the temperature-containing correlations we have been unable to conclude that any of the models performed distinctly better than the other two. The study does not necessarily weaken the case for the two modelling ideas considered herein but it does suggest that on their own they are insufficient. Various further explorations are prompted that may lead to more definitive improvements; among these the question of adopting a single turbulent time scale (k/ε) as characterizing the rates of progress of all the pressure interactions is one of the more serious. In the case of $\overline{u_k \theta^2}$, the thermal time scale $\overline{\theta^2}/\varepsilon_\theta$ would certainly have been preferable. Alternative modelling suggestions by Lumley (1978) also merit testing.

A major concern must be the use of the Gaussian approximation for the quadruple moments in intermittent regions. The only reason for retaining it is that there is no alternative available that seems convincingly better. Lumley (1978) has correctly commented that conceptually the most satisfactory treatment of intermittent regions is to model them as precisely that. One might envisage either adopting a transport equation for the intermittency itself (Libby, 1975) or developing a scheme based on probability density functions. Before such approaches would attract interest amongst those making practical calculations, however, they need to be shown to be necessary. Whether this is so can be most effectively determined by incorporating the third-moment models considered into a second-moment closure study of the present and other asymmetric flows.

Acknowledgements. The experimental work was undertaken at IMST, Marseille, and is published with the permission of the Director. Especial thanks go to Dr. C. Béguier for his material help and suggestions on all aspects of the experimental programme. Miss L. Towers at UMIST prepared the original camera-ready version of the paper. Authors' names appear alphabetically.

References

André, J. C., de Moor, G, Lacarrère, P., Therry, G., du Vachat, R.: The Clipping Approximation and Inhomogeneous Turbulence Simulations, in *Turbulent Shear Flows 1*, ed. by F. Durst, B. E. Launder, F. W. Schmidt, J. H. Whitelaw (Springer, Berlin, Heidelberg, New York 1979) p. 307

André, J. C., Lacarrère, P., Traoré, K.: Pressure Effects on Triple Correlations in Turbulent Convective Flow, in *Turbulent Shear Flows 3*, ed. by L. J. S. Bradbury, F. J. Durst, B. E. Launder, F. W. Schmidt, J. H. Whitelaw (Springer, Berlin, Heidelberg, New York 1982) p. 243

Béguier, C., Fulachier, L., Keffer, J. F. (1976): The turbulent mixing layer with an asymmetrical distribution of temperature. J. Fluid Mech. **89**, 561

Cormack, D. E., Leal, L. G., Seinfeld, J. H. (1978): An evaluation of mean Reynolds stress turbulence models: the triple velocity correlations. J. Fluids Eng. **100**, 47

Daly, B. J., Harlow, F. H. (1970): Transport equations in turbulence. Phys. Fluids **13**, 2634

Deardorff, J. W.: Three-Dimensional Numerical Modelling of the Planetary Boundary Layer, in *Proc. Workshop on Micrometeorology*, **271**, (Am. Met. Soc. Boston, Massachusetts, 1973)

Dekeyser, I. (1982): Etude d'un jet plan dissymmetrique chauffé en régime turbulent incompressible. Thèse Docteur-es-Sciences, Université d'Aix-Marseille II

Dekeyser, I. (1983): Journal de Mécanique Théorique et Appliquée. Vol 2, N° 6, 1983, 915

Gibson, M. M., Launder, B. E. (1978): Ground effects on pressure fluctuations in the atmospheric boundary layer. J. Fluid Mech. **86**, 491

Hanjalić, K., Launder, B. E. (1972): A Reynolds stress model of turbulence and its application to thin shear flows. J. Fluid Mech. **52**, 609

Keffer, J. F., Kawall, J. G., Giralt, F., Béguier, C.: Analysis of Turbulence Structure in Complex Shear Flows, in *Proc. 2nd Symposium on Turbulent Shear Flows* (Imperial College London 1979) p. 2

Launder, B. E. (1975): On the effects of a gravitational field on the transport of heat and momentum. J. Fluid Mech. **67**, 569

Launder, B. E., Reece, G. J., Rodi, W. (1975): Progress in the development of a Reynolds stress turbulent model. J. Fluid Mech. **68**, 537

Launder, B. E.: Heat and Mass Transport, in *Topics Appl. Phys.*, Vol. 12, 2nd ed., ed. by P. Bradshaw (Springer, Berlin, Heidelberg, New York 1978) p. 232

Launder, B. E., Samaraweera, D. S. A. (1979): Application of a second-moment closure to heat and mass transport in thin shear flows. Int. J. Heat Mass Trans. **22**, 1631
Launder, B. E. (1982): Second-moment closure: methodology and practice. EDF Summer School on Modélisation Numérique, Le Bréau, France
Libby, P. A. (1975): On the prediction of intermittent turbulent flows. J. Fluid Mech. **68**, 272
Lumley, J. L. (1978): Computational modeling of turbulent flows. AdV Appl. Mech. **18**, 123
Lumley, J. L., Zeman, O., Siess, J. (1978): The influence of buoyancy on turbulent transport. J. Fluid Mech. **84**, 581
Owen, R. G. (1973): An analytical turbulent transport model applied to non-isothermal fully-developed duct flows. Ph.D. Thesis. Penn. State Univ.
Schon, J.-P., Baille, A. (1972): Méthode d'isolement des fluctuations turbulentes cinématiques et thermiques au moyen d'une sonde anémométrique à trois fils. C.R. Acad. Sci. Paris **A 274**, 116
Wyngaard, J. C., Coté, O. (1974): The evolution of a convective planetary boundary layer. A higher order closure model study. Boundary Layer Met. **7**, 289
Wyngaard, J. C., Sundararajan, A.: The Temperature Skewness Budget in the Lower Atmosphere and Its Implications for Turbulence Modeling, in *Turbulent Shear Flows 1*, (Springer, Berlin, Heidelberg, New York 1979) p. 319

Part II

Free Flows

Introduction to Papers on Free Turbulent Flow

I. S. Gartshore

Department of Mechanical Engineering, University of British Columbia, Vancouver, B.C., Canada

Several convenient classifications of the papers in this section can be made. We find simple flows (which of course are not simple at all!); more complex flows, made up of simple flows in combination with each other; simple flows with curvature effects; and simple flows with swirl. All of the flows studied experimentally involve wakes: wakes in isolation, wakes in combination with boundary layers or with other wakes, wakes with curvature, and wakes with swirl. The one numerical study considers a plane mixing layer.

The wake of a circular cylinder, and circular cylinders were used in most cases considered here, is sensitive to both Reynolds number and free stream turbulence. (Why not use a normal flat plate which is sensitive to neither, over broad ranges?) There is a limited distance behind the cylinder, $x/d \gtrsim 15$ for simple cases at typical laboratory Reynolds numbers, in which large well-organized vortices appear. This "near wake" region is of interest in practical applications – wind engineering, heat exchange design and so on – but it is not a region in which turbulence itself is easy to study, measurements and visual studies being dominated by the large relatively regular vortex pattern. Beware then of interpreting all of these papers as studies of turbulence! Even for $x/d < 25$, we may expect a wake to contain relics of the vortex street, and to be changing its non-dimensional characteristics very rapidly; further downstream, only the cylinder drag coefficient determines the wake characteristics, which change slowly if at all.

Three combinations of simple flows are considered in these papers, and it is of interest to determine when these might be considered as superpositions of simpler cases. The time averaged equations lend support to simple superposition only if the velocity defects in the constituent flows are small enough to allow linearization of the acceleration terms. Even then, there remains the problem of the turbulent stress terms which are additive only after very restrictive assumptions. Considering vortex streets or large turbulent motions in the flow, a time sharing coexistence of structures can be postulated only near the outer edges of each participant flow, again where velocity defects are small. In consequence, the boundary layer/wake interaction examined by Tsiolakis, Krause and Muller, can be treated plausibly as a combination of its constituents only when the wake and boundary layer edges overlap slightly, and not when the wake is buried deep within the boundary layer. Similarly, the wake of cruciform circular cylinders, (Osaka, Yamada and Nakamura) be cannot (and is not) considered as a combination of two wakes, but must be examined as an interesting but complicated flow, in which vortex shedding effects appear only off the wake centre line.

In the asymmetric turbulent wake studied by Shokr, Keffer and Kawall, the vortex structures from the larger of the two parallel circular cylinders creating the wake, rapidly dominate the flow and disrupt or entrain coherent vortices shed from the smaller one. The combination leads, as the authors point out, to regions of apparent negative turbulent energy production in the wake. Vortex streets are not necessary for this phenomenon; a similar situation, involving turbulent motion but not vortex streets, occurs in a turbulent wall jet, where maximum velocity and zero stress positions do not coincide, again due to the overlap and interaction of shear flows with different length scales and different signs of strain rate.

Koyama investigates curvature effects in wakes. When the body generating the wake is normal to the curvature ("spanwise"), wake symmetry is destroyed due to the destabilizing effects on the inner side and stabilizing effects on the outer side. Two dimensionality is not necessarily affected. When the wake generator is placed along the radius of curvature, however, symmetry is maintained and two dimensionality is destroyed. In the latter case, the radial pressure gradient produces secondary flow which draws fluid along the wake, from a destabilized boundary layer on the outer tunnel wall, toward one which is stabilized on the inner wall. Measurements made at the wake mid span position are, at best, only representative of the wake development in this case, particularly as the length-to-diameter ratio of the wake-generating body is now only 10. The secondary flow disrupts the vortex street to some extent, a point which could prove useful in some practical applications, such as the case in which regular vortex shedding excites unwanted resonant vibrations. Spanwise flows induced by various means (e.g., air bubbles rising around a vertical cylinder in a water flow) should be kept in mind in this regard.

A final complicating effect considered in these papers is that of swirl (Leuchter and Solignac). Here the turbulent diffusion of axial vorticity is of particular importance because of the way the radial pressure gradients, and through them the axial pressure gradients, are affected. Axial pressure gradients near the centre line can be particularly adverse, more so than those at large radii, and because the dissipation is greatest near the centre-line, reverse flow can be produced near $r = 0$ similar to the flow reversal following separation from a solid surface in two-dimensional flow. This phenomenon in the swirling wake, which can occur with or without an externally imposed adverse pressure gradient is a major (if not *the* major) cause of "vortex bursting", studied extensively elsewhere. Radial diffusion of vorticity is critical to the phenomenon (see Mager, 1974 for example).

Two papers in this sequence consider simple flows, with or without perturbations. Fabris and Nakayama, by making measurements in a well-developed wake ($x/d \approx 500$), show that bursts and sweeps occur in this flow like those already observed in the outer region of boundary layers. It is encouraging to find similarities in these two cases, but perhaps not surprising that similar diagnostic techniques produce qualitatively similar conclusions. Comparisons are still useful, however. For example, the results show again that large structures in wakes are more effective in transferring momentum than those in other shear flows. It would be useful to know what differences in form or orientation of the large structures are responsible for these differences.

The numerical study of turbulence suppression in a shear layer (Nallasamy and Hussain) represents the flow by a series of vortices in a rotational inviscid process. Whether this adequately represents the major turbulent structures depends on the use being made of the model. The effect of sinusoidal perturbations on spanwise two-dimensional concentrations of vorticity can certainly be investigated. The results show that, consistent with some experimental observations, particular frequencies are most effective in suppressing cross-stream growth, provided the amplitude of excitation is in a suitable range. Very large or very small amplitudes are ineffective, a conclusion that remains to be verified experimentally.

Turbulence still represents an unpleasant black box to many able research workers and the reasons for this lie partly in the complexity of the motions, and partly in the way the equations are formulated or the measurements made. Almost all equations for turbulent flow use time-averaged quantities and time-averaging procedures, thereby making the equations indeterminate. In a parallel way, and for similar reasons of simplicity, experiments measure time-averaged quantities despite increasing evidence that a very important feature of turbulence is the unsteadiness arising from the irregular presence of large, more or less coherent structures which pass through growth-decay-renewal cycles (see Townsend, 1974). To average through active coherent regions and the intervening relatively passive periods, introduces experimental ambiguity similar to the indeterminacy so familiar in the corresponding equations. Conditional sampling based on multiple correlations has been introduced successfully (Mumford, 1982) and seems a promising way out of the wilderness of ambiguity in which we have been wandering. Such measurements are not easy or inexpensive, but with modern data acquisition and reduction systems they are increasingly practical. Flows with simple history effects, those that are demonstrably self-preserving, need to be compared with each other first, to document the forms and orientations of the large structures in each one. Similarities would be welcome! The distant objective would be to formulate a turbulent shear flow model containing a limited number of essential large structures with characteristics and relative strengths which change in response to the imposed strains. Recent steps in this direction are reported from Townsend's group in Cambridge by Savill (1982) and papers on subgrid models (see Bertoglio and Mathieu in the present symposium) also have this goal. Certainly the concept of large structures is here to stay and it would be useful to have direct descriptions of their forms rather than indirect inferences made through conventional averages.

References

Mager, A. (1974): Steady, incompressible swirling jets and wakes. AIAA J. **12**, 1540

Mumford, J. C. (1982): The structure of the large eddies in fully developed turbulent shear flows. Part 1. The plane jet. J. Fluid Mech. **118**, 241

Savill, A. M. (1982): Zonal Modelling – a response to the paper by Professor Kline, Volume 11, Proceedings of the 1980–81 Stanford AFOSR-HTTM Conference

Townsend, A. A. (1974): Flow patterns of large eddies in a wake and in a boundary layer. J. Fluid Mech. **95**, 515

Statistical Characteristics of the Turbulent Wake Behind an Intersecting Cruciform Circular Cylinder

Hideo Osaka, Hidemi Yamada

Dept. Mech. Eng. Yamaguchi Univ., Ube, Japan

Ikuo Nakamura

Dept. Mech. Eng. Nagoya Univ., Nagoya, Japan

Abstract

In order to clarify the periodic structure in the near field and the turbulent eddy structure in the far field of the intersecting cruciform circular cylinder wake, the detailed measurements have been performed on the statistical characteristics such as one-point and two-point time correlations, power spectra, triple velocity correlations and probability density profiles.

The experimental results are summarized as follows: The quasi-periodicity of the fluctuating velocity appears to different degrees at the different locations in the cross section of the near wake field, and is not found at the wake center. The turbulent large eddy structure conjectured from the contours of the space-time correlation is markedly different at each location in the cross section, and its decay rate is also slower at the center of the quasi-two-dimensional region than at the wake center. The magnitude of triple velocity correlation profiles on the bisector has the same order as that of the two-dimensional wake, but some of the shapes are significantly different.

Nomenclature

x	Distance in the streamwise direction from the origin of the cruciform circular cylinder, Fig. 1
y, z	Distances along two cylinder axes from the origin respectively, Fig. 1
y', z'	Distances along the bisector of $y = z$ and normal to the bisector from the origin respectively, Fig. 1
r_x, r_y, r_z	Distances between the fixed probe position and the moved probe position in x, y and z directions respectively
d	Diameter of circular cylinder, Fig. 1
$b_{y'}$	Half-width for the bisector, Fig. 1
L_u	Integral length scale
λ	Taylor's micro scale
l_ε	Kolmogorov's dissipative scale
t	Arbitrary time
τ	Time delay
U	Mean velocity in x direction, Fig. 1
U	Free stream velocity, Fig. 1
U_d	Mean velocity defect ($U_d = U_1 - U$), Fig. 1
U_{d_0}	Mean velocity defect at the wake center
u, v, w	Fluctuating velocity components in x, y and z directions respectively
v', w'	Fluctuating velocity components in y' and z' directions respectively
$(1/2)\,\overline{q^2}$	Turbulent kinetic energy ($\overline{q^2} = \overline{u^2} + \overline{v^2} + \overline{w^2}$)
$R_{uu}, R_{vv}, R_{ww}, R_{v'v'}, R_{w'w'}$	One-point auto-correlation coefficients of u-, v-, w-, v'- and w'-fluctuating velocities respectively
$R_{uv'}, R_{uv'}$	One-point cross correlation coefficients between u- and v-fluctuating velocities and between u- and v'-fluctuating velocities respectively
E	Power spectra of fluctuating velocity components
S	Skewness factors of fluctuating velocity components
F	Flatness factors of fluctuating velocity components
ν	Kinematic viscosity of air

1. Introduction

The intersecting cruciform circular cylinder which consists of two effectively long cylinders joined perpendicular to one another, produces the complex flow field due to the three-dimensional interference of each intersecting cylinder wake (Fig. 1). This three-dimensional flow interference is closely related to that found in various flow fields of interest to fluid engineers, for example the flow around heat exchanger tubes, on-shore and off-shore structures and damping screens formed from interwoven wires which are utilized as a means to generate isotropic turbulence.

The mean velocity profiles for the turbulent wake of an intersecting cylinder and two cylinders forming a cross, were investigated by Ukhanova [1] for the turbulent wake behind an intersecting cruciform circular cylinder with a finite length. His turbulent wake becomes approximately equal to the axisymmetric wake with increasing streamwise distance. The pressure distribution on the surface of two circular cylinders forming a cross was measured by Zdravkovich [2] and the variation of local drag coefficient, base pressure and Strouhal number along the span of two cylinders were discussed. The strong secondary flows on the surface behind the upstream cylinder and in front of the downstream cylinder were also visualized by using the oil-film technique.

A series of experiments in the turbulent wake behind an intersecting cruciform circular cylinder with an effective infinite length have been conducted by the present authors [3]–[5]. These results showed that the decay rate of the center line defect velocity was considerably slower than that found in either two-dimensional or axisymmetric wakes. Even far downstream the wake did not behave as a simple superposition of two perpendicular two-dimensional wakes, as inferred from Fig. 2(a) and (b) [3], [4]. In addition, the present authors found that the diffusion term of this wake in the turbulent energy budget was different from that of a two-dimensional wake, and that there existed the eight streamwise vortices which were symmetric with respect to the geometrical symmetric plane of this wake. Some unsolved problems in the turbulent wake behind an intersecting cruciform circular cylinder are the turbulent eddy structure, the feature of the large eddy structure and the behavior of vortex shedding from this obstacle; the former dominates the transport of momentum and the higher-order velocity correlation controls the configuration of the diffusion transport, while the latter acts on the flow-induced vibration.

The aims of this paper are to examine, through the measurements of various statistical quantities, the periodic structure of this three-dimensional wake in the near field and the detail of the turbulent eddy structure in the downstream region where the present wake is hardly affected by the pronounced periodic velocity fluctuation due to the vortices shed from

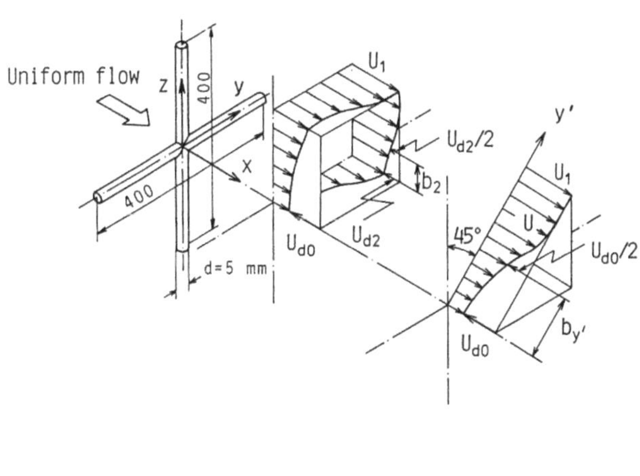

Fig. 1. Configuration of flow field

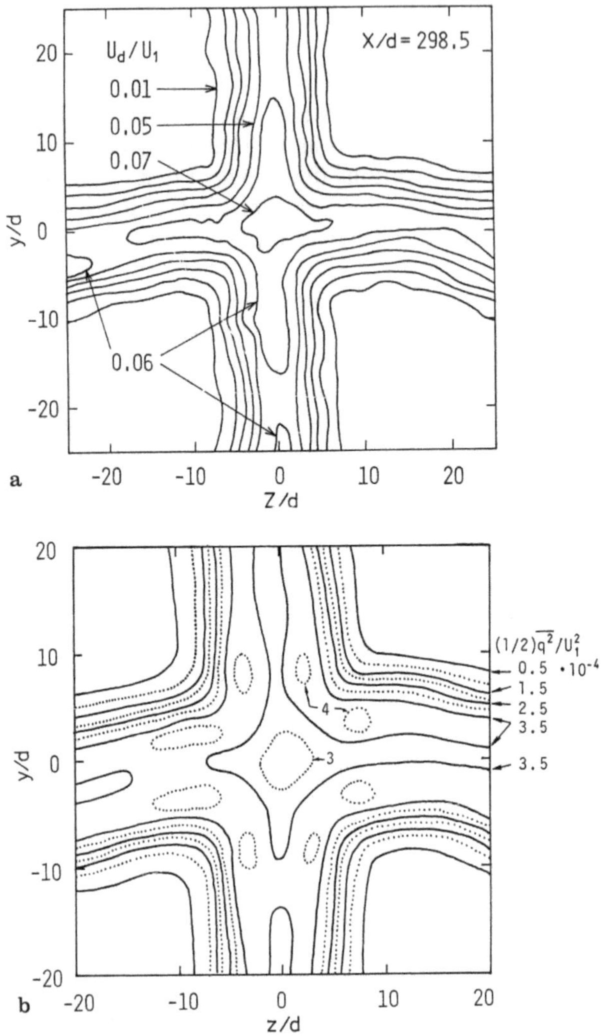

Fig. 2. Contours of the turbulent wake at the section of $x/d = 298.5$; (a) the mean velocity, (b) the turbulent kinetic energy

the cruciform cylinder. The measured quantities were one-point and two-point time correlations, power spectra, triple velocity correlations and probability density profiles.

2. Experimental Apparatus and Techniques

The wind tunnel used in this experiment has a working section 0.4 m square and about 4 m long [3]. The intersecting cruciform circular cylinder consists of two sufficiently long brass cylinders of 5 mm in diameter. The ratio of half length to diameter is 40. The configuration of the flow field and nomenclature are shown in Fig. 1.

The measurement of the mean and fluctuating velocities in the crisscross turbulent wake under a zero-pressure gradient has been made using a constant temperature hot-wire anemometer with I-type or X-type probes. A X-type probe was used for the measurement of one-point statistical quantities such as triple velocity correlations, power spectra and probability density profiles. On the other hand for the measurement of the two-point statistical quantities such as space-time correlations between two separated points, two *I*-type probes were employed. The value of the turbulent intensities was read directly from a rms meter and the triple velocity correlations were obtained from an analogue circuit with an integrator. The difference of the output voltage of the analogue circuit to the input one was not larger than 10% and the phase shift between the input signal and the output can not be observed in the frequency band below 10 kHz. For the measurement of one- and two-point time correlations, power spectra and probability density profiles, a San-ei 7T08 digital correlator was employed. The Reynolds number, $R_d = U_1 d/\nu$, was held constant at 4×10^3 for all tests.

3. Experimental Results and Discussions

3.1 Correlation Coefficients for Near Wake

It is generally known that the turbulent fluctuating velocity in the near wake region contains a contribution from the quasi-periodic fluctuating velocity component due to the vortex shedding from a bluff body. Figure 3 shows the auto-correlation coefficients of *u*-fluctuating

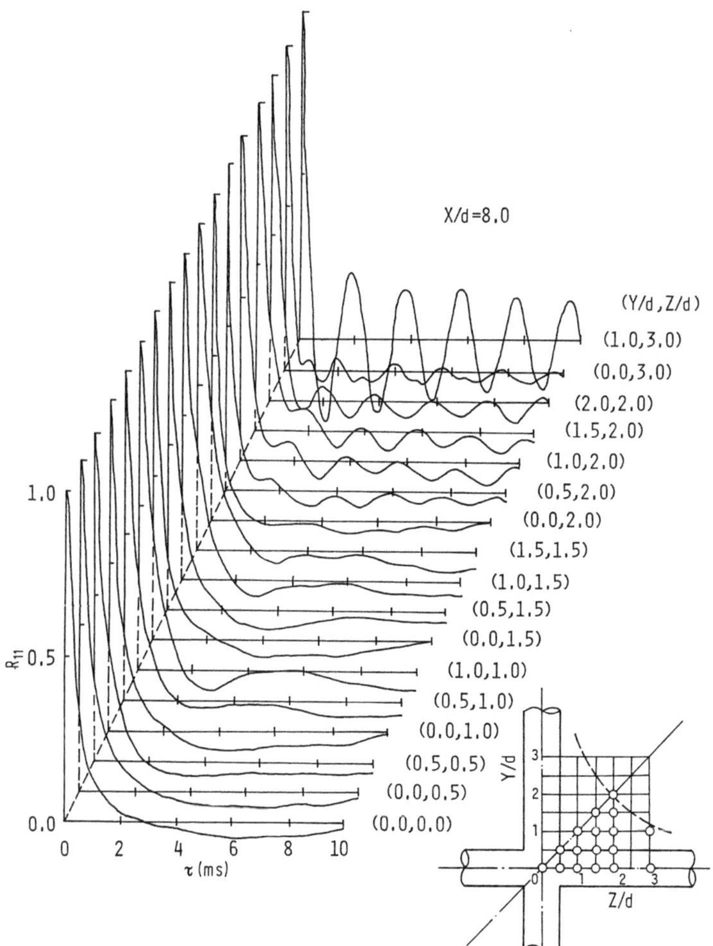

Fig. 3. Auto-correlation coefficients monitored at the section of $x/d = 8$

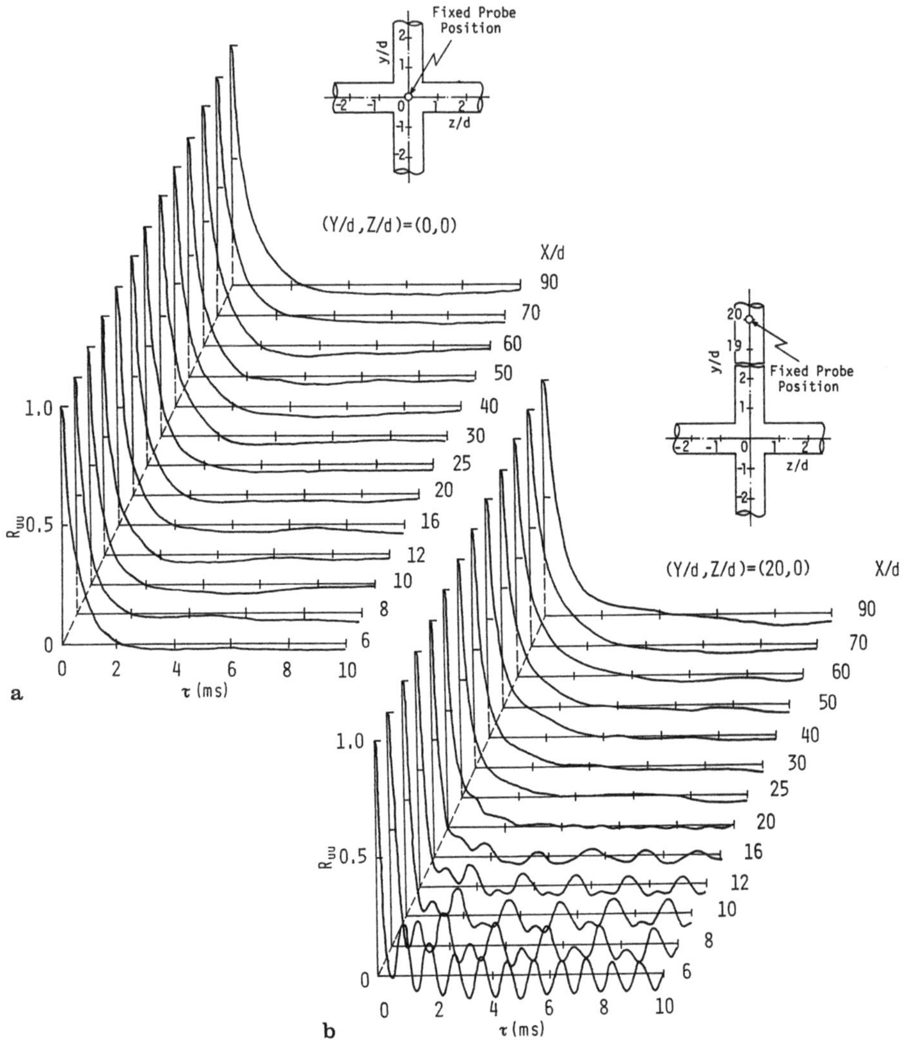

Fig. 4. Streamwise variation of auto-correlation coefficients; (a) along the wake center line, (b) along the center line in the quasi-two-dimensional region

velocity at various positions in the section of $x/d = 8$. The contribution of the quasi-periodic fluctuating velocity appears in the graphs of R_{uu} taken both on the bisector and on the cylinder axis away from the wake center, while they do not appear in the wake center region. This suggests that the vortex lines of the vortices shed from the present cruciform circular cylinder have the shape indicated by the broken line shown in Fig. 3.

Figure 4(a) and (b) shows the streamwise variation of the auto-correlation coefficients along both the wake center line and the center line in the quasi-two-dimensional region. In the whole range along the wake center line the contribution of the quasi-periodic fluctuating velocity can not be found. While, in the quasi-two-dimensional region the periodicity in the auto-correlation curve is clearly observed within the range of about $x/d = 20$. The contribution of the quasi-periodic fluctuating velocity also becomes gradually smaller as x increases.

It is usually assumed that the rate of reduction of each fluctuating velocity component differs from one another. Figure 5 shows the auto-correlation coefficients of u-, v- and

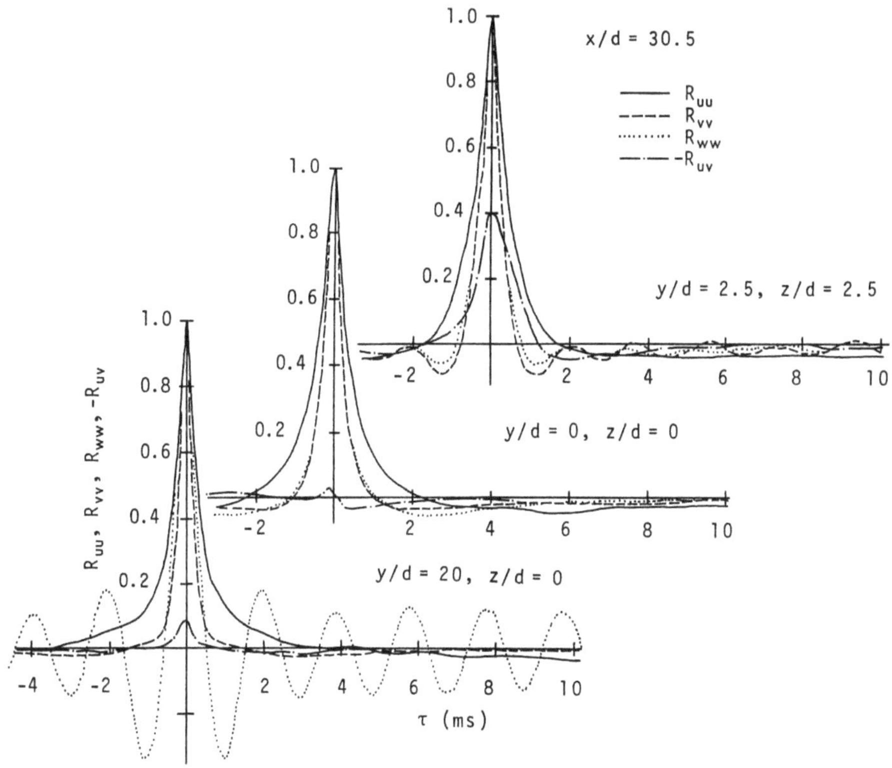

Fig. 5. Auto-correlation coefficients of fluctuating velocity components at the section of $x/d = 30.5$

w-fluctuating velocities at three typical locations in the section of $x/d = 30.5$. The influence due to the vortex shedding is not found for the auto-correlation coefficients of u-fluctuating velocity in the entire region of this section. While, at the center in the quasi-two-dimensional region far away from the wake center ($z/d = 0$, $y/d = 20$), the contribution due to vortex shedding can be recognized in the auto-correlation coefficient of w-fluctuating velocity whose direction is perpendicular to the y cylinder axis. In addition, the contribution due to vortex shedding have a wake effect on the auto-correlation curve of v- and w-fluctuating velocity components at the location of $y/d = z/d = 2.5$ away from the wake center along the bisector. The distance that the periodicity of v- and w-fluctuating velocity components are identifiable was not measured, but these influences clearly disappear in the far wake region ($x/d \geqq 298.5$) as described below (Fig. 11).

3.2 Turbulent Eddy Structure

As shown in Fig. 2, the present authors found that the turbulent kinetic energy at the wake center was smaller than that on the two cylinder axes [4]. It may be assumed that this result is closely related to the magnitude and the streamwise variation of the turbulent large eddy shown in Figs. 6 and 7.

The structure of the turbulent eddy was investigated using the following space-time correlation between two locations separated by a small distance in x direction

$$R_{uu}(r_x, r_y, r_z, \tau) = \frac{\overline{u(x_0, y_0, z_0, t_0) \cdot u(x_0 + r_x, y_0 + r_y, z_0 + r_z, t_0 + \tau)}}{[\overline{u^2(x_0, y_0, z_0, t_0) \cdot u^2(x_0 + r_x, y_0 + r_y, z_0 + r_z, t_0 + \tau)}]^{1/2}}$$

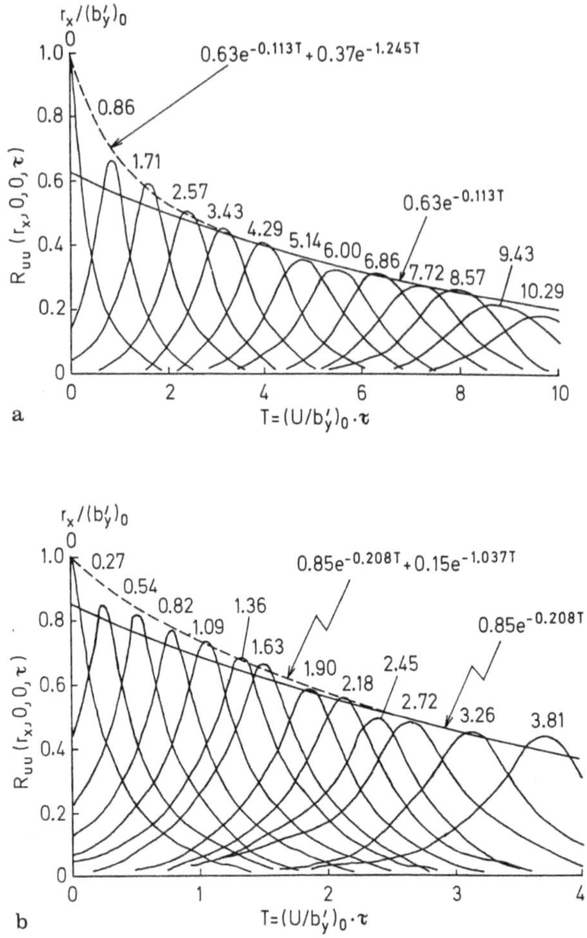

Fig. 6. Space-time correlation coefficients along the wake center line: Fixed probe position; (a) $x/d = 30.5$, (b) $x/d = 298.5$

where subscript 0 denotes the fixed probe position. The streamwise space-time correlations were measurd for two cases one; with the fixed probe set at the wake center and the other with the probe at the center of the quasi-two-dimensional region in the sections of $x/d = 30.5$ and 298.5 respectively. The results are shown in Fig. 6(a), (b) and Fig. 7(a), (b). It is well known that the envelopes of correlation coefficient curves indicate the decay of the turbulent eddy and they are described by the sum of the two decay exponential curves which represent the decay of the large eddy and small one respectively [6]. Considering this concept, we may say that the decay rates of the large eddy, taken as the exponential curves of slow decay, are smaller by one order of magnitude when compared with that of the small eddy at all the locations shown in the figures as well as those in the turbulent boundary layer, as is well known. The decay rate of the large eddy is slower at the center of the quasi-two-dimensional region than at the wake center. It is also slower in the upstream region than in the downstream region, if the non-dimensional time delay $T = (U/b_{y'})_0 \cdot \tau$ based on the local velocity U and the half width $b_{y'}$ for the plane including the fixed probe position is used. For example, the non-dimensional time delay T at which the exponential envelope for the large eddy structure approaches e^{-1} is 8.85 in the upstream region and 4.81 in the downstream region at the wake center. The value is 10.53 in the upstream region and 7.58 in the downstream region at the center of the quasi-two-dimensional region. It is also known that the inter-

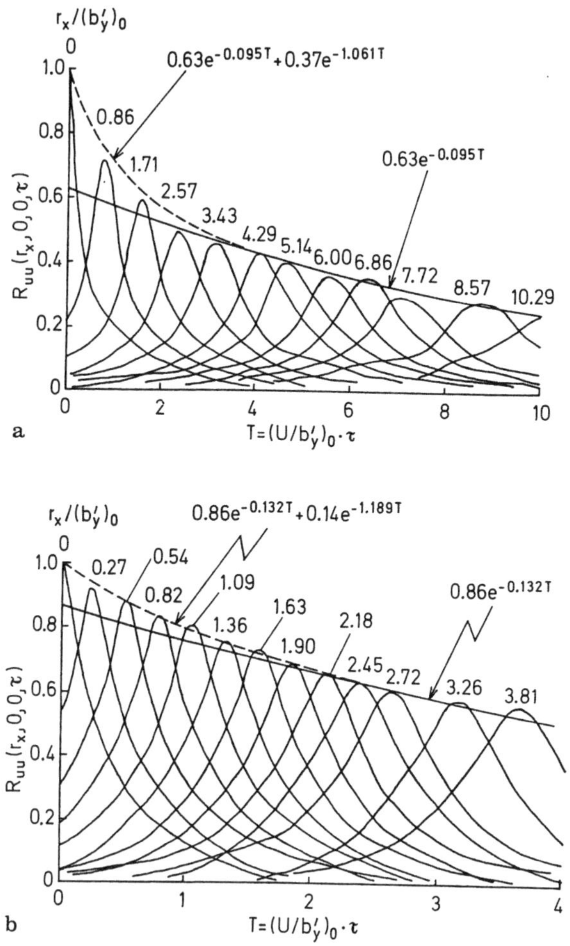

Fig. 7. Space-time correlation coefficients along the center line in the quasi-two-dimensional region: Fixed probe position; (a) $x/d = 30.5$, (b) $x/d = 298.5$

section of the slowly decaying envelope curve of the correlation coefficients at $r_x = 0$ or $\tau = 0$ represents the ratio of the contribution of the large eddy structure to the total turbulent energy [6]. The contribution of the large eddy is almost equal everywhere in the section of $x = \text{const.}$, and is larger in the downstream region than in the upstream region. This contribution is above 80% in the downstream region and is almost the same value as the outer region of the turbulent boundary layer [6].

Figure 8(a) to (d) shows contours of the space-time correlation, which briefly indicates the spatial structure of the turbulet large eddy, at the four typical locations in the section of $x/d = 298.5$. The shape of the turbulent large eddy, is stretched by two times compared the streamwise length with the transverse length, as well as that for the intermittent region in the turbulent boundary layer [7]. It is also seen that the entire scale of the turbulent eddy at the center of the quasi-two-dimensional wake region is fairly large compared with those at the other locations. Furthermore, the turbulent eddy scale at $y/d = z/d = 5$ in the bisector, where the turbulent kinetic energy takes the larger value, is not as large as expected and this configuration has the shape of a chestnut in the plane normal to the flow. While, at the wake center the shape of the turbulent eddy is axisymmetric and shows the spindle shape. In the location where the turbulent kinetic energy takes the maximum value, the turbulent eddy shows the elongated shape in the r_y direction when compared with the r_z direction. At the

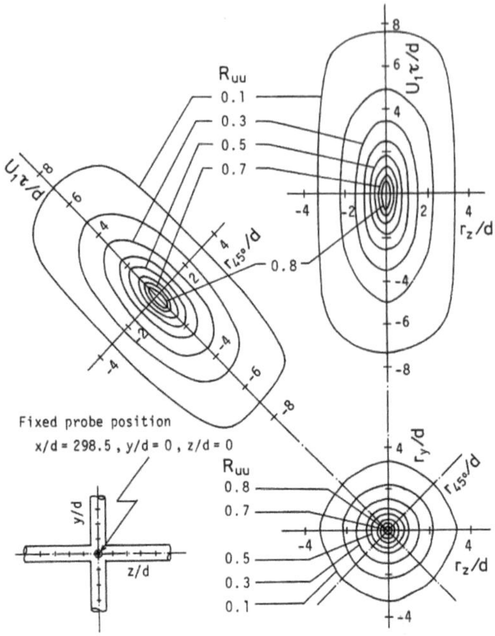

a

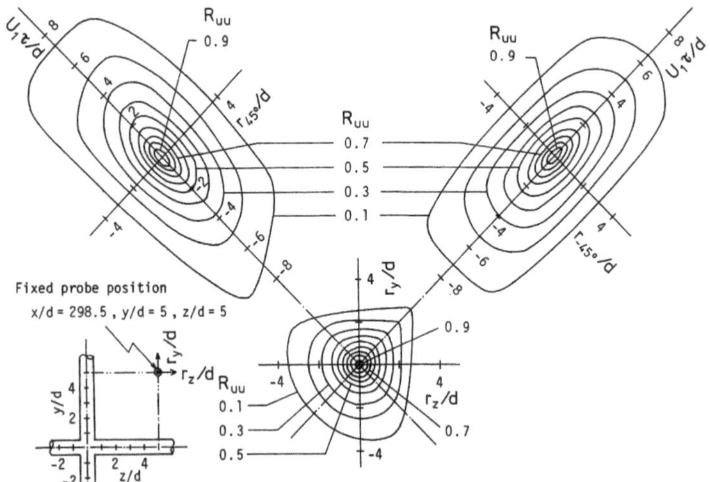

b

Fig. 8a–d. Contours of auto-correlation coefficient at the section of $x/d = 298.5$; (a) wake center location, (b) maximum turbulent kinetic energy location in the bisector, (c) maximum turbulent kinetic energy location in the cross section, (d) center location in the quasi-two-dimensional region

wake center the shape of the turbulent eddy is axisymmetric and shows the spindle shape. Therefore, the association of the configuration of turbulent large eddy with the local structure of turbulence is not necessarily evident.

The magnitude of the large eddy scale is often identified by the integral length scale estimated from the streamwise auto-correlation coefficient in the following manner,

$$L_u = U \int_0^{\infty} R_{uu}(\tau)\, d\tau,$$

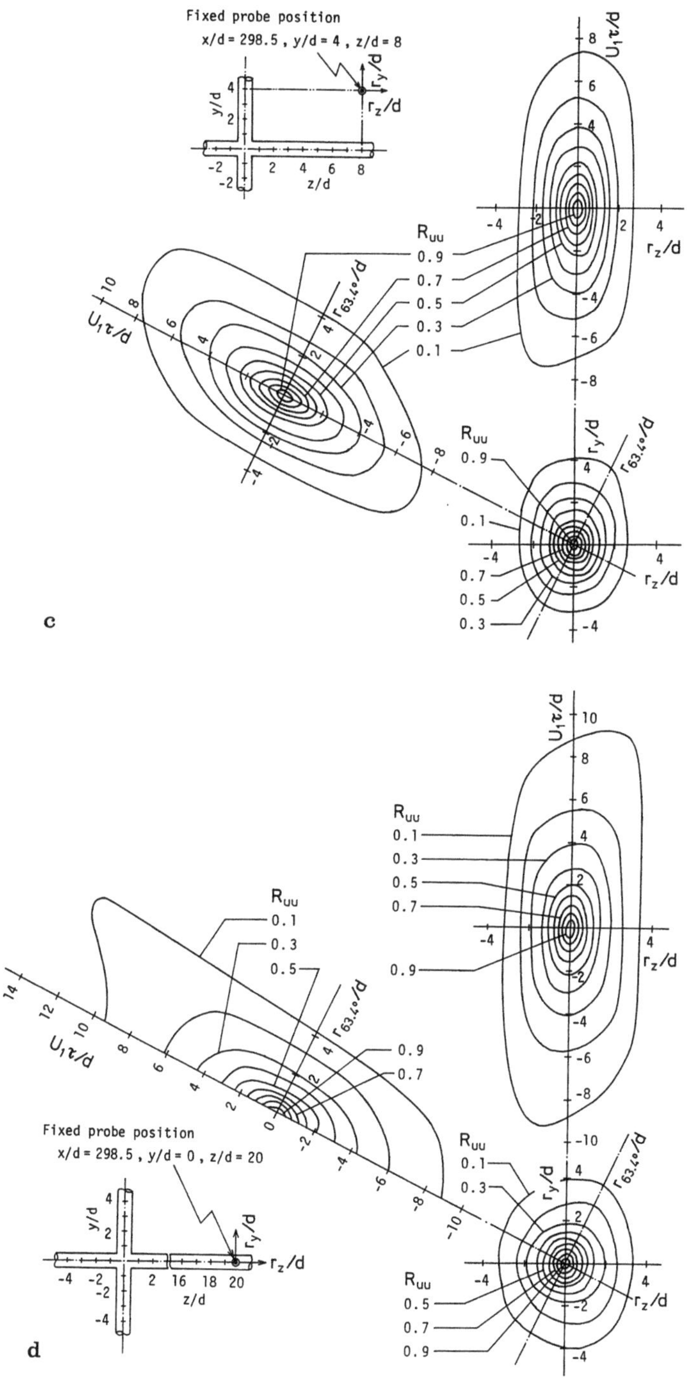

Fixed probe position
x/d = 298.5 , y/d = 4 , z/d = 8
y/d
z/d
r_y/d
r_z/d
$U_1\tau/d$
R_{uu}
0.9
0.7
0.5
0.3
0.1
$r_{63.4°}/d$
c
Fixed probe position
x/d = 298.5 , y/d = 0 , z/d = 20
d

Fig. 8c, d

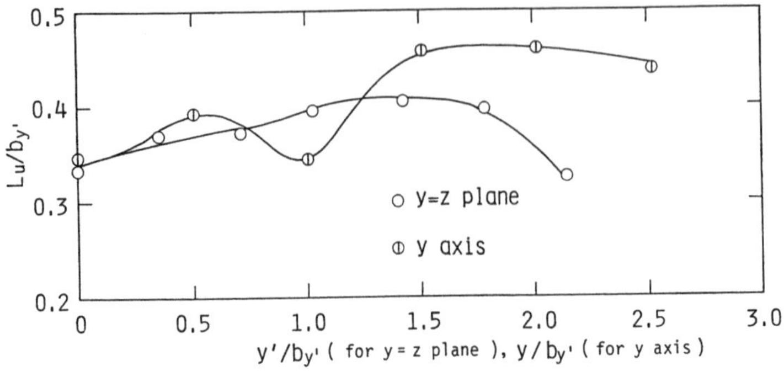

Fig. 9. Integral scale profile

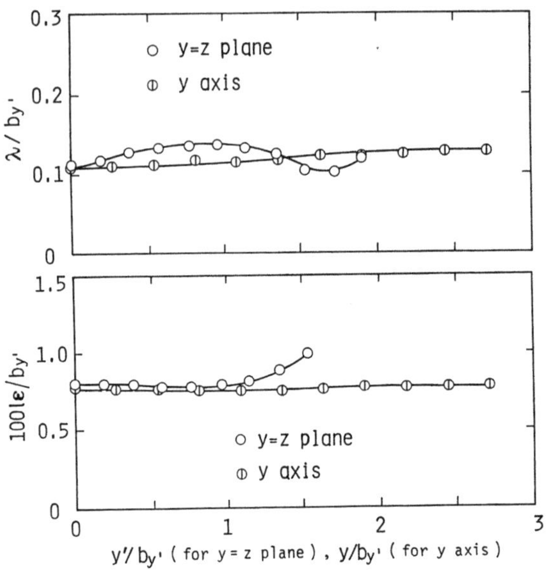

Fig. 10. Micro and dissipation scale profiles

where U is the local velocity. The ratio of L_u to half-width $b_{y'}$ in the far downstream region is shown in Fig. 9. The behavior of $L_u/b_{y'}$ with $y'/b_{y'}$ is relatively mild, while the profile of $L_u/b_{y'}$ on the cylinder axis has the distinct kink near $y/b_{y'} = 1$ where the diffusion term in the turbulent kinetic energy budget contributes a significant gain where compared with it contribution at other locations on the cylinder axis [5].

Figure 10 shows the profiles of the micro-scale and the dissipation scale defined as

$$\lambda = (15\nu \cdot \overline{u^2}/\varepsilon)^{1/2}, \quad l_\varepsilon = (\nu^3/\varepsilon)^{1/4}$$

respectively, where ε is the dissipation rate of the turbulent kinetic energy evaluated by the following equation, $\varepsilon = 15\nu\overline{(\partial u/\partial t)^2}/U^2$. The magnitude of the micro-scale is roughly 1/3 times the integral length scale, while the value of the dissipation scale is extremely small. Both behaviors of the micro-scale and the dissipation scale vary slightly along the bisector and the cylinder axis respectively, and the value of each turbulent scale almost equal on the symmetric planes.

It is very interesting to note that the spatial variation of the turbulent large eddy scale does not affect the scale of the small eddies such as λ and l_ε as shown in Fig. 9.

The turbulent eddy structure can be understood by the one-dimensional power spectrum analysis, which is one of the simple methods. Figure 11 (a) and (b) shows the power spectrum of each fluctuating velocity component both on the bisector and on the cylinder axis defined as following equation,

$$\overline{\chi^2} = \int_0^\infty E_{\chi\chi}(k)\, dk,$$

where χ is each fluctuating velocity component and k is wave number ($k = 2\pi f/U$). The profiles of $E_{uu}/\overline{u^2}$ have almost the same shape at all locations along the bisector and the cylinder axis. The inertia subrange for the present wake exists in the range of $1 \lesssim k \lesssim 10$ which is close to that of the two-dimensional wake in $R_d = 4320$ [8], and this wave number range is shifted toward the high wave number side compared with that of the outer region in the turbulent boundary layer ($0.5 \lesssim k \lesssim 5$) [9]. Both inertia subrange of $E_{vv}/\overline{v^2}$ and $E_{ww}/\overline{w^2}$ at all locations along the cylinder axis have the narrow range and these spectra also have the high power in the high wave number side compared with the profile of $E_{uu}/\overline{u^2}$. The

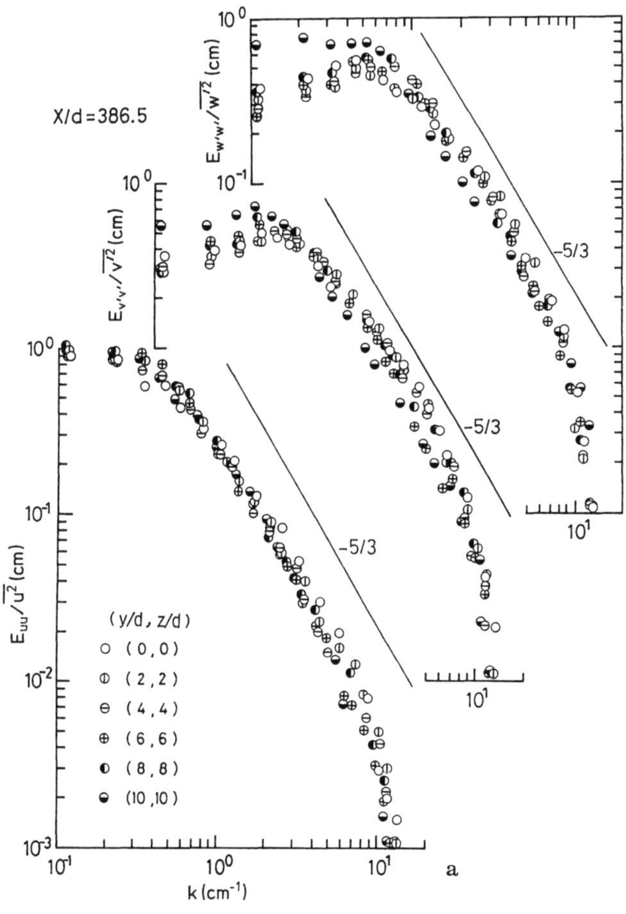

Fig. 11 a and b. One-dimensional power spectrum densities of fluctuating velocity components; (a) on the bisector, (b) on the cylinder axis

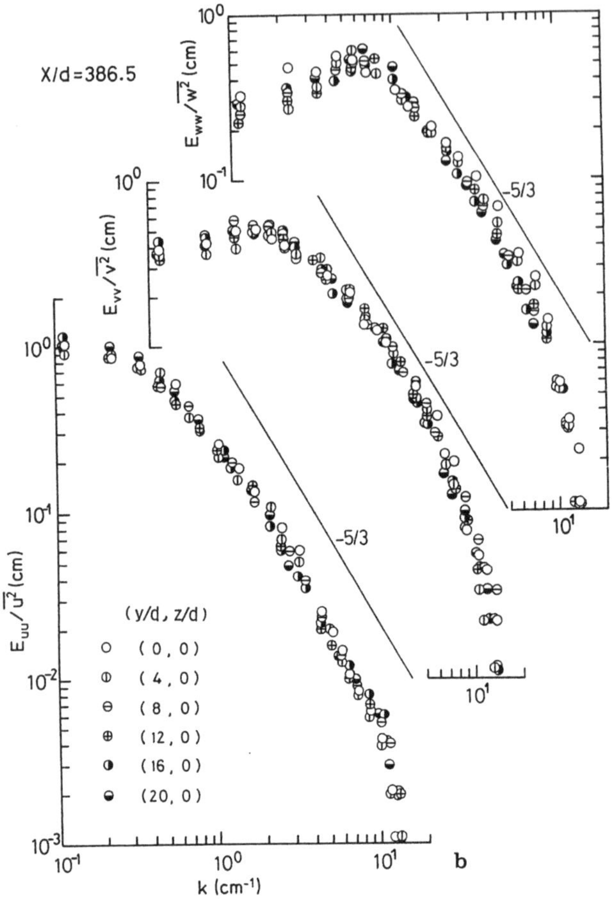

b

Fig. 11 b

profiles of $E_{v'v'}/\overline{v'^2}$ and $E_{w'w'}/\overline{w'^2}$ at any locations along the bisector have roughly the same feature as $E_{vv}/\overline{v^2}$ and $E_{ww}/\overline{w^2}$ described above. But, as the location approaches the main stream, the power in the low wave number side becomes large as well as in the case of the equilibrium boundary layer [10].

3.3 Triple Velocity Correlations and Distortion of Turbulent Eddy

The significance of the diffusion term in the trubulent kinetic energy budget was pointed out for the turbulent wake behind an intersecting cruciform circular cylinder by the present authors [5]. In addition, the data on the triple velocity correlation profiles may be expected to play a significant role in establishing the turbulent prediction method for the three-dimensional turbulent wakes. The diffusion transport terms are estimated from the gradient of the triple velocity correlation profiles. Figure 12 shows the triple velocity correlation profiles on the bisector at the section of $x/d = 386.5$. In the figure, the results of the two-dimensional wake [11] represented by the broken line, are non-dimensionalized by multiplying them by the ratio of $U_{d_2}/U_{d_0} = 0.7$ obtained for the present wake. They are of the same order of the magnitude as those on the bisector in the present wake. This means that the magnitude of the triple velocity correlation coefficient on the bisector has the same order as that in the quasi-two-dimensional region, although the velocity defect at the wake center is larger than that at the center in the quasi-two-dimensional region. The positions of the

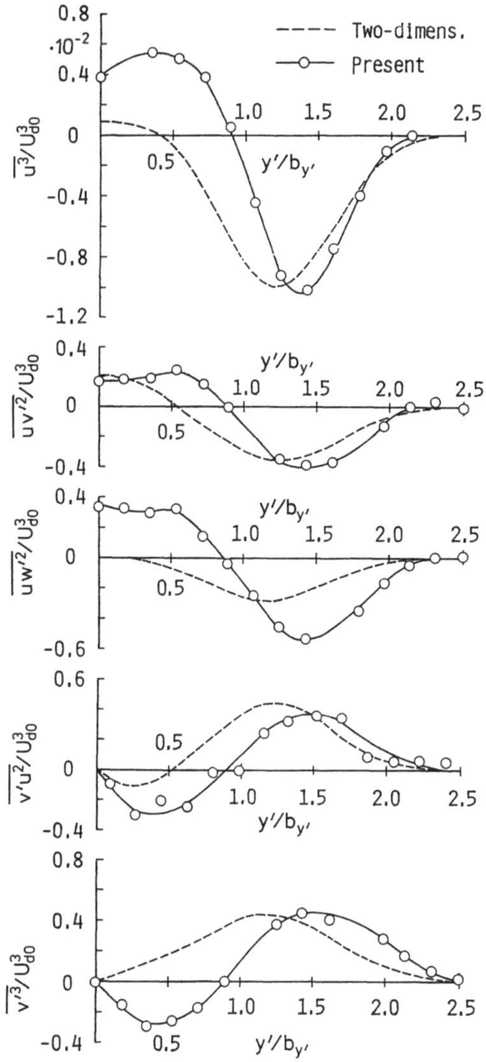

Fig. 12. Triple velocity correlation profiles in the bisector

maximum and minimum peaks for the velocity correlation profile are shifted towards the outer edge of the wake when compared with those for the two-dimensional wake. Both profiles of $\overline{v'u^2}$ and $\overline{v'^3}$ are markedly different from those for the two-dimensional wake respectively, but almost agree with each other on the bisector for the present wake.

Turbulent diffusion transport is associated with the distortion of the turbulent large eddy. The distortion of the eddy may be observed in the profiles of the probability density function of each fluctuating velocity. Figures 13 and 14 show the skewness and the flatness factors both on the bisector and on the cylinder axis defined as following equations,

$$S(\chi) = \int_{-\infty}^{\infty} \chi^3 p(\chi)\, d\chi \Big/ \left[\int_{-\infty}^{\infty} \chi^2 p(\chi)\, d\chi \right]^{3/2}$$

$$F(\chi) = \int_{-\infty}^{\infty} \chi^4 p(\chi)\, d\chi \Big/ \left[\int_{-\infty}^{\infty} \chi^2 p(\chi)\, d\chi \right]^{2}$$

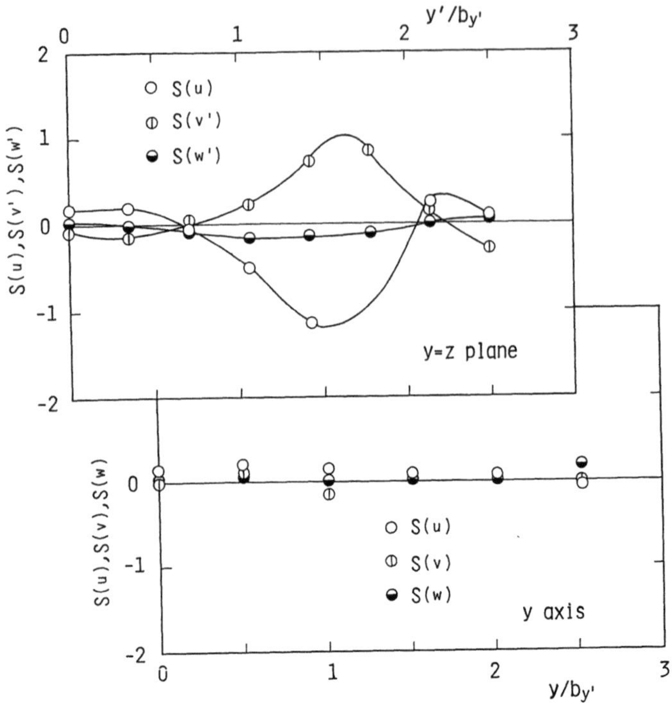

Fig. 13. Skewness factor profiles in the bisector and the cylinder axis

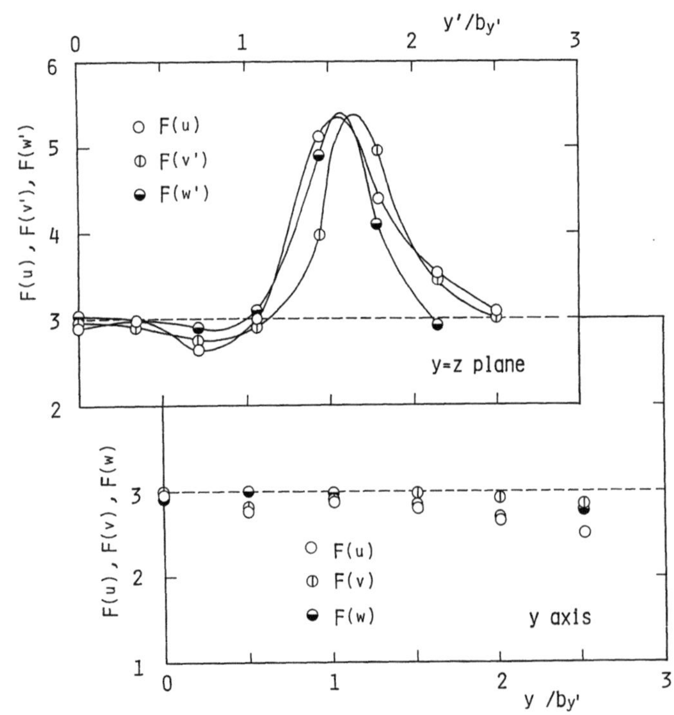

Fig. 14. Flatness factor profiles in the bisector and the cylinder axis

respectively, where $p(\chi)$ is the probability density function for each fluctuating velocity component. The values of $S(u)$ and $S(v')$ are negative and positive in the range of $y'/b_{y}' = 1 \sim 2$ respectively while $S(w')$ is nearly zero everywhere on the bisector as are $S(u)$, $S(v)$ and $S(w)$ on the cylinder axis. It may thus be said that the turbulent large eddy on the cylinder axis has very little spatial distortion. The flatness factors of every fluctuating velocity component on the bisector are larger than three in the range of $y'/b_{y'} = 1 \sim 2$, while they are zero everywhere on the cylinder axis. The profile behavior of the skewness and flatness factors on the bisector is roughly similar to those of the two-dimensional wake [11] as is the mean velocity profile.

4. Concluding Remarks

The turbulent eddy structure and various statistical quantities of the turbulent wake behind an intersecting cruciform circular cylinder were examined. The main results can be summarized as follows:

(1) The quasi-periodic fluctuating velocity due to the vortex shedding from a cruciform circular cylinder appears to different degrees at the different locations in the cross section, and is not found at the wake center. The magnitude of the contribution to the various turbulent quantities depends on each fluctuating velocity component.

(2) The ratio of the turbulent kinetic energy in the large eddy to total energy is almost equal everywhere in the section of x = const., but is larger in the downstream region than in the upstream region. The decay rate of the large eddy is slower in the downstream region than in the upstream region, and also slower at the center of the quasi-two-dimensional region than at the wake center.

(3) The turbulent eddy structure interpretated from the contours of the space-time correlation is markedly different at each location in the cross section.

(4) The magnitude of triple velocity correlation profiles on the bisector has the same order as that in a two-dimensional wake, but some of the shapes are significantly different. The profiles of the skewness and the flatness factors on the bisector are nearly similar to those of the two-dimensional wake, while the skewness and the flatness factors are respectively zero and three everywhere on the cylinder axis.

Acknowledgement. The authors wish to thank Professor Y. Kageyama for his valuable comments.

References

1. Ukhanova, L. N. (1973): Flow structure in three-dimensional turbulent wake. Eng. Phys. J. **25**, 893 (in Russian)
2. Zdravkovich, M. M. (1983): Interference between two circular cylinders forming a cross. J. Fluid Mech. **128**, 231
3. Osaka, H., Nakamura, I., Yamada, H., Kuwata, Y., Kageyama, Y. (1983): The structure of a turbulent wake behind a cruciform circular cylinder. 1st report: The mean velocity field. Bull. Jpn. Soc. Mech. Eng. **26**, 356
4. Osaka, H., Yamada, H., Nakamura, I., Kuwata, Y., Kageyama, Y. (1983): The structure of a turbulent wake behind a cruciform circular cylinder. 2nd report: The streamwise development of turbulent flow field. Bull. Jpn. Soc. Mech. Eng. **26**, 521
5. Osaka, H., Yamada, H., Nakamura, I.: Three-Dimensional Structure of the Turbulent Wake behind an Intersecting Cruciform Circular Cylinder, in *Three-Dimensional Turbulent Boundary Layers*, ed. by H. H. Fernholz, E. Krause (Springer, Berlin, Heidelberg, New York 1982) p. 221

6. Blackwelder, R. F., Kovasznay, L. S. G. (1972): Time scales and correlations in a turbulent boundary layer. Phys. Fluids **15**, 1545
7. Kovasznay, L. S. G., Kibens, V., Blackwelder, R. F. (1970): Large-scale motion in the intermittent region of a turbulent boundary layer. J. Fluid Mech. **41**, 283
8. Uberoi, M. S., Freymuth, P. (1969): Spectra of turbulence in wakes behind circular cylinders. Phys. Fluids **12**, 1359
9. Klebanoff, P. S. (1955): Characteristics of turbulence in a boundary layer with zero pressure gradient. NACA Report No. 1247, 1135
10. Bradshaw, P. (1967): The turbulent structure of equilibrium boundary layers. J. Fluid Mech. **29**, 625
11. Yamada, H., Kuwata, Y., Osaka, H., Kageyama, Y. (1980): Multiple correlation of a two-dimensional turbulent wake. Tech. Rep. Yamaguchi Univ. **2**, 341

Effects of Streamline Curvature on Laminar and Turbulent Wakes

Hide S. Koyama

Department of Mechanical Engineering, Tokyo Denki University, Tokyo, Japan

Abstract

Stabilizing, destabilizing, and secondary flow effects of streamline curvature on laminar and turbulent wakes behind a circular cylinder have been examined qualitatively and quantitatively. Experiments were carried out using a flow visualization method, a hot-wire anemometer and a signal analyzer. The results indicate that development of a turbulent wake is promoted by the destabilizing effect of streamline curvature and suppressed by the secondary flow effect of curvature.

Introduction

Coriolis, centrifugal and buoyancy forces in shear flows provide interesting problems in physics, engineering and the environmental sciences. Examples in which these body forces occur are the turbulent boundary layer on a curved wall as well as on a rotating wall; a curved mixing layer; and a turbulent shear flow in a thermally stratified field. The analogy between the buoyancy force and those due to curvature has been discussed by Bradshaw [1]. From a review of several published experimental studies [2–16] in which these forces are discussed, one may deduce that the principal effects of Coriolis, centrifugal and buoyancy forces are:

(1) to promote or suppress the development of two-dimensional laminar and turbulent boundary layers;
(2) to generate Taylor-Görtler type longitudinal vortices in a turbulent boundary layer;
(3) to generate large-scale vortices in a channel.

These are generally referred to as the stabilizing or destabilizing effects of streamline curvature. For example, in a curved shear flow, the motion of fluid particles is stabilized if there is a positive transverse velocity gradient. On the other hand, if the velocity gradient is negative, the fluid motion would be destabilized. If a velocity gradient exists in the spanwise direction, that is, normal to the plane containing the radius of curvature, secondary motion of the fluid particles arises as a result of the imbalance between the centrifugal force due to streamline curvature and the pressure-gradient force acting in the direction of the center of curvature. The resulting motion is referred to as the secondary flow effect of streamline curvature.

In experiments on boundary-layer development on curved walls, the majority of the results obtained exhibits coupled effects of stability and secondary flow, in particular on a concave wall. On a concave wall, Taylor-Görtler vortices generated in the boundary layer break down [16] and contribute substantially to the development of the turbulent boundary layer.

The aim of this investigation is to clarify the stabilizing, destabilizing and secondary flow effects of streamline curvature on turbulent shear flows. With this objective in view, qualitative and quantitative studies have been made on both laminar and turbulent wakes behind a circular cylinder which could be located either transversely or spanwise in a curved potential flow field. Similar work has been reported, for example, Margolis and Lumley [17] on a curved turbulent mixing layer; by Fette [18] on flow around a sphere placed in a rotating curved potential flow; by Raj and Lakshminarayana [19] on the turbulent wake behind a cascade of blades; and by Sarpkaya [20] and Perry et al. [21] on the Karman vortex street.

Apparatus and Experimental Method

A longitudinal cross section of the wind tunnel employed in the investigation is shown schematically in Fig. 1. Air, delivered to the tunnel by a fan blower with a capacity of 0.5 m^3/s at a delivery pressure only slightly greater than atmospheric pressure (980 Pa gauge), flowed through the rectification section and the convergent section of 14.8:1 contraction ratio into

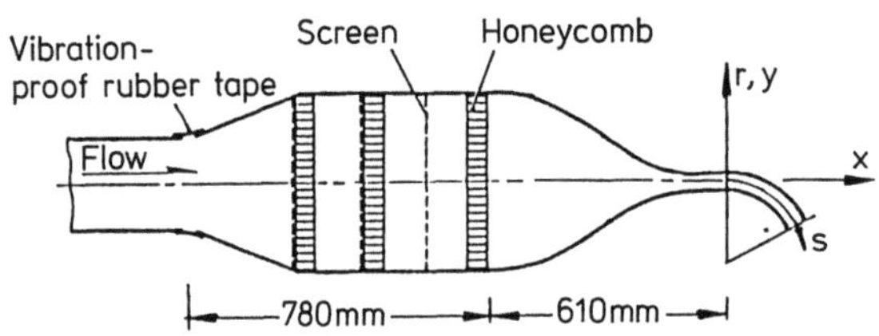

Fig. 1. Schematic of experimental apparatus

the test channel. Rectification was effected by means of layers of honeycomb flow straighteners interspersed with screens. The test channel, which could be either straight or curved, had a cross section 280 mm high × 50 mm wide and a length of 210 mm along the axis. The sides of the channel were of an optical quality transparent plastic material. The center line radius of curvature was 200 mm. Straight circular cylinders of different diameters were located 5 mm upstream of the entrance to the test channel in order to generate a wake. Such a cylinder can be fixed transversely (see Fig. 2a), that is, parallel to the short axis (50 mm), or spanwise (see Fig. 2b), that is, normal to the plane containing the radius of curvature. At low Reynolds numbers, it was found that wake development was sensitive to the manner in which the cylinder was fixed. Therefore, it was more convenient to leave the cylinder fixed in either the transverse direction or spanwise, changing the test channel instead. The effects of streamline curvature on the wake behind a cylinder were deduced by comparing the results obtained with a straight test channel with those from a curved test channel.

Figure 3 is a schematic diagram of the measuring system used in the investigation. A constant-temperature anemometer (KANOMAX, 7000) with 5 μm tungsten hot-wire probe was used to obtain time-mean and fluctuating velocity measurements. The hot-wire probe could be moved continuously in the streamwise, transverse and spanwise directions by a suitable traversing mechanism. The output of a multiple-turn potentiometer indicated the position of the hot wire, and the dc signal and ac signal (root mean square of the fluctuating velocity) from the hot-wire anemometer were recorded by $X - Y$ plotters (YEW, 3063). The ac signal was also fed into a 12-bit signal analyzer (IWATSU, SM-2100A) and the frequency spectrum obtained. These results were stored on a magnetic disc and afterwards recorded by $X - Y$ plotter (IWATSU, DPL-3321).

Mean velocity components and Reynolds stresses were measured by using six hot-wire probes, each inclined at a known angle to the mean flow, that could be traversed spanwise at different streamwise positions downstream of the cylinder. The signals from these probes were reduced the equations derived by Fujita and Kovasznay [22].

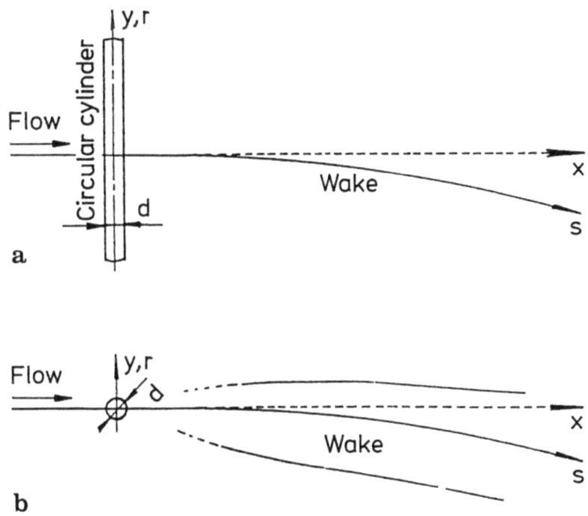

Fig. 2. Wakes in curved potential flows

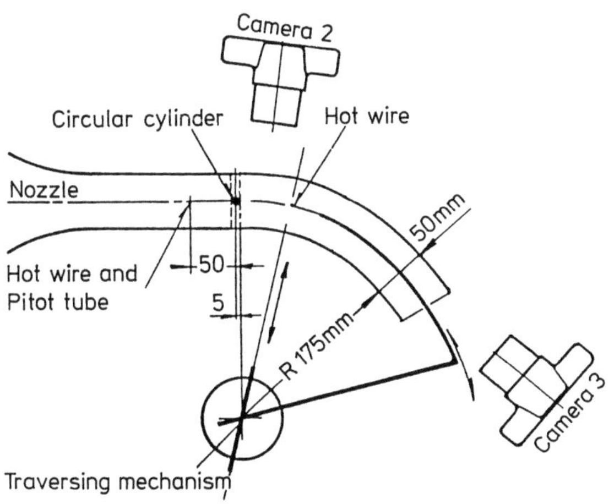

Fig. 3. Schematic diagram of measuring system

The smoke-wire method of flow visualization was used to reveal the pattern in the wake behind the cylinder. A nichrome wire 0.1 mm in diameter was stretched spanwise in the test channel and painted with paraffin wax. A dense white smoke was produced by passing a strong current through the wire. A camera located at either positions 2 or 3, Fig. 3, and at position 1 (not shown in Fig. 3) above the plane of the paper enabled photographs to be taken from the side, front, and top, respectively, of the test channel. The free stream velocity was measured by a Pitot tube and by a hot wire 50 mm upstream from the fixed cylinder. The velocity was varied by changing the rotational speed of the blower. In calculating the flow Reynolds number, the free stream velocity and the diameter of the fixed cylinder were used.

Results and Considerations

Figure 4 shows the variation of the Strouhal number St with Reynolds number of the free stream. The measurement was taken in a straight potential flow. The Strouhal number was calculated on the basis of the shedding frequency f_k of the Karman vortex street. The points shown were obtained with fixed cylinders of three different diameters, as given in the figure,

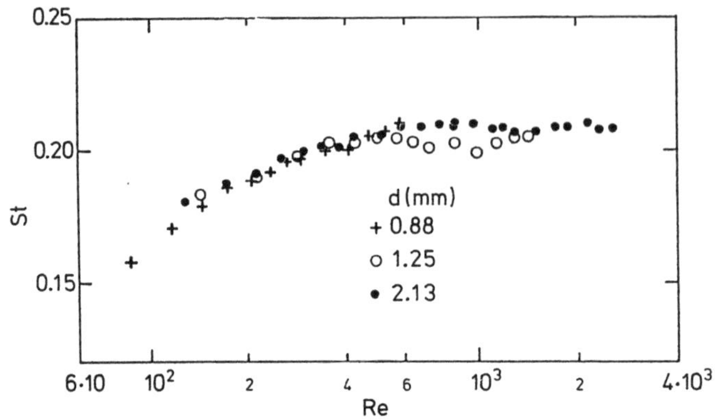

Fig. 4. Variation of Strouhal number with Reynolds number

and for different velocities in the range of Reynolds numbers between about 90 and 2570. The shedding frequency was measured by the signal analyzer. The result shown in Fig. 4 agree with those obtained by Roshko [23].

Secondary Flow Effect

For inviscid, incompressible flow the equation of motion in cylindrical coordinates is

$$\frac{DV}{Dt} - \frac{u^2}{r}\,\boldsymbol{n} + \frac{u\,v}{r}\,\boldsymbol{t} = -\frac{1}{\varrho}\nabla p, \tag{1}$$

where V is the velocity vector with components u, v and w in the azimuthal, radial and axial directions, respectively; $\boldsymbol{n}$, $\boldsymbol{t}$ and $\boldsymbol{k}$ are unit vectors in the corresponding directions. The second and third terms in (1) are the centrifugal and Coriolis accelerations and appear automatically upon transformation from rectangular to cylindrical coordinates.

On calculating the rotation of (1), the vector equation of vorticity is obtained

$$\frac{D\omega}{Dt} - \frac{\partial}{\partial z}\left(\frac{u\,v}{r}\right)\boldsymbol{n} - \frac{\partial}{\partial z}\left(\frac{u^2}{r}\right)\boldsymbol{t} + \left[\frac{\partial}{\partial r}\left(\frac{u\,v}{r}\right) + \frac{1}{r}\frac{\partial}{\partial \theta}\left(\frac{u^2}{r}\right)\right]\boldsymbol{k} = 0. \tag{2}$$

It is clear from (2) that the vorticity components will be changed by the velocity gradients in a curved flow [24]. Assuming there is a two-dimensional vortex in a curved potential flow as in Fig. 5, the vorticity vector which initially was directed at right angles to the stream is now inclined at an angle to it so that a streamwise component of vorticity ω_s is generated. In other words, a secondary flow shown by the dotted lines will be set up in the vortex. This is referred to as the secondary flow effect of streamline curvature.

The smoke-wire method of flow visualization was used to reveal the secondary flow effect of curvature. Figure 6 shows photographs of instantaneous smoke streak-line patterns behind a fixed cylinder in straight and curved potential flows at a Reynolds number of 150, calculated with a cylinder diameter of 3.0 mm. The cylinder was located in the transverse position and the smoke wire was stretched spanwise 10 mm downstream of it. In the side-view photograph of the straight wake, the pattern known as a Karman vortex street is clearly seen. The vortices were observed to move more slowly than the free stream. From many similar photographs the ratio of spanwise spacing to streamwise spacing of the vortices in the wake close to the fixed cylinder was found to be about 0.3, but this tended to increase further downstream. In contrast, in the curved wake it is not at all clear that the Karman vortex street exists. The confusion is apparently due to breakdown of individual vortices. However,

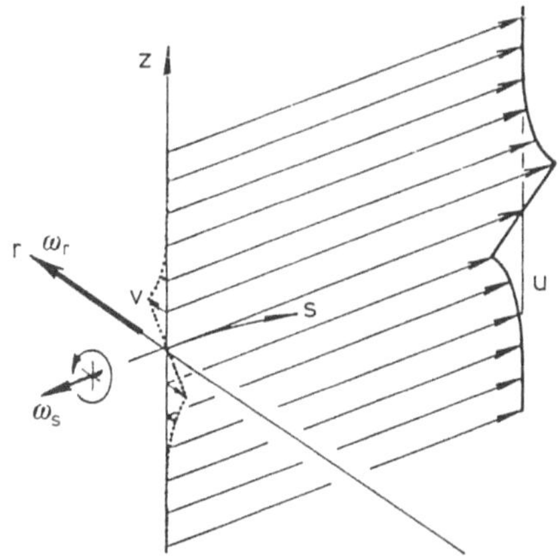

Fig. 5. Generation of streamwise component of vorticity

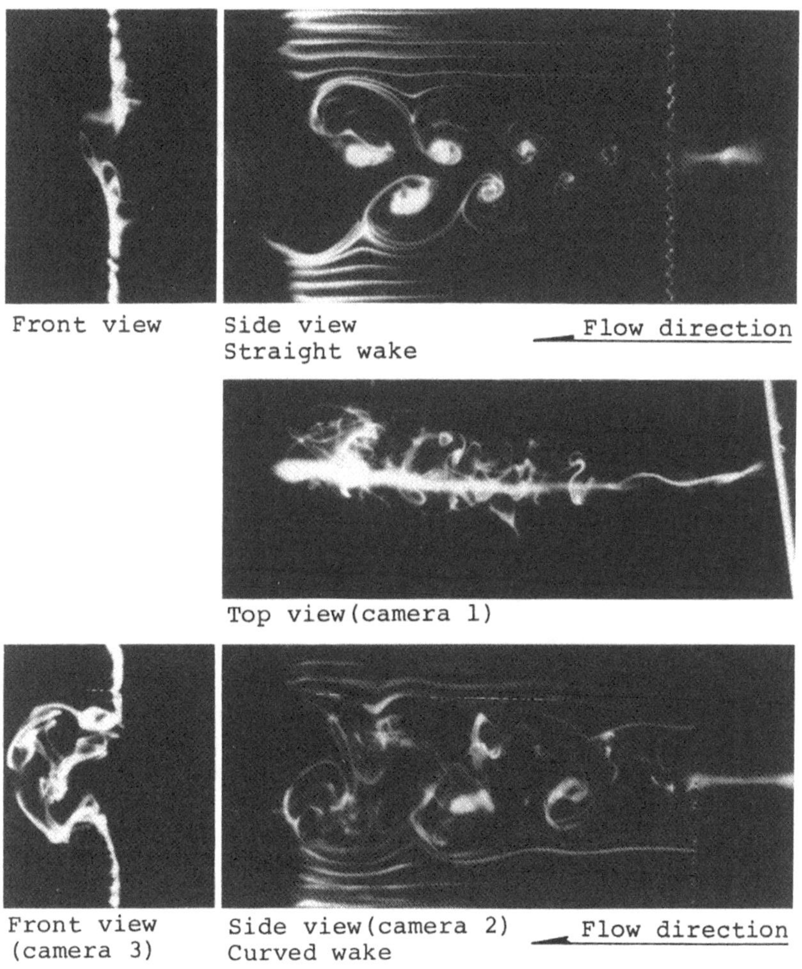

Fig. 6. Photographs of instantaneous smoke-streakline patterns behind fixed cylinder in straight and curved potential flows at Reynolds number of 150

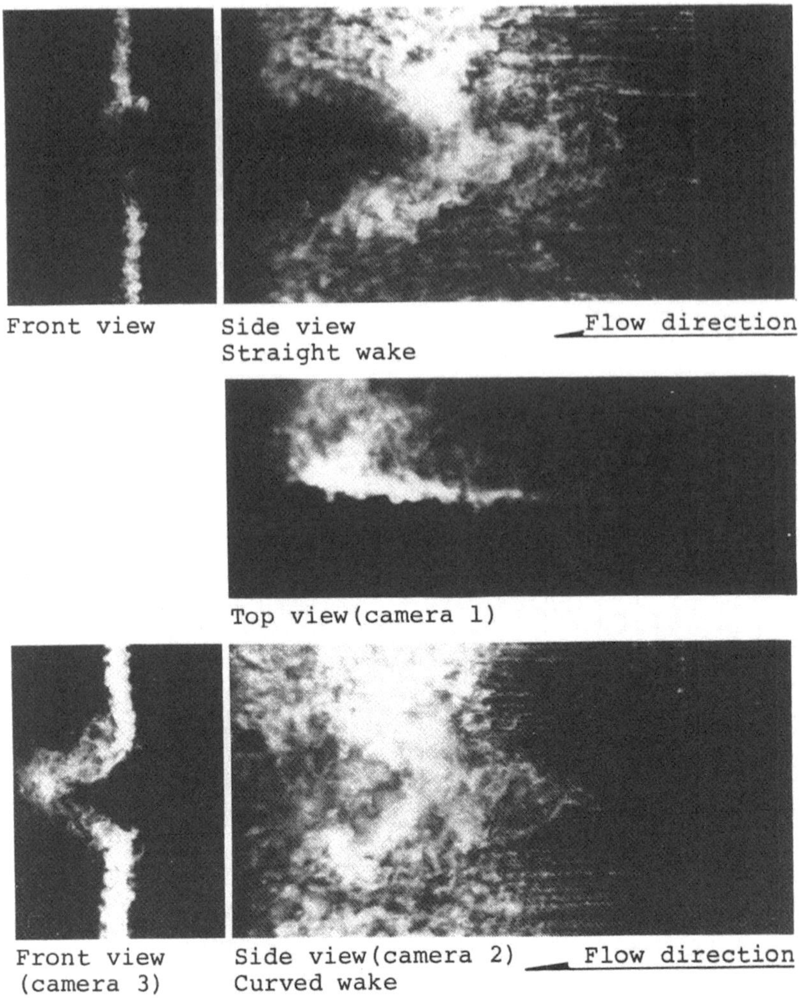

Fig. 7. Photographs of instantaneous smoke-streakline patterns behind fixed cylinder in straight and curved potential flows at Reynolds number of 1500

a large-scale secondary flow towards the center of the curvature may be seen in the top-view and the front-view photographs. In addition, in the former, small-scale secondary flows emanating from the vortices can be identified.

At higher Reynolds numbers, the regular vortex pattern in the straight wake becomes turbulent and discrete vortices cannot be distinguished by the smoke-wire method, as is evident in Fig. 7. The velocity defect region is clearly seen in the straight wake but not in the curved wake. Also, the width of the straight wake is greater than that of the curved wake. Secondary flow in the curved wake, predicted by the vorticity equation (2), is confirmed by flow visualization, as may be seen in the top- and front-view photographs in that figure. The maximum velocity of the secondary flow is found at the wake centerline.

Quantitative information concerning discrete vortices in the wake were obtained by spectral analysis. For straight and curved wakes at Reynolds numbers of 150, 200, 300 and 500, power spectra of the ac signals from the hot-wire anemometer are shown in Fig. 8. The hot-wire probe was located at a fixed streamwise position x/d or s/d equal to 2 and at the spanwise position z/d equal to unity behind the cylinder. An output of 10 V from the

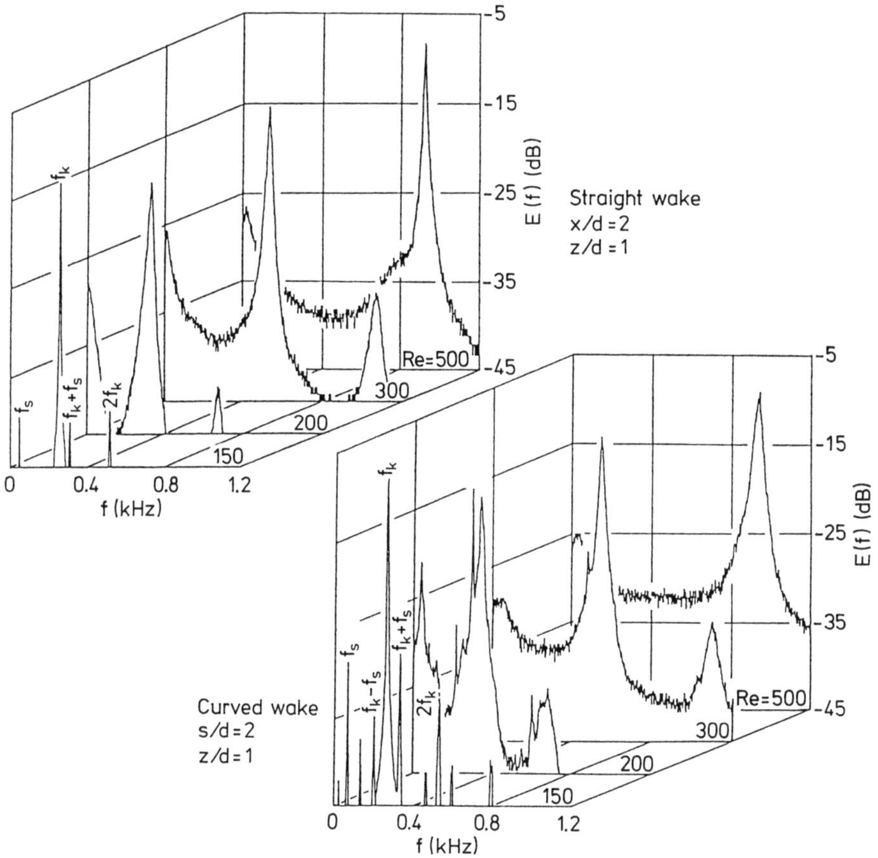

Fig. 8. Power spectra of hot-wire signals in straight and curved wakes at Reynolds number of 150, 200, 300 and 500

anemometer was equivalent to 10 m/s for the power-spectrum analysis. The results obtained were analyzed in the frequency range 0–2 kHz. The peaks of the principal and harmonic frequencies of the Karman vortex street appear at f_k, $2f_k$, and $3f_k$, respectively. The small peak at f_s may be due to a disturbance emanating from the surface roughness of the fixed cylinder and its method of fixing. It appears only downstream of the cylinder, not upstream of it. In the curved wake, many peaks with sidebands can be seen when the Reynolds numbers are 150 and 200. The full widths at half-maxima of the principal and harmonic frequencies in the curved wake are wider than in the straight wake. The results obtained when the Reynolds number is 150 are analogous to the frequency spectrum of an amplitude-modulated signal e_{out}, shown in Fig. 9, given by

$$\begin{aligned} e_{out} &= K[E_s(1 + m\cos(2\pi f_s t))] \cdot [E_c \cos(2\pi f_c t)] \\ &= 1/2 \cdot K E_s E_c m[\cos 2\pi(f_c + f_s) + \cos 2\pi(f_c - f_s)t] + K E_s E_c \cos(2\pi f_c t), \end{aligned} \quad (3)$$

where K is a constant, E_s and E_c are the amplitudes of the modulating signal and carrier signal, respectively, and m is the degree of modulation.

In this case, the frequency band of the carrier signal is f_c and the two-sideband frequencies are $(f_c + f_s)$ and $(f_c - f_s)$, so that the carrier frequency and modulating frequencies correspond to those of the vortex street and disturbance, respectively.

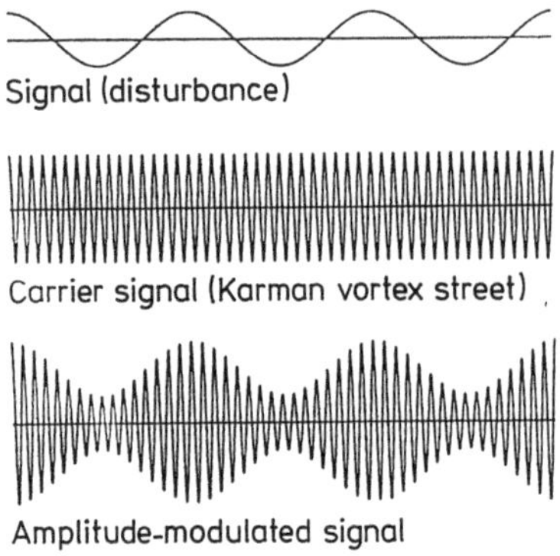

Fig. 9. Modulating, carrier and modulated signals

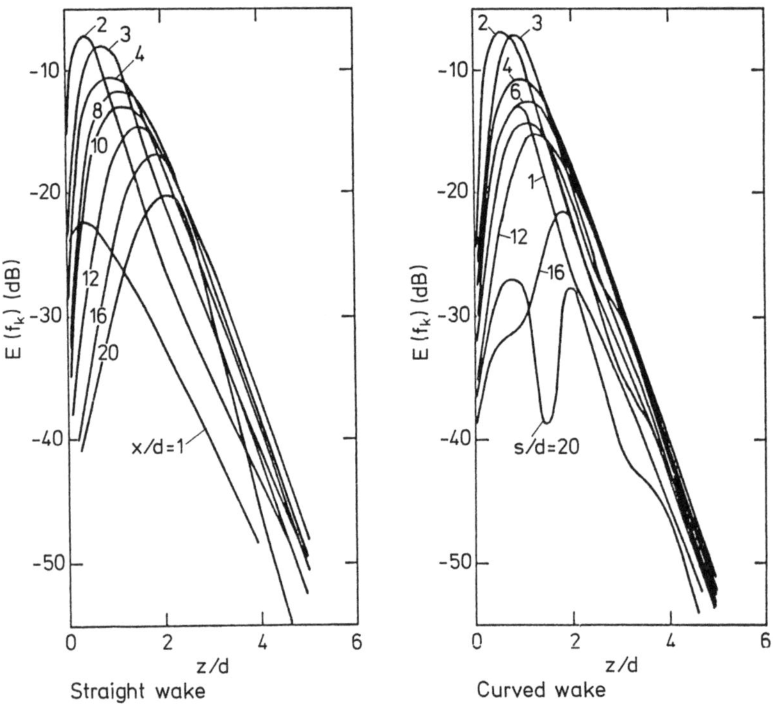

Fig. 10. Power distribution $E(f_k)$ at shedding frequency f_k of Karman vortex street for straight and curved wakes at Reynolds number of 150

In the curved wake, flow visualization revealed breakdown of the shed vortices and, hence, it was important to clarify the behavior of the flow in proximity to, and immediately behind, the fixed cylinder. For this purpose, detailed measurements of the power spectrum of the fluctuating velocity were made in the wake close to the cylinder.

Figure 10 shows the power distribution $E(f_k)$ at the shedding frequency f_k of a Karman vortex street for straight and curved wakes at a Reynolds number of 150. The peaks of the power profiles coincide with the center of the vortices at increasing distances x/d or s/d from the fixed cylinder. It is evident that there is no difference in the absolute maximum power in the two wakes although the spanwise displacement of the shed vortices is slightly sup-

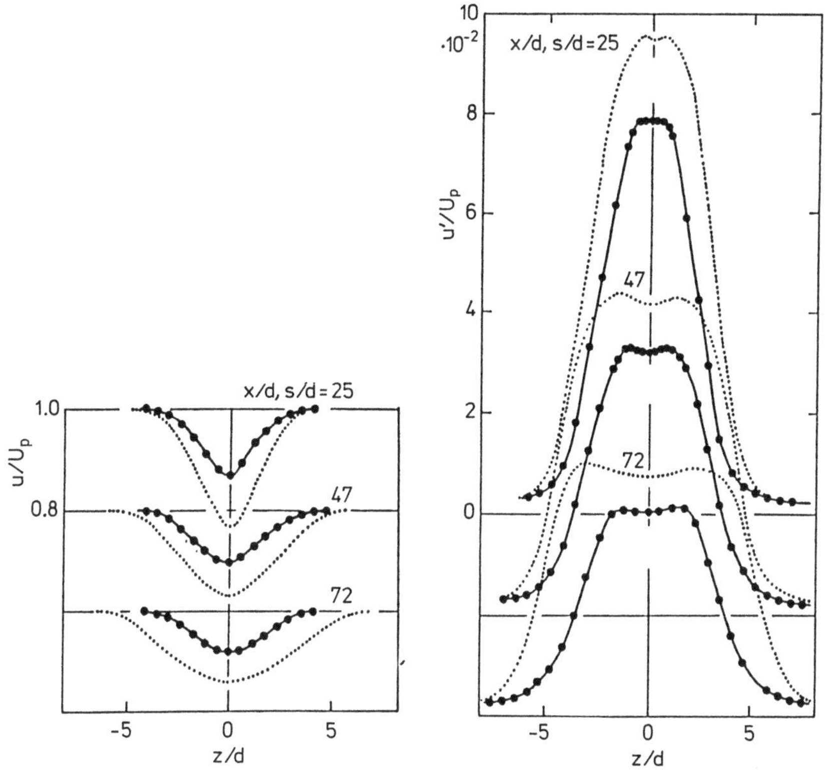

Fig. 11. Mean velocity and turbulence intensity distributions in straight and curved wakes at Reynolds number of 1500

pressed in the curved wake. Breakdown of the vortices apparently occurs at about s/d equal to 16 in the curved wake.

The mean velocity and turbulence intensity distributions in straight and curved wakes at a Reynolds number of 1500 behind a fixed transverse cylinder of diameter 2.13 mm are shown in Fig. 11, where, as in Figs. 12, 16, 17, the two wakes are compared, the solid line' representing the curved wake and the dotted line the straight wake. Measurements were made at three streamwise locations x/d or s/d equal to 25, 47, and 72. The mean velocity profiles and turbulence intensity profiles are symmetrical about the wake centerline for both the straight and curved wakes. However, the width of the straight wake increases with distance from the fixed cylinder while for the curved wake it is constant. The velocity defect, the difference between the hypothetical potential wake velocity and the actual velocity, is also smaller in the curved wake. Turbulence intensities are also smaller in the curved wake.

The spanwise distribution of the transverse velocity component v in the curved wake at the location s/d equal to 72 is shown in Fig. 12 for a Reynolds number of 1500. Also shown is a comparison of the Reynolds stresses $\overline{uw}/Up^2$ and $\overline{vw}/Up^2$ in the spanwise direction at the same streamwise location x/d or s/d equal to 72 for the two wakes. The Reynolds number for the two cases is 1500. In the curved wake, there is a secondary flow directed towards the center of curvature of the primary flow due to an imbalance between the centrifugal force arising from curvature of the particle pathline and the pressure gradient force. The maximum transverse velocity occurs at the wake centerline, a result which conforms to the photograph in Fig. 7, taken in frontal view using the smoke-wire method. The Reynolds stresses are seen

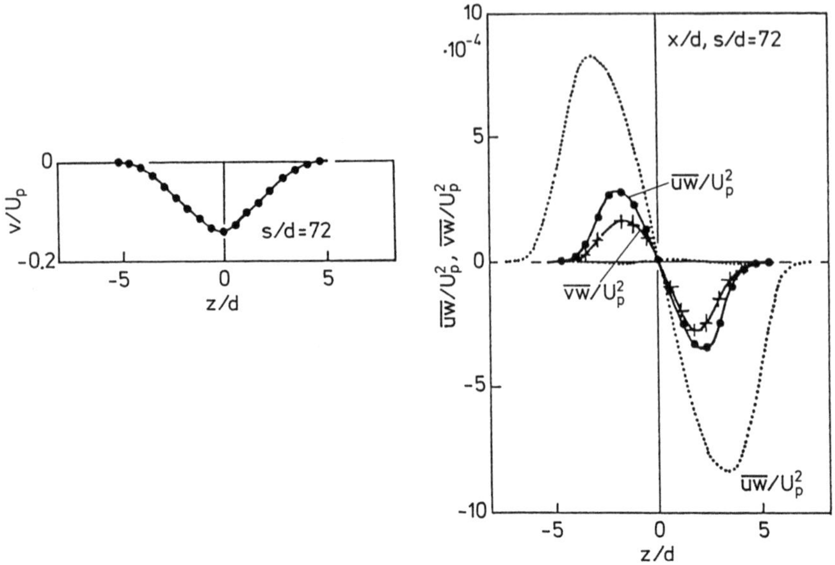

Fig. 12. Spanwise distribution of transverse velocity component and Reynolds stresses at Reynolds number of 1500

to change sign in Fig. 12 at the centerline of both the straight and curved wakes, with mirror-image distributions about the horizontal line of zero stress. The stress $\overline{uw}/U_p^2$ in the curved wake is considerably smaller than in the straight wake. The stress $\overline{vw}/U_p^2$ is found in the curved wake only because there is a spanwise gradient of transverse velocity v.

The power spectrum of the fluctuating velocity in a turbulent straight wake at a Reynolds number of 1500 behind a fixed cylinder is shown in Fig. 13. It is generally believed that in a two-dimensional wake behind a cylinder, a relatively large two-dimensional wave corresponding to the Karman vortex street and small three-dimensional disturbance waves propagate along the stream. A principal frequency f_k corresponding to a Strouhal number

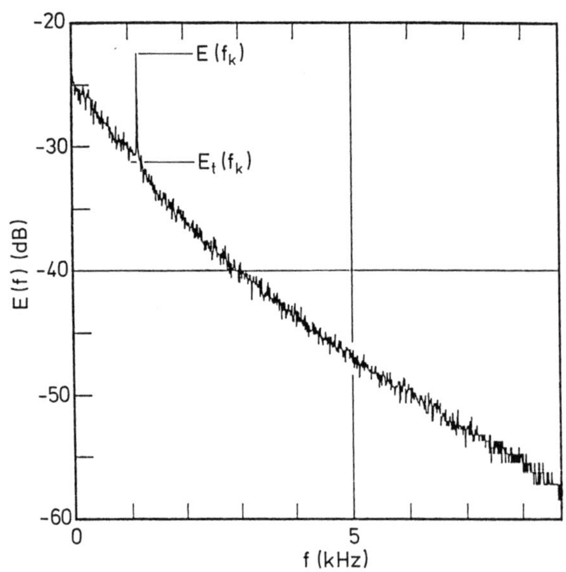

Fig. 13. Power spectrum of fluctuating velocity in turbulent straight wake behind cylinder at Reynolds number of 1500

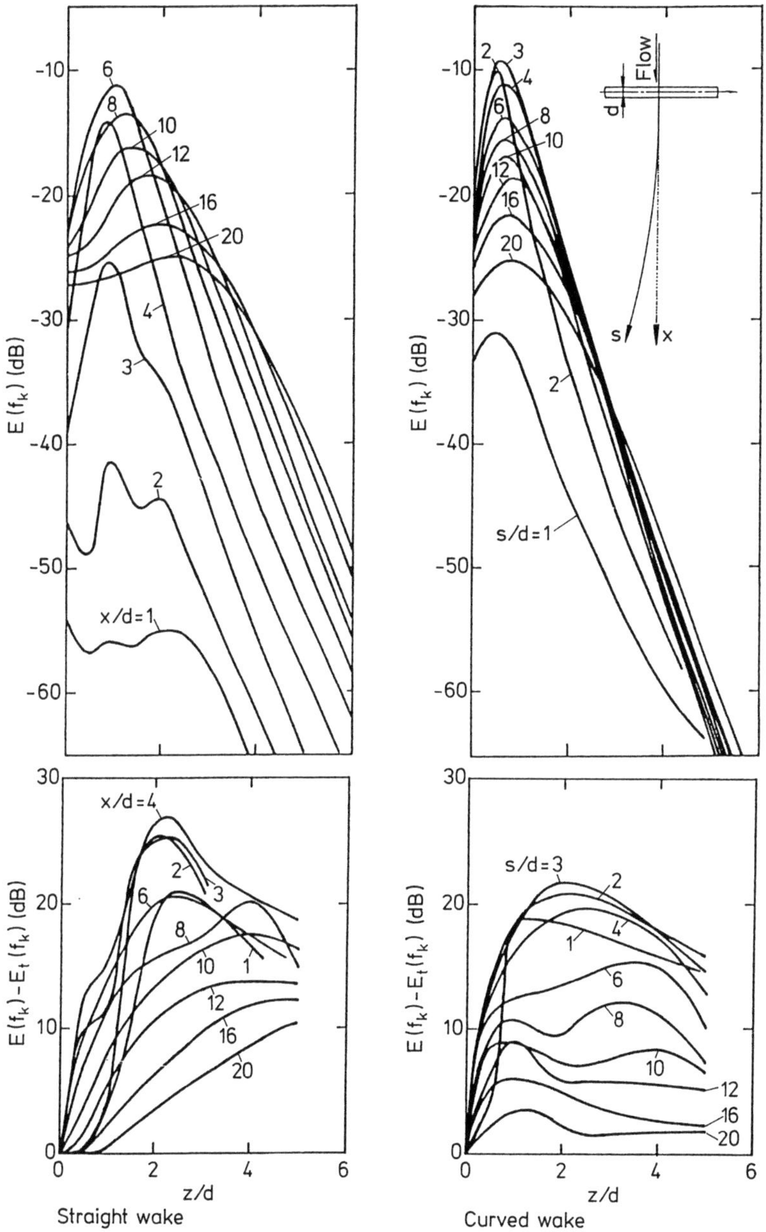

Fig. 14. Spanwise distributions of power $E(f_k)$ at principal frequency f_k of Karman vortex street and power difference $E(f_k) - E_t(f_k)$ at Reynolds number of 1500

equal to 0.22 for the Karman vortex street is observed in the power spectrum. The spectrum also contains a continuous power distribution due to the turbulence which is superimposed on the vortices.

The spanwise distributions of power $E(f_k)$ at the principal frequency f_k of the Karman vortex street and also of the power difference $E(f_k) - E_t(f_k)$ are shown in Fig. 14 for straight and curved wakes, where $E_t(f_k)$ is an interpolated imaginary power at f_k. The Reynolds number of the flow is 1500. It is evident from this figure that in the curved wake the spanwise spacing of shed vortices is smaller and their displacement in that direction is also more restricted than in the straight wake.

Stability

The two-dimensional wake in the transverse direction of a curved potential flow involves stability (Fig. 15). In the inner region of a curved turbulent wake the motion of fluid particles becomes destabilized while in the outer region of the wake, it stabilizes because of the existence of the imbalanced force. The mean velocity and turbulence intensity distributions in straight and curved wakes behind a cylinder fixed in the spanwise direction are shown in Fig. 16. The cylinder was fixed in the long axis of symmetry of the test channel and measure-

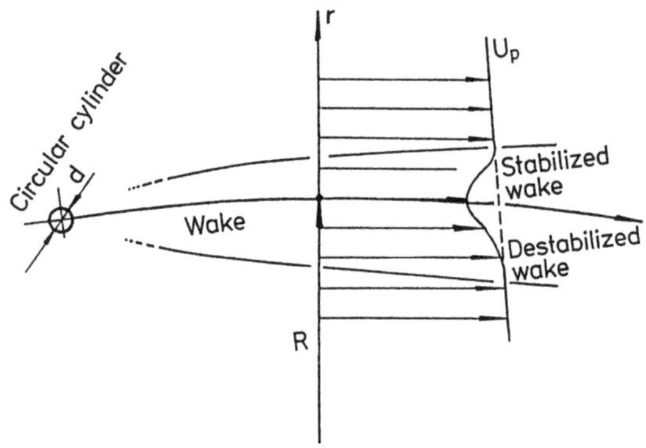

Fig. 15. Two-dimensional wake in the transverse direction of curved potential flow

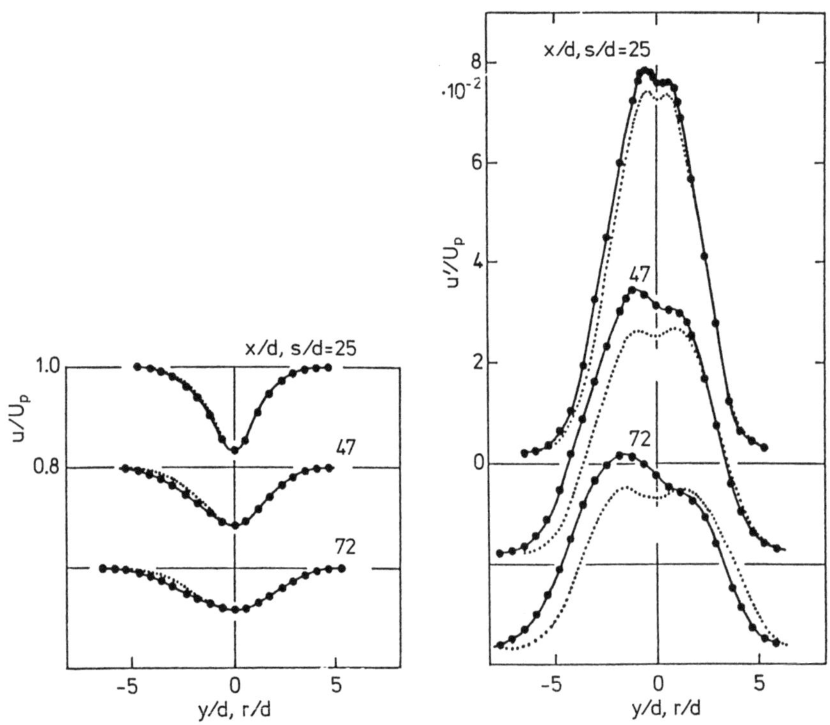

Fig. 16. Mean velocity and turbulence intensity distributions in straight and curved wakes behind cylinder fixed in spanwise direction at Reynolds number of 1500

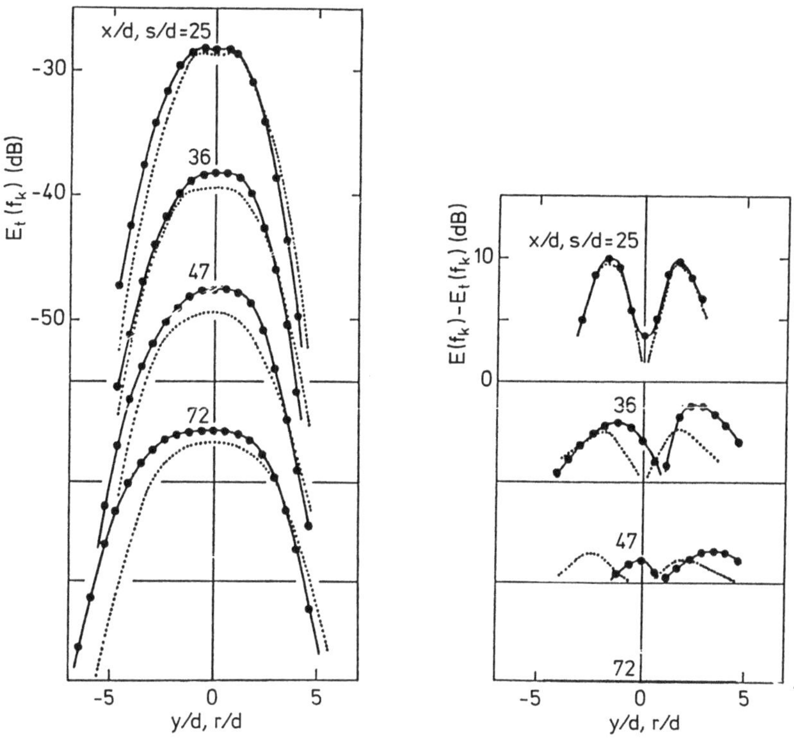

Fig. 17. Transverse distributions of imaginary interpolated power $E_t(f_k)$ at principal frequency f_k of Karman vortex street and power difference $E(f_k) - E_t(f_k)$ at Reynolds number of 1500

ments were made at streamwise locations x/d or s/d equal to 25, 47, and 72 at a Reynolds number of 1500. The distributions are asymmetrical in the curved wake and the asymmetry is attributed to the net destabilizing effect of curvature of the primary flow. In the curved wake, the turbulence intensity in the unstable region is higher than in the neutral-stable straight wake. But in the stable region of the curved wake, the turbulence intensity becomes smaller than in the straight wake as the ratio x/d or s/d is increased. The distinction between stable and unstable regions of a curved wake is based on the comparison between the mean velocity distributions in the two wakes.

Figure 17 shows the transverse distributions at streamwise locations x/d or s/d equal to 25, 36, 47 and 72 of the imaginary interpolated power $E_t(f_k)$ at the principal shedding frequency f_k of the Karman vortex street at a Reynolds number of 1500. Also shown are the power difference distributions, namely, $[E(f_k) - E_t(f_k)]$. The asymmetry of $E_t(f_k)$ about the line r/d equal to zero may be compared with that of the turbulence intensity in Fig. 16. Peaks in the power difference distributions in the two wakes were found to disappear at about x/d or s/d equal to 60. In the curved wake peaks in the unstable region decayed more rapidly than in the stable region.

Conclusions

To clarify stabilizing, destabilizing, and secondary flow effects of streamline curvature in a shear flow, a comparative experimental study was undertaken on laminar and turbulent wakes behind a cylinder in straight and curved potential flow fields. The conclusions reached as a result of this study may be summarized as follows.

1. For a turbulent wake behind a cylinder located spanwise, profiles of mean velocity, turbulence intensity, and power spectrum of the fluctuating velocity are not symmetrical about the wake centerline. However, two-dimensionality of the wake is preserved. Development of turbulence is promoted in the inside wake region but is suppressed in the outside region. The destabilizing effect of curvature in a turbulent shear flow is slightly greater than the stabilizing effect.
2. For a turbulent wake behind a cylinder located transversely, profiles of mean velocity, turbulence intensity and power spectrum of the fluctuating velocity are symmetrical about the wake centerline. A secondary flow, which has a maximum velocity at the wake centerline and is directed towards the center of curvature of the primary flow, arises from an imbalance between the centrifugal and pressure-gradient forces.
3. In a curved wake, breakdown of shed vortices occurs at lower Reynolds numbers than in a straight wake because of secondary flow.

Acknowledgements. I express my sincere thanks to Prof. B. E. L. Deckker, University of Saskatchewan, for his advice and kind reading of the manuscript, and also to Messrs. T. Endo and M. Ohuchi, Tokyo Denki University, for design and manufacture of the electrical equipment used. This work was supported in part by a grant-in-aid for scientific research from the Ministry of Education and also in part by the Center for Research, Tokyo Denki University.

References

1. Bradshaw, P. (1969): The analogy between streamline curvature and buoyancy in turbulent shear flow. J. Fluid Mech. **36**, 177
2. Hill, P. G., Moon, I. M. (1962): Effects of Coriolis forces in the turbulent boundary layer in rotating machines. Rep. No. 69, Gas Turbine Laboratory, MIT
3. Moore, J. (1973): A wake and an eddy in a rotating, radial-flow passage (part 1 and 2). ASME J. Eng. Power **95**, Ser. A, 205
4. Johnston, J. P., Halleen, R. M., Lezius, D. K. (1972): Effects of spanwise rotation on structure of two-dimensional fully developed, turbulent channel flow. J. Fluid Mech. **56**, 533
5. Wagner, R. E., Velkoff, H. R. (1972): Measurements of secondary flows in a rotation duct. ASME J. Eng. Power **94**, Ser. A, 261
6. Murakami, M., Kikuyama, K. (1976): Effects of rotation on diffuser flow. Jpn. Soc. Mech. Eng. **42**, 828
7. Hart, J. E. (1971): Instability and secondary motion in a rotating channel flow. J. Fluid Mech. **45**, 341
8. Rothe, P. H., Johnston, J. P. (1976): Effects of system rotation on the performance of two-dimensional diffusers. ASME J. Fluid Eng. **98**, Ser. I, 422
9. Koyama, H., Masuda, S., Ariga, I., Watanabe, I. (1979): Stabilizing and destabilizing effects of Coriolis force on two-dimensional laminar and turbulent boundary layers. ASME J. Fluid Eng. Power **101**, 23
10. Koyama, H., Masuda, S., Ariga, I., Watanabe, I.: Turbulence structure and three-dimensionality of a rotating two-dimensional turbulent boundary layer, in *Turbulent Shear Flows 2* (London, England 1979) pp. 4.22–4.27
11. Tani, I. (1962): Production of longitudinal vortices in the boundary layer along a concave wall. J. Geophys. Res. **67**, 3075
12. Bradshaw, P. (1973): Effects of streamline curvature on turbulent flow. AGARD-ograph **169**
13. Honami, S., Abe, T., Ariga, I., Watanabe, I. (1976): Investigation of turbulent flows in curved channels. ASME Paper, No. 75-FE-32
14. Meroney, R. N., Bradshaw, P. (1975): Turbulent boundary layer growth over a longitudinal curved surface. AIAA J. **13**, 1448
15. Ramaprian, B. R., Shivaprasad, B. G. (1978): The structure of turbulent boundary layers along mildly curved surfaces. J. Fluid Mech. **85**, Part 2, 273
16. So, R. M. C., Mellor, G. L. (1975): Experiment on turbulent boundary layers on a concave wall. Aeronaut. Q. **26**, 25
17. Margolis, D. P., Lumley, J. L. (1965): Curved turbulent mixing layer. Phys. Fluids **8**, 1775

18. Fette, von H. (1933): Strömungsversuche im rotierenden Laboratorium. Z. Tech. Phys. **7**, 257
19. Raj, R., Lakshminarayana, B. (1973): Characteristics of the wake behind a cascade of airfoils. J. Fluid Mech. **61**, 707
20. Sarpkaya, T. (1979): Vortex-induced oscillations. J. Appl. Mech. **46**, 241
21. Perry, A. E., Chong, M. S., Lim, T. T. (1982): The vortex-shedding process behind two-dimensional bluff bodies. J. Fluid Mech. **116**, 77
22. Fujita, H., Kovasznay, L. S. G. (1968): Measurement of Reynolds stress by a single rotated hot-wire anemometer. Rev. Sci. Instrum. **39**, 1351
23. Roshko, A. (1954): On the development of turbulent wakes from vortex streets. NACA Rep. 1191
24. Squire, H. B., Winter, K. G. (1951): The secondary flow in a cascade of airfoils in a nonuniform stream. J. Aeronaut. Sci. **8**, 271

Experimental Investigation of the Turbulent Structure of Vortex Wakes

O. Leuchter and J. L. Solignac

Office National d'Études et de Recherches Aérospatiales (ONERA), BP 72-92322 Châtillon Cedex, France

Abstract

Two configurations of vortical wake flows are investigated. The first (flow A) corresponds to a double-branched trailing vortex, the second (flow B) to the leading edge vortex of a delta wing. Both flows are studied in the presence of a destabilizing adverse pressure gradient leading to vortex breakdown. For various degrees of destabilization, the flow field is explored by means of a two-color laser velocimeter that yields mean velocity and Reynolds stress distributions across the flow. Breakdown is characterized by a strong deceleration of the flow near the axis of the vortex, by a significant decrease of the rotation rate in the center and consequently by a large production of turbulence. When breakdown occurs, both types of flow reveal similar characteristics of the flow structure. However, there are also significant differences, the most important being the abruptness of the breakdown phenomenon and its instability character which is much more pronounced for vortex flow B.

Nomenclature

E	turbulent energy, $1/2(\overline{u^2} + \overline{v^2} + \overline{w^2})$
K	circulation parameter, rV_t/U_0
K_p	pressure coefficient, $(p - p_0)/q_0$
p	pressure
q	dynamic pressure
r	radial distance from the axis
$U, V,$	mean velocity components in the x, y, z directions
u, v, w	fluctuating velocity components in the x, y, z directions
V_r, v_r	mean and fluctuating velocity in the radial direction
V_t, v_t	mean and fluctuating velocity in the tangential direction
x, y, z	coordinates in the streamwise, horizontal and vertical directions
α	angle of incidence
β	adverse pressure gradient parameter, $(d\,Kp_e/dx) \cdot K_e$

Subscripts

o	reference upstream flow conditions
e	external flow conditions
a	conditions on the axis of the vortex
c	conditions in the test chamber (flow A)

Introduction

Vortical type flows play a major role in modern aerodynamics. Of special practical interest are vortex wakes generated by lifting surfaces or by rotor blades. In particular, the phenomenon of vortex bursting that arises as a consequence of destabilization in the presence of an adverse pressure gradient strongly affects flight qualities at high angles of incidence. Also in many other fields, rotating flows play a dominant practical role. For instance, in the field of combustion, swirling flows are commonly used to enhance mixing rates and to stabilize combustion.

A great deal of effort is presently being devoted to the development of numerical predictive methods for rotating and recirculating viscous flows. Turbulence plays a dominant role there and the success of the prediction depends strongly on the degree of accuracy of the model used in the calculations. While satisfactory progress has been achieved so far as the numerics are concerned, the progress in the field of turbulence modeling in swirling and recirculating flows is much less satisfactory and our knowledge concerning the effects of rotation on turbulence is still incomplete and fragmentary.

Several basic research projects on vortex flows have been recently initiated at ONERA with the aim of providing a better physical understanding of the fundamental processes involved and of obtaining quantitative information concerning the structure of the mean and the turbulent fields, especially under breakdown conditions. The present study, which belongs to this basic research activity, concerns two types of subsonic vortex flows: the first (flow A) corresponds to a double-branched trailing vortex. This yields a quasi-axisymmetrical swirling flow configuration which has been found convenient for fundamental investigations of turbulence structure, especially under destabilizing conditions. The second (flow B) corresponds to a leading edge vortex of a delta wing. It was used for a parametric study of the conditions under which breakdown occurs and to establish practical breakdown criteria. Flow B yields a stronger vortex and is therefore used for studying the instability features of the breakdown. In both cases, destabilization is achieved by means of an imposed positive pressure gradient.

Laser Doppler anemometry has been extensively used in the present study. The advantages of this method lie in its nonintrusive character and in the ability to detect the direction of the velocity, thus allowing recirculating flow regions due to breakdown to be explored accurately. Some typical results concerning mean and turbulent fields are presented for both types of vortex flows, especially under destabilizing conditions. Common features as well as typical differences between the two configurations are discussed.

Experimental Arrangements

Flow A is generated by two identical airfoils spanning the test section and assembled together on a slender cylindrical body, as indicated in Fig. 1. The two airfoils (NACA 0012 profile) have opposite incidences so that the vortex sheets from the blades roll up into a double-branched trailing vortex. The vortex strength can be varied by modifying the angle of attack of the airfoils. In the present study, fixed values of $\pm 12°$ have been used, yielding a maximal tangential velocity of $0.7\ U_0$. The cylindrical test section has a diameter of 0.3 m and the upstream reference velocity is 32.5 m/s.

The positive pressure gradient in the flow is created by means of secondary suction which acts on the open part of the test section (between the locations $x = 153$ mm and $x = 287$ mm

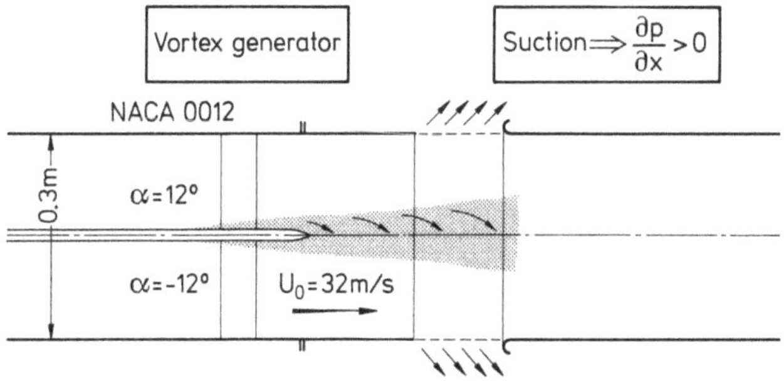

Fig. 1. Schematic of vortex flow A

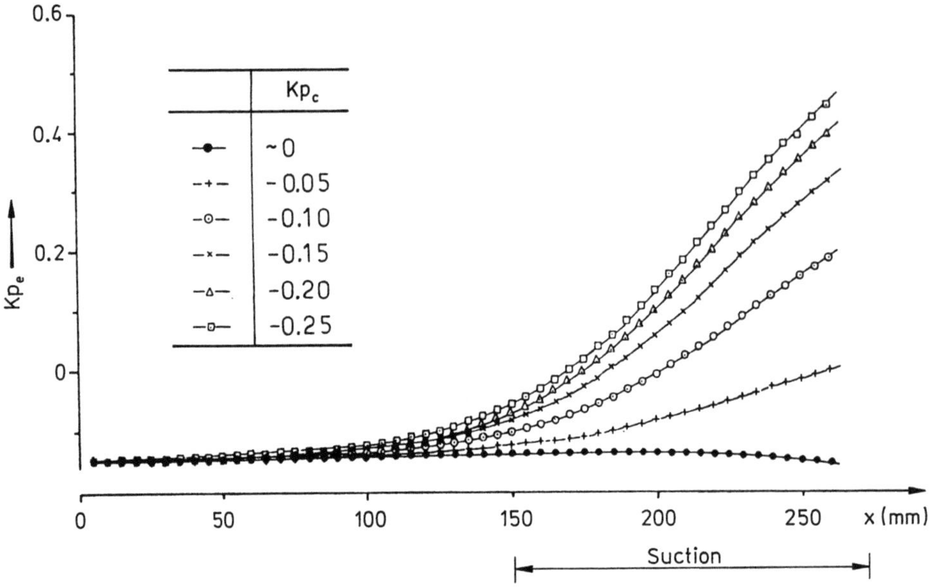

Fig. 2. Evolution of this static pressure in the outer flow (flow A)

downstream from the vortex generator). The pressure gradient is easily adjustable by varying the secondary mass flow rate extracted from the external test chamber. The longitudinal evolution of the static pressure in the outer part of the flow (measured at a radial distance of $r = 80$ mm) is shown in Fig. 2 for different aspiration rates (expressed here by the pressure coefficient Kp_c relative to the pressure in the test chamber). The positive pressure gradient corresponding to the middle of the open test section is expressed in terms of the nondimensional parameter $\beta = (d\,Kp_e/dx).\ K_e$, where the circulation parameter K_e represents a characteristic dimension of the vortex [1]. The corresponding values of the pressure gradient parameter β will be given further below in the text.

The longitudinal velocity in the flow is maintained constant by means of a sonic throat located downstream of the test chamber. Stable flow conditions are thus achieved even when breakdown occurs.

Flow B is generated by means of a slender delta wing (sweep angle 75°) mounted at incidence in the test section of a subsonic wind tunnel (Fig. 3). The diameter of the test section is 1 m and the free stream velocity U_0 about 15 m/s.

The vortex strength can be easily adjusted by varying the angle of attack of the wing. In the present study, an angle of 20° has been adopted, yielding a maximal tangential velocity of the order of U_0.

One of the two vortices generated by the wing enters into an air intake, the flow rate of which can be adjusted by means of moving flaps. The resulting adverse pressure gradient

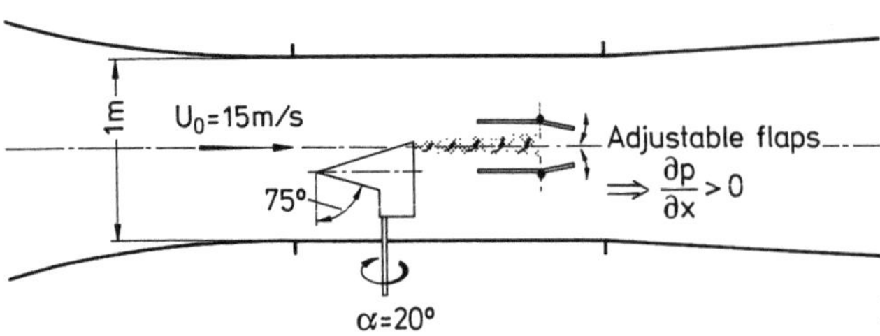

Fig. 3. Schematic of vortex flow B

acting on the vortex can thus be easily varied. In the present study, the position of the flaps has been adjusted so that breakdown of the vortex occurs between the trailing edge of the wing and the air intake. The corresponding pressure gradient parameter β is of the order of 0.04.

Instrumentation

The flow field measurements presented below are obtained with a two-component laser velocimeter system developed at ONERA. It will only be briefly described here, as a more detailed description is given in [1].

The velocimeter uses a 15 W argon laser as the light source. The laser beam is split into two beams of blue (4880 Å) and green (5145 Å) light. Both optical circuits have Bragg cell frequency shifting, allowing the direction of the velocity to the detected. Seeding of the flow is generally provided by naturally occurring particles. In some extreme cases, incense smoke is employed. The system provides simultaneoulsy the longitudinal and the vertical velocity components measured on the same (validated) particle. Statistical averages are calculated from an ensemble of at least 2000 samples. The velocimeter system operates in the forward scattering mode. The optics are mounted on rigid tables ensuring computer controlled displacements of the measuring volume along the three orthogonal axes. The visualization technique with a laser light sheet has also been used in preliminary stages of the study in order to characterize qualitatively the flow structure during the destabilizing process leading to breakdown. Examples of this kind of optical investigation can be found in [1].

Results and Discussion

Flow A. In this case, the region corresponding to the open part of the test section can be easily explored by the laser velocimeter, due to the particular arrangement shown in Fig. 1. Horizontal and vertical explorations are made in a longitudinal cross section located at $x = 200$ mm, for the whole set of adverse pressure gradients as defined in Fig. 2. The corresponding values for the pressure gradient parameter β are: $\beta \approx 0$, 0.008, 0.020, 0.027, 0.032 and 0.035. Only for the last four values are the pressure gradients strong enough to destabilize the vortex wake. Therefore, part of the results, especially those concerning the turbulent field, are limited to these four cases.

As the velocimeter measures the two velocity components U and W (longitudinal and vertical), the horizontal explorations provide (for axisymmetrical flow) the tangential velocity profiles, and the vertical explorations that of the radial ones. Comparison of the longitudinal mean and rms velocity profiles measured in both transverse directions y and z showed indeed that axisymmetry properties are well established in this flow [1]. It will therefore be meaningful to associate the profiles of the three velocity components $U(y)$, $V_t(y)$ and $V_r(z)$. Simultaneous measurements of all three components in the same transverse direction would imply the use of a three-dimensional velocimeter system. These measurements are planned for a later stage of this study with an extended version of the velocimeter.

The profiles of the three mean velocity components U, V_t and V_r measured at the location $x = 200$ mm are given in Fig. 4 for the six different values of the pressure gradient parameter β.

The undisturbed configuration (no secondary aspiration, $\beta \approx 0$) yields for the longitudinal component U an almost uniform velocity profile except in the viscous core near the axis and in the external flow region where the tracks of the vortex sheets are clearly

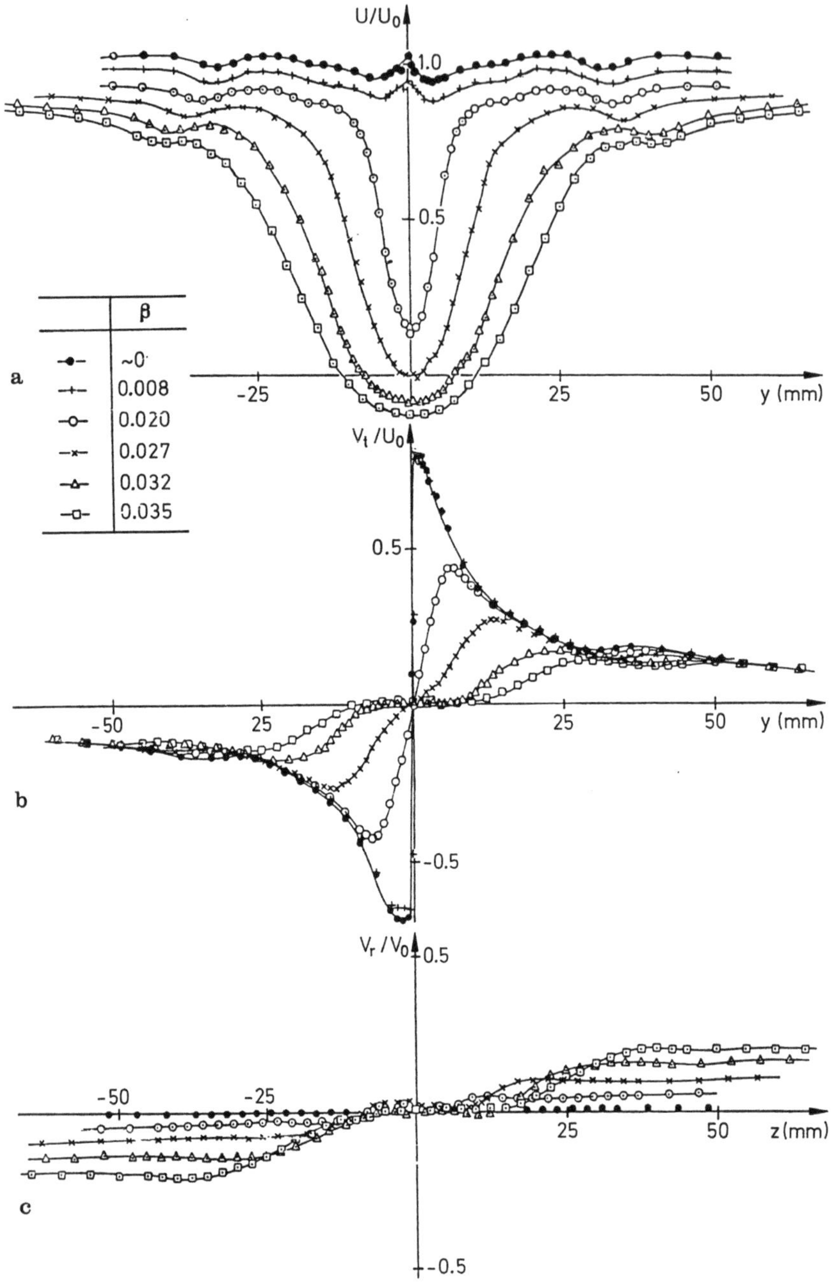

Fig. 4. Mean velocity profiles (flow A). a) axial component; b) tangential component; c) radial component

evidenced. The tangential velocity profile shows behavior typical for a potential vortex, except in the region near the axis where viscous effects predominate. The radial component is for this case approximately zero, justifying the approximation of quasi-cylindrical flow.

For a moderate compression rate ($\beta = 0.008$), the vortex wake remains stable in the sense that the U profile undergoes only a slight reduction without changing its form and that the tangential velocity profile remains approximately invariant.

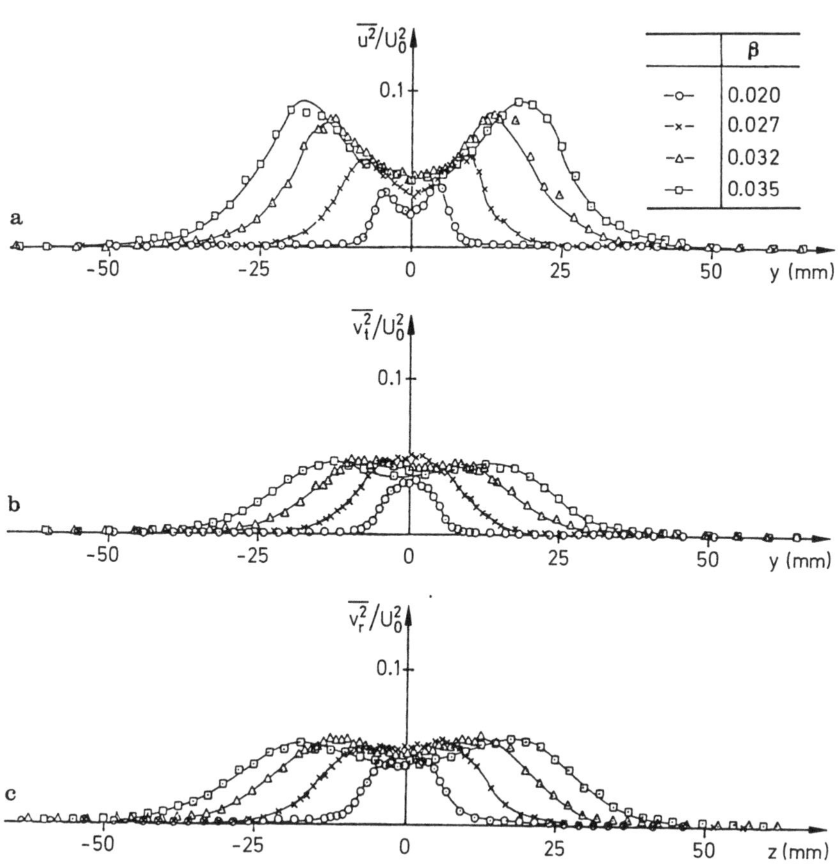

Fig. 5. Profiles of the normal stresses (flow A). a) axial component; b) tangential component; c) radial component

For higher values of β, the flow can no longer sustain the externally imposed pressure rise and the vortex breaks down. This fact is clearly evidenced in Fig. 4 by the deep defect appearing in the longitudinal velocity profiles near the axis and by the concomitant decrease of the angular velocity. The radial component remains negligibly small in the central region where the vortex is destabilized, but increases noticeably in the outer field when β grows. This is a consequence of the increasing divergence of the flow for higher aspiration rates. For the two highest values of β, the longitudinal velocity component becomes negative in the center of the wake due to the formation of a bubble with recirculation flow. The tangential component there is approximately zero.

Laser sheet visualizations have shown that the destabilization process is accompanied by a strong helical distortion of the vortex core with the helix diameter growing when the pressure gradient is increased [1]. This results in a strong radial expansion of the dissipative region as indicated by the velocity profiles in Fig. 4. High-speed cinematography has revealed that this process is highly turbulent.

This fact is clearly evidenced by the profiles of the three normal stresses presented in Fig. 5. The rapid expansion of the turbulent zone with increasing β corresponds to that of the velocity defect appearing in the profiles of Fig. 4a. The individual stresses show typical differences, in so far as those relative to the tangential and to the radial velocity components tend to become rather flat for growing destabilization, whereas the longitudinal stress

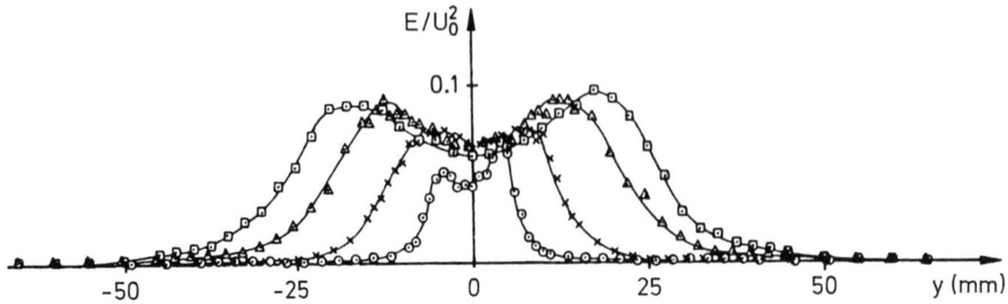

Fig. 6. Turbulent energy profiles (flow A)

profiles show very distinct peaks of increasing level, reaching twice that on the axis. The position of the peaks corresponds roughly to the maximum slope of the mean velocity profiles (Fig. 4a). Figure 5 shows moreover that the $\overline{v_r^2}$ stresses expand radially farther than the $\overline{v_t^2}$ stresses and that at the center of the vortex the three normal stresses are of the same order of magnitude (corresponding to a rms level with respect to U_0 of roughly 20%).

If axial symmetry is assumed, the turbulent energy E can be deduced directly from the normal stresses of Fig. 5. Its distribution along the horizontal direction is given in Fig. 6 (with $\overline{v_r^2}$ taken from the vertical exploration). The evolution of this quantity, which is of central importance in turbulence modeling, shows roughly the same behavior during the destabilization process as that of the individual stresses previously discussed.

Only two of the three off-diagonal terms of the Reynolds stress tensor are available from the present laser velocimeter measurements. The longitudinal-tangential velocity correlation $\overline{uv_t}$ is obtained from the horizontal exploration, whereas the longitudinal-radial velocity correlation $\overline{uv_r}$ results from the vertical one.

The $\overline{uv_r}$ profiles of Fig. 7a are representative of axisymmetrical wake type flows, whereas the $\overline{uv_t}$ stresses are zero in nonrotating flows. Both stresses expand radially with increasing destabilization conditions in the same manner as normal stresses. At the locations of maximum shear stress, the shear stress to energy ratio reaches values of about 0.3 for $\overline{uv_r}$ and of about 0.2 for $\overline{uv_t}$. A slight lack of axial symmetry can be observed for both quantities although the mean velocity profiles are apparently symmetric. This can be due in part to the strong mutual contamination of the two transverse components V_r and V_t if the horizontal or vertical explorations do not go exactly through the center of the vortex.

Flow B. In contrast to flow A, only one destabilized vortex configuration has been investigated. However, the measurements can be made here in a more extended domain including regions upstream and downstream of the breakdown position. Explorations have been done in the longitudinal (x) direction on the axis of the vortex wake as well as horizontally (y) and vertically (z) for several longitudinal positions.

Figure 8 shows the axial evolution of the mean velocity U and the rms values relative to the longitudinal and transverse components. As in flow A, breakdown is characterized by a rapid decrease of the mean velocity on the axis. The initial value, about twice the free stream velocity, is much higher than in flow A. A bubble is formed downstream of the breakdown position, as evidenced by negative axial velocities.

The rapid decrease of the mean velocity is accompanied by a sharp increase of the longitudinal fluctuation intensity whereas the vertical fluctuation decreases. In the bubble both rms values are approximately of the same magnitude (as was observed in flow A), of the order of 0.35 U_0. The particular behavior of both fluctuations in the breakdown region can be qualitatively explained from the joint probability density function (JPDF).

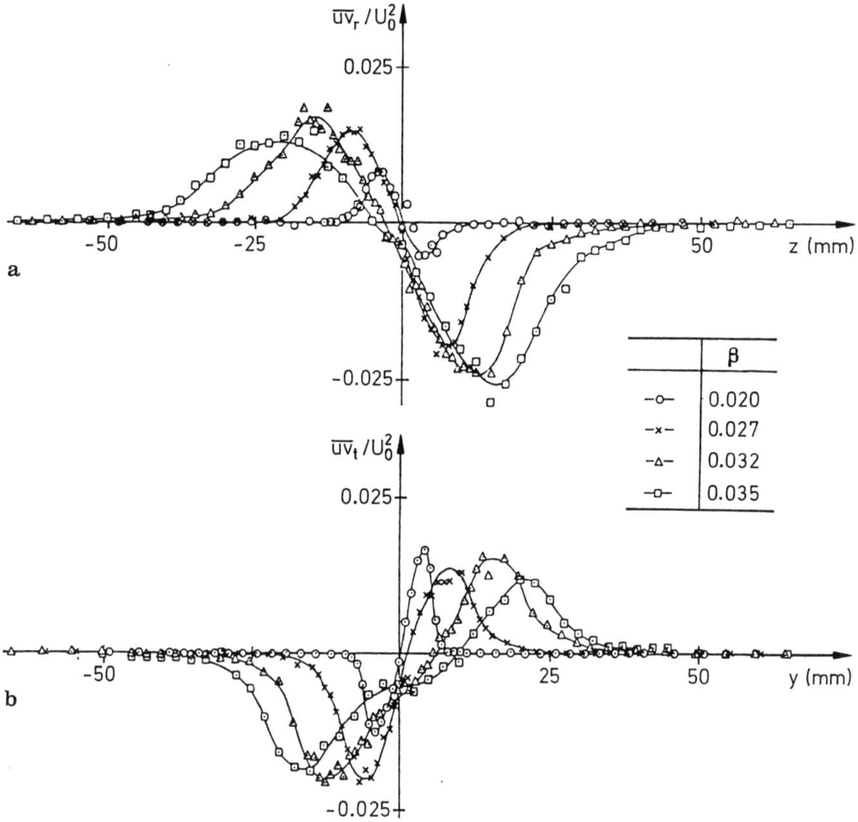

Fig. 7. Profiles of the off-diagonal Reynolds stresses (flow A). a) axial-radial; b) axial-tangential

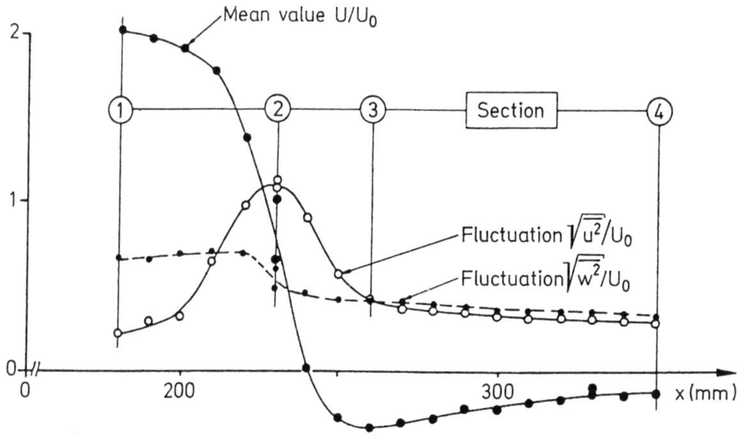

Fig. 8. Axial evolution of the mean velocity and the rms values (flow B)

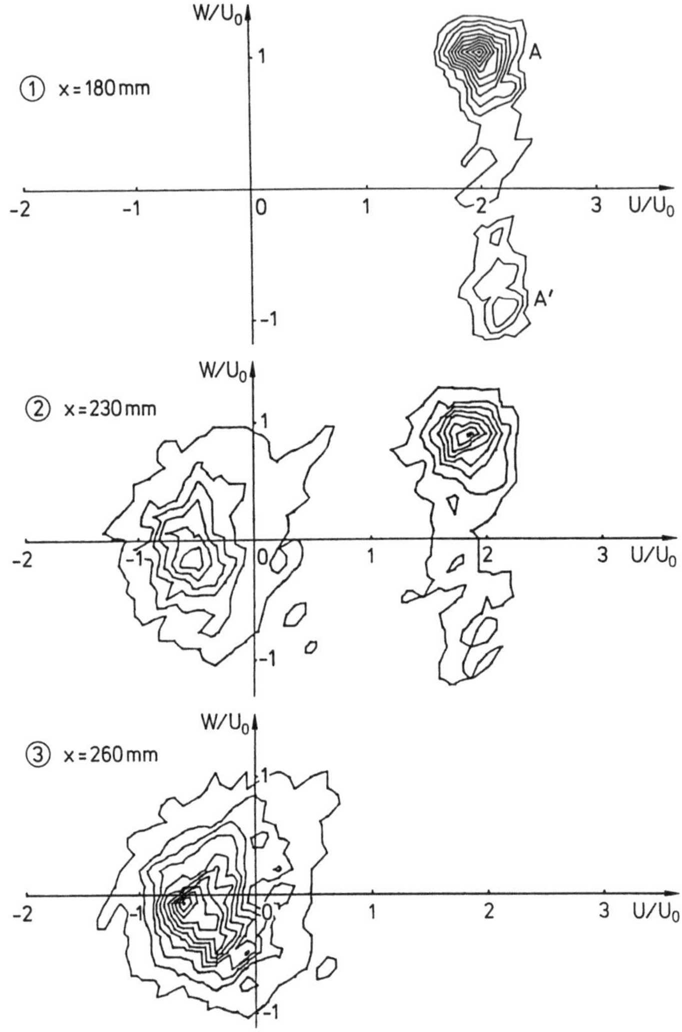

Fig. 9. Contours of the joint probability density function (flow B)

Computer-plotted contours of the JPDF in the U, W plane are shown in Fig. 9 for the three axial positions 1, 2 and 3. In the upstream Section 1 only positive U values occur, whereas two peaks A and A' are evident. Their symmetrical positions with respect to the U axis suggest the existence of small fluctuations of the vortex as a whole, which could explain the relatively high level of $\sqrt{\overline{w^2}}$ before breakdown.

In the breakdown region (Section 2), the plots clearly show the existence of an intermittent state between two distinct flow configurations, the first corresponding to the upstream position 1, the second to the downstream position 3. This strongly intermittent state explains the high level of $\sqrt{\overline{u^2}}$ shown in Fig. 8.

In the downstream Section 3, the single-peaked JPDF indicates a nonintermittent state, with a negative mean value for U. Moreover, the rms values with respect to U and W are of the same order, as shown in Fig. 8. The longitudinal and vertical mean velocity profiles are given in Fig. 10 for the four Sections 1 to 4 indicated in Fig. 8. The vertical component

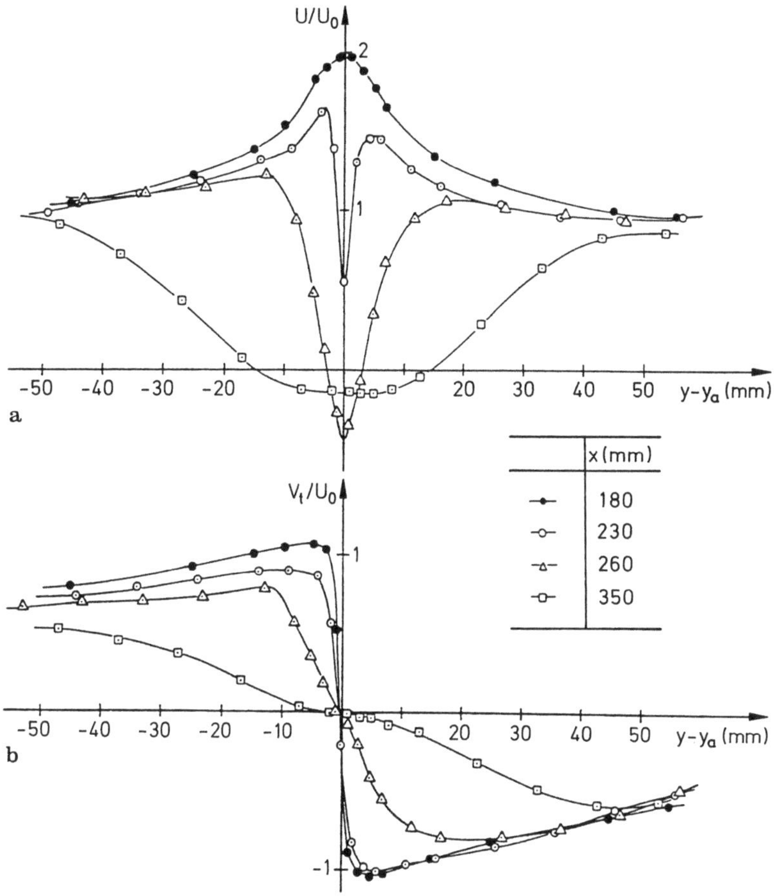

Fig. 10. Mean velocity profiles (flow B). a) axial component; b) tangential component

can be interpreted as in flow A as the tangential velocity, provided axial symmetry is assumed. The profiles show typical evolution for destabilized vortex flows, analogous to those shown in Fig. 4 for flow A. However, it must be noted that the initial profiles (Section 1, $x = 180$ mm) are quite different from those of case A: a higher overspeed on the axis for U and a flatter evolution for V_t. These characteristics are typical of Euler vortices generated by delta wings [1, 2].

The profiles of the rms values for the three velocity components are given in Fig. 11. In contrast to Fig. 5, very high levels are obtained in the region of breakdown (Section 2, $x = 230$ mm and Section 3, $x = 260$ mm). This is due to the high degree of instationarity of the breakdown. Only in Section 4 do the profiles become more regular and qualitatively similar to those of flow A (Fig. 5).

The profiles of the velocity correlation $\overline{uv_t}$ are given in Fig. 12. As in Fig. 7, they are approximately antisymmetrical and $\overline{uv_t}$ has the same sign as V_t, except in the region near the axis. As for the rms values, very high levels are obtained in the sections nearest to breakdown.

The formation of both vortex flows results from spiral rolling up of the vortex sheets coming from the vortex generators. This leads to a more or less stratified structure of the flow, as can be seen from visualizations [1], especially in undisturbed vortex configura-

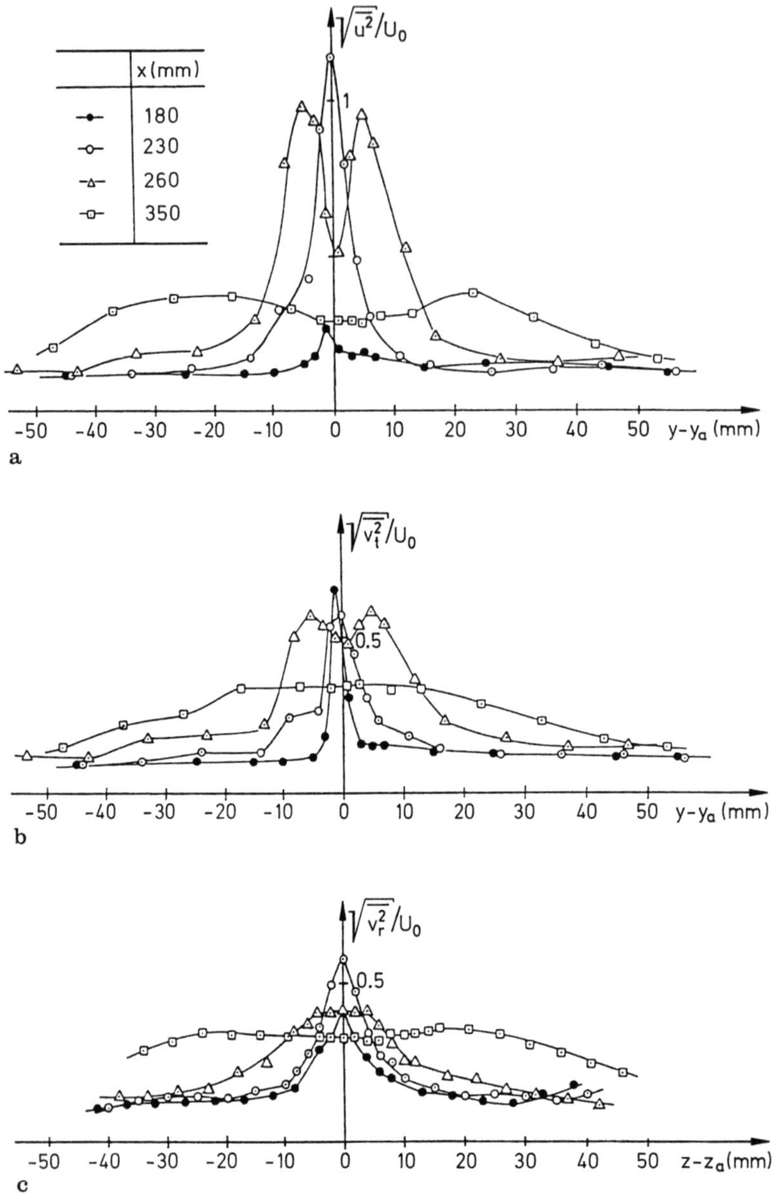

Fig. 11. Profiles of the rms values (flow B). a) axial component; b) tangential component; c) radial component

tions. A detailed exploration of Section 1 upstream of the breakdown position has been done in order to construct lines of equal mean and rms values in the $y - z$ plane and to characterize the properties of axisymmetry for this flow. Figure 13 shows that the highest rms values are located at the center of the vortex and that a ridge of locally high values prolongs the track of the wake issuing from the trailing edge of the wing. The rolling up of the sheet, which can be easily recognized on this figure, is similar to that given by the visualizations (see Ref. [1] and Fig. 14).

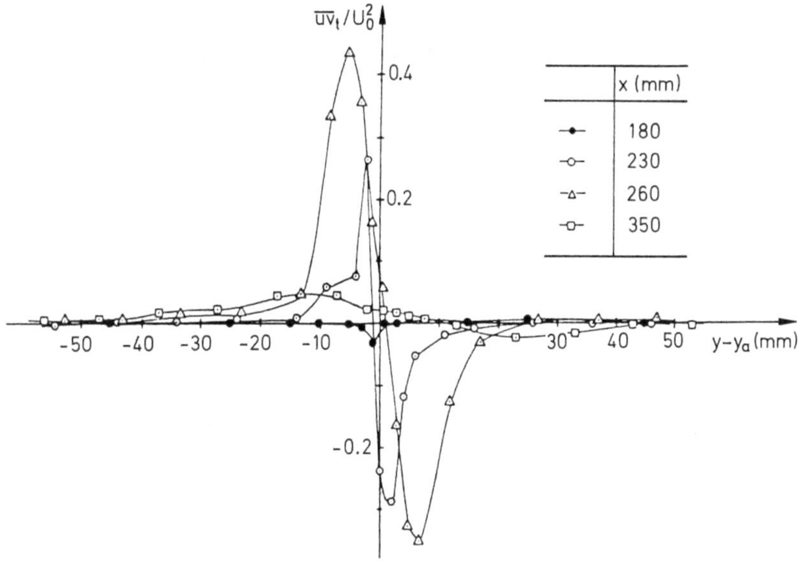

Fig. 12. Profiles of the $\overline{uv_t}$ correlation (flow B)

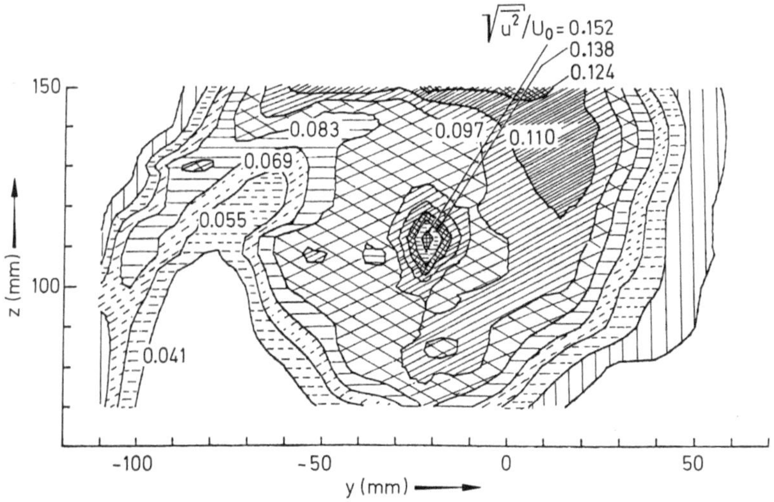

Fig. 13. Contours of equal rms values in the $y - z$ plane for station 1 (flow B)

Concluding Remarks

The mean and turbulent velocity fields of two different vortex flows have been investigated by laser anemometry, under destabilizing conditions produced by adverse pressure gradients. The two flow configurations present close similarities as far as the general features of destabilization are concerned: breakdown originates in both flows in the region close to the axis, with the axial velocity and the rotation rate decreasing rapidly and with a strong

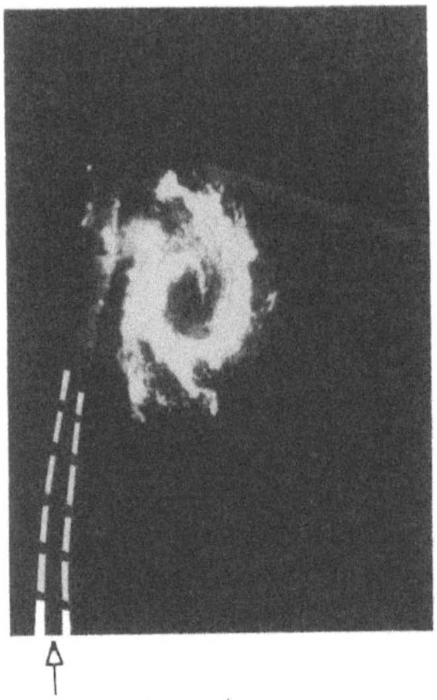

Fig. 14. Vortex visualized by laser sheet (flow B)

increase of turbulence production. Fundamental differences between the two flows lie in the internal structure of the vortex wakes as a result of the distinct generation modes, in the vortex strength which is higher with the delta wing generator and in the abruptness of the destabilization process, which is more pronounced in flow case B. This destabilization results in a highly unstable character of flow B with much higher rms levels. The same trend can be observed from previously published results; compare for instance [3] concerning the tip vortex and [4] for the leading edge vortex.

Flow B will be used further for studying the breakdown conditions for different flow parameters (vortex strength and adverse pressure gradient) and for working out practical breakdown criteria. The instationary character of breakdown will be more thoroughly investigated by means of conditional sampling techniques. Flow A, which is more stable, will be used for more fundamental studies of the turbulence structure and of the influence of the rotation on turbulence. The results given here for both flows should be considered only as the first step of a more exhaustive study on the general topic of turbulent flows with strong rotation effects.

References

1. Solignac, J. L., Leuchter, O. (1983): Études expérimentales d'écoulements tourbillonnaires soumis à des effets de gradient de pression adverse, AGARD-FDP Aerodynamics of Vortical Type Flows in Three Dimensions. Rotterdam, April 25–27
2. Hall, M. G. (1961): A theory for the core of leading edge vortex. J. Fluid Mech. **11**, 209
3. Singh, P. I., Uberoi, M. S. (1976): Experiments on vortex stability. Phys. Fluids **19**, 1858
4. Hayashi, Y., Nakaya, T. (1978): Flow field in a vortex with breakdown above sharp delta wings. NASA TM **75**, 339

Numerical Study of the Phenomenon of Turbulence Suppression in a Plane Shear Layer

M. Nallasamy and A. K. M. F. Hussain

ED 42, Systems Dynamics Lab., NASA Marshall Space Flight Center Huntsville, Alabama 35812, USA

Abstract

The phenomenon of turbulence suppression in a free shear layer due to controlled excitation has been studied using vortex-in-cell method. When the amplitude of the sinusoidal excitation is low, the maximum suppression occurs for excitation at the maximally unstable frequency ($\mathrm{St}_\theta = 0.017$) of the shear layer, consistent with experimental data. An interesting new result of the numerical simulation is that the above Strouhal number preference is lost at high amplitudes of excitation. Also, it is found that the level of random fluctuations superimposed on the initial sinusoidal disturbance affects the extent of suppression significantly.

Introduction

The phenomenon of turbulence suppression due to controlled excitation of free shear layers was experimentally investigated by Zaman and Hussain (1981) in a number of experimental facilities – circular jets, a plane jet and a single-stream plane mixing layer. They identified the conditions for the occurrence of maximum suppression in free shear flows. The main findings of their study are: (a) The turbulence suppression depends on the laminar exit shear layer characteristics (no suppression has been detected when the exit boundary layer is turbulent). (b) The suppression effect is maximum when the shear layer is forced at the maximally unstable frequency, i.e. $\mathrm{St}_\theta = 0.017$ (Michalke 1965). Here $\mathrm{St}_\theta (\equiv f\theta_e/U_e)$ is the excitation Strouhal number, f is the forcing frequency, θ_e and U_e are the exit boundary layer momentum thickness and free stream velocity, respectively. Since the phenomenon is associated with the initial shear layer instability, it occurs not only in circular jets but also in plane jets and plane mixing layers. (c) The suppression occurs over the entire thickness of the shear layer. Suppression is observed in longitudinal and transverse fluctuations as well as in the Reynolds stress. (d) The maximum suppression characterized by u'_{ex}/u'_{ux} (the ratio of longitudinal fluctuation intensities with and without excitation) occurs at an axial location of about 400 θ_e downstream from the separation point (i.e., $x = 0$). The axial extent of suppression observed in jets was in the range of $0.75\,D < x < 8\,D$. In a single-stream plane mixing layer, they observed the suppression effect to persist as far downstream as 6000 θ_e. From flow visualization and conditional sampling studies of excited and unexcited shear layers, they conclude that excitation at $\mathrm{St}_\theta = 0.017$ produces the fastest growth and roll up of the shear layer, resulting in early saturation and transition (breakdown) of the structures, and inhibits the formation of large energetic vortices. As a consequence, the large fluctuation intensity, otherwise caused by the passage and interaction of the energetic vortices is reduced. Thus, Zaman and Hussain established that the turbulence suppression occurs in a range of Strouhal numbers (0.009–0.021) due to the laminar instability characteristics of the shear layer. Note that Vlasov and Ginevskiy (1974) and Petersen et al. (1974) independently reported turbulence suppression at jet excitation Strouhal numbers $\mathrm{St}_D = 2.75$ and 3.0, respectively, even though they neither focused on this observation nor attempted to explain the phenomenon. In an effort to further investigate and understand the phenomenon of turbulence suppression in shear flows, a numerical study of turbulence suppression was undertaken. The attractiveness of the numerical simulation is that the amplitude and frequency of the sinusoidal excitation as well as the exit boundary layer momentum thickness can be varied at will. Furthermore, controlled initial turbulence can be introduced numerically. In an experiment these characterizing parameters cannot be independently varied, limiting the choices available.

This paper reports the results of numerical simulation of excitation of a plane two-dimensional mixing layer. First, a brief description of the numerical simulation technique is given. Then the results of the simulation are discussed in relation to experimentally observed suppression along with the limitations of the simulation.

Numerical Simulation

Mixing layers are known to be dominated by quasi-two-dimensional structures which develop initially from the Kelvin-Helmholtz instability of a vortex sheet (Batchelor 1967). Controlled excitation studies of jets and mixing layers suggest that the sensitivity of the normally turbulent flows to external forcing is related to the rotational inviscid behavior of these flows. Then, an inviscid two-dimensional computation (with no vortex stretching and dissipation which are crucial to the energy cascade argument) may shed light on at least the dynamics of large structures. To this end, several shear layer simulations which essentially solve two-dimensional Euler equations have been reported (Acton 1976; Delcourt and Brown 1979; Ashurst 1979; Aref and Siggia 1980). In these simulations, the vorticity, originally confined in a thin layer, is concentrated further into a finite number of point vortices for simulation purposes. In other words, the piecewise continuous distribution of vorticity ω is replaced by a finite sum of N delta functions,

$$\omega(\boldsymbol{X}, t) = \sum_{n=1}^{N} \Gamma_n \delta(\boldsymbol{X} - \boldsymbol{X}_n(t)), \tag{1}$$

where δ is the Dirac delta function, $\boldsymbol{X}_n = (X_n, Y_n)$ is the location of the Nth vortex of circulation Γ_n. The motion is followed by integrating the system of ordinary differential equations,

$$\frac{d\boldsymbol{X}_n}{dt} = \boldsymbol{U}(X_n, t). \tag{2}$$

The solution of the Eqs. (1) and (2) presumably represents the solutions of the two-dimensional Euler equations:

$$\frac{\partial \omega}{\partial t} + (\boldsymbol{U} \cdot \nabla) \quad \omega = 0. \tag{3}$$

$$\nabla^2 \psi = -\omega, \tag{4}$$

$$U_x = \frac{\partial \psi}{\partial y}; \quad U_y = -\frac{\partial \psi}{\partial x}. \tag{5}$$

No precise answer can be given for the question of the number N of vortices required for a simulation. For a given N and t, confidence in the method is based on comparison with experimental results. The selection of a reasonable N is based on the requirement of the simulation and is many a time limited by the computer time and memory requirements (Saffman and Baker 1979; Leonard 1980; Aref 1983).

The velocity field induced by the vorticity distribution may be calculated in two different ways. One way is the classical direct summation procedure in which one calculates the velocity field by directly summing the velocity fields of the individual point vortices. The other method is the vortex-in-cell method (Christiansen 1973) in which a grid is superimposed on an array of point vortices and the stream function at the grid points is computed by solving the Poisson equation for ψ (Eq. 4) after ω has been redistributed onto the mesh points. The velocity at X_n is then found by interpolation. The inversion of Poisson Eq. (4) can be performed very efficiently by using the fast Fourier transform method. The advantage of the vortex-in-cell method is that the amount of computation is linear in N, whereas in direct summation it is quadratic in N. Also, the viscous effects can easily be incorporated in the vortex-in-cell method.

The mixing layer in a laboratory is a spatially developing flow. However, simulation of spatially developing mixing layer is computationally expensive. Only Ashurst's (1979) simulation is that of a space-developing mixing layer. All other simulations consider time-developing mixing layers. In the time-developing perturbed shear layer simulation, Riley and Metcalfe (1980) solve directly the fully nonlinear low pass-filtered equations of motion using pseudo-spectral methods. This is called direct numerical simulation (sometimes referred to as large eddy simulation when sub-grid scale modeling is employed).

We consider a time-developing two-dimensional shear layer and use the vortex-in-cell method. The method is briefly discussed below. The superimposed grid has mesh spacings Δx and Δy, assumed uniform in x and y directions, respectively. To obtain the stream function ψ, a finite difference (central difference) approximation is made to Eq. (3).

$$(\psi_{i+1,j} - 2\psi_{i,j} + \psi_{i-1,j})/(\Delta x)^2 + (\psi_{i,j+1} - 2\psi_{i,j} + \psi_{i,j-1})/(\Delta y)^2 = -\omega_{i,j}. \tag{6}$$

The vorticity is represented at the points (X_n, Y_n) and so a redistribution scheme is introduced to assign values at the grid points and the Eq. (6) is solved. Figure 1 shows the redistribution scheme,

$$\omega(k) = A_k \Gamma_n / \Delta x \cdot \Delta y, \tag{7}$$

where the A's are the areas shown. Equation (6) is solved using a fast Poisson solver (Hockney 1970) using a fast Fourier transform method. To determine the velocity of a point vortex, the velocities at the nearest four grid points are calculated using a central difference formula,

$$U_{i,j} = (\psi_{i,j+1} - \psi_{i,j-1})/2\Delta y \tag{8}$$

$$V_{i,j} = -(\psi_{i+1,j} - \psi_{i-1,j})/2\Delta x, \tag{9}$$

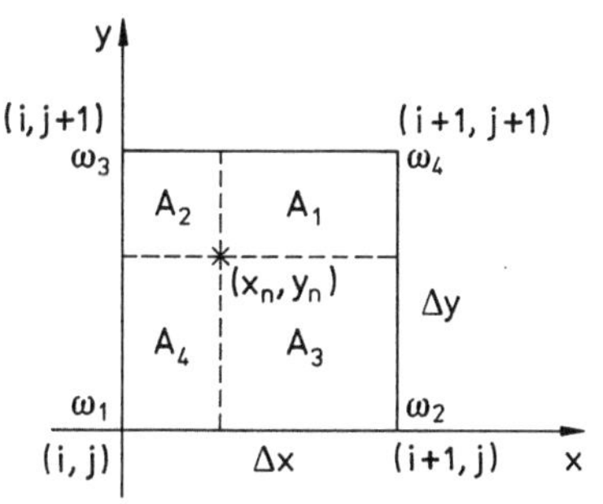

Fig. 1. Definition sketch of the vortex-in-cell redistribution procedure

and then interpolated,

$$U_n = \left(\sum_{k=1}^{4} U(k) A_k \right) \Big/ \Delta x \, \Delta y. \tag{10}$$

The vortices are moved forward in time using a stable second order accurate method,

$$X_n(t + \Delta t) = X_n(t) + 0.5(U_n(X_n) + U_n(\bar{X}_n)) \, \Delta t, \tag{11}$$

$$Y_n(t + \Delta t) = Y_n(t) + 0.5(V_n(Y_n) + V_n(\bar{Y}_n)) \, \Delta t, \tag{12}$$

where

$$\bar{X}_n = X_n(t) + U_n(X_n) \, \Delta t, \tag{13}$$

and

$$\bar{Y}_n = Y_n(t) + V_n(Y_n) \, \Delta t. \tag{14}$$

In this way the vorticity distribution at $(t + \Delta t)$ is computed, and the procedure is repeated to study the evolution of the flow in time.

In the present simulation, we superimpose a 128×128 grid on the flow field of the shear layer represented by 4096 point vortices with periodic boundary conditions in the x-direction. 4096 point vortices have been found to be adequate in representing the shear layer in the present simulation. The computation was carried out on the AS/9000N computer system. Preliminary calculations were done on a 64×64 grid with 2048 vortices. The adequacy of the time stepping accuracy has been verified on both the solution of a single shear layer with sinusoidal disturbance and the evolution of two shear layers with symmetric sinusoidal disturbance (simulating "puffing" instability of a plane jet). Although smoothing procedures (Moore 1981) could be employed to extend the useful evolution time, no such procedure was attempted in the present simulation.

The shear layer in the present simulation is given an initial thickness by representing it by four rows of point vortices. The distance between the rows was arrived at by examining the velocity profile and momentum thickness in a number of configurations. It was found that a transverse distance between the rows of 0.2 times the mesh size results in a velocity profile which compares well with the tanhyperbolic velocity profile (Fig. 2). The particular configuration chosen yields a momentum thickness which enables the study of excitations at various St_θ on the chosen grid.

The excitation frequency St_θ in the numerical simulation is expressed as $\mathrm{St}_\theta = 0.5\, \theta_i/\lambda$; θ_i is the initial momentum thickness of the undisturbed shear layer and λ is the wavelength of the perturbation. Thus, in the numerical simulation we choose the wavelength λ to obtain the desired frequency St_θ of excitation. This is equivalent to the experimental procedure adopted by Zaman and Hussain (1981) for a single-stream plane mixing layer, where they vary the frequency of excitation at constant values of the exit velocity U_e and the momentum thickness θ_e.

The excitation frequency St_θ in the numerical simulation is expressed as $\mathrm{St}_\theta = 0.5\, \theta_i/\lambda$; the chosen frequency to the initial distribution of point vortices representing the shear layer. An extremely small random disturbance is superimposed on the sinusoidal disturbance for enabling different realizations for ensemble average. The unforced case is obtained by computing the roll up and evolution of the large-scale vortex structures without any si-

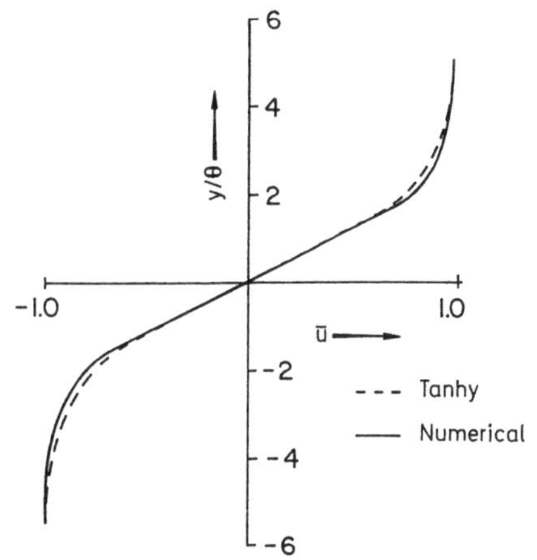

Fig. 2. Initial velocity profile

nusoidal initial perturbation. The number of frequencies investigated is limited by the grid size and the periodic boundary conditions. The amplitude of excitation has been varied over a large range. The effect of superimposing random fluctuations (representing the high frequency random fluctuations that exist in a real flow) on the regular sinusoidal disturbance has also been studied with different, levels σ of the random fluctuations.

Results and Discussion

For any initial disturbance specified, the temporal evolution of the shear layer is computed. At any instant of time the position of vortices and the velocities on the grid points are known in the simulation. The mean velocity $\bar{u}$ is obtained as x average,

$$\bar{u}(y,t) = \int u(x,y,t)\,dx. \tag{15}$$

The mean velocity $\bar{u}$ goes to a constant value $\Delta U/2$ well above the shear layer and $-\Delta U/2$ well below the shear layer. We can compute the momentum thickness as,

$$\theta(t) = \int (0.25 - (\bar{u}/\Delta U)^2)\,dy. \tag{16}$$

The momentum thickness and the width of the shear layer $B(\equiv y_{0.95} - y_{0.1})$ are computed at every time step; $y_{0.95}$ and $y_{0.10}$ are the transverse locations where $\bar{u}$ is 95% and 10% of the free-stream velocity difference ΔU. At regular intervals the profiles of turbulence intensities and Reynolds stress are obtained. The typical interval for the time averages were such that the momentum thickness increases by about 5% during the interval. The ratio u'_{ex}/u'_{ux} of the peak longitudinal fluctuation with excitation u'_{ex} to the peak longitudinal fluctuation without excitation u'_{ux} is defined as the suppression factor. Preliminary investigations showed no noticeable change in the evolutions of the shear layer for amplitudes of excitation A/λ between 0.003 and 0.0125. Figure 3 shows the width of the shear layer as a function of time for two Strouhal numbers, 0 and 0.017 for an excitation amplitude of 0.0125. We observe near linear variation of the width with time, with little difference between the forced and unforced cases. In this amplitude range of excitation, no recognizable suppression occurs. However, in the range of amplitude of excitation 0.025 to 0.2, significant changes in flow development occur, and we observe turbulence suppression as discussed below. A major

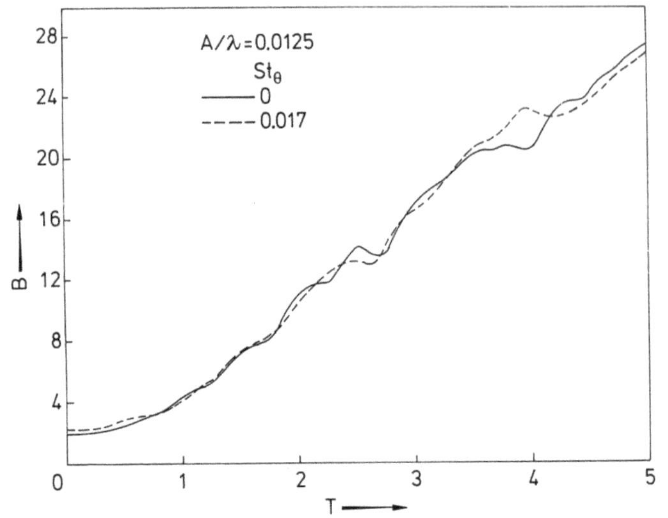

Fig. 3. Low amplitude excitation

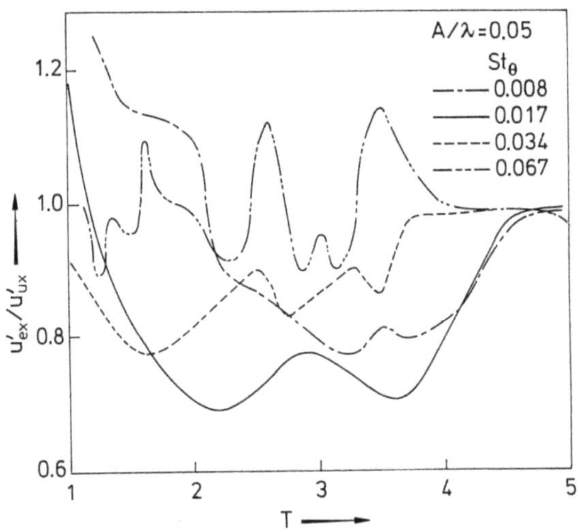

Fig. 4. Effect of St_θ

part of the results of suppression reported below has been obtained for an excitation amplitude of 0.05. In the experiments of Zaman and Hussain, the excitation amplitude (u_e'/U_e) was in the range of 0.3 to 1%. The higher forcing levels required in the simulation compared to those used in the experiments may be an inherent constraint of the simulation. The inadequacy may partly stem from the grid size and number of vortices employed. Acton (1980) found similar requirements of high forcing levels in her numerical simulation compared to experimental levels. The suppression factor u_{ex}'/u_{ux}' as a function of time is shown in Fig. 4 for four Strouhal numbers for an excitation amplitude of 0.05. It is seen that the maximum suppression occurs for a Strouhal number of 0.017, the theoretical "maximally" unstable frequency. The suppression for $St_\theta = 0.034$ and 0.008 is less pronounced. This is in qualitative agreement with the experimentally observed results. For a Strouhal number (0.067) much higher than the maximally unstable frequency, we observe turbulence augmentation rather than suppression. The maximum suppression observed at $St_\theta = 0.017$ is about 35%. It may be instructive to note that the maximum suppression observed by Zaman and

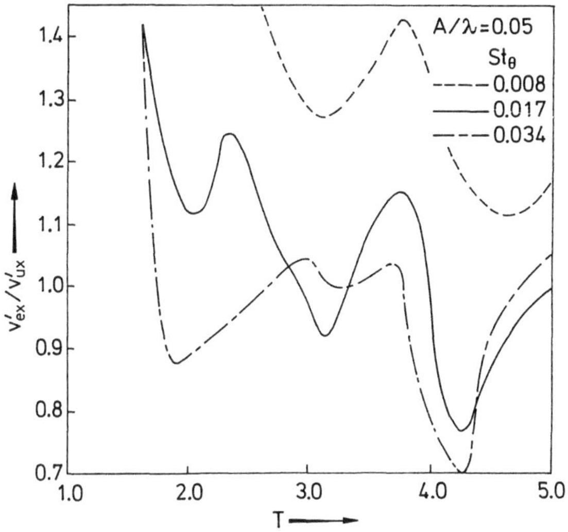

Fig. 5. Suppression in transverse component

Hussain was about 42% at $U_e = 20$ m/sec and about 58% at $U_e = 10$ m/sec for a plane mixing layer. The measurements were made at a constant transverse distance $y(= 1.27$ cm from the lip) whereas the suppression factor reported here is the ratio of peak longitudinal fluctuations which occur at a different y. Another aspect of the numerical simulation is that the results show oscillations which are not observed in acoustic excitation measurements. Crighton and Gaster's (1976) analyses showed that small differences in the mean velocity profile could result in significant changes in instability characteristics. The observed oscillations in the numerical simulation could be due to the difference in the simulation initial velocity profile (Fig. 2). Also, a larger simulation and an ensemble average of large samples will reduce the oscillations.

Figure 5 shows the transverse peak velocity fluctuation ratio v'_{ex}/v'_{ux} against time. We notice that the observed Strouhal number preference ($\mathrm{St}_\theta = 0.017$) for suppression in longitudinal fluctuations does not exist for transverse velocity fluctuations. Zaman and Hussain obtained the distribution of the longitudinal and transverse components of the turbulence intensities u' and v' over the entire width of the shear layer at $\mathrm{St}_\theta = 0.017$ with and without excitation. They found that for the excitation case both longitudinal and transverse components of turbulence intensities are small, over the entire cross-section compared to the corresponding values in the unexcited case. But it is not clear if the Strouhal number preference for the maximum suppression will be true for the transverse component of the turbulence intensity too. The experiments show that $\overline{uv}/U_e^2$ also suffer suppression with excitation at $\mathrm{St}_\theta = 0.017$. The ratio of $\overline{uv_{ex}}/\overline{uv_{ux}}$ obtained in the simulation is shown in Fig. 6. We see that the Reynolds stress shows large excursions due to the small number of rolled up vortices in the simulation. That is, the excursions are large statistical fluctuations in a small sample, and this can be demonstrated numerically (Aref and Siggia 1980). Because of cancellations due to fluctuations in the sign of uv, we need a larger simulation length of the shear layer and a larger ensemble size for this quantity than for the longitudinal and transverse velocity fluctuations which are positive definitive quantities. To obtain a smooth average one needs a large system of vortices and an ensemble average of large samples. In the present case because of the small number of rolled-up vortices considered an ensemble average of 3 realizations was found to result in only a marginal improvement of the results. The evolution of the momentum thickness [Eq. (16)] and width of the shear layer are shown

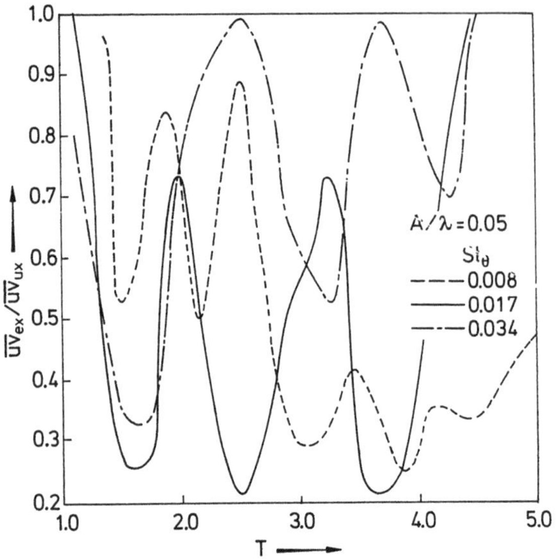

Fig. 6. Suppression in Reynolds stress

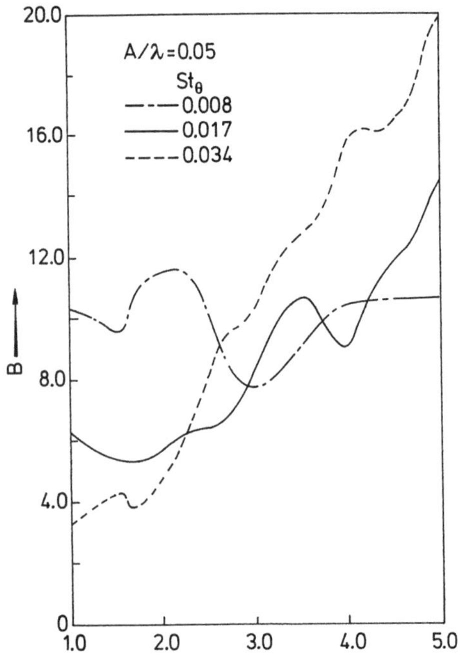

Fig. 7. Variation of halfwidth with St_θ

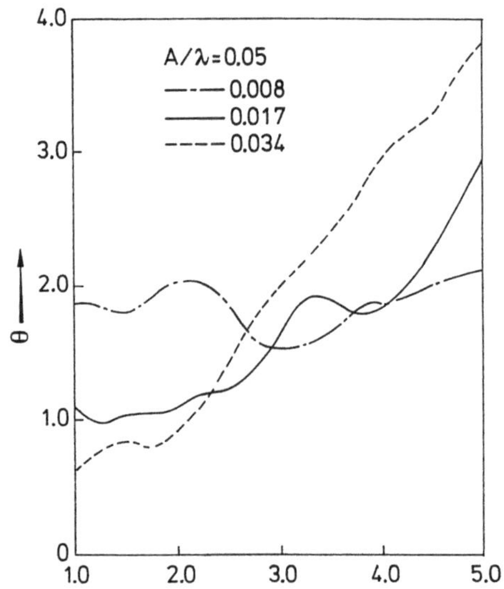

Fig. 8. Variation of θ with St_θ

in Figs. 7 and 8 respectively, for the Strouhal numbers, 0.008, 0.017 and 0.034. We see that the growth rates for Strouhal numbers 0.017 and 0.034 are not very much different.

We need to digress a little here and discuss the success of the numerical methods in predicting the velocity fluctuations. Ashurst's computations incorporated viscosity twice, using both the random walk and the viscous increase of size (aging of the discrete vortices) in an effort to match the simulation fluctuation velocity components to the experimental ones. Even after inclusion of viscous effects he could match the longitudinal and transverse fluctuation intensities of the simulation with those of experiments only for low Reynolds numbers. For high Reynolds numbers he could match only the longitudinal intensities. He conjectured that the pressure strain correlations ($\overline{p\,\partial u'/\partial x}$), which are responsible for maintaining the transverse velocity fluctuations, may not be correctly simulated. Aref and Siggia found that the peak velocity fluctuations in their simulation (256 × 256 grid with 4096 vortices) were high compared to the experimental values. Also, the transverse velocity fluctuations were larger than the longitudinal ones though the reverse is true at high Reynolds numbers. Acton (1980) also found that the values of fluctuation components were much higher than in the experiments, which is true with the present simulation too. Since we are considering only the ratios of the fluctuation intensities, it is believed that the results of suppression have been reasonably correctly predicted in the simulations. No attempt was made to include viscous effects in the present simulation.

Next, it is of interest to see if the Strouhal number dependence of turbulence suppression is true for any amplitude of excitation. We obtained the suppression factor u'_{ex}/u'_{ux} for three Strouhal numbers for an excitation amplitude of 0.2. The results are shown in Fig. 9. We see that at this amplitude of excitation the strong Strouhal number preference of the maximum suppression is lost. An experimental study of suppression at high amplitude ($u'_f/U_e = 4.5\%$) carried out by us in an axisymmetric mixing layer clearly showes that the Strouhal number preference is lost at high amplitudes. The maximum suppression was found to occur at $St_\theta = 0.022$ (maximum frequency studied in the experiment) and not at 0.017, the maximally unstable frequency. At high amplitudes of excitation nonlinear effects dominate the flow evolution. A high initial amplitude together with the fast transition (inferred from the longitudinal velocity spectra at $U/U_e = 0.7$ along x) for excitation frequencies larger than the maximally unstable frequency produces less energetic smaller structures earlier in x. The saturation amplitude at a high frequency is also less. The combined effect of these is to produce a higher suppression.

The dependence of the extent of suppression on the amplitude of excitation is shown in Fig. 10 for $St_\theta = 0.017$. The maximum suppression is found to increase with increase in the amplitude of excitation. However, one would expect the suppression to saturate at some maximum amplitude of excitation. This does not occur in the present simulation since the disturbance is in the position of the point vortices. For amplitudes of excitation greater than

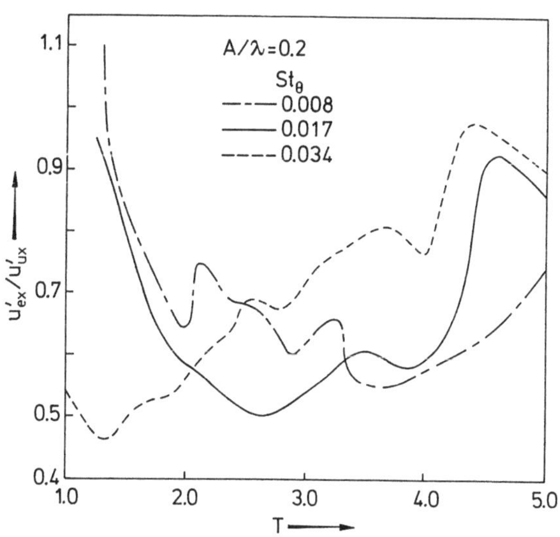

Fig. 9. High amplitude excitation ($A/\lambda = 0.2$)

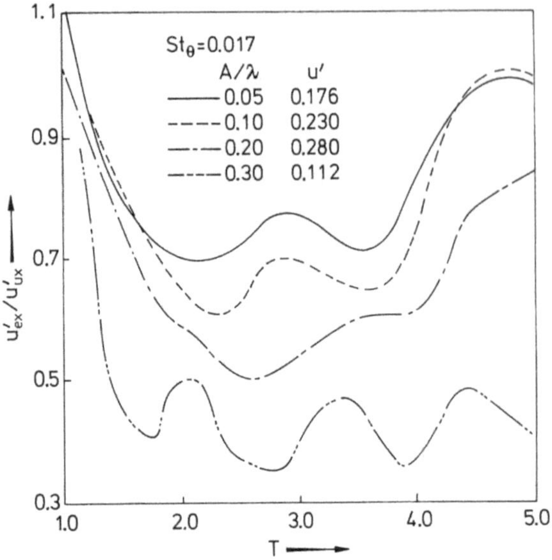

Fig. 10. Effect of amplitude of excitation

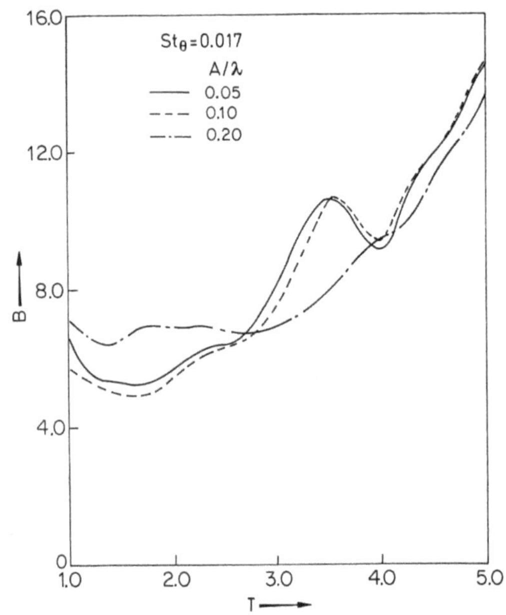

Fig. 11. Variation of halfwidth with amplitude

0.2, the shear layer becomes so thick that it does not evolve and grow. This is shown in Figs. 11 and 12, which show the variations of B and θ with time at different amplitudes of excitation. We see that as the excitation amplitude increases, the growth rate decreases. Such behavior at high amplitudes of excitation in point vortex simulation has been reported by Acton (1976). The initial velocity disturbance introduced corresponding to each amplitude of excitation is shown in the figure. We see that for amplitudes of excitation greater than 0.2, the disturbance velocity decreases, indicating the non-physical nature of the shear layer representation. Thus the simulation is meaningful only up to an amplitude of excitation of 0.2.

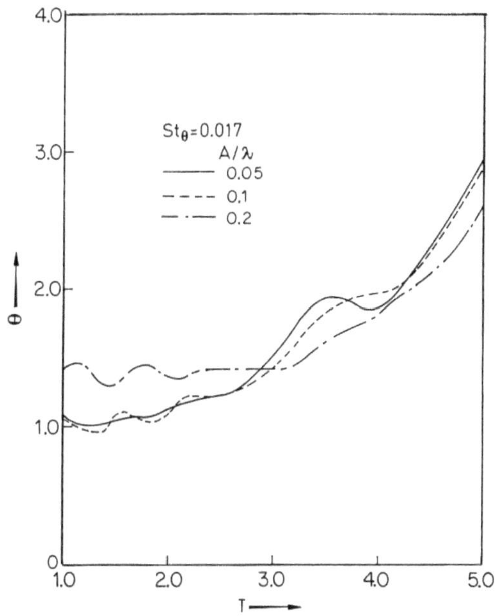

Fig. 12. Variation of θ with amplitude

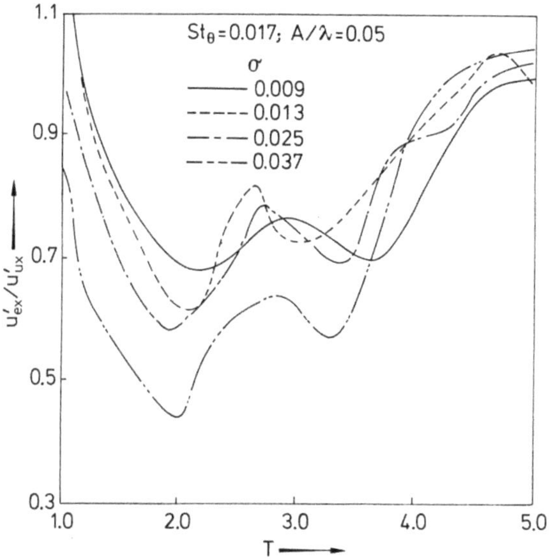

Fig. 13. Effect of random fluctuation

The disturbance introduced into intrinsically unstable shear layer, appears as high frequency random fluctuations superimposed on low ones, depending on the frequency of the disturbance (Ho and Nossier 1981, Oster and Wygnanski 1982). Also in a real flow situation there always exist random (incoherent) velocity fluctuations, whose effect on the flow depends on the level of these fluctuations. To incorporate these random fluctuations in the simulation, we superimposed random fluctuations (of positions of vortices) over the regular sinusoidal fluctuations of the specified frequency. The level (σ, standard deviation) of the random fluctuations was varied and its effect on the suppression of longitudinal turbulence intensity studied. The results are shown in Fig. 13. We see that the maximum suppression

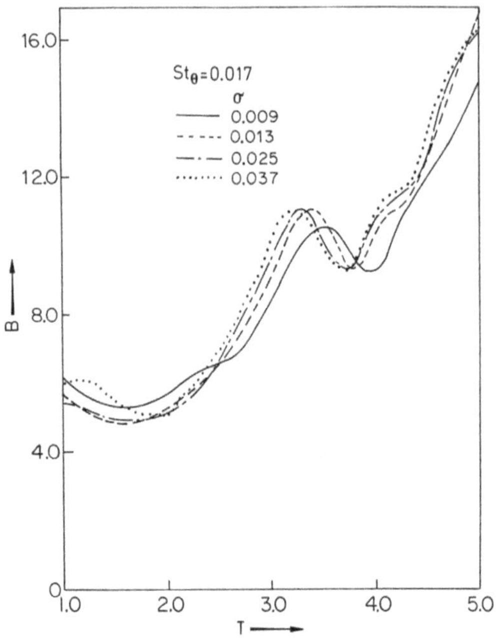

Fig. 14. Variation of halfwidth with random fluctuation

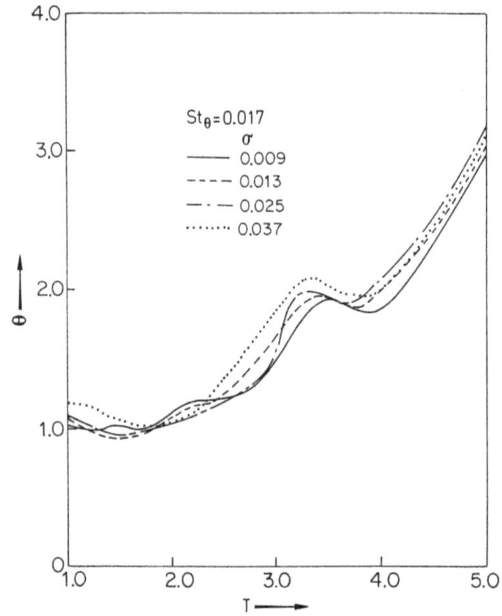

Fig. 15. Variation of θ with random fluctuation

increases with increase in the level of random fluctuations. As with the amplitude of excitation, the maximum suppression increases with increase in the level of random fluctuation without attaining a saturation level. For values of σ beyond 0.037, the general shape of the curve (u'_{ex}/u'_{ux} against time) changes, and the suppression increases drastically which is not physically meaningful. The reason for this is the excessive initial thickening of the shear layer with such high levels of random fluctuation as shown in Figs. 14 and 15.

Conclusions

The present numerical simulation results may be summarized as follows:

1. No noticeable suppression occurs in the range of amplitude of excitation 0.003–0.0125. Turbulence suppression occurs in the range of amplitude of excitation 0.025–0.2.
2. At an amplitude of excitation of 0.05, the simulation shows that the maximum suppression occurs at the maximally unstable frequency of $St_\theta = 0.017$, consistant with the experimental results.
3. At high amplitudes of excitation ($A/\lambda = 0.2$) this Strouhal number preference is lost. (This has been verified experimentally by us recently.)
4. The level of random fluctuations superimposed on the sinusoidal fluctuations has significant effect on the extent of suppression.

The research is supported by the National Science Foundation under Grant MEA 81-11676.

References

Acton, E. (1976): J. Fluid Mech. **76**, 561
Acton, E. (1980): J. Fluid Mech. **98**, 1
Aref, H. (1983): Ann. Rev. Fluid Mech. **15**, 345
Aref, H., Siggia, E. D. (1980): J. Fluid Mech. **100**, 705
Ashurst, W. T.: In: *Turbulent Shear Flows 1*, ed. by F. Durst, B. E. Launder, F. W. Schmidt, J. H. Whitelaw (Springer, Berlin, Heidelberg, New York 1979) p. 402
Batchelor, G. K.: *An Introduction to Fluid Mechanics* (Cambridge University Press, Cambridge 1967)
Christiansen, J. P. (1973): J. Comput. Phys. **13**, 423
Crighton, D. G., Gaster, M. (1976): J. Fluid Mech. **77**, 397
Delcourt, B. A. G., Brown, G. L.: In: *Proc. of the 2nd Int. Symp. on Turbulent Shear Flows* (Imperial College, London, July 2–4, 1979) p. 14.35
Ho, C. M., Nossier, N. S. (1981): J. Fluid Mech. **105**, 119
Hockney, R. W.: In: *Methods in Computational Physics*, Vol. 9, ed. by B. Alder, S. Fernberg, M. Rotenberg (Academic Press, New York 1970) p. 135
Leonard, A. (1980): J. Comput. Phys. **37**, 289
Michalke, A. (1965): J. Fluid Mech. **22**, 351
Moore, D. W. (1981): SIAM J. Sci. Stat. Comput. **2**, 65
Oster, D., Wygnanski, I. (1982): J. Fluid Mech. **123**, 91
Riley, J. J., Metcalfe, R. W. (1980): AIAA Paper No. 80-0274
Saffman, P. G., Baker, G. R. (1979): Ann. Rev. Fluid Mech. **11**, 95
Zaman, K. M. B. Q., Hussain, A. K. M. F. (1981): J. Fluid Mech. **103**, 133

Structural Features of the Near Region of an Asymmetric Turbulent Wake

M. Shokr*, J. F. Keffer and J. G. Kawall

Department of Mechanical Engineering, University of Toronto, Toronto, Canada

Abstract

Statistical properties of the turbulent/non-turbulent interfaces in the near region of an asymmetric wake generated by a pair of closely-spaced, unequal-diameter cylinders have been determined by means of digital image analysis of ciné flow visualization pictures of the flow. The results, comprising moments, burst rates, correlations and auto-spectra, indicate that the lower interface of the flow, which is associated with the large cylinder, has markedly steeper slopes, on the average, than the upper interface and its motion is considerably less random than that of the latter. Moreover, they establish that a "Kármán vortex street", centred in the lower part of the wake and with a characteristic frequency corresponding to a Strouhal number significantly less than that pertaining to the symmetric cylinder generated wake, is produced.

Nomenclature

D_1	lower cylinder diameter, mm
D_2	upper cylinder diameter, mm
D	equivalent diameter equal to $(D_1 + D_2)/2 + L$, mm
E	auto-spectrum, s
F	flatness factor
$\hat{f}$	burst rate, Hz
$\hat{f}_0$	burst rate at the mean interface position, Hz
L	spacing between cylinder centres
N	number of ciné picture frames
n	frequency, Hz
n_s	vortex shedding frequency, Hz
S	skewness factor
t	time, s
U_1	free-stream velocity, ms^{-1}
x	streamwise coordinate
Y	instantaneous lateral interface position, mm
$\bar{Y}$	mean value of Y, mm
y	lateral coordinate, or instantaneous lateral interface position relative to $\bar{Y}$, mm
y'	root-mean-square value of Y, mm
z	spanwise coordinate
α	reference variable
β	reference variable
Δt	time interval between ciné pictures frames, s
ϱ	auto- or cross-correlation function
τ	time lag, s
Φ	instantaneous interface slope, equivalent to $\partial Y/\partial x$
$\bar{\Phi}$	mean value of Φ, deg
ϕ	instantaneous interface slope relative to $\bar{\Phi}$
ϕ'	root-mean-square of Φ

Subscripts

U	upper
L	lower
y	position
ϕ	slope

Introduction

Recent experimental results (e.g., see Keffer, Kawall, Giralt and Béguier, 1979) have shown that the turbulent transport mechanism associated with certain "asymmetric" turbulent shear flows is fundamentally different from that associated with classical "symmetric" flows,

* Present address: Atmospheric Environment Service, Toronto, Canada

viz., wakes and jets with symmetrical mean profiles, and mixing layers and boundary layers with monotonically-varying mean profiles. Such asymmetric flows have been found to contain well-defined localized regions within which the net production of turbulence kinetic energy (or its thermal equivalent) is negative. One such flow is the turbulent wake generated by a pair of closely-spaced, unequal diameter circular cylinders, which was first studied experimentally by Palmer and Keffer (1972).

In a review paper on negative production, Béguier, Giralt, Fulachier and Keffer (1977) suggested that the existence of the negative production zone in the Palmer-Keffer wake was linked to the presence of "Kármán vortex street" (KVS) activity, i.e., the quasi-periodic occurrence of relatively highly-organized motion, or coherent structures, within the near region of this flow. We note that Diaz, Gavaldà, Kawall and Giralt (1982) have found zones produced by a rotating circular cylinder, as did Gavaldà, Diaz, Giralt, Kawall and Keffer (1983) in the more complicated situation of a wake generated by a pair of closely-spaced, equal diameter circular cylinders, one of which is rotating. In both instances, the results indicate that the occurrence of these zones is related to KVS activity.

In an effort to gain further insight into the structure of the Palmer-Keffer asymmetric wake, Shokr and Keffer (1982) analysed ciné flow visualization pictures of the near region of this flow, using a digital image processing technique. Specifically, they measured the lateral positions of the upper and lower turbulent/non-turbulent interfaces of the flow as functions of time, for several streamwise locations, and hence, determined various interface statistics. In the present work we have carried out an extended analysis of the aforementioned ciné pictures in order to obtain additional quantitative information on the interfaces. In particular, we have determined a number of statistical properties of the interface slopes. It should be pointed out that these properties cannot be readily obtained by means of either hot-wire anemometry or laser-Doppler anemometry.

Experimental Details

The flow visualization pictures, employed herein were obtained from the Institut de Mécanique Statistique de la Turbulence (IMST), Marseille. The flow, a definition sketch of which is shown in Fig. 1, was generated by means of two cylinders of diameters $D_1 = 20$ mm and $D_2 = 12$ mm mounted horizontally in the IMST flow visualization water tunnel. The axes of the cylinders were separated in the lateral direction by a distance, L, of 23 mm so that the quantity $(D_1 + D_2)/2 + L$, which is defined as the equivalent diameter, D, was 39 mm. This particular cylinder configuration resulted in a relatively strong "source interaction", as in the case of the Palmer-Keffer experiments. The free-stream velocity was 0.17 ms^{-1}, which produced a Reynolds number based on D of 7000. Approximately 1000 frames and a time

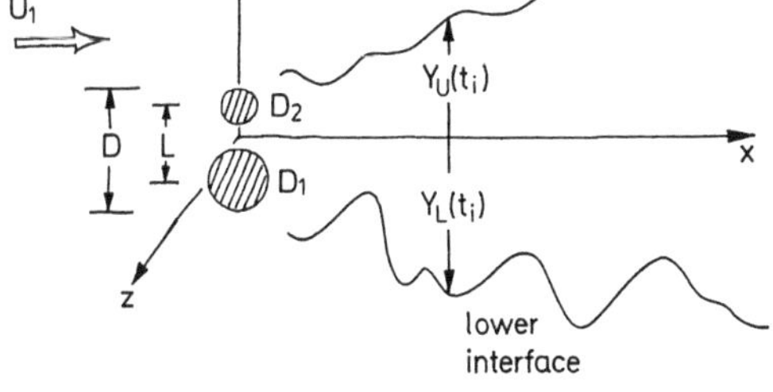

Fig. 1. Definition sketch

interval between successive frames of 1/12 s were used in the analysis, and results pertaining to five selected downstream locations (i.e., x/D values) were computed.

The pictorial information was digitized, and the resulting digital data were processed by means of a Bausch and Lomb Omnicon digital image analysis system, in conjunction with an IBM 3033 computer, at the University of Toronto. The turbulent and non-turbulent regions within the pictures were identified, and the upper and lower interface positions relative to the centreline of the flow, Y_U and Y_L, and the upper and lower interface slopes, $\Phi_U(\cong \partial Y_U/\partial x)$ and $\Phi_L(\cong \partial Y_L/\partial x)$, were then measured as functions of time for the selected x/D values. (The technique used to determine $Y_{U,L}$ and $\Phi_{U,L}$ is described by Shokr and Keffer, 1982.)

Interface Statistics

The statistical properties of $Y_{U,L}$ and $\Phi_{U,L}$ that were examined in the present work are as follows.

(1) The first four moments of $\Phi_{U,L}$, which are given by

$$\overline{\Phi^m_{U,L}} = \frac{1}{N}\sum_{i=1}^{N} \Phi^m_{U,L}(t_i), \qquad m = 1, 2, 3, 4.$$

In this expression, N represents the number of frames used; $t_i = (i-1)\,\Delta t$, where Δt is the time interval between frames; $m = 1$ yields the upper and lower mean interface slopes, $\bar{\Phi}_U$ and $\bar{\Phi}_L$; $m = 2$ yields the root-mean-square (RMS) values,

$$\phi'_{U,L} = \sqrt{\overline{\phi^2_{U,L}}},$$

where $\phi = \Phi - \bar{\Phi}$; $m = 3$ yields the skewness factors,

$$S_{U,L} = (\overline{\phi^3}/\phi'^3)_{U,L};$$

and $m = 4$ yields the flatness factors,

$$F_{U,L} = (\overline{\phi^4}/\phi'^4)_{U,L}.$$

(2) The burst rate, $\hat{f}$, which represents the average frequency of occurrence of turbulent bursts (or zones) at a given point in the flow, and is equal to one-half the average frequency at which the interface crosses that point. We note that if Y and Φ are uncorrelated Gaussian random variables then $\hat{f}$ will be given by

$$\hat{f} = \hat{f}_0 \exp(-\, y^2/2\, y'^2),$$

where $\hat{f}_0$ is the burst rate at the mean interface position, $\bar{Y}$.

(3) The auto-correlation function, which is given by

$$\varrho(\tau_j) = \left[\frac{1}{(N-j)}\sum_{i=1}^{N-j} \alpha(t_i)\,\alpha(t_{i+j})\right]\Big/\alpha'^2,$$

where $\tau_j = j\,\Delta t$, $j = 0, 1, 2, \ldots J$, $J \ll N$, α denotes y or ϕ, and α' is the RMS value of α.

(4) The auto-spectrum, $E(n)$, which is defined as

$$E(n) = \int_{-\infty}^{\infty} \varrho(\tau) \exp(-i 2\pi n \tau)\, d\tau.$$

(5) The cross-correlation function, which is given by

$$\varrho_{\alpha\beta}(\tau_j) = \begin{cases} \left[\dfrac{1}{(N-j)} \sum\limits_{i=1}^{N-j} \alpha(t_i)\, \beta(t_{i+j})\right] \Big/ \alpha'\beta', & \text{for } \tau_j \geqq 0, \\ \left[\dfrac{1}{(N-j)} \sum\limits_{i=1}^{N-j} \beta(t_i)\, \alpha(t_{i+j})\right] \Big/ \alpha'\beta', & \text{for } \tau_j < 0, \end{cases}$$

where α and β denote y_U and ϕ_U, y_L and ϕ_L, or y_L and y_U.

Results and Discussion

The streamwise variations of the mean values, the rms values, the skewness factors and the flatness factors of the upper and lower interface slopes are presented in Figs. 2, 3, 4 and 5, where it is seen that these quantities all remain essentially constant as x/D increases. One can infer from the results for the mean values (Fig. 2) that, on the average, (a) both interfaces are fairly flat, $|\bar{\Phi}_U|$ and $|\bar{\Phi}_L|$ being less than 13°, and (b) the slopes of the lower interface, associated with the large cylinder (see Fig. 1), are significantly steeper than those of the upper interface, $|\bar{\Phi}_L|$ being roughly twice as large as $|\bar{\Phi}|_U$. These inferences are consistent both with visual observations of the ciné film, which reveal that the intermittent bulges in the lower part of the wake are larger than those in the upper part, and with the results for the mean interface positions, $\bar{Y}_U$ and $\bar{Y}_L$ (see Shokr and Keffer, 1982), which show that $\bar{Y}_L$ spreads more rapidly than does $\bar{Y}_U$ – so that the rate of entrainment of potential fluid is greater across the *steeper* lower interface.

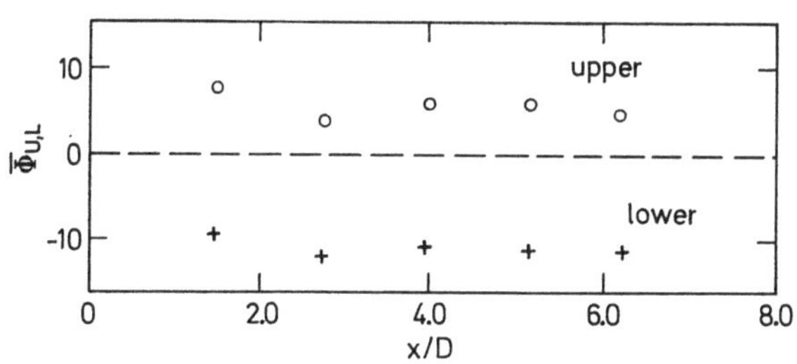

Fig. 2. Mean values of interface slopes

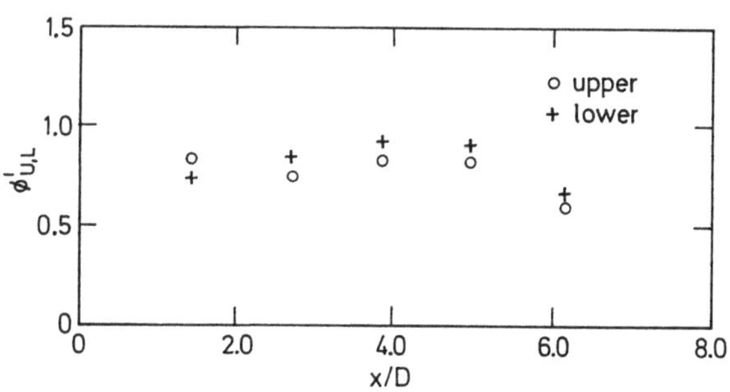

Fig. 3. rms values of interface slopes

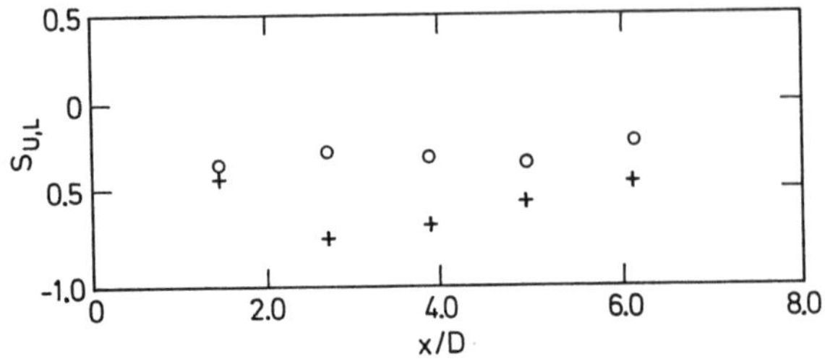

Fig. 4. Skewness factors of interface slopes

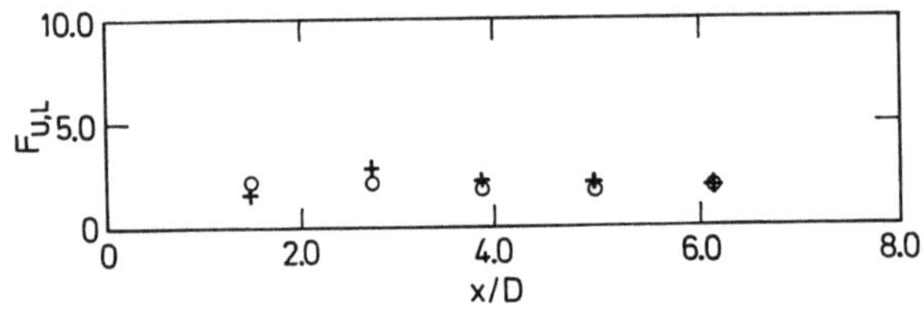

Fig. 5. Flatness factors of interface slopes

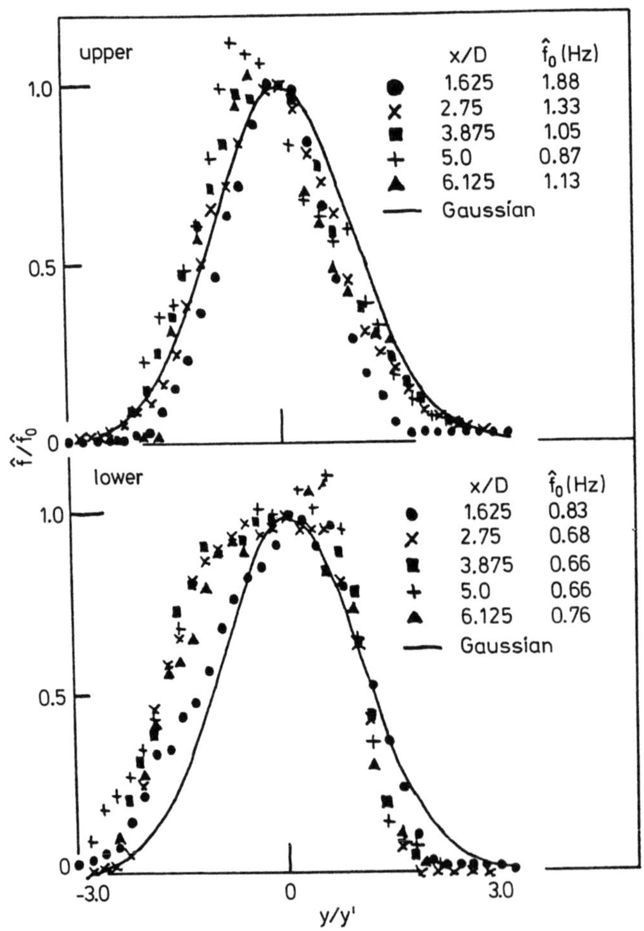

Fig. 6. Burst rate distributions

Figure 3 demonstrates that the rms values of the upper and lower interface slopes, ϕ'_U and ϕ'_L, are more or less the same. The magnitude of these quantities is approximately 45°, which, together with the fact that both $|\bar{\Phi}_U|$ and $|\bar{\Phi}_L|$ are less that 13°, leads to the conclusion that interface folding does not occur to any significant extend (or equivalently, that the interface positions are single-valued functions). This is, in fact, the case, as evidenced by visual observations of the ciné film. The skewness factors of both the upper slopes and the lower slopes, S_U and S_L (Fig. 4), deviate significantly from zero, and $|S_L|$ is consistently greater than $|S_U|$. Hence, it follows that the interface bulges are asymmetric in shape, the more so in the case of the lower interface. The upper and lower flatness factors, F_U and F_L (Fig. 5), are essentially equal, and their magnitude is roughly 2.0, so that these results reinforce the above conclusion that interface folding is negligible in the present situation.

Normalized upper and lower burst rate distributions pertaining to the various x/D locations examined here are depicted in Fig. 6. The scatter associated with these results is seen to be large. Nonetheless, because the extent of the scatter depends upon N (the number of frames used), decreasing as N increases, we can conclude that the burst rate profiles for the present flow are self-similar. Furthermore, it appears that the upper profiles tend to follow the Gaussian curve; the lower burst rate profiles, however, deviate markedly from this curve. It is interesting to note that the intermittency factor profiles (see Shokr and Keffer, 1982) display a similar trait: the upper profiles conform to the Gaussian distribution function, but the lower profiles show little agreement with this function. The implication here is that in the near region of the wake, the process governing the motion of the upper interface (i.e., the fluctuations of its lateral position) is predominantly Gaussian in nature, whereas that governing the motion of the lower interface is not.

In Figs. 7 and 8, we present upper and lower interface-position and interface-slope auto-correlation functions pertaining to $x/D = 3.875$, and in Figs. 9 and 10 the corresponding auto-spectra. As can be seen, the auto-correlation functions display distinct regular oscillations, establishing the presence of quasi-periodic coherent structures within the flow. The oscillations do not occur, however, until τ exceeds certain limiting values, and they decay (albeit slowly) with increasing τ. It is clear that a "Kármán vortex street" (KVS), somewhat similar to the KVS within the symmetric cylinder-generated wake, exists within the asymmetric wake; moreover, the process that gives rise to this KVS is not strictly periodic but rather is randomly modulated, as in the case of the symmetric wake (see Budny, Kawall and Keffer, 1979). Notice that the lower auto-correlations exhibit a more pronounced oscillatory behaviour than do the upper auto-correlations, suggesting that the KVS is centred in the lower part of the flow. The auto-spectra (Figs. 9 and 10) serve to show that this is, in fact, so – the dominant peaks in these spectra, which correspond to the oscillations in the auto-correlations, being both larger and noticeably sharper in the case of the lower spectra than in the case of the upper spectra. Evidently, the frequency at which the aforementioned peaks are centred is the nominal frequency at which vortices are shed from the two cylinders (n_s). This frequency yields a Strouhal number ($n_s D/U_1$) of 0.14, which is significantly less than that associated with the symmetric wake (viz., 0.21).

It is interesting to note that whereas the upper spectra display only a single peak centred at n_s – the "fundamental", the lower spectra display several "harmonics" as well as the "fundamental". This trait is also exhibited by the auto-spectra of hot-wire signals from the upper and lower parts of the asymmetrical wake, that were measured by Palmer and Keffer (1972). It is interesting to note as well that a comparison of the interface-position auto-spectra and the interface-slope auto-spectra (Figs. 9 and 10) suggests that the interface slope, $\phi(t)$, is closely related to the time derivative of the interface position, $\partial y(t)/\partial t$, in as much as the auto-spectra of $\phi(t)$ are more sensitive to the KVS activity than are the auto-spectra of $y(t)$.

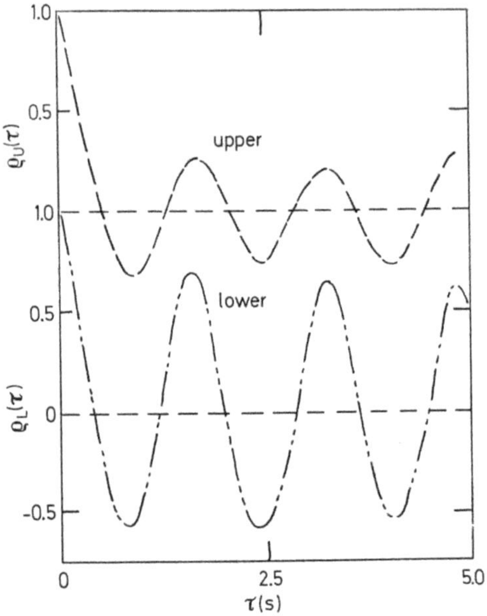

Fig. 7. Auto-correlations of interface positions $x/D = 3.875$

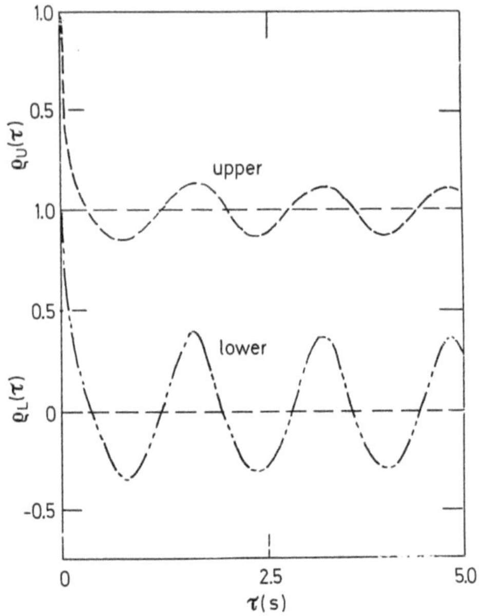

Fig. 8. Auto-correlations of interface slopes, $x/D = 3.875$

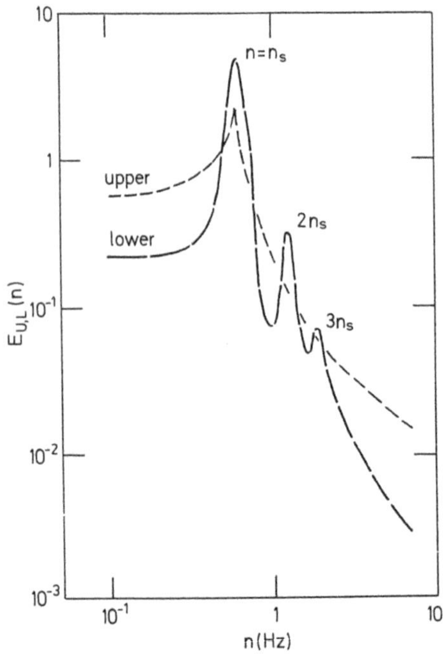

Fig. 9. Auto-spectra of interface positions, $x/D = 3.875$

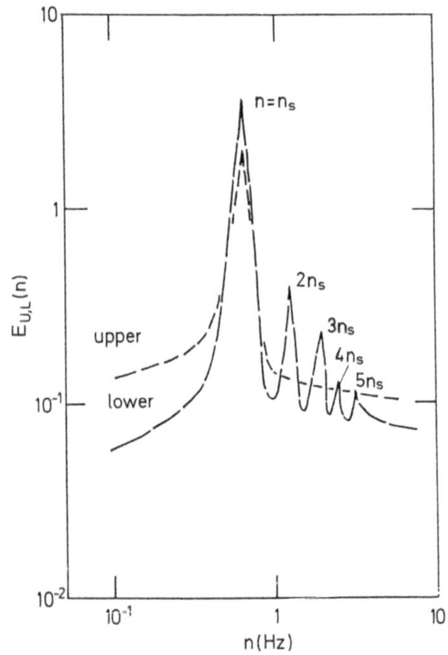

Fig. 10. Auto-spectra of interface slopes, $x/D = 3.875$

On the basis of the above results, we can regard the motion of the upper and lower interfaces in the near region of the flow as comprising a large-scale quasi-periodic component (caused by KVS activity) and a small-scale random component, with the former dominating the motion of the lower interface, once the KVS is well-established, and the latter dominating the motion of the upper interface. This picture is consistent both with the fact that the values of $\hat{f}_0$ pertaining to the upper interface are everywhere larger than those pertaining to the lower interface (Fig. 6), and with visual observations of the ciné film which reveal that the upper interface contains numerous small random convolutions, while the lower interface contain relatively few such convolutions.

Figure 11 depicts the upper and lower cross-correlation functions of the interface position and the interface slope for $x/D = 3.875$. It is clearly evident from this figure that the random components of $y(t)$ and $\phi(t)$ are uncorrelated, which implies, in the light of the foregoing results, that these components are uncorrelated Gaussian random variables, and hence, are statistically independent. (The fact that the burst rate and intermittency factor profiles pertaining to the lower interface deviate from the Gaussian models can be attributed to the relatively strong effect of the quasi-periodic components of $y_L(t)$ and $\phi_L(t)$.) We note that in the case of the upper interface, there is a phase difference between $y(t)$ and $\phi(t)$ of about 90°, the cross-correlation function being approximately zero at $\tau = 0$, and in the case of the less random lower interface, this phase difference is somewhat greater than 90°, the value of the cross-correlation function lying between zero and its minimum value. These cross-correlation results, therefore, lend support to the view that there is a strong relationship between ϕ and $\partial y/\partial t$. Indeed, with respect to the upper interface, this relationship may be taken as: $\partial y/\partial t = U_c\,\phi(t)$, where U_c is a characteristic (local) bulge convection velocity. It may be remarked that if this relationship holds, then for sufficiently small Δt, U_c will be given by, $(\overline{\Delta y^2}/\overline{\phi \Delta y}/\Delta t)$, where

$$\Delta y = y(t + \Delta t) - y(t).$$

In Fig. 12, we present the cross-correlation functions of the upper and lower interface positions for selected x/D locations in the near region of the flow. These results establish, not surprisingly, that as in the case of the symmetric wake (see Budny et al., 1979), only the quasi-periodic components of the interface positions on opposite sides of the centreline of

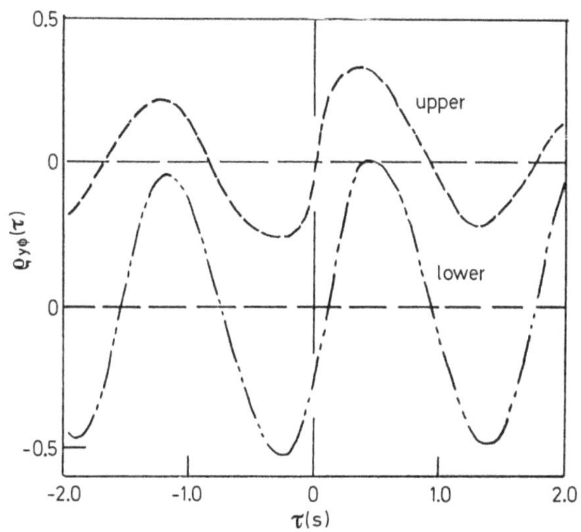

Fig. 11. Cross-correlations of interface positions and slopes, $x/D = 3.875$

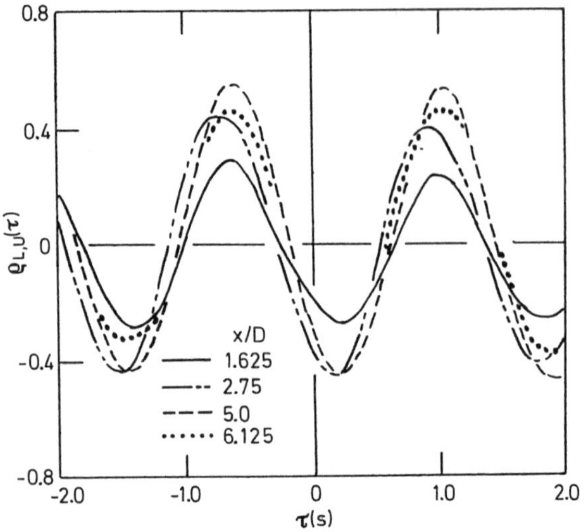

Fig. 12. Cross-correlations of interface positions

the flow are correlated. Furthermore, they demonstrate that these components are not 180° out-of-phase as they are for the symmetric wake, since the minimum values of the correlations do not occur at $\tau = 0$. This is a consequence of the KVS being centred in the lower part of the wake, and implies that the coherent structures comprising the KVS are *not* arranged as in the classic KVS configuration. Finally, it should be noted that the peak values of the cross-correlation curves vary with x/D, increasing initially for x/D less than about 5, then decreasing as x/D increases beyond 5. From this, in conjunction with the fact that $\hat{f}_0$ pertaining to both the upper interface and the lower interface decreases initially then increases as x/D increaees from 1.625 to 6.125 (see Fig. 6), we conclude that the KVS is well-established by x/D equal to about 4, so that its formation takes place within a streamwise distance of about $3\,D$ from the cylinders, and that it starts to decay for x/D in excess of about 5.

Summary and Concluding Remarks

The present findings indicate that large vortices shed from the lower (large) cylinder coalescence with or entrain small vortices shed from the upper cylinder and thus form coherent structures that are significantly larger than the large vortices. This "vortex-merging" process is rapid (taking place within about $3\,D$), more or less stable and quasi-periodic, and results in the occurrence of a "Kármán vortex street" (KVS) with a characteristic frequency corresponding to a Strouhal number considerably less than 0.21 and centred in the lower part of the wake. In consequence, the lower interface is, on the average, considerably steeper than the upper interface and its motion is significantly less random than that of the latter. It is probable that such a "skewed" KVS is indirectly responsible for the existence of the zones of negative production detected in the asymmetrical wake by Palmer and Keffer (1972).

Acknowledgement. This research was supported by the Natural Sciences and Engineering Research Council of Canada through Grant A-2746 and the Atmospheric Environment Services of Canada. The ciné film was produced by the Institut de Mécanique Statistique de la Turbulence, Marseille.

References

Béguier, C., Giralt, F., Fulachier, L., Keffer, J. F.: Negative production in turbulent shear flows, in *Structure and Mechanisms of Turbulence II*, ed. by H. Fiedler, Lecture Notes Phys. Vol. 76 (Springer, Berlin, Heidelberg, New York 1978) pp. 22–35

Budny, R. S., Kawall, J. G., Keffer, J. F.: Vortex street evolution in the wake of a circular cylinder, in *Proc. 2nd Symp. on Turbulent Shear Flows* (Imperial College, London 1979)

Diaz, F., Gavaldà, J., Kawall, J. G., Giralt, F.: Interpretation of the complex turbulent flow generated by a rotating circular cylinder, in *Structure of Complex Turbulent Shear Flow*, IUTAM Symposium Marseille 1982, ed. by R. Dumas, L. Fulachier (Springer, Berlin, Heidelberg, New York 1983) pp. 175–184

Gavaldà, J., Diaz, F., Giralt, F., Kawall, J. G., Keffer, J. F. (1984): Analysis of the wake generated by the interaction of a rotating cylinder and a stationary cylinder. To be submitted to J. Fluid Mech.

Keffer, J. F., Kawall, J. G., Giralt, F., Béguier, C.: Analysis of turbulent structures in complex shear flows, in *Proc. 2nd Symp. on Turbulent Shear Flows* (Imperial College, London 1979)

Palmer, M. D., Keffer, J. F. (1972): An experimental investigation of an asymmetrical turbulent wake. J. Fluid Mech. **53**, 593

Shokr, M., Keffer, J. F.: Digital image analysis of a complex turbulent wake, in *Structure of Complex Turbulent Shear Flow*, IUTAM Symposium Marseille 1982, ed. by R. Dumas, L. Fulachier (Springer, Berlin, Heidelberg, New York 1983) pp. 175–184

An Investigation of Reynolds Stress Structures in Plane Turbulent Wakes

G. Fabris*

University of California, Los Angeles, USA

A. Nakayama

California State University, Long Beach, California

Abstract

Instantaneous u-v signals in the turbulent far wake of a cylinder have been analysed using an extended "quadrant analysis" in order to investigate the Reynolds stress structure and the features of large eddies. The results indicate that the region between the location of maximum shear and the edge of the wake is similar to the outer layer of boundary layers and bursts of slow moving flow ejecting outward contribute the most to the Reynolds shear stress. In the wake centerline region, fast moving fluid crossing the centerline contributes significantly but with opposite sign. There is a special type of flow near the centerline: flow moving upstream along the centerline slower than the mean velocity. This type of motion does not contribute to the shear stress but contains kinetic energy.

Introduction

The Reynolds stress ($-\overline{uv}$) in a turbulent shear flow is the key quantity in the mutually linked mechanisms of production of turbulent kinetic energy and transfer of momentum. It is well established that large eddies contribute the most to the Reynolds stress perhaps in an orderly quasi-periodic manner.

Features of these large eddies have been studied during the last decade using flow visualization and conditional sampling techniques. In the case of boundary layers [1–10] and pipe flow [11] quantification of features of the violent eddies has been made using so-called "quadrant analysis" in which "events" in the sequence of bursting mechanism were identified by the signs and magnitude of the instantaneous u and v velocity components.

In the case of turbulent wakes, the mechanism of the governing large eddies is not as clear. Gerrard [12] discussed observations of large-scale motions in wakes and concluded that most questions about growth and characteristics of these motions remain unanswered. As early as 1947 Townsend [13] concluded one of his classical wake papers with the statement "the jet structure (large eddies) can only be surmised at present, and many interesting questions remain to be answered. It is possible that useful information may be obtained from measurements of quadruple velocity products of the form $\overline{u^2 v^2}$ ... In this way more exact information about jet (turbulent burst) widths and rates of development could be collected and clearer understanding of the details of the motion attained." There appears to be no follow up to date on this visionary suggestion though Bradshaw et al. [14] demonstrated how revealing it is to study $\overline{v^2 w^2}$ in an axisymmetric jet. Recent studies of wakes behind cylinders by Townsend [15] and Wlezian [16] suggest patterns of large eddies resembling those of Kármán vortex street.

The present work is an attempt to investigate large structures in fully-developed cylinder wake making use of the quadrant analysis. This method classifies the instantaneous motion by directions in which the flow is temporarily moving and whether the particular motion contributes substantially to the Reynolds stress. It may be contrasted with the conditional

* Also at Transamerica Delaval, Inc., Santa Monica, California

sampling measurements of the near wake of a flat plate made by Andreopoulos and Bradshaw [17] who discriminated the instantaneous flow based on the origin of the fluid and whether the fluid has mixed with the fluid coming from the other side of the plate (discrimination conditioned on temperature threshold). The method employed here might prove to be more suitable for studying interacting shear layers than the temperature tagging. Namely the heat has the tendency of rapid diffusion into unheated turbulent flow permitting unambiguous conditional sampling only in the near interaction region. It is believed that the quadrant analysis may shed new light on instantaneous Reynolds stress mechanisms especially in the central region of the wake. The conventional Reynolds stress is low and changes sign in this region providing no information about the underlining physics.

Conditional Amplitude Analysis

Instantaneous *u-v* Signals and Structure of Turbulence

Measurements of instantaneous *u-v* signals were made using a hot-wire probe with *x*-array sensors at 500 diameters downstream of a circular cylinder of diameter of 1/16 in. in an otherwise uniform flow of speed of 50 fps. Signals were analysed by digital methods by sampling and digitizing them at 20 kHz.

Figure 1 displays simultaneous instantaneous traces of u, v, uv and $u^2 + v^2$ quantities. The figures are ordered starting with the intermittent wake edge and ending at the centerline. The structure of turbulence appears to be much more deterministic than in the pipe flow [11]. Existence of occasional large uv bursts is evident. At the wake edge the large excursions are one sided almost exclusively due to the slow moving fluid (negative u) ejecting outward (positive v). As the centerline is approached the uv peaks are more and more double sided with both negative and positive contributions to $\overline{uv}$. While this is expected, careful examination of u, v and uv traces also indicates that, at the maximum-shear location and at the centerline, some of the large negative excursions of uv signals are due to fast moving fluid (positive u) moving towards the centerline. These two types of motions contributing to $\overline{uv}$ are similar to the "ejections" and "sweeps", respectively, in boundary layers [2]. They may correspond to the downstream and upstream sides of a large vortex-like structure seen in the smoke pictures of Wlezian [16] or the Kármán vortex street like pattern discussed by Townsend [15]. At the centerline there is an ambiguity in the sense that positive v may mean either moving towards the centerline or away from the centerline. Similar ambiguity has been noted by Sabot and Comte-Bellot [11] near the centerline of a pipe flow. However, if they are associated with large scale motion, it is likely that when a (large) positive v occurs with a positive u it is due to a large eddy belonging to the lower half of the wake sending fast moving fluid towards and through the centerline. When a positive v is associated with a negative u it is the returning of such fluid originated from the upper half. These events are generally called "interactions" in the quadrant analysis of boundary layers.

The accompanying $u^2 + v^2$ traces reveal that they contain occasional "bursts" with no corresponding high excursions in the uv signal, indicating only one of u or v has high values. This is seen more clearly near the centerline and they are mainly associated with flows downstream along the centerline. Townsend [15] observed recently that eddies contributing most to the shear stress are significantly different in shape and in orientation from those predominantly contributing to the turbulent energy. For this reason these events are also depicted and analysed here. The above analysis is limited of course and may even be biased in the sense that the spanwise velocity component is not included. However it does imply certain characteristics and some orderlines of the large-scale motion.

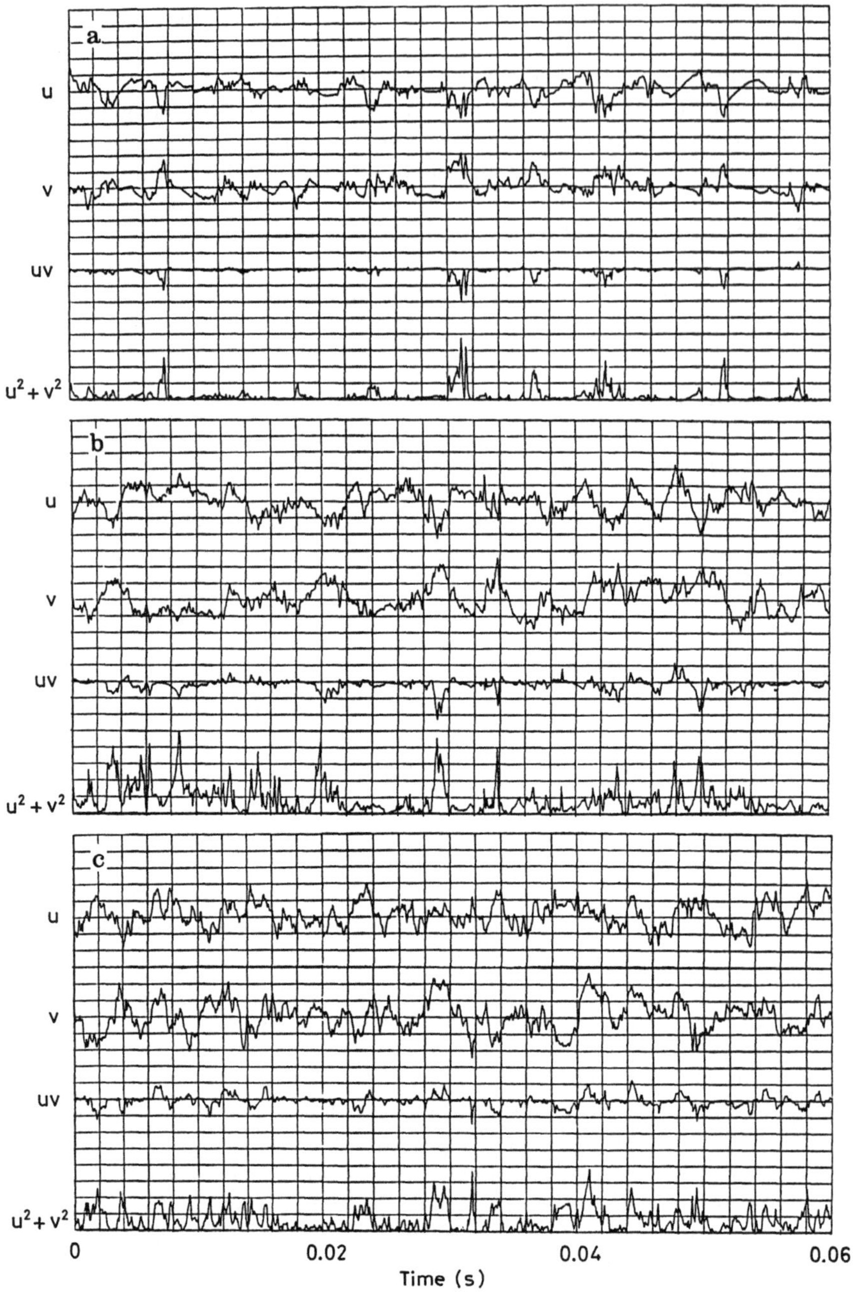

Fig. 1. Instantaneous u, v, uv and $u^2 + v^2$ traces; (a) near the wake edge, (b) at the maximum-shear, (c) at the centerline

Classification of Events

For the reasons mentioned in the preceeding section the following quantitative analysis has been attempted. As in the quadrant analysis of boundary layer flows the instantaneous motion was classified into different events based on the signs and magnitudes of the instanta-

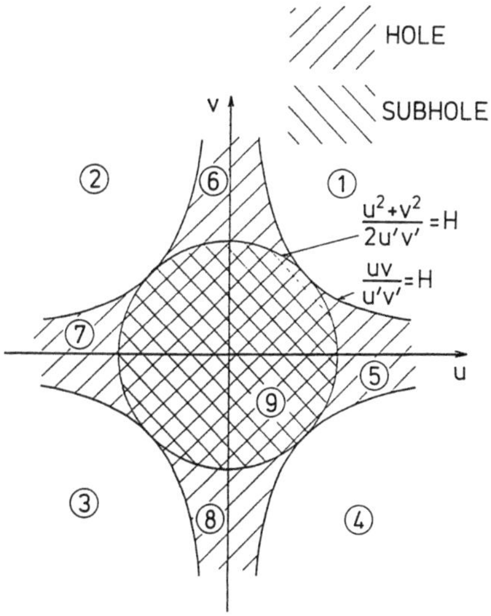

Fig. 2. Classification of the events in the *u*-*v* plane

neous *u* and *v*. Figure 2 illustrates the classification of the events in the instantaneous *u*-*v* plane. The method is close to the one introduced originally by Lu and Willmarth [2]. Events containing significant Reynolds stress of either sign are identified by the condition

$$uv > H u' v'$$

where u' and v' are the rms intensities and H is a parameter. These events correspond to regions 1, 2, 3 and 4. In order to depict and analyse the energy containing eddies that do not contribute to the Reynolds stress, the "hole" ($uv < H u' v'$) was subdivided by the circle

$$u^2 + v^2 = 2 H u' v'$$

to identify such events.

Although this flow differs from boundary layers we call the event represented by region 4 "sweep" and region 2 "ejection" since the former is likely to be the sweep of recently entrained potential flow, while the latter is likely to be the ejection of fully turbulent eddy originating in the wake central region. We also call the regions 1 and 3 "interactions", with a reservation near the centerline where the interactions can be sweeps penetrating through the centerline or ejections originating on one side emerging into the other side of the wake. The events represented by regions 5, 6, 7 and 8 are called respectively the turbulence moving downstream, outwards, upstream and inwards. The region 9, called subhole, represents motions that do not contribute appreciably to either the Reynolds stress or the turbulent kinetic energy.

Time Fractions

The first part of the analysis consists of an investigation of the effects of the threshold level H on time fractions the *u*-*v* trace spent in different regions (events). The fraction of the time spent in regions 1 through 4, hole and subhole, are shown in Fig. 3(a) and Fig. 3(b) as function of H for near-maximum-shear location and the wake center respectively.

Most of the quadrant analysis studies of boundary layers do not provide values of this quantity except for the paper by Brodkey et al. [3] where it is given as a function of y for

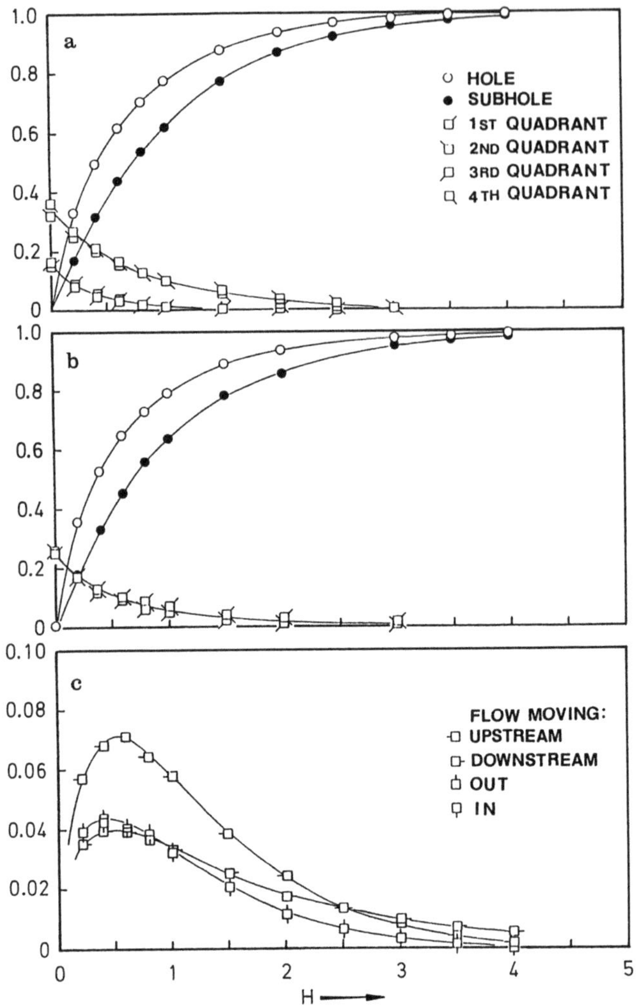

Fig. 3. Fractions of time occupied by the events at; (a) the maximum-shear, (b) the centerline, (c) by subregions at the centerline

$H = 0$ only. Levels of these quantities shown in [3] are very close to the present results at $H = 0$. Lu and Willmarth [2] reported fraction of time spent in the hole at several locations in the boundary layer. Again the level and shape are quite similar. The ejection-like motions and the sweep-like motions take almost equal time at the maximum-shear location. It should be noted, however, that the curve for ejections cross the curve for sweeps at about $H = 1.0$ and the former is higher than the latter at $H > 1.0$. Although not shown, the data taken at positions between the wake edge and the maximum-shear location indicate that the cross-over points exist at progressivley smaller H values as the edge is approached. It is reasonable to say that in this region the ejection-like motions are likely to be fully turbulent eddies originating probably in the wake central region while the sweep-like motions are likely to be recently entrained potential flow. Diminishing contributions of sweeps with increasing threshold H indicate that they are considerably less violent than ejections. The first and third quadrant can be considered to have either the interaction-type motions or sweeps and ejections associated with the large structure on the other side of the wake. At the maximum-shear position these motions take a considerably smaller fraction of time than the ejections and sweeps and become negligible at about $H = 1.5$. At the wake centerline the events in all four quadrants take about equal time but the fast moving motions (first and fourth quadrants) appear to be more violent.

The fraction of time spent in the regions inside the hole but outside the subhole (regions 5, 6, 7 and 8) is not negligible. The distributions are shown in Fig. 3(c) for the wake centerline only in order to show that the fluid moving upstream takes nearly twice as much time as other modes of motions. At the maximum-shear location time fractions of fluid moving in the upstream and downstream directions are almost the same, while at the near-edge position downstream eruptions dominate these energy containing motions with small contributions to the shear stress.

Fractional Contributions to the Reynolds Shear Stress

Figure 4 displays the contributions to the shear stress correlation coefficient $-\overline{uv}/u'v'$ from four quadrants for $H = 0$ at various positions across the wake. The total $-\overline{uv}/u'v'$ is also shown. The subscript $J = 1$, 2, 3 and 4 indicates respective quadrants. The reason for showing the contributions to the shear correlation coefficient instead of the shear stress $-\overline{uv}$ is that the shear stress is zero at the centerline and near the wake edge. Since the total $-\overline{uv}/u'v'$ is nearly constant over the shear zone, the present plots are nearly proportional to the fractional contributions to $-\overline{uv}$ in that zone.

First, Fig. 4 indicates that the contribution from the ejection-like event (second quadrant) dominates in the outer half of the wake and accounts for about 80 percent of the total shear stress which is similar to the results obtained by Brodkey et al. [2] in the boundary layer. The contributions from the sweep-like event (fourth quadrant) is generally lower but is higher in the wake central region. The events represented by the first and third quadrants which transfer momentum in the "wrong" direction account for only 20 percent in the shear zone which is only half as much as contributions of the interactions in boundary layers. This is another indication of the fact that eddies in the wake are more efficient than those in boundary layers. In the central region, as discussed earlier the first quadrant and fourth quadrant contributions balance out and the second and third quadrants are also of nearly equal magnitude. It is noted, however, that accelerated motions contribute more than the decelerated motions.

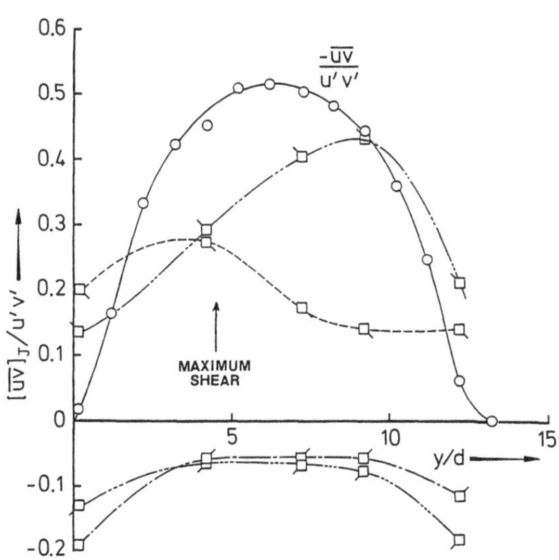

Fig. 4. Contributions to the Reynolds stress correlation coefficient for $H = 0$

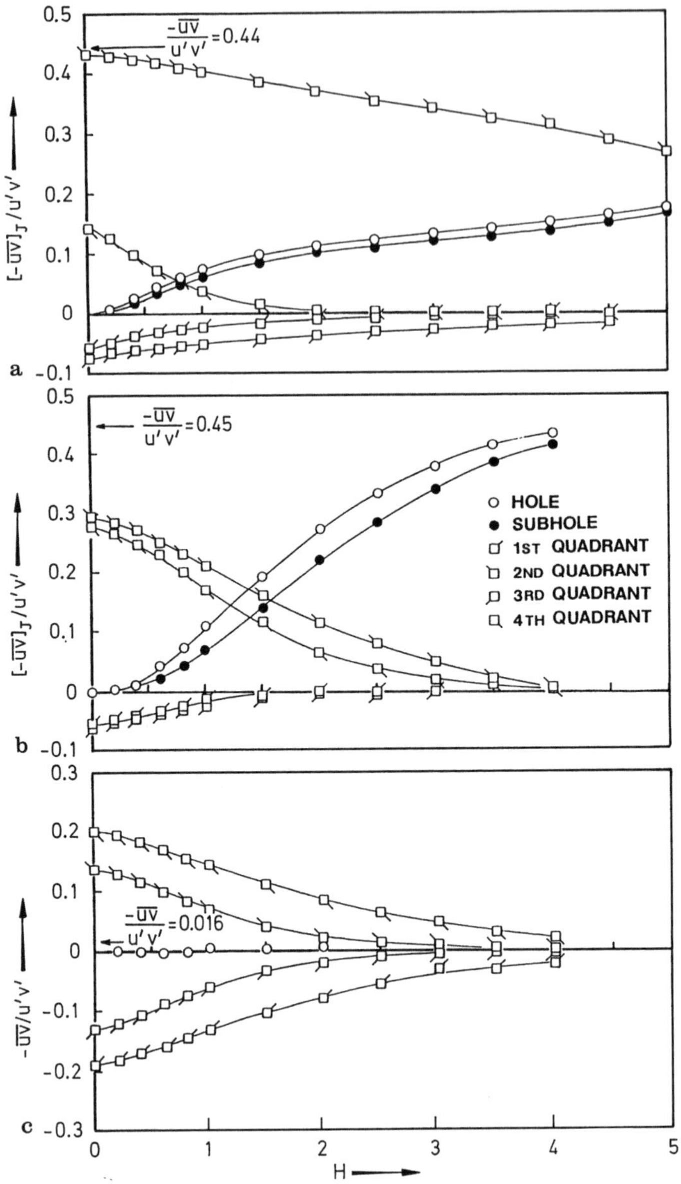

Fig. 5. Contributions to the Reynolds stress; (a) near the wake edge, (b) at the maximum-shear, (c) at the centerline

Figures 5(a), (b) and (c) show variation of the contributions by different events as function *a* of *H*. Figure 5(a) is for a position near the edge, (b) is for the maximum-shear location and (c) is for the wake centerline. These diagrams can be directly compared with those for the outer part of boundary layers [2] and pipe flow [11]. The boundary layer data appear to fit between the near-wake-edge data of Fig. 5(a) and maximum-shear position shown in Fig. 5(b). Near the wake edge there are surprisingly large contributions from very violent eddies moving out and very small contributions from the sweep-like events which are likely to be potential. This was expected from the instantaneous traces of *u*, *v*, and *uv*, but Fig. 5(a) confirms this quantitatively. At the maximum-shear location the ejection-like events and sweep-like events contribute most and almost equally significant while the interactions have very small contributions and are negligible at about $H = 2$. It is interesting to observe that at the centerline the fast moving fluid (first and fourth quadrants) contributions to the shear at high H are even larger than the ejections and sweep-like events at the maximum-shear location. It suggests that these may be quite violent motions that cancel out in most of the conventional averages.

As expected the contributions from the region between the hole and the subhole is very small. Though not shown, our data do indicate that the individual contributions are small and the canceling between the different subregions (5, 6, 7 and 8) is not the reason for this.

Fractional Contributions to the Kinetic Energy

Fractional contributions to $\overline{u^2} + \overline{v^2}$ (representative of the turbulent kinetic energy) at the position of maximum-shear are shown in Fig. 6(a). Trends are similar to the Reynolds stress shown in Fig. 5(a). Figure 6(b) offers the same contributions at the wake centerline. The significance of the regions within the hole but outside of the subhole is now more apparent

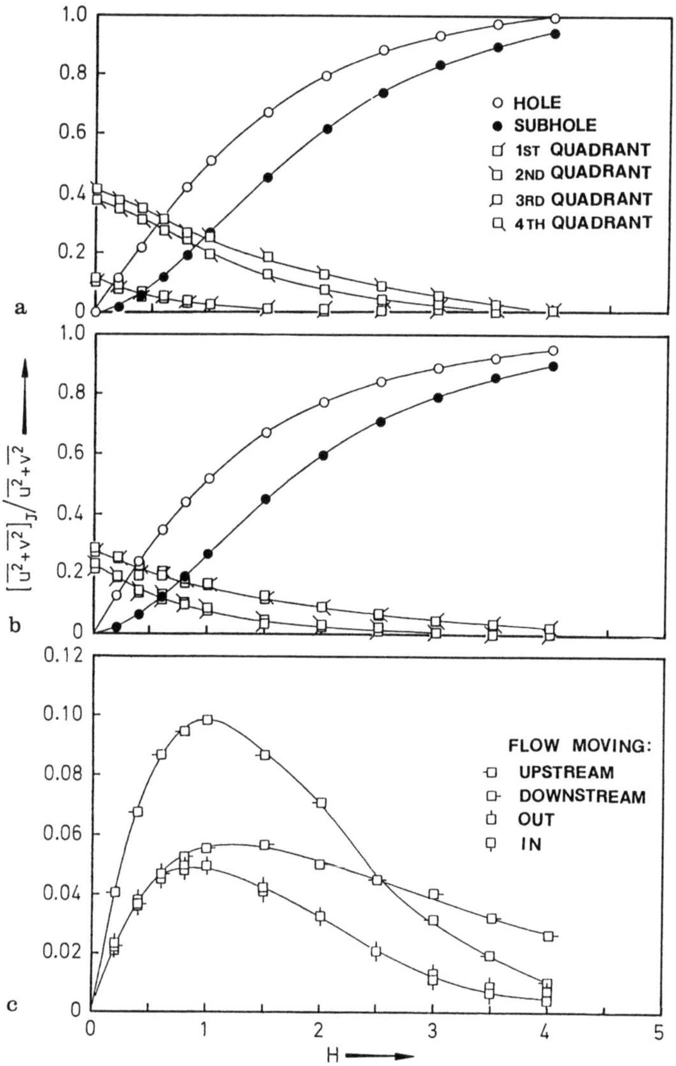

Fig. 6. Contributions to $u^2 + v^2$; (a) at the maximum-shear, (b) at the centerline, (c) at the centerline from subregions

with wide difference a between the curves for the hole and subhole. The contributions from the subhole regions (5, 6, 7 and 8) are further displayed on Fig. 6(c). By far the most contributing subregion is 7 where the streamwise velocity is lower than the mean and the lateral velocity is small.

Conditional Velocities of the Events

Conditional mean velocities U_J and V_J for four events (1, 2, 3 and 4) at the position of maximum-shear and the centerline are given in Fig. 7(a) and (b), respectively. The values are connected by equal H lines. This diagram is very informative of flow directions and intensities of different events. Points on the graph indicate the average direction and the magnitude of the velocity in the x-y plane of all the motions in appropriate events with the instantaneous Reynolds stress larger than $Hu'v'$. The plot for the maximum-shear location reveals that in generall, ejection-like and sweep-like motions are stronger than interaction-type motions. However, this tendency is less pronounced if averaged over large Reynolds stress

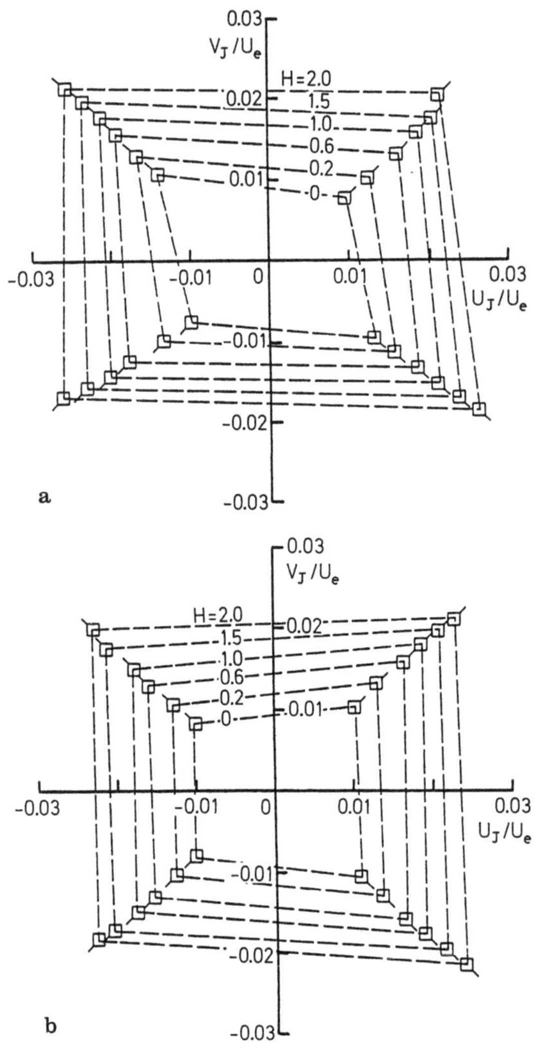

Fig. 7. Conditional velocities of the events; (a) at the maximum-shear, (b) at the centerline

carrying motions (large H). The figure for the wake centerline shows that faster moving motions (first and fourth quadrants) are stronger.

Durations and Intervals

Lu and Willmarth [2] and Sabot and Comte-Bellot [11] determined the periods of duration of and intervals between large bursts and sweeps in the boundary layer by taking a characteristic value of H determined from the plots of contributions to $\overline{uv}$ from different events. On the other hand Brodkey et al. [3] considered $H = 0$ only. We have tried an approach similar to the former to obtain information on the durations and intervals of various events.

Figure 8 is a plot of mean intervals between consecutive occurrences of events of the same type at the location of maximum-shear and the wake centerline. They are normalized by the similarity variable for time, $(x - x_0)/U_\infty$, where x_0 is the virtual origin of the wake taken here as $x_0 = -40D$. The trend is very similar to that given by Lu and Willmarth [2] in the sense that the intervals depend very strongly on H. At the location of maximum-shear the ejection-like event occurs most frequently and the interaction type motions least frequently. If $H = 3$, for which the contributions to $-\overline{uv}$ from interactions become negligible at this position, is taken as a characteristic threshold, the mean interval between the ejection-like events is $0.2(x - x_0)/U_\infty$. The characteristic interval between sweep-like events is about $0.5(x - x_0)/U_\infty$. The ratio is not very different from that of the boundary layer. At the wake centerline, the flow symmetry requires that the events associated with quadrants one and four occur at the same frequency and those of quadrants two and three also occur at the same rate. The diagram shows that the events associated with flow moving downstream (quadrants one and three) occur much more often than those moving upstream. Figure 5(c) indicates that $H = 3$ may also be taken as a characteristic threshold for determining the frequency of violent events moving faster than the mean velocity. It is slightly larger than that for the ejections at the maximum-shear location.

Figure 9 shows averages of durations of the ejections and sweeps. The durations, as in the boundary layers, exhibit a slow variation over a range of H. The durations of fast moving

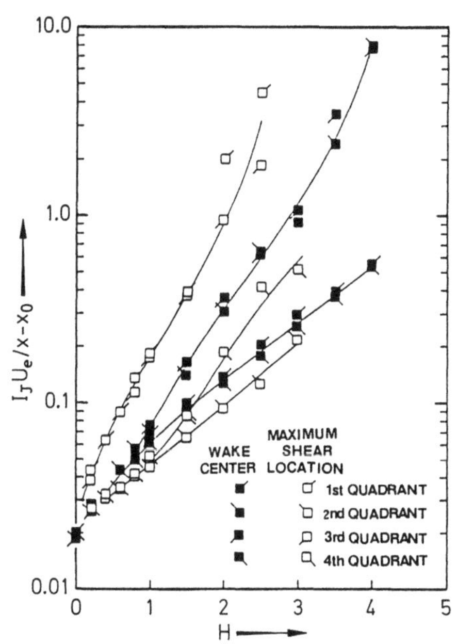

Fig. 8. Mean interval between large bursts; open symbols at the maximum-shear, closed symbols at the centerline

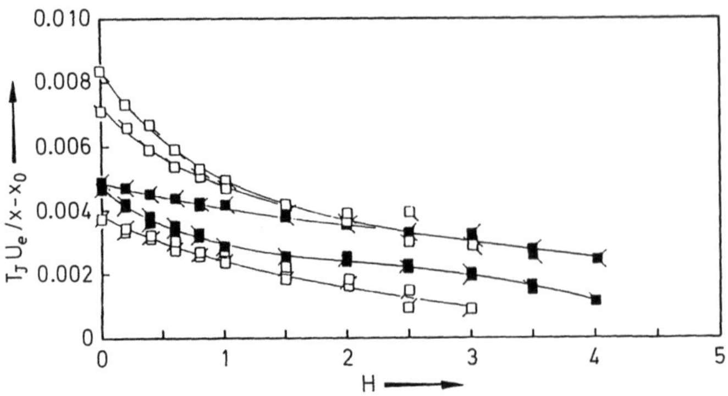

Fig. 9. Mean period of large burst; symbols as on Fig. 8

events at the wake center are very close to the durations of sweep and burst events at the maximum-shear position. The durations of slow moving events are shorter than these but longer than the interaction type motions at the locations of maximum-shear. This implies that the larger structure near the centerline consists more of sweeps coming from both sides of the centerline than of the structures that develop into ejection-like motions. The upstream moving flow has shorter durations and is smaller scaled. These events are moving streamwise slower than the mean flow along the centerline can be named "old turbulence" being convected along the central region of the wake. This fluid was entrained a long time ago.

Summary

The instantaneous *u-v* signals in the far wake of a cylinder have been analysed using an extension of the quadrant analysis method which is often employed to investigate bursting phenomena in boundary layers. Considerable qualitative and quantitative features of the large and violent eddies in the wake of a cylinder were obtained by this method. The present analysis by sorting the instantaneous signal on the basis of Reynolds stress contribution, even without clear visualization of the large structures in the wake, does show structures similar to boundary layers in the shear zone and a kind of coexistence of two opposite-sided large structures and smaller scaled "old turbulence" near the centerline. Future work which incorporates other type of conditioning together with the present quadrant type analysis will further clarify the Reynolds stress structure in the far wake.

References

1. Willmarth, W. W., Lu, S. S. (1972): J. Fluid Mech. **55**, 65
2. Lu, S. S., Willmarth, W. W. (1973): J. Fluid Mech. **60**, 481
3. Brodkey, S. R., Wallayce, M. J., Eckelmann, H. (1974): J. Fluid Mech. **63**, 209
4. Wallace, M. J., Brodkey, S. R., Eckelmann, H. (1977): J. Fluid Mech. **83**, 673
5. Eckelmann, H., Nychas, S. G., Brodkey, S. R., Wallace, J. M. (1977): Phys. Fluids **20**, S 225
6. Wallace, J. M., Brodkey, S. R. (1977): Phys. Fluids **20**, 351
7. Kreplin, P. H., Eckelmann, H. (1979): Phys. Fluids **22**, 1233
8. Panton, R. L., Goldman, A. L., Lowery, R. L., Reischan, M. M. (1980): J. Fluid Mech. **97**, 299
9. Raupach, R. M. (1981): J. Fluid Mech. **108**, 363
10. Nakagawa, H., Nezu, I. (1977): J. Fluid Mech. **80**, 99
11. Sabot, J., Comte-Bellot, G. (1976): J. Fluid Mech. **74**, 767

12. Gerrard, J. H.: Proc. of the Third Symp. of *Turbulent Shear Flows* (University of California, Davis, Davis California 1981) p. 11
13. Townsend, A. A. (1947): Proc. Roy. Soc. A **197**, 124
14. Bradshaw, P., Ferris, D. H., Johnson, R. F. (1964): J. Fluid Mech. **19**, 591
15. Townsend, A. A. (1979): J. Fluid Mech. **95**, 515
16. Wlezian, R. (1981): Ph.D. Thesis, Illinois Institute of Technology
17. Andreopoulos, J., Bradshaw, P. (1980): J. Heat Transfer **102**, 755
18. Fabris, G. (1979): J. Fluid Mech. **94**, 673

Turbulent Boundary Layer-Wake Interaction

E. P. Tsiolakis, E. Krause and U. R. Müller

Aerodynamic Institute of Technical University Aachen, 5100 Aachen, Federal Republic of Germany

Abstract

The interaction of a two-dimensional, incompressible, turbulent boundary layer on a flat plate with a plane wake of a circular cylinder was investigated experimentally and theoretically. The cylinder was positioned at various distances above the plate, with the axis normal to the free-stream direction and parallel to the surface of the plate. The time-averaged velocity components, the Reynolds stresses, the wall-pressure distribution and the wall shear stresses were measured downstream from the cylinder in the wake and in the boundary layer. Mean and fluctuating velocities were measured with hot-wire probes, conventional data reduction was used. It could be shown that the wake retains its self-preserving characteristics in the outer part and approximately in the inner layer as well. The profiles of the Reynolds stresses in the interacting part of the flow were used to deduce a mixing-length closure assumption, which was incorporated in a numerical finite-difference simulation of the relaxing boundary layer.

Experimental Setup and Hot-Wire Data Reduction

The experiment was carried out in the incompressible, two-dimensional, turbulent boundary layer on a smooth plate placed in the open test section (∅ 1 m) of the return-circuit low-speed wind tunnel of the Aerodynamisches Institut. The flow was tripped at the leading edge with a wire of 2 mm diameter. A distance of 500 mm downstream, the boundary layer was disturbed by a circular cylinder with diameter $D = 5$ mm or 10 mm, Fig. 1. The cylinder was mounted with its axis normal to the free-stream direction and parallel to the plate at various wall distances y_c. In this paper discussion is limited to conditions, for which the cylinder ($D = 10$ mm) is fully immersed in the boundary layer ($y_c = \delta_0/2$), and for which the distance between axis and wall is equal to $y_c = 2\delta_0$. The quantity $\delta_0 = 26$ mm is the thickness of the undisturbed boundary layer at the locus of the cylinder. The details of all measurements, also including the cases $y_c = 5$ mm and 26 mm, were reported by Tsiolakis (1982). The free-stream Reynolds number per unit length was $1.8 \cdot 10^6\,\mathrm{m}^{-1}$. Downstream of the cylinder

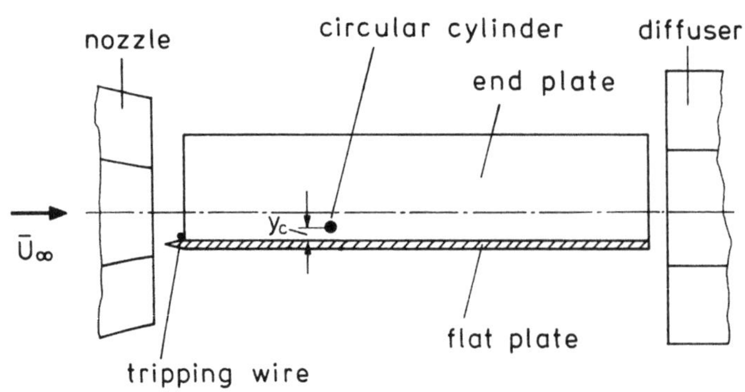

Fig. 1. Experimental setup

profiles of mean velocities and all Reynolds stresses were measured by means of miniature hot wires. The wall shear stress distribution, inferred from Preston tube measurements, and the pressure distribution, obtained from pressure taps (∅ 1 mm) at the wall, were also described in the investigation quoted. For sake of comparison all measurements were carried out in the unperturbed boundary layer as well.

Mean and fluctuating velocities were measured by means of rotating hot wires. The influence of the magnitude and direction of the local velocity vector on the signal interpretation were taken into account in the calibration according to Müller (1982). For a hot-wire fixed coordinate system, the cooling law was described with an effective cooling velocity U_c

$$E^2/S^2 = U_c^2 = U_N^2 + k^2 U_T^2 + h^2 U_Q^2 . \tag{1}$$

The velocity components normal and tangential to a hot-wire in the plane of the prongs were denoted by U_N and U_T, respectively, and U_Q was perpendicular to both. The directional sensitivities k and h were determined from separate calibrations for each hot wire used, the quantity S was the constant of proportionality between the electrical output E of an analogue linerarizer circuit and U_c. Transformation of the instantaneous turbulent velocities $\bar{U}_i + u_i$ of a laboratory fixed coordinate system onto the hot-wire fixed frame of reference yielded the cooling law in the form

$$E(\psi, \gamma, k, h)/S = U_c = [L(\bar{U}_i \bar{U}_j) + l(u_i) + f(u_i u_j)]^{0.5}, \tag{2}$$

where L and f contain all double velocity correlations of mean and fluctuating velocity components, respectively; the function l includes the linear fluctuation terms. With the hot-wire angle γ between wire and prong and the sensitivities k and h known, rotation of the probe around its longitudinal axis with various angles ψ yields a sufficiently large number of electrical signals for evaluating all mean velocities $\bar{U}_i$ and the Reynolds stresses $\overline{u_i u_j}$.

As was pointed out already by Müller (1982), large scatter can result in the data reduction, if the squared time averaged signal

$$\overline{E^2}(\psi)/S^2 = \bar{L} + \bar{f} \tag{3}$$

is used to solve the corresponding set of linear equations as proposed by Durst (1971) or Acrivlellis (1979):

$$d_{ij}(\psi, \gamma, k, h)\, \vec{W} = \vec{B} \tag{4}$$

with $B_i = \overline{E_i^2}/S^2$ and $\vec{W} = (\overline{U^2} + \overline{u^2}, \overline{V^2} + \overline{v^2}, \overline{w^2}, \overline{UV} + \overline{uv})$.

One reason for the scatter is the unequal order of magnitude of the terms $\bar{L}$ and $\bar{f}$, with $\bar{L}$, in general, being much larger than $\bar{f}$. Another reason is that the coefficient matrix is ill-conditioned: Hadamard's criterion, see Jordan-Engeln and Reutter (1976), for an $(n \times n)$-type matrix

$$K(d_{ij}) = |\det d_{ij}|/(a_1 a_2 \ldots a_n), \tag{5}$$

with a_m being defined as the square root of the sum of all squared elements of the m-th row of d_{ij}, yielded a value of $K(d_{ij}) = 1.5 \cdot 10^{-5} \ll 1$, indicating that Eq. (3) cannot be used for accurate data evaluation, if the turbulence level is low. However, for high-intensity flows locally reversed velocities become significant and may falsify the hot-wire signal $\overline{E^2}$ because of rectification, Müller (1983).

In the present investigation the conventional method for calculating the Reynolds stresses from the rms-measurements of the hot-wire output was used. Root-expansion of Eq. (2) and time-averaging yielded

$$\overline{[\bar{E}(\psi) - E(\psi)]^2}/S^2 = \overline{e^2}(\psi)/S^2 = a_{ij}(\psi)\, \overline{u_i u_j}$$

and

$$\overline{[e(\psi) \pm e(\psi + 180°)]^2}/S^2 = b_{ij}(\psi)\, \overline{u_i u_j}. \tag{6}$$

For this set of equations Hadamard's condition yielded $K = 0.9978 \approx 1$ indicating a well-conditioned system.

After measuring all Reynolds stresses, the mean velocities were evaluated from Eq. (4) as proposed by Müller (1982), yielding an accuracy of order $(\overline{u^3})$ compared to the conventional method, which has an inherent error of order $(\overline{u^2})$.

Discussion of Results

Measurements

Figures 2 and 3 show the measured mean velocities for the test cases $D = 10$ mm and $y_c = 52$ mm and 13 mm, respectively, compared to the smoothed results of the undisturbed boundary layer. The data show the spreading of the wake and the decay of the velocity defect. The near-wall region of the perturbed boundary layer recovers over a shorter distance than the outer layer. Within the downstream measuring range $20 \leqq x/D \leqq 86$, relative minima of the velocity profiles were only measured with the cylinder positioned outside of the boundary layer and for $y_c = 26$ mm and $x/D = 20$. Mean velocity data of boundary layer-cylinder wake interactions similar to those discussed here were reported by Klebanoff and Diehl (1952), Clauser (1956) or Marumo et al. (1978).

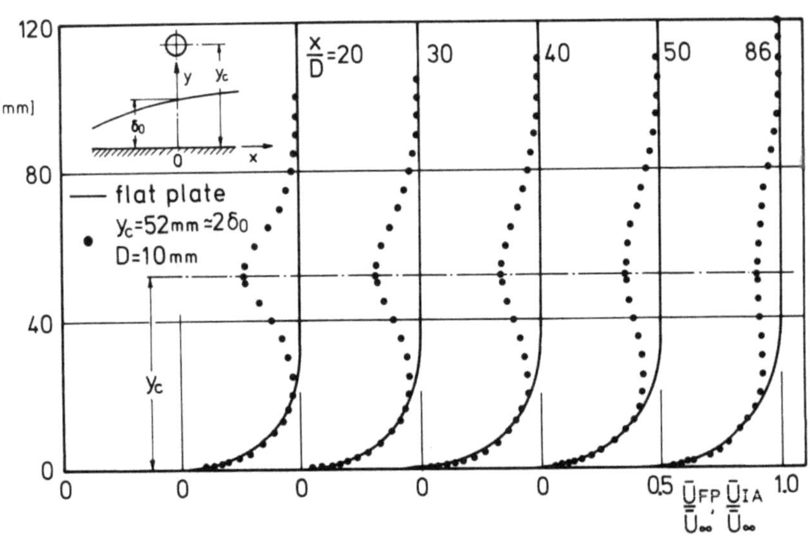

Fig. 2. Measured mean velocites of interacting flow (. . .) compared to undisturbed, flat-plate flow (—)

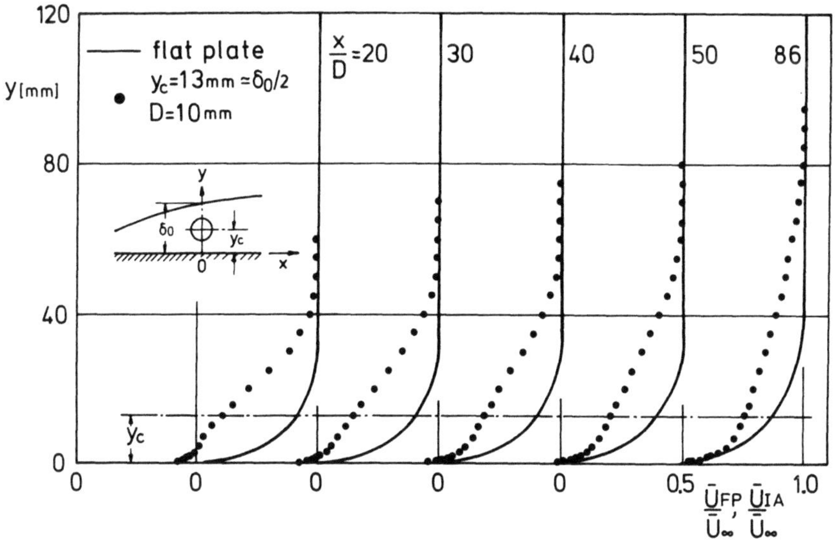

Fig. 3. Measured mean velocities of interacting flow (...) compared to undisturbed, flat-plate flow (—)

The present investigation also includes measurements of the Reynolds stress tensor as shown in Figs. 4–7. For $y_c = 2\delta_0 = 52$ mm the normal stresses are close to those of an undisturbed boundary layer ($\overline{u^2} > \overline{w^2} > \overline{v^2}$) for $y \leqq \delta_0/2$, Fig. 4. In the outer part of the flow the measured profiles correspond to those of Townsend (1956) obtained with a cylinder in uniform free-stream ($\overline{v^2} > \overline{w^2} > \overline{u^2}$). Because of vortex shedding the $\overline{v^2}$-correlation is the largest one in the wake flow. The profiles of $\overline{w^2}$ exhibit a relative minimum at $y = y_c$ which disappears at $x/D \approx 50$. At the measuring station farthest downstream ($x/D = 86$), the normal stresses measured in the wake region, Fig. 5, approach equal orders of magnitude, while in the boundary layer the turbulence structure remains unchanged. The Reynolds shear stress $\overline{uv}$ approaches zero with increasing distance from the wall, in the outer region it has a symmetrical shape about y_c and correspondingly changes its sign. The measured results of mean velocities and Reynolds stresses for this test case with $y_c = 2\delta_0$ indicate negligible interaction between boundary layer and wake flow. In the other example chosen for discussion, $y_c = 13$ mm (Fig. 6), boundary layer and wake do interact with each other. In contrast to the measurements discussed above, the maximum value of $\overline{v^2}$ is reduced considerably yielding $\overline{u^2} > \overline{w^2} > \overline{v^2}$. The $\overline{uv}$-correlation does not change its sign in this case. At $x/D = 86$, Fig. 7, the measured profiles of the Reynolds stresses are close to those of an undisturbed boundary layer except for the large values of $\overline{w^2}$.

For those test series, in which the profiles of the mean velocity indicate relative minima, the wall distances $y(\partial\bar{U}/\partial y = 0)$ and those of $y(\overline{uv} = 0)$ are equal and the sign of the turbulent shear stress is determined by that of the velocity gradient. Such results are consistent with gradient-type closure assumptions and therefore the mixing length distributions were evaluated. The length scale of the free shear layer was found to be constant across the wake with a magnitude being about three times the value obtained in the outer layer of the wall-bounded undisturbed shear flow. The mixing length distributions of the interacting flow of the other test cases were also evaluated; the results obtained for $y_c = 13$ mm are shown in Fig. 8. The near-wall slope $\varkappa = \partial l/\partial y$ decreases with increasing downstream distance ($0.8 \gtrsim \varkappa \gtrsim 0.5$). The average value is about 0.6. In the outer layer l/δ increases from 0.1 at $x/D = 20$ to $l/\delta \cong 0.14$ for $x/D \geqq 40$. The measured near-wall and outer-layer asymptotes

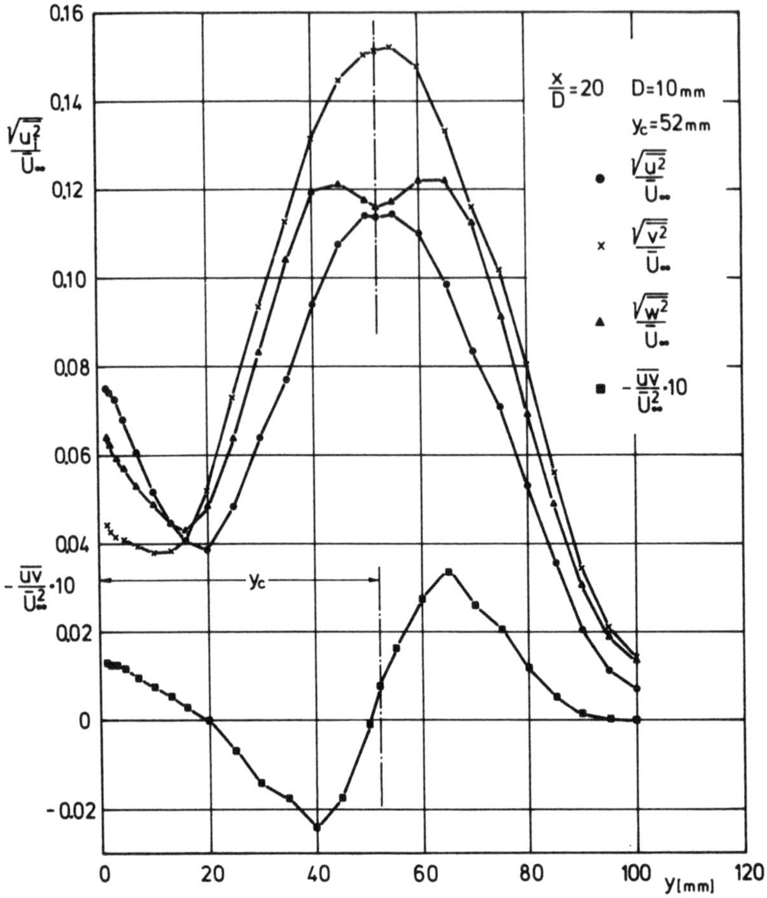

Fig. 4. Measured Reynolds stresses

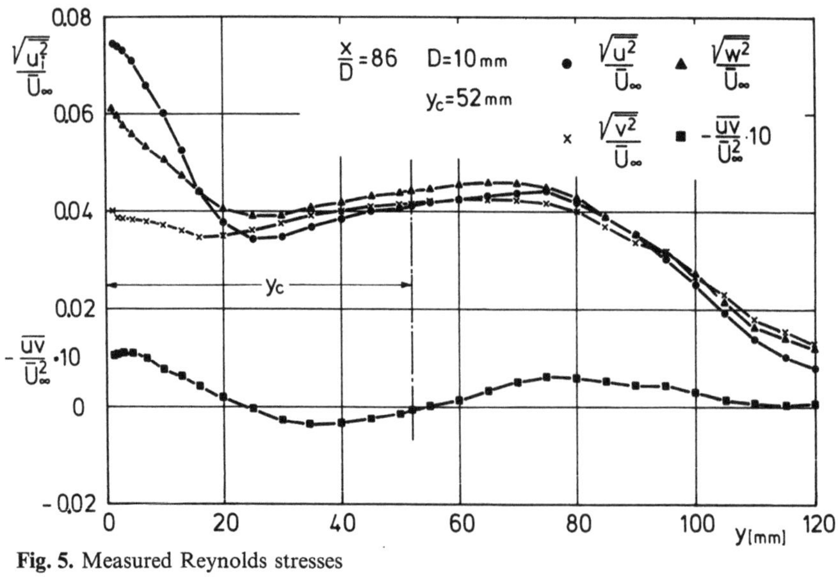

Fig. 5. Measured Reynolds stresses

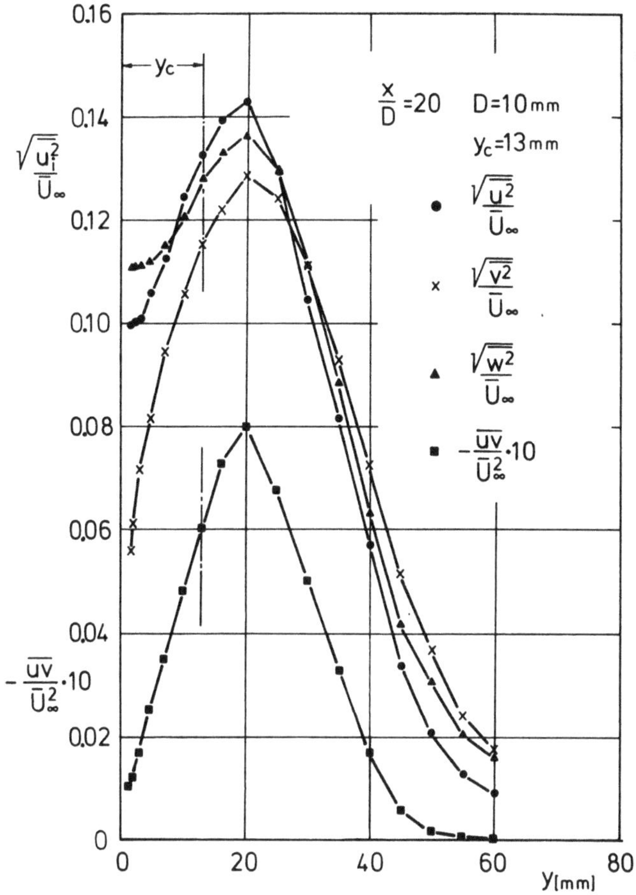

Fig. 6. Measured Reynolds stresses

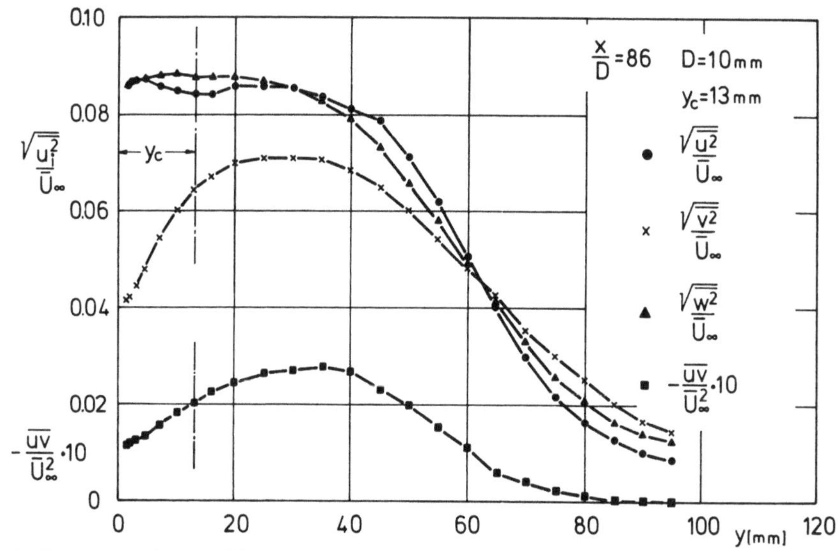

Fig. 7. Measured Reynolds stresses

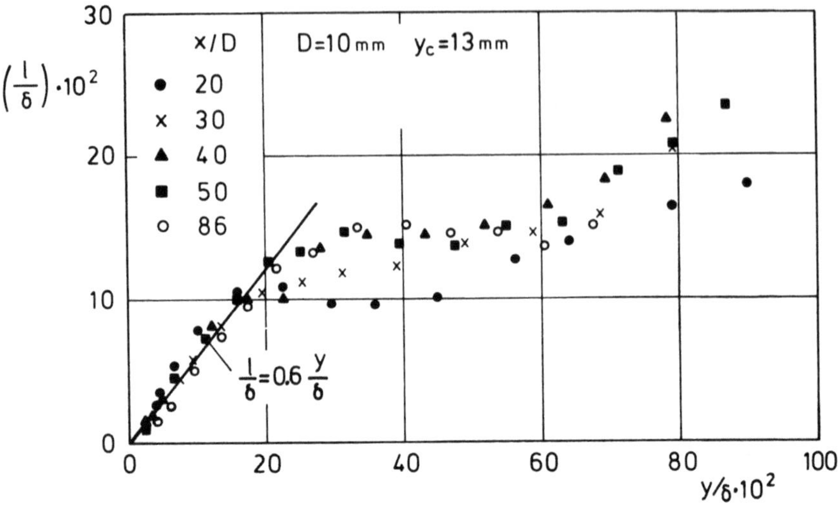

Fig. 8. Measured mixing length distributions normalized by boundary layer thicknesses

are substantially larger than the corresponding ones of equilibrium boundary layers, $\varkappa = 0.41$ and $l/\delta = 0.085$, respectively. The mixing length distributions evaluated for the test case $y_c = 5$ mm closely agree with those of Fig. 8. The same results were also obtained for the inner layer of test case $y_c = 26$ mm, while in the outer layer a constant length scale identical to that of case $y_c = 52$ mm was evaluated.

Numerical Simulation and Analysis of Self-Similarity

All flow fields investigated were simulated numerically by solving the boundary-layer equations with an implicit finite-difference method developed at the Aerodynamisches Institut. Eddy viscosities were prescribed with a two-layer mixing-length closure assumption adjusted to the measurements. Using the smoothed velocity profiles measured at $x/D = 20$ as initial conditions and neglecting downstream pressure gradients according to the measurements of Tsiolakis (1982), good agreement between computed mean velocity profiles and experimental data was obtained. For the test case $y_c = 13$ mm, Fig. 9, the calculations show a slight overshoot in the outer layer compared to the measurements. In the transition region between law-of-the-wall and interactive flow, $y \approx 5$ mm, the measurements indicate large curvatures $\partial^2 \bar{U}_{IA}/\partial y^2$, which could not be simulated correctly in the computations. Probably a more sophisticated turbulence model would be suited better for predicting the relaxing boundary layer, though for the present flow the definition of the initial conditions is expected to give rise to problems.

The experimental data were also analyzed with regard to self-preserving characteristics of the flow. The velocity defect created by the cylinder was obtained by subtracting the mean velocity profiles of the perturbed, interacting flow from those of the undisturbed, flat-plate boundary layer, see Figs. 2 and 3:

$$\Delta\bar{U} = (\bar{U}_{FP} - \bar{U}_{IA}). \tag{7}$$

Following Schlichting (1930) and Reichardt (1942), the existence of self-similarity of the wake requires $\Delta\bar{U} \sim (x/D)^{-0.5}$ and $L \sim (x/D)^{0.5}$ as scaling laws for characteristic velocity and length scales. Therefore the maximum velocity defect $\Delta\bar{U}_{max} = \max(\bar{U}_{FP} - \bar{U}_{IA})$ as well

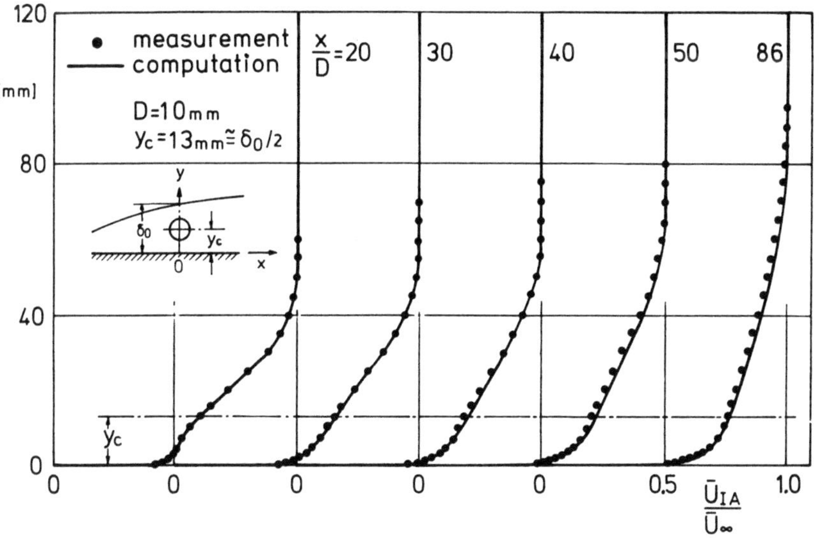

Fig. 9. Comparison of measured mean velocities with results of finite-difference solution

as the outer-layer half width

$$b_{1/2} = y(0.5\,\Delta\bar{U}_{max}) - y(\Delta\bar{U}_{max}), \tag{8}$$

evaluated for all test cases with $D = 10$ mm, are plotted in Fig. 10 and compared with the empirical far-field scaling laws of the investigations quoted above. Since scaling with $(x/D)^{0.5}$ approximately exists, except for $y_c = 5$ mm, the normalized velocity defect $\Delta\bar{U}/\Delta\bar{U}_{max}$ was plotted for $y_c = 52$ mm and 13 mm, e.g., in Figs. 11 and 12 versus the normalized wall distance $\eta/\eta_{1/2}$, with

$$\eta = [(y - y(\Delta\bar{U}_{max}))/D]/(x/D)^{0.5} \tag{9}$$

and

$$\eta_{1/2} = b_{1/2}/(x/D)^{0.5}. \tag{10}$$

The correlations indicate that the self-preserving wake characteristics are retained for $y_c = 52$ mm and for the outer shear layer of $y_c = 13$ mm. In the interacting region of the flow, similarity exists for $-0.5 \gtrsim \eta/\eta_{1/2} \leqq 0$, but deviations become larger as $\eta/\eta_{1/2}$ approaches -1. Steep velocity gradients in the inner part of the wake region are typical for all correlations evaluated in contrast to vanishing gradients at the outer edge.

The computation of outer and inner-layer similar mean velocity profiles was carried out by assuming a constant mixing length or a constant eddy viscosity across the wake, as proposed by Schlichting (1930) and Reichardt (1942), respectively. Since both results agreed closely, discussion will be limited to the latter assumption. For a cylinder in uniform free-stream, substitution of

$$\Delta\bar{U}/\bar{U}_\infty = f(\eta)/(x/D)^{0.5} \tag{11}$$

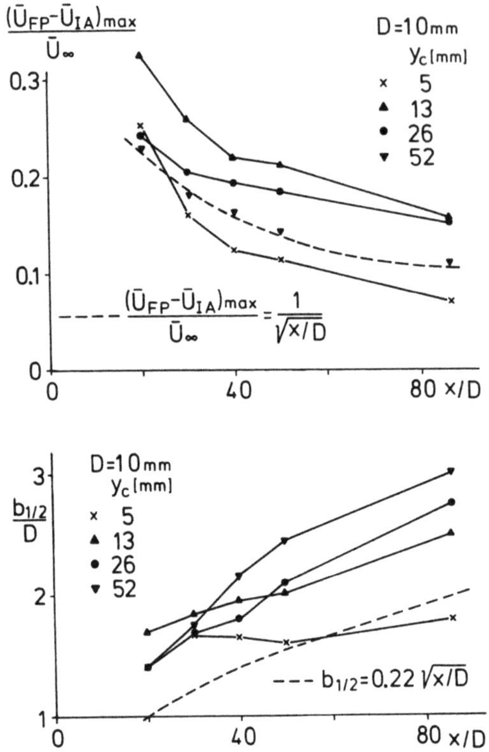

Fig. 10. Measured maximum velocity defect of cylinder wake and outer-layer half width

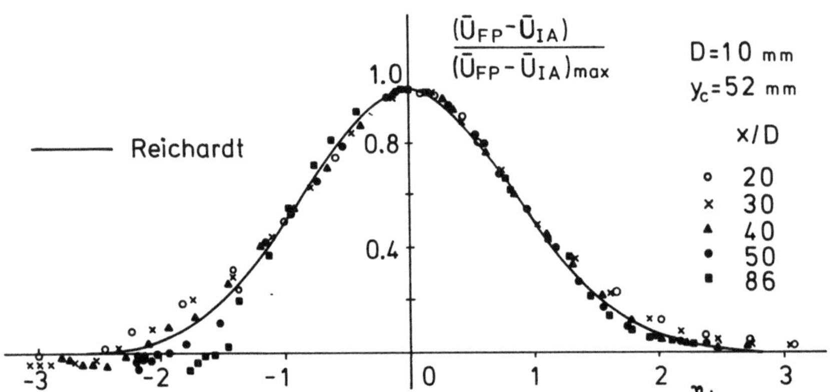

Fig. 11. Self-preservation of wake flow

into the momentum equation

$$\bar{U}_\infty \frac{\partial \Delta \bar{U}}{\partial x} = \hat{\nu}_t \frac{\partial^2 \Delta \bar{U}}{\partial y^2} \tag{12}$$

yields the similar solution

$$f/f_{max} = \exp[-\eta^2/(4\nu_t)]. \tag{13}$$

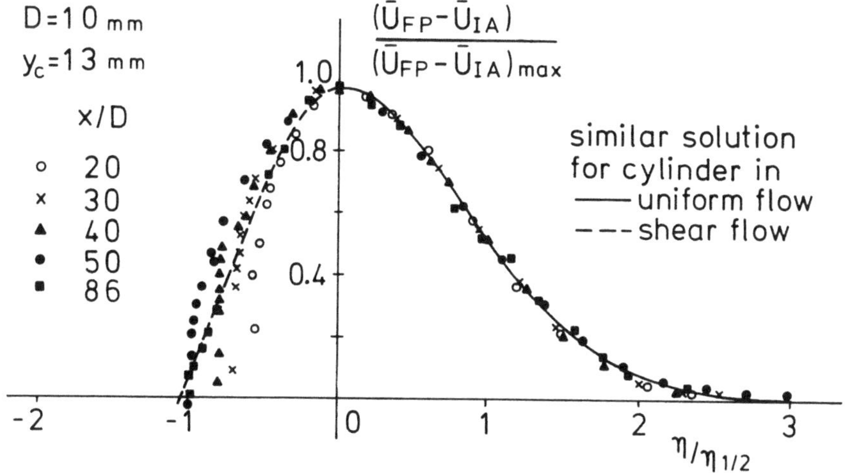

Fig. 12. Measured self-preservation of wake flow compared to similar solutions for cylinder in uniform and sheared free-stream

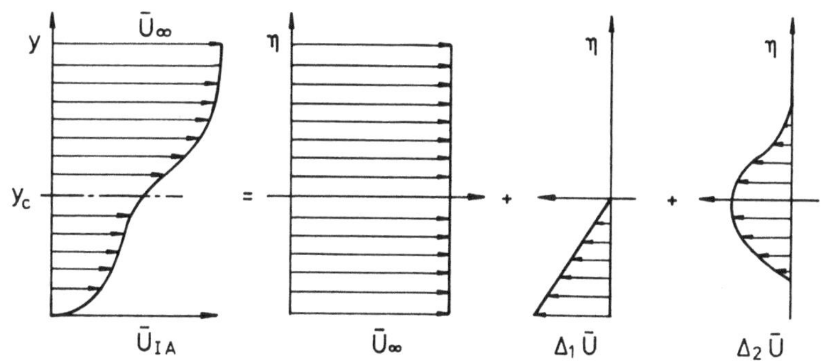

Fig. 13. Decomposition of boundary layer-wake interaction for analyzing self-similarity of wake flow

Using Reichardt's experimental far-field results for the half width ($\eta_{1/2} \cong 0.22$) and the non-dimensional eddy viscosity ($\nu_t = \hat{\nu}_t/(\bar{U}_\infty D) = 0.0164$), results in the profile plotted versus $\eta/\eta_{1/2}$ in Fig. 11 and in the outer half of Fig. 12. The measured correlations agree excellently with this solution, since with the measured half width being approximately $1.5 \cdot 0.22$, see Fig. 10, and with the average eddy viscosity being about $2.5 \cdot 0.0164$, the same solution for $f/f_{\max}$ versus $\eta/\eta_{1/2}$ is obtained.

The inner-layer similarity of the cylinder wake ($\eta/\eta_{1/2} \leqq 0$) was analyzed by assuming the boundary layer flow to be created by a velocity defect $\Delta_1\bar{U}$, Fig. 13. The flow of the wake of the cylinder, $\Delta_2\bar{U}$, can be superimposed in a first approximation to yield the resultant velocity defect

$$\Delta\bar{U} = \Delta_1\bar{U} + \Delta_2\bar{U}. \tag{14}$$

Substituting $\Delta_1\bar{U}$, approximated by

$$\frac{\Delta_1\bar{U}}{\bar{U}_\infty} = \frac{f_1(\eta)}{(x/D)^{0.5}} = \frac{c\eta}{(x/D)^{0.5}} \quad \text{with } c \cong -9.5 \tag{15}$$

and

$$\frac{\Delta_2 \bar{U}}{\bar{U}_\infty} = \frac{f_2(\eta)}{(x/D)^{0.5}} \tag{16}$$

into the momentum Eq. (12), yields a differential equation for the similar profile $f_2(\eta)$ of the inner-layer. With the eddy viscosity prescribed by $v_t \cong 0.04$ as found for all wake flows investigated, Runge-Kutta integration resulted in the profile $f_2/f_{2\,max}$ plotted for $\eta/\eta_{1/2} \leqq 0$ in Fig. 12; according to the measurements of Fig. 10, the half width was set to $\eta_{1/2} = 0.33$. The calculated self-similar profile reasonably approximates the measured correlations in the range $-0.5 \leqq \eta/\eta_{1/2} \leqq 0$. For $\eta/\eta_{1/2} \to -1$ the computed velocity gradient considerably overestimates the measured values. Obviously in the near-wall region self-similarity breaks down.

The measured profiles of the Reynolds shear stress $\overline{uv}$ were analyzed in the same way as described for the mean velocities. In the outer-layer wake self similarity was found to correspond to that of a cylinder in uniform flow. The results obtained for the inner-layer failed to indicate acceptable agreement between theory and experiment because of inaccuracies of the measurements and due to the coarse simplifications inherent in the theoretical similarity analysis.

Concluding Remarks

The interaction of an incompressible, two-dimensional turbulent boundary layer with a cylinder wake was investigated experimentally and theoretically. Cylinders with different diameters D were positioned at various distances y_c from the wall; the paper discusses typical results obtained from the test series with $D = 10$ mm and $y_c = \delta_0/2$ and $y_c = 2\delta_0$ with $\delta_0 = 26$ mm being the thickness of the undisturbed boundary layer at the position of the cylinder. In addition to the wall shear stress and the pressure distributions, profiles of mean velocities and Reynolds stress tensors were measured by means of hot wires. The method of Durst (1971) or Acrivlellis (1979) for evaluating the mean and fluctuating velocities from the squared hot-wire signals was shown to yield an ill-conditioned set of equations for turbulence levels below 50%. Therefore the conventional data reduction method based on a root-expansion of the hot-wire cooling law was used.

The results obtained from the measurements and their analysis may be summarized as follows:

- Interaction of turbulent boundary layer-wake flow occurred for the test series with $y_c \leqq \delta_0/2$ as well as for $y_c = \delta_0$ in the range $y < y_c$; in the outer layer of the latter case and for $y_c = 2\delta_0$ the wake was created by uniform free-stream.
- The mean velocities perturbed by the wake recovered over a shorter distance in the near-wall region that in the outer layer.
- At the measuring station farthest downstream ($x/D = 86$) the boundary layer had not recovered completely.
- Vortex shedding generated large velocity correlations $\overline{v^2}$, as well $\overline{w^2}$ was found to be larger than $\overline{u^2}$. The downstream relaxation reestablished the order of magnitude $\overline{u^2} > \overline{w^2} > \overline{v^2}$ typical for undisturbed boundary layers.
- The mixing length distributions determined from the measurements differed markedly from those of an undisturbed boundary layer. In the interacting flow region the near-wall slopes $\varkappa = \partial l/\partial y$ decreased with increasing downstream distance ($0.8 \gtrsim \varkappa \gtrsim 0.5$). The

average value was about $\varkappa = 0.6$. In the outer layer l/δ increased from 0.1 at $x/D = 20$ to $l/\delta \cong 0.14$ for $x/D \geqq 40$. The turbulent length scale for the non-interacting wake flow was found to be constant across the shear layer with a magnitude being about two and a half times the value measured in the outer region of the undisturbed boundary layer. Consistent with a gradient-type turbulence model, the wall distances $y(\partial\bar{U}/\partial y = 0)$ agreed with $y(\overline{uv} = 0)$. At these coordinates the $\overline{uv}$-shear stress changed its sign.

- Numerical simulations of all flows investigated were carried out by solving the boundary-layer equations with a finite-difference method. Prescribing the eddy viscosities with a two-layer mixing-length model adjusted to the measurements, the computed mean velocities indicated good agreement with the experimental data.
- The velocity defect measured in the cylinder wake could be correlated – except for the near-wall region – in terms of similarity variables. The outer-layer correlations agreed closely with classical similarity theories, for the interacting inner-layer flow a modified similar solution was developed.

References

Acrivlellis, M. (1979): Auswertung von Hitzdrahtmessungen mehrdimensionaler Strömungen beliebiger Turbulenzintensität. Habil.-Schrift, TH Karlsruhe

Clauser, F. H. (1956): The turbulent boundary layer. Adv. Appl. Mech. **4**, 1

Durst, F. (1971): Evaluation of hot-wire anemometer measurements in turbulent flows. Imperial College, Dept. of Mech. Eng. Rept. ET/TN/A/9

Jordan-Engeln, G., Reutter, F.: Formelsammlung zur numerischen Mathematik mit Fortran IV-Programmen (Bibliographisches Institut, Mannheim 1976) p. 58

Klebanoff, P. S., Diehl, Z. W. (1952): Some features of artificially thickened fully developed turbulent boundary layers with zero pressure gradient. NACA Report 1110

Marumo, E., Suzuki, K., Sato, T. (1978): A turbulent boundary layer disturbed by a cylinder. J. Fluid Mech. **87**, 121

Müller, U. R. (1982): On the accuracy of turbulence measurements with inclined hot wires. J. Fluid Mech. **119**, 155

Müller, U. R. (1983): A hot-wire method for high-intensity turbulent flows. Proc. 10th International Congress on Instrumentation in Aerospace Simulation Facilities, St. Louis (France), IEEE Publication 83 CH 1954-7

Reichardt, H. (1942): Gesetzmäßigkeiten der freien Turnbulenz. VDI-Forschungsheft 414, 2. Auflage 1951, (Deutscher Ingenieur-Verlag GmbH, Düsseldorf)

Townsend, A. A.: *The Structure of Turbulent Shear Flow* (Cambridge University Press, New York 1956)

Tsiolakis, E. P. (1982): Geschwindigkeiten und Reynoldssche Spannungen in einer mit einem Kreiszylinder gestörten turbulenten Plattengrenzschicht. Ph.D. Thesis, RWTH Aachen

Schlichting, H. (1930): Über das ebene Windschattenproblem. Ph.D. Thesis, Göttingen, Ing. Arch. **1**, 533

Part III

Boundary Layers

Boundary Layers

M. M. Gibson

Mechanical Engineering Department, Imperial College, London, England

Introductory Remarks

The papers presented at the Fourth International Shear Flows Symposium show that the measurement and prediction of the flow in turbulent boundary layers continue to engage the attention of researchers. The practical relevance of the study of boundary layers needs no explanation; at a more basic level many of the fundamental features of turbulent shear flow are conveniently studied in thin shear layers, while the mathematical simplifications of the boundary layer hypothesis have not yet been made unnecessary by advances in numerical methods. It can be argued that the uncertainty involved in the numerical solution of elliptic differential equations for complex flows with recirculation makes the measurements from thin shear layers which can be described by parabolic equations a more reliable test for turbulence model predictions. The acquisition of suitable data for this purpose is one of the two main objectives of experimental turbulence research; the other is still to achieve a more complete understanding of the physical nature of turbulence. The paper by Iritani, Kasagi and Hirata falls into this category. It is concerned with a subject of continuing research interest: the details of turbulence in the immediate vicinity of the wall. A convenient entry to this topic is provided by Hinze in the second edition of his book, which contains a section summarizing the status and observations to 1975. Some of the later work is cited by the present authors who have now made a further contribution by studying the heat transfer mechanism and associated turbulence structure in the near wall region of a boundary layer in a water channel. The results show that the turbulent temperature field near the wall resembles closely the flow structure in the viscous wall region and are compatible with what is known about the streamwise sublayer vortices from previous investigations.

A decade which has seen a huge increase in computing power, and a number of new methods of solving the turbulence closure problem, has also seen a shift of interest in boundary layer research, from the relatively simple type of flow documented in the 1968 Stanford Conference Proceedings [1] to "complex" boundary layers whose structure and development are affected by a variety of complicating factors. These factors include buoyancy interaction in density stratified flow, extra mean rates of strain in the flow over curved walls, body forces in rotating boundary layers, unsteadiness of the mean flow, and the effects of compressibility and interactions with shock waves. The diversity of topics treated in the remaining six conference papers selected for inclusion in this chapter reflect this research activity though it is, perhaps, significant that only one of them is directly concerned with turbulence modelling and flow prediction. Work in this field seems now to have settled down to valuable but unspectacular improvements in detail to the closure methods devised in the early 1970s. A basic need is for reliable experimental data to guide model development because the passage of time has done little to narrow what Bradshaw [2] in 1972 called the "fact gap" between potential computing power; and the experimental database on which models are based. The 1981–82 Stanford Conferences [3] on complex flow prediction revealed how slender is the data base currently available. From this point of view the quantitative data contained in the papers by Muck and Smits, Nakagawa et al., and Ramaprian et al., from compressible, three-dimensional, and periodic boundary layers respectively, exemplify the sort of material that is required.

In the first of these papers Muck and Smits present the results of hot-wire measurements made in a two-dimensional compressible boundary layer disturbed to the point of separation by an oblique shock wave generated in a compression corner. The main features of turbulence are known to be unaffected by compressibility in high speed flow when the free stream Mach number is less than about five. However, the passage of a turbulent boundary layer through a shock wave produces large increases in turbulent intensity, and correspondingly large decreases when the flow passes through an expansion. These changes, which are probably due to the direct influence of extra strain on the turbulence structure, are a demanding test for modelling methods [3, 4]. In discussing their results the authors surmise that the observed amplification of the turbulence may be due to the combined effects of compressibility and longitudinal destabilizing curvature, with the former outweighing the latter in the ratio of 5:2. An interesting observation is that the structure function, $-\overline{uv}/u'v'$, after increasing to about 0.35 in the inner part of the layer in the interaction zone, decreases in the recovery flow downstream to an extent that it actually undershoots the undisturbed layer value of about 0.25. The authors point out that similar trend was observed in an incompressible boundary layer relaxing to plane flow conditions from an impulse in concave wall curvature.

In a different area of boundary layer research, the work reported by Nakagawa and his colleagues stems from the possibility of multi-cellular secondary currents in straight wide rivers which is suggested by the periodic lateral distribution of sediment concentration. The authors have investigated the development of stress-driven secondary flows in the corners of smooth walled rectangular ducts of different aspect ratios without finding evidence for this supposition. The results show that the extent of the secondary flows is confined to the corner regions and becomes approximately constant irrespective of the duct aspect ratio. Cellular secondary currents are not produced in the central zone of the rectangular channel when the aspect ratio is greater than four; in the large aspect ratio ducts the flow in the centre appears to be remarkably two-dimensional. Previous studies of stress-driven secondary flow have nearly always been made in low aspect ratio ducts where the effects are most apparent. A merit of this contribution is that the effects of aspect ratio have been systematically investigated. The authors suggest that multi-cellular secondary currents may result from the formation of longitudinal sand ridges by scouring in regions of high shear stress. They have carried out some experiments with artificial ridges which seem to support this hypothesis but evidently more work needs to be done on this interesting problem.

The authors of the two papers on skin-friction measurement describe different ways of evaluating this basic quantity in a diversity of flows. The surveys of existing techniques by Winter [5] and Hanratty and Campbell [6] provide an entry to the subject, the former including an interesting brief description of the simultaneous use of two Preston tubes of different diameters. Nitsche et al. have now further developed this technique and have tested the results against measurements made in a number of different flows with sublayer fences, conventional Preston tubes, and surface hot film gauges. The calibration is based on an assumed van Driest function containing a number of disposable constants and good results are apparently obtained when only one of these constants, the von Karman constant, is varied. The method appears to be promising and experimenters will follow with interest the further developments promised by the authors. Acharya and Escudier have followed the example of numerous workers in addressing themselves to the difficult problem of measuring the wall shear stress directly. Winter [5] lists more than twenty previous descriptions of floating element skin-friction gauges, following the early work of Schultz-Grunow in 1940. The present authors appear to have overcome the difficulties associated with the null reading type of instrument to obtain satisfactory measurements of the skin friction in rough wall and transitional boundary layers. Their paper may usefully be read in conjunction with the two papers cited and that of Brown and Joubert [7].

The papers by Hanjalić and Stõsić and by Ramaprian et al. deal with unsteady boundary layers. Previous experimental work on this topic which has been reviewed by Carr [8, 9] includes several contributions to previous volumes in the Turbulent Shear Flows series. It appears from over forty experiments covered by the survey that the time-averaged mean velocity in a boundary layer is the same as the value expected for a steady flow having a velocity corresponding to the mean of the oscillatory free stream. Several of the experiments examined showed the turbulence structure to be unaffected by the unsteadiness of the flow field, in that although significant variations occurred in the shear stress and turbulent energy, their ratio remained constant at the steady flow value. Other results indicated that conditions exist for which substantial changes in the turbulence can occur. Quasi-steady behaviour is observed for low frequencies and low amplitudes of the periodic free stream velocity. As the frequency is increased, a critical value is reached which is related to that of turbulent bursts in the flow and at which there can be significant interaction between the periodic motion and the turbulence structure. Carr's findings can be used to place the results of the two present papers in context. Fig. 1, which has been redrawn from [9], shows the data in question plotted on a graph of dimensionless frequency, the Strouhal number, against percentage amplitude. f is the frequency of oscillation, δ is the boundary layer thickness or pipe radius for pipe flow, and U is the average velocity of the free stream. The shaded area shows the ranges of amplitude and frequency for which no unsteady effects have been reported. A Strouhal number of about 0.2 corresponds to the bursting frequency. The experiments presented in the present paper by Ramaprian et al. are represented by filled symbols: circles for the pipe flow and a triangle for the boundary layer. The two high frequency experiments fall in the area where marked unsteady effects are to be expected and where only a limited amount of data have been previously obtained. The results show that not only is the turbulence structure disturbed from equilibrium to varying degrees depending on the frequency, but also that the externally imposed periodicity produces a small but measurable effect on the time mean velocity.

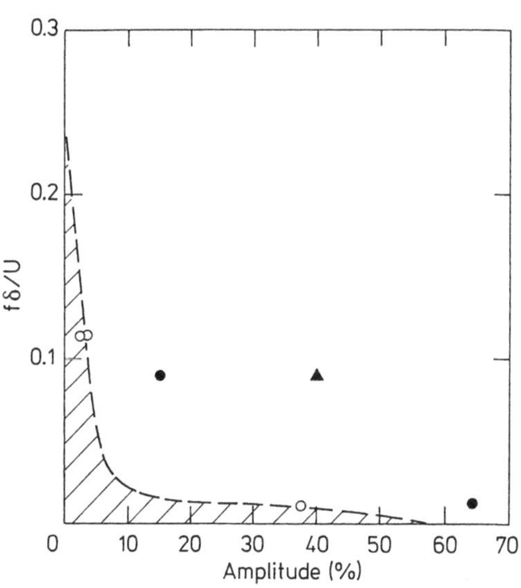

Fig. 1. Amplitudes and frequencies of unsteady boundary layer experiments. From Carr [9]

Hanjalić and Stõsić have calculated the response of boundary layers to unsteadiness in the free stream. The predictions are compared with the results of boundary layer experiments at the frequencies and amplitudes shown by the open symbols on the figure. These fall on the edge of the shaded region inside which, according to Carr [9], unsteady effects are not to be found. Comparisons are also made with a third set of data which are not shown on the figure; these had been obtained in a water channel at very low frequencies and amplitudes approaching 100 per cent. Quasi-steady behaviour in these conditions does not, however, preclude significant changes in the turbulence in each oscillation cycle. These changes lag behind those of the mean flow and, as the authors note, the time lags in the acceleration period are markedly different to those which occur in the deceleration phase, because the response of the turbulence structure varies across the wave number spectrum. The response of the turbulent shear stress and energy then shows pronounced hysteresis which is tolerably well predicted by the single- and two-scale energy-dissipation models and single-scale Reynolds stress model tested by the authors. The finding that the single-scale models are not noticeably inferior is probably attributable to the quasi-steady behaviour of the flows examined. The use of multiple scales may well be seen to greater advantage in the calculation of flows of larger amplitude and frequency, such as those documented here by Ramaprian et al. The paper also contains points of general interest to turbulence modellers in the further application of multiple-scale modelling, and in two areas where current methods are recognised as defective, namely the formulation of the source terms in the energy-dissipation rate equation and the treatment of the turbulence near the wall.

The variety of topics treated in these papers is impressive, as is the range of experimental techniques deployed in the experiments. The results from the complex flows documented here should be especially useful to researchers concerned with problems of turbulence modelling and flow calculation. In this context the usefulness of experimental research depends crucially on completeness in it's documentation. All too often one finds important items of information missing in accounts of interesting experiments, or that initial and boundary conditions or other essential data for the calculator have to be extracted with difficulty from small-scale graphs or are complicated by imprecise normalizing factors. This short introduction concludes with a plea to experimenters to remember the needs of flow calculators in this respect. It is anticipated that the contributions made in this group of papers may be of lasting value.

References

1. Coles, D. E., Hirst, E. A. (eds.) (1968): *Computation of Turbulent Boundary Layers*, Proc. AFOSR-IFP-Stanford Conference
2. Bradshaw, P. (1972): The understanding and prediction of turbulent flow. Aeronaut. J. **76**, 403
3. Kline, S. J., Cantwell, B. J., Lilley, G. M. (eds.) (1981): Proc. 1980–81 AFOSR-HTTM-Stanford Conference on Complex Turbulent Flows, Stanford
4. Viegas, J. R., Horstman, C. C. (1979): Comparison of multi-equation turbulence models for several shock boundary-layer interaction flows. AIAA J. **17**, 811
5. Winter, K. G. (1979): An outline of the techniques available for the measurement of skin friction in turbulent boundary layers. Progr. Aerosp. Sci. **18**, 1
6. Hanratty, T. J., Campbell, J. A.: Measurement of wall shear stress, in *Fluid Mechanics Measurements*, ed. by R. J. Goldstein (Hemisphere, New York 1983) Chap. 11
7. Brown, K. C., Joubert, P. N. (1969): The measurement of skin friction in turbulent boundary layers with adverse pressure gradients. J. Fluid Mech. **35**, 737
8. Carr, L. W. (1981): A compilation of unsteady boundary layer data. AGARDograph AG 265
9. Carr, L. W.: A review of unsteady turbulent boundary layer experiments, in *Unsteady Turbulent Shear Flows*, ed. by R. Michel, J. Cousteix, R. Houdeville (Springer, Berlin, Heidelberg, New York 1981)

Heat Transfer Mechanism and Associated Turbulence Structure in the Near-Wall Region of a Turbulent Boundary Layer

Yoichiro Iritani, Nobuhide Kasagi and Masaru Hirata

Department of Mechanical Engineering, University of Tokyo, Bunkyo-ku, Tokyo 113, Japan

Abstract

Heat transfer mechanism and associated turbulence structure in the near-wall region of a turbulent boundary layer have been investigated experimentally. With the aid of temperature-sensitive liquid crystal and hydrogen bubble method, the wall temperature fluctuation and the flow structure are visualized simultaneously to study directly the heat exchange between the fluid and the wall. The structure of the temperature field near the wall is further studied by the measurement of cross correlation of temperature fluctuations in the spanwise and streamwise directions. All of these experimental results show that the turbulent temperature field near the wall has a feature very much similar to the flow structure in the viscous wall region, e.g., the streamwise elongated structure with a quasi-periodicity in the spanwise direction, and are compatible with what is known about the streamwise sublayer vortices previously studied.

Nomenclature

Coh^2	square of coherence, Eq. (5)
F	bursting frequency, 1/ms
f	frequency, Hz
$f\upsilon$	$= u^{*2}/\nu$, Hz
k	mass transfer fluctuation
N	number of event
R	correlation coefficient
T_B	bursting period, s, Eq. (1)
t	time, s
U_∞	free stream velocity, m/s
u	velocity fluctuation in x-direction, m/s
u^*	friction velocity, m/s
v	velocity fluctuation in y-direction, m/s
w	velocity fluctuation in z-direction, m/s
x	streamwise coordinate
y	coordinate normal to the wall
z	transverse coordinate
δ	boundary layer thickness ($0.99\,U_\infty$), mm
δ_T	thermal boundary layer thickness, mm
δ^*	displacement thickness
θ	momentum thickness, mm, or temperature fluctuations, °C
λ	spanwise spacing between streaks
ν	kinematic viscosity, m²/s
σ	standard deviation

Superscripts

$(\ )^+$	non-dimensionalized with u^* and ν
$(\ )'$	root-mean-square value

Subscripts

$(\ )_f$	low-speed streaks
$(\ )_T$	high-temperature streaks
$(\ )\vert_w$	at the wall

Introduction

In the wall region of a turbulent boundary layer, characteristic intermittent turbulence motions are the ejection of low-momentum fluid from the wall and the inrush of high-momentum fluid towards the wall (sweep), i.e., the bursting phenomena [1–3]. In the region further close to the wall, the fluid motion is much affected by the molecular viscosity and there appear so-called low- and high-speed streaks. Characteristic spatial and time scales of such streaks are known to be statistically deterministic [1–4], but the turbulent motions associated with streaky structures are also highly intermittent and non-isotropic. While the ejection and sweep motions during the bursting process have an important role upon not only turbulent momentum flux [2, 5] but also turbulent heat flux [5, 6], the relationship between the coherent streaky structure and the heat transfer mechanism has not been clear enough, primarily due to the fact that up to the present the temperature field has been studied mostly by point type measurements. Meanwhile the visualization of wall pressure fluctuation has revealed many important features of wall turbulence [7] which are difficult to obtain by the measurement above.

In this study, with the use of the liquid crystal technique [8, 9] combined with the hydrogen bubble method, the wall temperature fluctuation as well as the flow structure is visualized simultaneously to study directly the heat exchange between the wall and the fluid with an emphasis on the streaky structures near the wall. In addition, the temperature fluctuations are measured by a conventional thermocouple probe to obtain the cross correlation in the spanwise and streamwise directions.

Experimental Apparatus and Procedures

All the experiments have been carried out in a low-speed closed-loop water channel [10] and the measurements have been made in the 2 m-long test section, shown in Fig. 1, which is attached to the end of the nozzle (contraction ratio is 9:1) of the water channel. The test section is made of transparent plexiglas and has a cross-section of 0.3 m × 0.3 m. One of the vertical side walls of the channel is used as a working wall on which the turbulent boundary layer is developed. The pressure gradient in the streamwise direction is removed by an additional roof plate installed in the test channel. The cylindrical rod of 4 mm in diameter is introduced at the inlet of the test channel to promote boundary layer transition.

The test plate for the visualization of wall temperature fluctuation shown in Fig. 2 is installed at the position 1600 mm downstream of the inlet of test section. Four stainless steel foils, each of them 300 mm in length, 45 mm in width and 30 μm in thickness, are bounded by epoxy adhesive to the surface of a bakelite plate and connected electrically in series. Constant heat flux heating is achieved by A.C. supply into these foils. A liquid crystal sheet, which is composed of polyester sheet of 50 μm thickness, a liquid crystal layer and black paint [9, 10], is attached flush on the heater, and the total thickness of these materials is about 100 μm. Thus, the wall temperature fluctuation is visualized as a color change of liquid crystal. The upper limit of frequency range in this experiment, where the color change can be recognized, is estimated as about 2 Hz [9], and is approximately equal to the frequency, where the power spectrum of streamwise velocity fluctuation is one tenth of its maximum value [10].

The hydrogen bubble method is used for flow visualization. The cathode is a tungsten wire and the pulse interval of D.C. supply is controlled as 0.1 sec. Presently, transverse and vertical wires are used in the region about 1750 mm downstream of the tripping rod. The transverse wire of 20 μm in diameter is positioned perpendicular to the flow direction and parallel to the wall, while the vertical wire of 50 μm in diameter is positioned perpendicular to the wall.

For the quantitative measurement of temperature fluctuation, copper-constantan thermocouples of 50 μm in diameter are used. The frequency response is fast enough and is estimated to be flat up to about 120 Hz. Signals obtained are recorded on a FM-type tape recorder, TEAC R210-B, through a DC amplifier and a 35 Hz low-pass filter. The resultant frequency response of this system is 35 Hz ($= fv/2.47$). The analogue signals are digitized at a sampling rate of 100 Hz, stored on the magnetic tapes, and successively processed by a mini-computer system, MELCOM 70/25. A "fixed probe" and a "moving probe" are used.

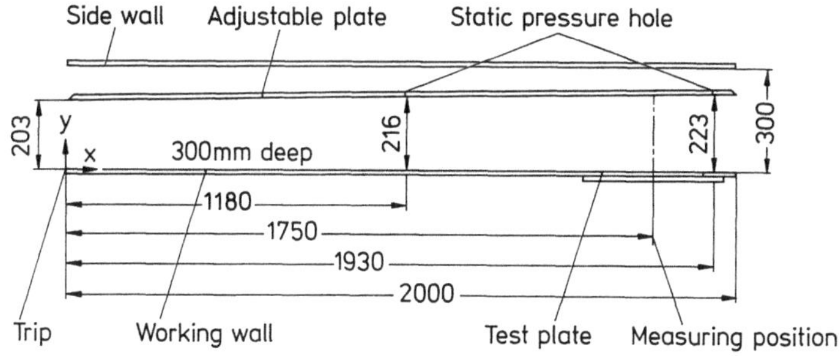

Fig. 1. Schematic of the test section

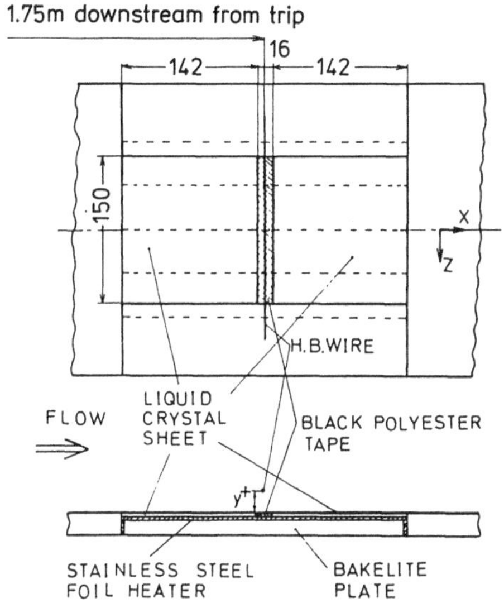

Fig. 2. Test plate for the simultaneous visualization of wall temperature fluctuation and flow field

The location of the fixed probe is 1.75 m downstream of the trip and right above the center line of the test plate. The moving probe is traversed relative to the fixed probe in the spanwise and streamwise directions, with the distance from the wall being the same as that of the fixed probe. A thermocouple is soldered on the backside surface of stainless steel foil heater to measure the wall temperature fluctuation. In this case, the liquid crystal sheet is not attached on the heater surface to assure a higher frequency limit than the liquid crystal technique.

Basic experimental conditions are shown in Table 1. The mean flow velocity is measured by the Pitot tube of a boundary layer type and an inclined U-tube manometer of two fluid type (water and benzene). The free stream turbulent intensity measured by a hot-film anemometer system, DISA 55M10, is about 2%, while the water temperature is kept at 20 °C. At the location of measurement 1.75 m downstream from the trip, the flow field is confirmed to be completely two-dimensional in the central 60 mm region and the turbulent

Table 1. Basic data of the turbulent boundary layer

U_∞ m/s	δ mm	δ^* mm	θ mm	$R_{\delta^*} = \frac{U_\infty \delta^*}{\nu}$	$R_\theta = \frac{U_\infty \theta}{\nu}$	$H = \frac{\delta^*}{\theta}$	u^* m/s	δ_T mm
0.204	50	6.8	4.9	1370	990	1.4	9.35×10^{-3}	9.4

boundary layer is fully developed [10]. The friction velocity is determined by the Clauser's chart. In the present experiment, only the streamwise 300 mm length of the test plate is heated with a uniform heat flux, and the thermal boundary layer thickness is 9.4 mm ($y^+ = 87$) at the half length of the heated plate. Thus the thermal contamination is confined deep into the turbulent boundary layer so that the structure of near-wall region is mainly studied. Heating is always slight and the temperature can be recognized as a passive contaminant.

Experimental Results and Discussion

Simultaneous Visualization of Flow Structure and Wall Temperature Fluctuation

In a preliminary experiment, the present technique using liquid crystal was tested in a laminar boundary layer along the flat plate similar to that shown in Fig. 2. The steady color pattern observed indicated that the wall temperature distribution was uniform in the spanwise direction and that the local heat transfer coefficient decreased gradually in the streamwise direction.

A typical photograph of the H_2-bubble/liquid crystal visualization in the turbulent boundary layer is shown in Fig. 3, where the location of transverse wire is at $y^+ = 6$. The streaky structures of high-speed fluid and low-speed fluid are observed appearing alternatively in the spanwise direction and being elongated in the streamwise direction. The wall temperature fluctuation visualized by the liquid crystal also indicates streaky pattern, and the thermal field consists of high-temperature streaks (black in Fig. 3 but actually blue) and low-temperature streaks (white but actually orange). These thermal streaks are observed more or less periodically in the spanwise direction and elongated in the streamwise direction much same as the momentum streaks. The streamwise extent of these thermal streaks is about 1000–2000 ν/u^* and is by an order of magnitude longer than the spanwise spacing. In the 16 mm motion pictures taken in this experiment, every thermal streak appears randomly in space and time and vanishes out after some timespan in which it moves downstream. It is qualitatively confirmed that the high-temperature streaks are mostly associated with the low-speed streaks and low-temperature streaks with the high-speed streaks.

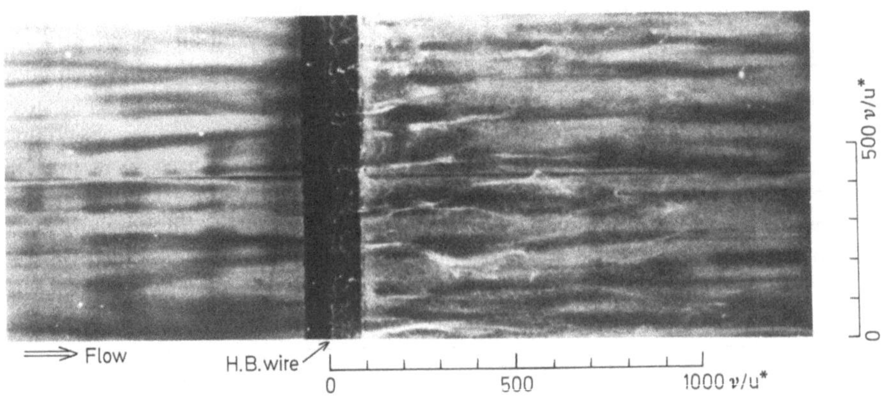

Fig. 3. Simultaneous visualization of wall temperature fluctuation and flow field (A hydrogen bubble wire as located at $y^+ = 6$)

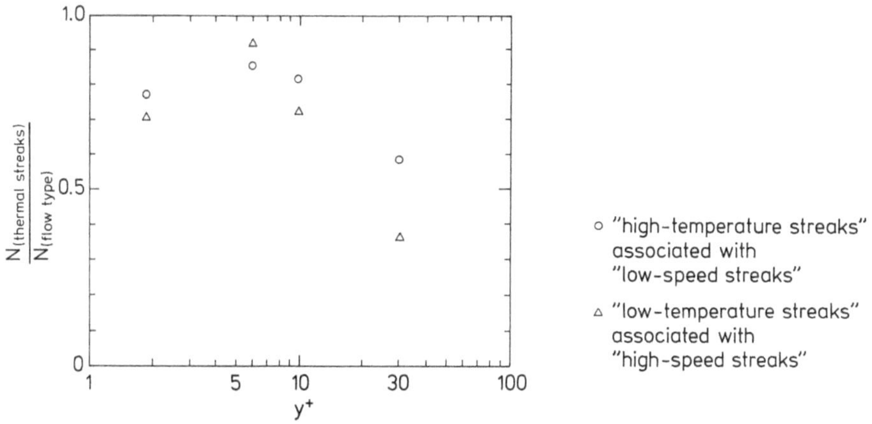

Fig. 4. Correspondence between thermal streaks and momentum streaks as a function of distance from the wall

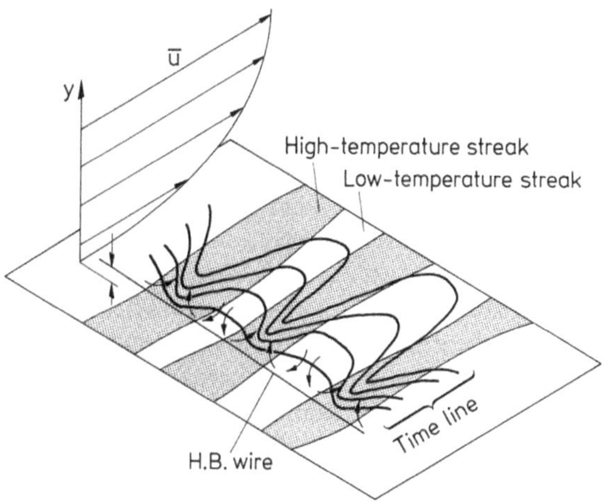

Fig. 5. Conceptual model of the relationship between thermal streaks and momentum streaks

The correspondence between the thermal and momentum streaks is shown in Fig. 4, as a function of a distance from the wall. These data have been obtained from 20 – 40 streaks in the 8 conventional photographs taken at each position over 2.16 sec ($= 187\ \nu/u^{*2}$) by using a motor-driven camera. The observed correspondence is remarkable in the region $y^+ \leqq 10$ and more than 70% of the thermal streaks are uniquely associated with the prescribed momentum streaks. But these tight relationship fades out in the region $y^+ \geqq 30$ due to the disappearance of streaky structures. These results agree with the observation that the center of the streamwise vortices associated with the bursting phenomena lies at approximately $y^+ = 20 - 30$ [11] (also see Fig. 10). A conceptual model of the relationship between the coherent streaky structures and thermal streaks is presented in Fig. 5.

Simulataneous observations of the wall thermal streaks and the flow pattern in the x-y plane are given by using the vertical hydrogen-bubble wire as shown in Fig. 6. Figure 6(a) indicates a typical pattern that the high-temperature streak is associated with the lift-up event [1], which is presently defined as the case when the lift-up motion of low-speed streak and the inflectional point of time lines appear within the area $x^+ = 100$ downstream of the

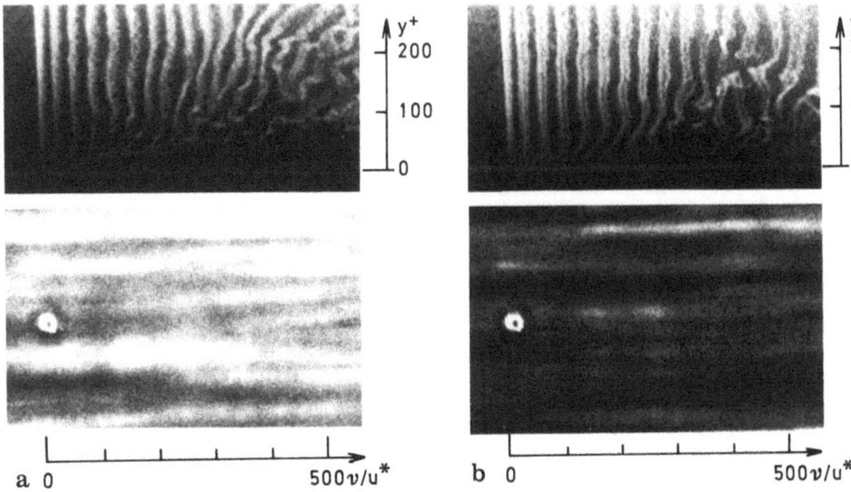

Fig. 6 a. Typical appearance of high-temperature streak associated with lift-up event. **b** Typical appearance of low-temperature streak associated with sweep event

wire. On the other hand, Fig. 6(b) indicates that the low-temperature streak is formed by the sweep event, i.e., the appearance of large fluid element toward the wall in the region $y^+ \leqq 50$ within $x^+ = 100$ downstream of the wire. While 82 couples of photograph have been obtained, the high-temperature streaks have appeared with 70% of the total lift-up events, and the low-temperature streaks with 72% of the total sweep events.

From the experimental results described above, the heat exchange between the wall and the fluid is considered primarily ruled by the streaky structures in the region at least $y^+ \leqq 10$. Under the present constant heat-flux condition, the instantaneous local heat transfer coefficient is relatively small and large at the high- and low-temperature streaks, respectively. When the local inrush towards the wall is induced by the streamwise vortices [11–13], the local wall temperature is rapidly decreased, and this results in the formation of the elongated low-temperature streak with the streamwise convection of low-temperature high-momentum fluid. During this streamwise convection, the fluid is decelerated by the wall shear and, at the same time, heated by the wall heat flux, and this results in the high-temperature low-momentum streak which is eventually ejected from the wall by the streamwise vortices.

Spanwise Spacing of Streaks

The 16 mm motion pictures with the transverse wire positioned at $y^+ = 4$ are analyzed on a screen of a motion analyzer. The spanwise spacing of low-speed streaks, λ_f, is sampled and measured in every 10 frames (at an interval of 0.417 sec = $36\,\nu/u^{*2}$). The total average spacing is $89\,\nu/u^*$, and the probability density distribution is well approximated by a log-normal distribution [3, 10]. These experimental results are in good agreement with the values obtained in previous studies [1–4]. (In Ref. [4], measured values of λ_f have been summarized.)

In order to determine the spanwise spacing of the thermal streaks quantitatively, high- and low-temperature streaks are defined in the following manner as shown in Fig. 7. A threshold is the border between green and orange-yellow, and is presently equal to 30.9 °C. Uncertainty of ± 0.35 °C, however, lies in identifying whether the local wall temperature is

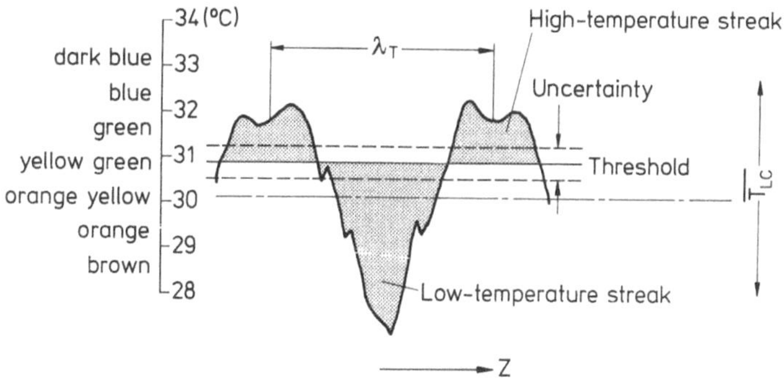

Fig. 7. High- and low-temperature streaks defined by the threshold of color change

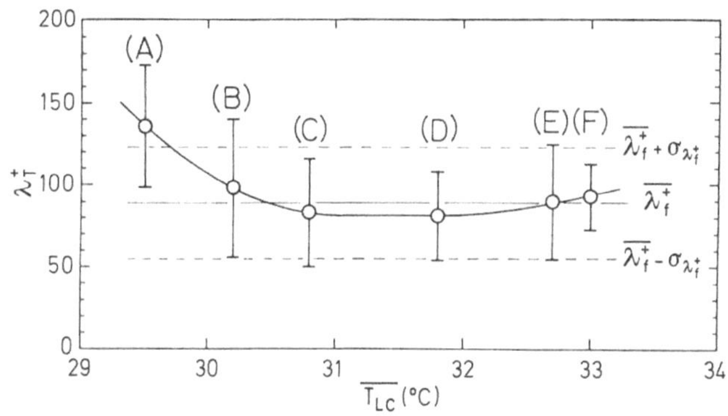

Fig. 8. Mean spacing and its standard deviation of high-temperature streaks

below or beyond this threshold. High-temperature streaks are defined as the regions where the temperature is higher than the threshold. The spanwise spacing of high-temperature streaks, λ_T, is defined as the distance between the centers of two neighboring high-temperature streaks. The time-averaged temperature of liquid crystal layer, $\overline{T}_{LC}$, as depends on the heat flux given at the wall, while the free stream temperature is kept constant, $T_\infty = 20\,°C$. Hence, the experimental results of the mean and the standard deviation of λ_T are obtained for various heat fluxes, i.e., various wall temperatures, as shown in Fig. 8. It is easily understood that the mean spacing between the thermal streaks thus defined is independent of the mean temperature level of the surface liquid crystal layer in the temperature range of $\overline{T}_{LC}$ between the case of (C) and (D). The value measured is $81\,\nu/u^*$ and agrees well with that of momentum streaks, $89\,\nu/u^*$. It is noted that the probability density distribution can also be approximated as a lognormal distribution. Under the present experimental condition of constant wall heat-flux, the wall temperature fluctuation can be recognized straightforward as the fluctuation of local heat transfer coefficient. In other words, the heat transfer coefficient also has a quasi-periodicity of $\overline{\lambda}_T^+ = 81$ in the spanwise direction, but this result is somewhat different from the result of Shaw and Hanratty [14], where the lateral wavelength characterizing the spatial variation of the mass transfer rate at high Schmidt number would be one-half the lateral wavelength of the flow oriented eddies (also see Fig. 11).

Frequency of the Occurance of Streaks

The visual break-up frequency of the low-speed streaks [1] is counted in the motion pictures with the transverse wire located at $y^+ = 4$. The result is shown as a straight line in Fig. 9, which indicates the frequency of $F = 80$ (1/ms), i.e. $F^+ = F\nu^2/u^{*3} = 99.8 \times 10^{-3}$. The pass-

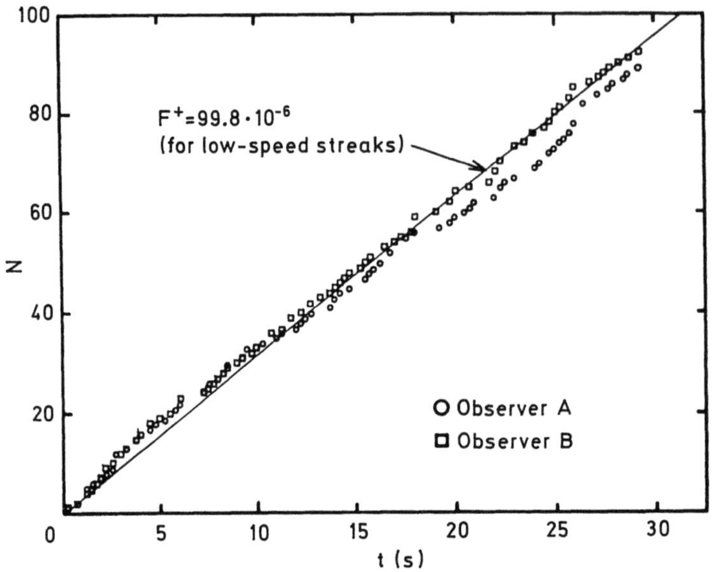

Fig. 9. Comparison of the passing frequency of low-temperature streaks with the break-up frequency of low-speed streaks

ing frequency of low-temperature streak is also counted with the wall temperature range, where the measured spanwise spacing is constant as described previously. The results are also included in Fig. 9, where a good agreement is obtained between the break-up frequency of the low-speed streak and the passing frequency of the low-temperature streak. These results again suggest that the turbulent motions of sublayer vortices, which lead to the formation of the low- and high-speed streaks, produce the similar pattern of wall tempera-ture fluctuation shown in Fig. 3.

If the bursting period is defined by the following equation [1];

$$T_B = 1/F\,\lambda_f \tag{1}$$

then it is 1.3 sec ($= 112\,\nu/u^{*2}$) in the present experiment and is in agreement with the values obtained previously [1, 2, 4].

Temperature Fluctuation in the Viscous Wall Region

The structure of the temperature fluctuation is investigated by the analyses of cross correla-tions. The cross correlation coefficient between wall temperature fluctuations and fluid temperature fluctuations is defined as follows;

$$R_{\theta_w\theta} = \overline{\theta_w\theta}/\theta'_w\theta'. \tag{2}$$

The result is shown in Fig. 10 as a function of a distance from the wall. This figure also include the correlation coefficient of spanwise velocity fluctuations measured by Blackwelder and Eckelmann [11]. The temperature correlation shows a rapid decrease in the region of $y^+ \leqq 10$ and gradually approaches zero without any negative value, in contrast to the spanwise velocity fluctuations. This discrepancy is considered due to the difference between the measured variables, i.e., a scalar and a spanwise vector, respectively.

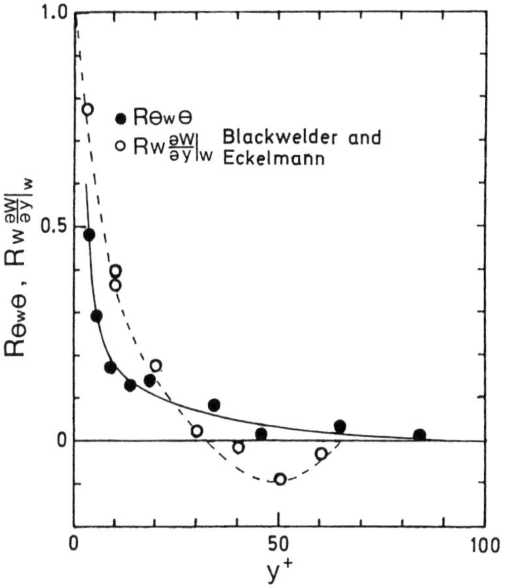

Fig. 10. Spatial correlation of the temperature fluctuations with the separation normal to the wall

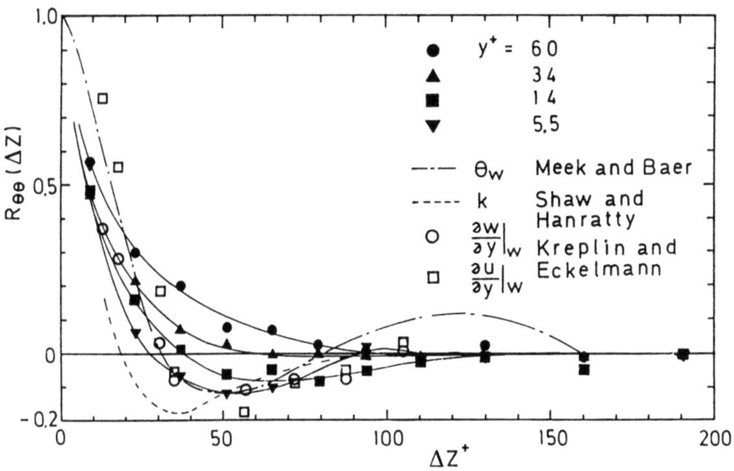

Fig. 11. Spanwise correlations of the temperature fluctuations

The spanwise and streamwise cross correlations are defined as;

$$R_{\theta\theta}(\Delta z) = \overline{\theta(z)\,\theta(z+\Delta z)}/\theta'(z)\,\theta'(z+\Delta z) \tag{3}$$

$$R_{\theta\theta}(\Delta x) = \overline{\theta(x)\,\theta(x+\Delta x)}/\theta'(x)\,\theta'(x+\Delta x). \tag{4}$$

The present rcsult of spanwise correlation is shown in Fig. 11 along with the related data obtained by previous researchers [12, 14, 15]. At the edge of the viscous sublayer, $y^+ = 5.5$, there is a negative peak at the distance of $\Delta z^+ = 50$, and a positive small peak around $\Delta z^+ = 100$. Thus the quasi-periodicity with a period of about $\Delta z^+ = 100$ can be confirmed. As the distance from the wall is increased, the distance which gives the first zero-crossing increases, and a negative correlation is hardly observed beyond $y^+ = 34$. The present data at $y^+ = 5.5$ are very much similar to the correlations of the spanwise velocity gradient

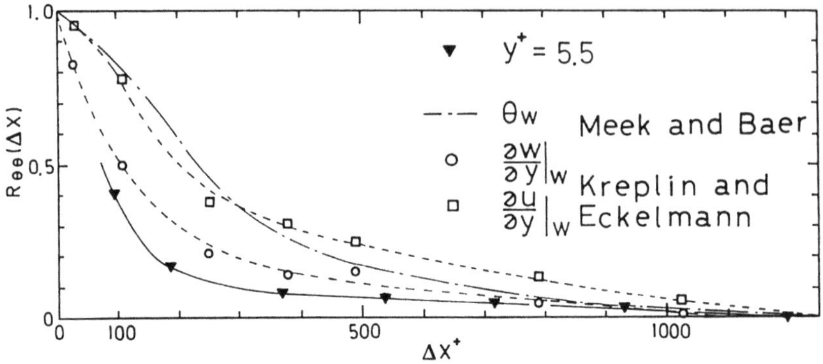

Fig. 12. Streamwise correlations of the temperature fluctuations

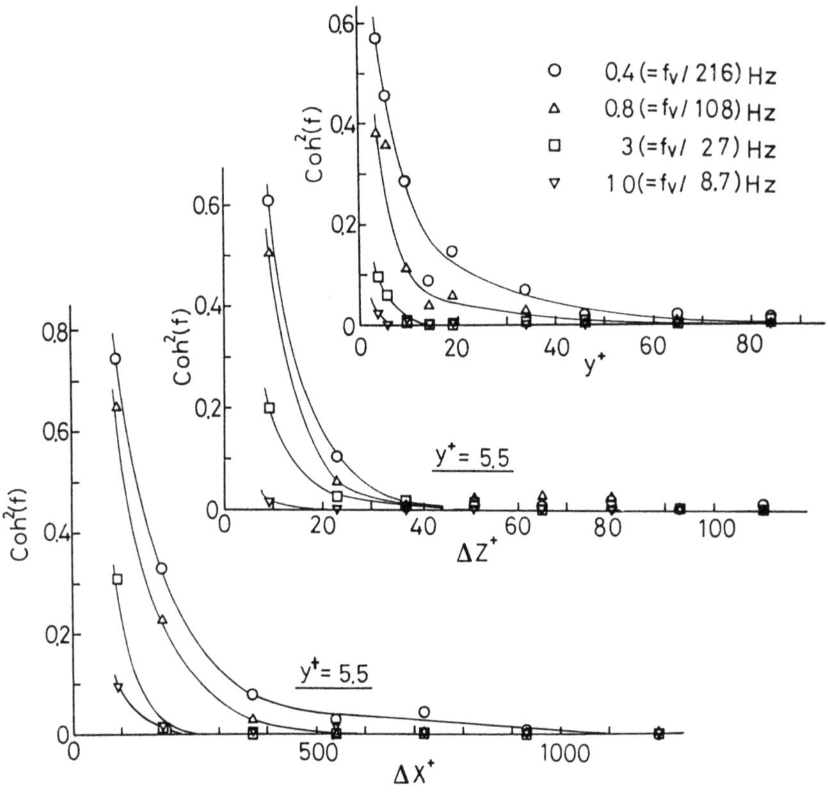

Fig. 13. Coherence of the temperature fluctuation

fluctuation at the wall, but not so much to those of the streamwise velocity gradient. The data of the wall temperature fluctuations by Meek and Baer [15] and the mass transfer fluctuations by Shaw and Hanratty [14] are also somewhat different from the present data.

The result of the streamwise correlation at $y^+ = 5.5$ is shown in Fig. 12, where a positive correlation can be found as far as $\Delta x^+ = 1100$. This length scale is qualitatively in agreement with the streamwise extent which has been visually confirmed by the liquid crystal technique (see Fig. 13). It also agrees with the length scale suggest by Kreplin and Eckelmann [12]. The curve for the temperature fluctuations at $y^+ = 5.5$ is again similar to that for the spanwise velocity gradient rather than the streamwise velocity gradient, even though data by Meek and Baer is considerably different from the present result. All of these results of spatial temperature correlations suggest not only the existence of streamwise sublayer vortices [11–14] but also their important role on the heat transfer mechanism in the viscous wall region.

In order to analyze the frequency range, of which the above cross correlations are composed, the coherence is obtained by using the following equation;

$$\mathrm{coh}^2(f) = |\Theta_{12}|^2/\Theta_1\Theta_2. \qquad (5)$$

In Eq. (5), Θ_{12} is a cross spectrum of θ_1 and θ_2, while Θ_1 and Θ_2 are spectra of θ_1 and θ_2, respectively. Numerical data reduction has been carried out by using FFT and Hanning filter. As a result, it can be said that the component of an order of the bursting frequency must be a main contributor to the coherent transport mechanism in the viscous sublayer.

Conclusions

The relationship between the coherent structures and the temperature fluctuations in the near-wall region of a turbulent boundary layer has been investigated experimentally.

Through the visual data obtained with the aid of the liquid crystal technique combined with the hydrogen bubble method, it is directly confirmed that the high-temperature streaks are mostly associated with the low-speed streaks and the low-temperature streaks with the high-speed streaks. In addition, quantitative agreements of the spacing and the frequency between thermal and momentum streaks are obtained.

In the viscous sublayer and the buffer layer, the fluctuating temperature field has a quasi-periodicity in the spanwise direction with a mean period of about $\bar{\lambda}_f^+$ and has a much elongated scale in the streamwise direction, about $1100\,\nu/u^*$, and these scales are consistent with the scale of streamwise vortices proposed by previous studies. These coherent structures are composed by components with a frequency of the order of the bursting frequency.

The turbulent motions of studied lead to the formation of streaky structures and these play a main role on the turbulent heat transfer in the viscous-wall region, at least for $y^+ \leqq 10$.

References

1. Kline, S. J., Reynolds, W. C., Schraub, F. A., Runstadler, P. W. (1967): The structure of turbulent boundary layers. J. Fluid Mech. **30**, 741
2. Kim, H. T., Kline, S. J., Reynolds, W. C. (1971): The production of turbulence near a smooth wall in a turbulent boundary layer. J. Fluid Mech. **50**, 133
3. Smith, C. R., Metzler, S. P. (1983): The characteristics of low-speed streaks in the near-wall region of a turbulent boundary layer. J. Fluid Mech. **129**, 27
4. Hirata, M., Tanaka, H., Kawamura, H., Kasagi, N. (1982): Heat transfer in turbulent flows. Proc. 7th Int. Heat Transfer Conf. Vol. 1, p. 32
5. Perry, A. E., Hoffmann, P. H. (1976): An experimental study of turbulent convective heat transfer from a flat plate. J. Fluid Mech. **77**, 355
6. Zaric, Z. (1975): Wall turbulence structure and convection heat transfer. Int. J. Heat Mass Transfer **18**, 831
7. Dinkelacker, A., Hessel, M., Meier, G. E. A., Schewe, G. (1977): Investigation of pressure fluctuations beneath a turbulent boundary layer by means of an optical method. Phys. Fluids **20**, S 216

8. Kasagi, N. (1980): Liquid crystal applications in heat transfer experiments. Report IL-27, Thermosciences Division, Mech. Eng. Dept. Stanford University
9. Kasagi, N. (1982): Application of temperature-sensitive liquid crystals. J. Flow Visual. Soc. Jpn **2**, 647 [in Japanese]
10. Iritani, Y., Kasagi, N., Hirata, M. (1982): Transport mechanism in a turbulent boundary layer. 1st Rept. Trans. JSME, Ser. B**48**, 2284 [in Japanese]
11. Blackwelder, R. F., Eckelmann, H. (1979): Streamwise vortices associated with the bursting phenomenon. J. Fluid Mech. **94**, 577
12. Kreplin, H. P., Eckelmann, H. (1979): Propagation of perturbation in the viscous sublayer and adjacent wall region. J. Fluid Mech. **95**, 305
13. Lee, M. K., Ekelman, L. D., Hanratty, T. J. (1974): Identification of turbulent wall eddies through the phase relation of the components of the fluctuating velocity gradient. J. Fluid Mech. **66**, 17
14. Shaw, D. A., Hanratty, T. J. (1977): Influence of Schmidt number on the fluctuations of turbulent mass transfer to a wall. AIChE J. **23**, 160
15. Meek, P. L., Baer, D. (1970): The periodic viscous sublayer in turbulent flow. AIChE J. **16**, 841

The Behaviour of a Compressible Turbulent Boundary Layer Under Incipient Separation Conditions

K. C. Muck and A. J. Smits

Gas Dynamics Laboratory, Mechanical and Aerospace Engineering Department,
Princeton University, Princeton, NJ 08544, USA

Abstract

This paper presents an experimental study of a turbulent boundary-layer/shock-wave interaction. The interaction was generated by a two-dimensional compression corner, and the flow was on the point of separating. Measurements were made using both normal and inclined hot wires, and the data include measurements of the longitudinal mass-flow fluctuation intensity and the mass-weighted Reynolds shear stress.

Introduction

When a shock wave interacts with a turbulent boundary layer, the boundary layer properties experience a sudden, severe perturbation. The resulting distortion of the mean flow properties has been clearly documented, and considerable data appears to be available (see, for instance Fernholz and Finley, 1977). In contrast, direct measurements of the turbulent stresses are still rather scarce and sometimes of dubious quality (see, for example Fernholz and Finley, 1981). Nevertheless, high quality turbulence measurements in a wide range of interactions are urgently required; calculation methods for shock-wave/boundary-layer interactions have generally not performed well, and only if accurate and extensive turbulence data become available will better turbulence models be developed.

At the Gas Dynamics Laboratory at Princeton University, we hope to fulfill partially this need by studying the behaviour of turbulence in a range of compression corner flows. The corners are two-dimensional, and the flows range from fully attached to fully-separated. The mean flow behaviour was extensively documented by Settles (1975) and Settles et al. (1979), and, using these results, four corner angles were selected for detailed turbulence measurements: 8°, 16°, 20°, and 24°.

In the current contribution, we present measurements in the 16° corner interaction, which, at a freestream Mach number of 2.87 corresponds to incipient separation (Settles et al., 1976). The measurements were made using hot wire anemometry, and both normal and inclined wires were used. The data therefore include longitudinal mass-flow fluctuations $\overline{(\varrho u)'^2}$ as well as measurements of the mass-weighted Reynolds shear stress $\overline{(\varrho u)' v'}$. To gain further insight into the turbulence structure, the longitudinal velocity fluctuation intensity, $\overline{u'^2}$, and the correlation $\overline{u'v'}$ were deduced using Morkovin's "Strong Reynolds Analogy".

We begin by describing the apparatus and experimental conditions, and follow this with a presentation and discussion of the results. A summary of the conclusions is given in the final section.

This work was supported by NASA Headquarters under Grant NAGW-240, monitored by Drs. Clint Brown and Gary Hicks.

Apparatus and Experimental Techniques

The experiment was performed in the Princeton University high Reynolds number 20 cm × 20 cm supersonic blow-down tunnel (see Fig. 1). The tunnel was operated at a stagnation pressure of 6.9×10^5 N/m^2, the wall conditions were near adiabatic and the freestream rms mass-flow turbulence level was about 1%. The upstream boundary layer developed in a region of essentially zero pressure gradient along the tunnel floor. Immediately prior to the interaction, the boundary layer thickness was about 28 mm and the Reynolds number based on the momentum thickness was approximately 77,600 (see Table 1).

The test model was a two-dimensional ramp with a corner angle of 16°. The width was 15.24 cm, and side fences were installed to isolate the corner flow from the tunnel wall boundary layers (see Fig. 1). The measurement stations are shown in Fig. 2.

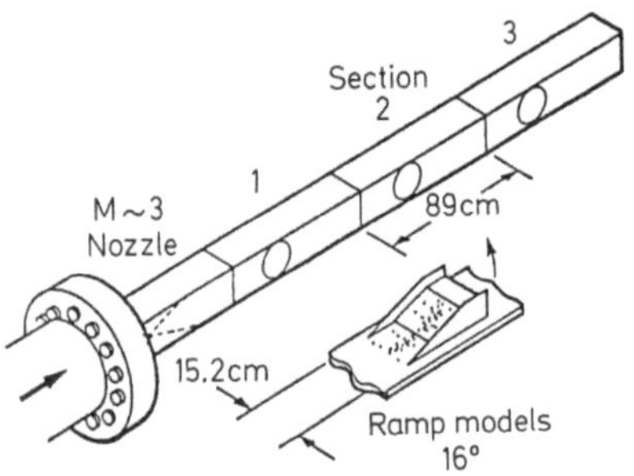

Fig. 1. Sketch of the 20 × 20 cm high Reynolds number channel and test model installation

Table 1. Incoming flow conditions. The boundary layer parameters were taken at $x = -51$ mm. The thickness δ_0 is defined as the distance where the mass flow rate reaches 99% of the freestream value

Re_∞/m	6.3×10^7/m
$M_{a\infty}$	2.87
p_0	6.9×10^5 N/m^2
T_w/T_0	1.04
δ_0	28.0 mm
δ^*_{ref}	6.7 mm
θ_{ref}	1.3 mm

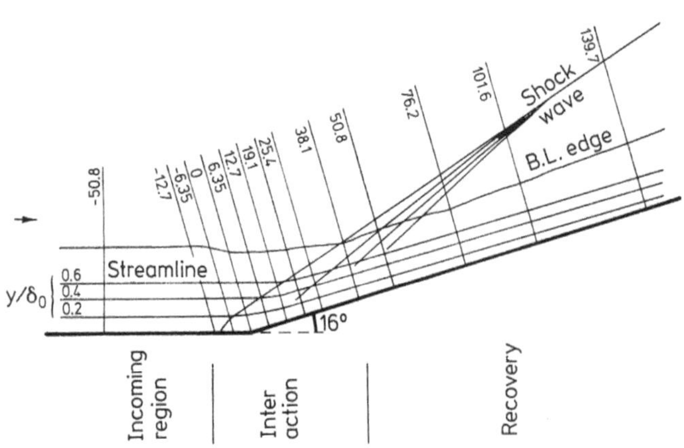

Fig. 2. The 16° two-dimensional compression corner flow field with indicated locations of the measuring stations. All dimensions are in mm

A DISA 55M 10 constant-temperature anemometer was used for the turbulence measurements. In an effort to improve the quality of these measurements, the authors recently undertook a critical study of hot-wire techniques in supersonic flows (Smits et al., 1983, and Smits and Muck, 1984) which resulted in a substantial improvement in both normal and inclined wire measurement accuracy.

The probes were constructed using 5 μm diameter tungsten wire with an active length of 0.8 ~ 1.0 mm, and they were operated at overheats of 1.0 ~ 1.1. At these high overheat ratios, the contribution due to the temperature fluctuations was small and it was therefore neglected. For the incoming boundary layer the uncertainties in $\overline{(\varrho u)'^2}/(\varrho U)^2_{ref}$ were − 10% to + 19%, and for $\overline{(\varrho u)' v'}/(\varrho U^2)_{ref}$ these were − 32% to + 21% (see Smits et al., 1983, and Smits and Muck, 1984). In addition, these estimates may be used as an indication of the measurement accuracy downstream of the interaction. It is important to note, however, that the results are only reliable where the instantaneous *normal* Much number exceeds 1.2 since the calibration was not extended into the transonic regime. This restriction applies in particular to the inclined wire measurements, and it may explain why the Reynolds shear stress levels in a supersonic boundary layer are often suspiciously low near the wall.

Results and Discussion

Mean flow

The properties of the incoming boundary layer at a position − 51 mm upstream of the corner are summarized in Table 1. A shadowgram of the interaction region is given in Fig. 3, and the distributions of static pressure and skin friction coefficient are shown in Fig. 4.

The pressure distribution indicates a considerable upstream influence. As measurements by Dolling and Murphy (1982) in a similar configuration suggest, this upstream influence is actually caused by relatively low frequency oscillations in the shock location. Downstream of the corner the pressure gradient decreases, and the pressure eventually reaches the inviscid level, as expected. There is no evidence of flow three-dimensionality, and the surface streak patterns observed by Settles (1975) did not reveal the presence of any longitudinal Taylor-

Fig. 3. Shadowgram of the shock wave/boundary layer interaction

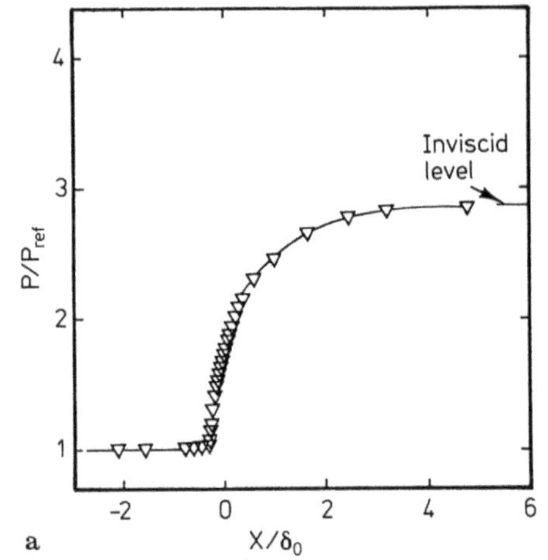

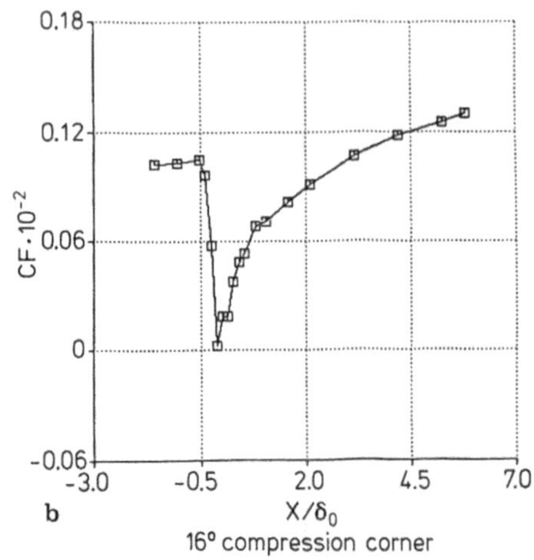

Fig. 4a. Surface pressure distribution (from Settles et al. 1979). **b** Distribution of the local skin-friction coefficient (from Settles et al. 1979)

Görtler-like vortices. (These vortices are commonly observed in subsonic boundary layers under the action of concave streamline curvature.)

The skin friction coefficient indicates that the separation zone, if it exists, is extremely small. Downstream of the corner the skin friction coefficient increases rapidly; this increase continues even after the pressure gradient has fallen to zero, and the level actually overshoots the equilibrium value (shown the right-hand-side of Fig. 4b).

Turbulence measurements

Normal wire results. The rms mass-flow fluctuation level $\langle(\varrho u)'\rangle$ is shown in Fig. 5. The incoming freestream mass flow rate $(\varrho U)_{\text{ref}}$ and the upstream boundary layer thickness δ_0 were used to non-dimensionalize the results, and these scales were chosen simply because it is not clear what scaling variables are appropriate within the interaction region. Figure 5 therefore presents information on the behaviour of the *absolute* mass-flow fluctuation level. Where applicable, the point at which the local normal Mach number equals 1.2 is shown in the figure; measurements below this point should be treated cautiously.

The most obvious feature of the turbulence behaviour is the large increase in the level of $\langle(\varrho u)'\rangle$. The initial rise occurs as the flow passes through the interaction zone and results in a doubling of the turbulence level, but what is in some ways more surprising is the continued rise of $\langle(\varrho u)'\rangle$ within the "recovery" zone. At the furthest downstream station, the peak value of $\langle(\varrho u)'\rangle$ actually reaches a level 2.8 times the upstream value. A similar, but smaller amplification was observed in the corresponding 8° corner flow (Hayakawa et al., 1982).

To investigate the approximate behaviour of the rms velocity fluctuation intensity $\langle u'\rangle$, we assumed that Morkovin's (1962) "Strong Reynolds Analogy" could be applied in this flow. In other words, we assumed that,

$$\frac{\sqrt{\overline{\varrho'^2}}}{\bar{\varrho}} = (\gamma - 1)\, M_a^2 \frac{\sqrt{\overline{u'^2}}}{\bar{U}}.$$

Thus,

$$\frac{\overline{u'^2}}{\bar{U}_{\text{ref}}^2} = \frac{\overline{(\varrho u)'^2}}{\bar{\varrho}^2 \bar{U}_{\text{ref}}^2} \{1 + 2 R_{\varrho u}(\gamma - 1) M_a^2 + [(\gamma - 1) M_a^2]^2\}^{-1}$$

where

$$R_{\varrho u} = \frac{\overline{\varrho' u'}}{\sqrt{\overline{\varrho'^2}} \sqrt{\overline{u'^2}}}.$$

$R_{\varrho u}$ was taken to be a constant across the boundary layer and equal to 0.8 (Dussauge and Gaviglio, 1981). Although these assumptions are always somewhat questionable in highly disturbed flows, even approximate information regarding the velocity fluctuations is valuable. The results are shown in Fig. 6.

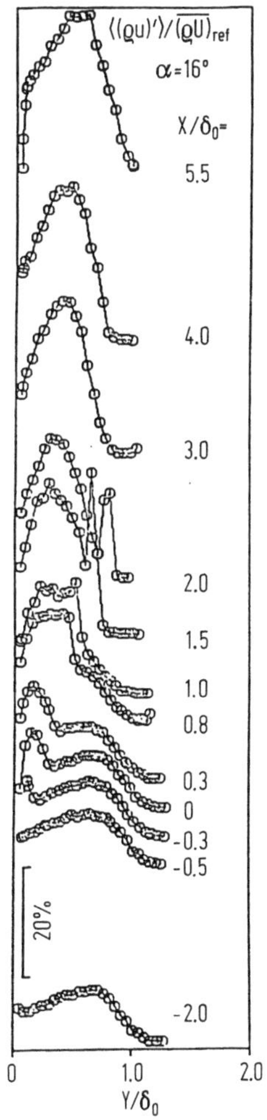

Fig. 5

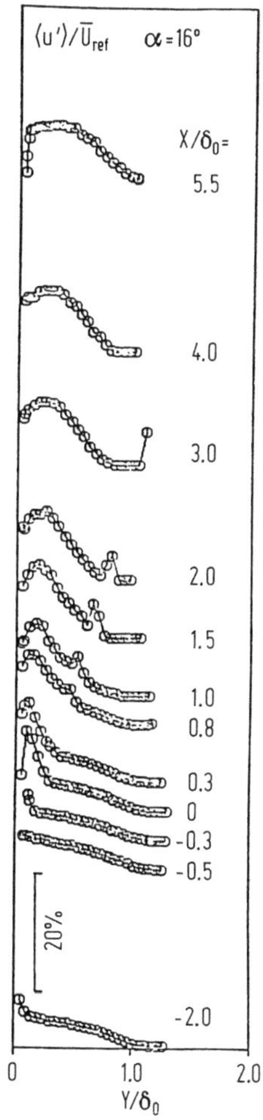

Fig. 6

Fig. 5. Profiles of the rms longitudinal mass flow fluctuations $\langle(\varrho u)'\rangle/(\overline{\varrho U})_{\text{ref}}$

Fig. 6. Profiles of the rms longitudinal velocity fluctuation $\langle u'\rangle/\bar{U}_{\text{ref}}$

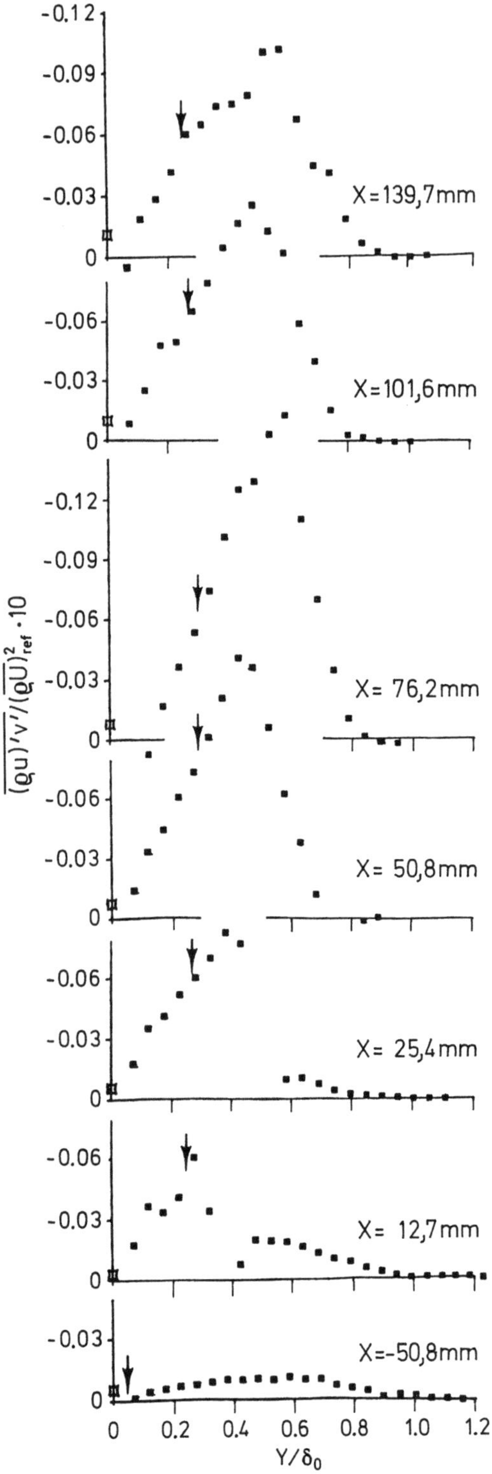

Fig. 7. Profiles of the mass weighted Reynolds shear stress $\overline{(\varrho u)' v'}/(\overline{\varrho U^2})_{\mathrm{ref}}$. The arrows indicate the locations at which the mean local *normal* Mach number is 1.2. The values given at $y = 0$ were taken from the mean flow data

The initial behavior of $\langle u' \rangle$ is broadly similar to the behavior of $\langle(\varrho u)'\rangle$ in that there is a sudden increase within the interaction zone. Downstream, in contrast to the increase observed in $\langle(\varrho u)'\rangle$, the level of $\langle u' \rangle$ reaches a plateau and eventually begins to decrease.

Inclined wire results. The mass weighted Reynolds shear stress $\overline{(\varrho u)' v'}/(\overline{\varrho U^2})_{ref}$ is shown in Fig. 7. Before discussing these results, the accuracy and limitation of the hot-wire technique should be considered.

It is becoming increasingly clear that one of the major limitations on using inclined hot-wires in low Mach number supersonic flows is the requirement that the instantaneous *normal* Mach number should be greater than 1.2 (Smits and Muck, 198). Within the transonic range, the hot-wire behavior changes drastically, and a sharp fall-off occurs in the inferred turbulence intensity. For this reason, results for which the normal Mach number is below 1.2 are expected to be in error and should be used cautiously, although it may be possible to use the shear stress at the wall as a guide for interpreting the results near the surface.

Consider now the results shown in Fig. 7. The behavior of $\overline{(\varrho u)' v'}$ is obviously rather different from that displayed by $\overline{(\varrho u)'^2}$. In passing through the interaction region, $\overline{(\varrho u)' v'}$ increases by as much as 16 times, whereas $\overline{(\varrho u)'^2}$ increases by less than 10 times over the same distance. In additional contrast, $\overline{(\varrho u)' v'}$ decreases significantly further downstream.

To determine the behavior of the turbulent shear stress $\tau = \overline{\varrho u' v'}$, we again assumed that the "Strong Reynolds Analogy" could be applied. That is,

$$\tau/\tau_w = \frac{\bar{\varrho}\,\overline{u'v'}}{\varrho_w u_\tau^2} = \frac{2}{c_f}\,\frac{\bar{\varrho}\,\bar{U}^2}{\bar{\varrho}_{ref}\,\bar{U}^2_{ref}}\,\frac{1}{[1 + (\gamma - 1)\,M_a^2]}\,\frac{\overline{(\varrho u)' v'}}{\bar{\varrho}\,\bar{U}^2}$$

where the subscript "*w*" indicates at the wall. In the case of the undisturbed, upstream boundary layer (see Fig. 8), the results agree closely with Sandborn's "best fit", which does not *prove* that the data are accurate, but does give some confidence in our method.

Although the behavior of the Reynolds stress $\bar{\varrho}\,\overline{u'v'}$ is interesting, the behavior of the kinematic shear stress $\overline{u'v'}$ is in some ways even more interesting. The results are shown in Fig. 9.

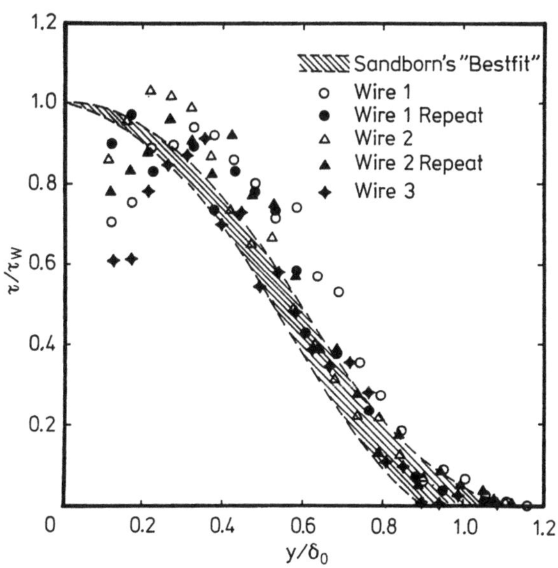

Fig. 8. Profiles of the Reynolds shear stress normalized by the surface shear stress, τ/τ_w for different wires at the upstream station ($x = -50.8$ mm). Wire 1 was used for all the measurements presented in this paper. The region of "best fit" is taken from Sandborn (1974)

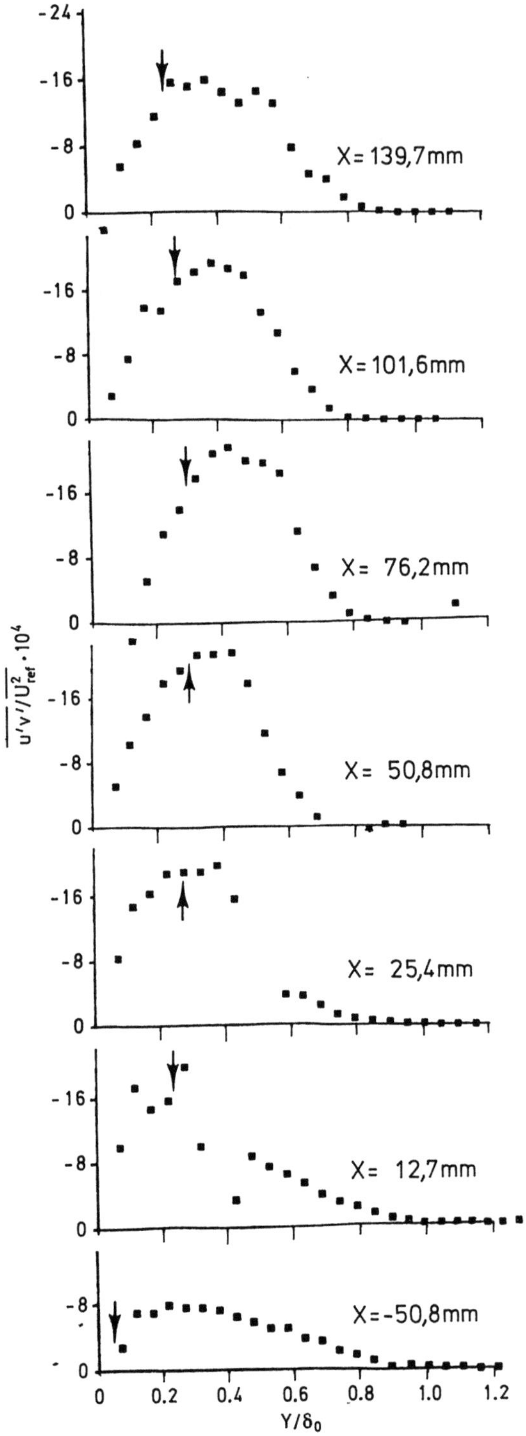

Fig. 9. Profiles of the kinematic shear stress $\overline{(u'v')}/\bar{U}_{ref}^2$. The arrows indicate the locations at which the mean local *normal* Mach number is 1.2

The streamwise trend displayed by $\overline{u'v'}$ is qualitatively similar to that observed for $\overline{(\varrho u)'v'}$; a sudden increase immediately downstream of the shock wave is followed by a more gradual rise which then levels off and subsequently begins to decrease. The magnitude of the increase in $\overline{u'v'}$, however is considerable smaller than the observed increase in $\overline{(\varrho u)'v'}$. At $x = 76$ mm, for instance, $\overline{u'v'}$ has increased by a factor of 2.6 whereas $\overline{(\varrho u)'v'}$ increased about 16 times. At the same station, the corresponding amplification factor for $\overline{u'^2}$ and $\overline{(\varrho u)'^2}$ were 4 and 8 respectively, which suggests a considerable change in the turbulence structure.

For instance, the ratio $-\overline{u'v'}/\overline{u'^2}$ can be taken as a structure parameter. In the undisturbed boundary layer at $x = -51$ mm, this ratio has a value of about 0.25 (at $y/\delta_0 = 0.6$), which agrees well with the value commonly quoted for incompressible boundary layers (see Townsend, 1976, p. 107). Through the interaction zone, the ratio $\overline{u'v'}/\overline{u'^2}$ increases to about 0.32, although it must be stated that the measurements of $\overline{u'v'}$ in this region are probably not too reliable. Downstream in the recovery region, however, this ratio decreases significantly, and it reaches a value of $0.16 \sim 0.18$ at the furthest downstream station ($x = 140$ mm). A similar trend was observed by Smits et al. (1979) in a subsonic boundary layer subjected to an impulse in concave curvature. It is interesting to note that in the present experiment, where both curvature and compressibility effects are important, the turbulence appears to respond in a fashion similar to that observed in an incompressible curved flow.

Conclusions and Final Discussion

All the turbulence parameters considered in this study, including $\overline{(\varrho u)'^2}$, $\overline{u'^2}$, $\overline{(\varrho u)'v'}$ and $\overline{u'v'}$ have shown a similar trend. All of these parameters increase through the interaction over a distance which corresponds approximately to the region of non-zero pressure gradient and the increases are not insignificant; amplification factors vary from 2.6 to 16. It was observed that the turbulence at the furthest downstream station is still far from equilibrium, and it appears that the relaxation process is very slow. In addition, the turbulence structure is considerably distorted by the interaction. For instance, the ratio $\overline{u'v'}/\overline{u'^2}$ first increases by about 30% and then decreases by approximately the same amount.

To interpret these results, we can begin by considering the theoretical work of Zang et al. (1982) and Anyiwo and Bushnell (1982). Using the two-dimensional Euler equations, these authors showed that several possible turbulence amplification or generation mechanisms may occur during an interaction between a plane shock wave and an incident turbulence field. These mechanisms include (1) direct amplification, (2) "generation" of turbulence from incident acoustic and entropy fluctuations, and (3) "focussing" caused by distortions of the shock front.

When a shock wave interacts with a turbulent boundary layer, however, the flow field is rather different from the ideal case considered by these authors. The shock wave within the boundary layer curves and is followed by a system of instantaneous compression waves (see, for instance, Fig. 3). All these waves are unsteady and appear to be in constant, low frequency motion (Dolling and Murphy, 1982). As Zang et al. (1982) point out, the direct conversion of mean flow energy into turbulence by shock oscillation can serve as a powerful turbulence amplification mechanism. In our particular experiment, however, it does not appear that shock oscillation is an important amplification mechanism. This conclusion is based on the results given in Figs. 5 and 6; the amplification begins very close to the mean shock location, and if shock oscillation was important we would expect to see a greater upstream influence.

Of the other three mechanism, the first two are probably more important than the third (Anyiwo and Bushnell, 1982). We can consider these two mechanisms together for our purposes, and simply call them "direct" shock effects. In any shock-wave/boundary layer interaction, however, the compression occurs over some distance, and compressibility effects occur downstream of the shock front. These compressibility effects are similar to the compressibility effects identified by Bradshaw (1973, 1974) in supersonic boundary layers subjected to adverse pressure gradients.

Alternatively, the interaction zone may be viewed simply as a region of impulsively applied pressure gradient. The physical mechanism by which turbulence is amplified across an interaction zone can then be identified as a change in the cross-sectional area of fluid elements, caused by the total compression. A useful measure of the strength of the interaction is given by the integral of the divergence of the velocity field over the time it acts, and this is approximately equal to $\theta_p = (1/\gamma)\ln(p_2/p_1)$, where p_2/p_1 represents the pressure rise across the shock wave (Hayakawa et al., 1982).

In the case of a compression corner, the streamlines experience concave longitudinal curvature as they pass through the interaction zone. Concave curvature is an unstable extra strain rate in that it causes turbulence levels to increase (Bradshaw, 1973), and therefore it must be considered as an additional amplification mechanism. We can similarly characterize the strength of the curvature effect by evaluating the integral of the extra strain rate over the zone it acts, and we find that this is equal to the total turning angle θ_c (Smits et al., 1979).

For the present experiment, $\theta_p \approx 0.74$ and $\theta_c = 0.28$ which suggests that the effects of compressibility outweigh the effects of curvature, if no interaction between these effects occurs.

We have remarked that the turbulence structure appears to be significantly distorted by the interaction. This observation is in accord with the analysis presented by Debieve et al. (1980). By using an approach based on rapid distortion theory, and by modelling the shock wave as a discontinuity, they showed that the shock wave rotates the Reynolds stress tensor. As a consequence, a structure parameter such as $\overline{u'v'}/\overline{u'^2}$ must change, and their analysis predicts an increase, as observed in our measurements. The subsequent decrease indicates that this analysis apparently breaks down in the recovery region.

A final note should be added regarding the time-dependent response of the turbulent motions. It seems highly likely that the full impact of the distortion is not felt until some distance downstream of the interaction zone. At first sight, the results do not show evidence of such a lagged response. This aspect of the turbulence behavior, however, is under further investigation.

References

Anyiwo, J. C., Bushnell, D. M. (1982): Turbulence amplification in shock-wave boundary layer interaction. AIAA J. **20**, 893

Bradshaw, P. (1973): Effects of streamline curvature on turbulent flow. AGARDograph 169

Bradshaw, P. (1974): The effect of mean compression or dilatation on the turbulent structure of supersonic boundary layers. J. Fluid Mech. **63**, 449

Debieve, J.-F., Gouin, H., Gaviglio, J. (1982): Evolution of the Reynolds stress tensor in a shock-wave/turbulence interaction. Indian J. Technology **20**, 90

Dolling, D. S., Murphy, M. (1982): Wall pressure in a supersonic separated compression ramp flow field. AIAA Paper 82-0986

Dussauge, J. P., Gaviglio, J. (1981): Bulk dilatation effects on Reynolds stress in the rapid expansion of a turbulent boundary layer at supersonic speed. Proc. Third Symp. on *Turbulent Shear Flows*, Univ. of Calif., Davis, pp. 2.33

Fernholz, H. H., Finley, P. J. (1977): A critical compilation of compressible turbulent boundary layer data. AGARD-AG-223

Fernholz, H. H., Finley, P. J. (1981): A further compilation of compressible boundary layer data with a survey of turbulence data. AGARD-AG-263

Hayakawa, K., Smits, A. J., Bogdonoff, S. M. (1982): Hot-wire investigation of unseparated shock-wave/turbulent boundary layer interaction. AIAA Paper No. 82-0985

Morkovin, M. V. (1962): Mecanique de la turbulence. Favre, A. ed., C.N.R.S., Paris, pp. 367–380

Sandborn, V. A. (1974): A review of turbulence measurements in compressible flow. NASA TM X-62,337

Settles, G. S.: An experimental study of compressible turbulent boundary layer separation at high Reynolds numbers. Ph.D. Thesis, Aerospace and Mechanical Sciences (Department, Princeton University, Princeton, NJ 1975)

Settles, G. S., Bogdonoff, S. M., Vas, I. E. (1976): Incipient separation of a supersonic turbulent boundary layer at high Reynolds numbers. AIAA J. **14**, 50

Settles, G. S., Fitzpatrick, T. J., Bogdonoff, S. M. (1979): Detailed study of attached and separated compression corner flow fields in high Reynolds number supersonic flow. AIAA J. **17**, 579

Smits, A. J., Young, S. T. B., Bradshaw, P. (1979): The effect of short regions of high surface curvature on turbulent boundary layers. J. Fluid Mech. **94**, 209

Smits, A. J., Hayakawa, K., Muck, K. C. (1983): Constant temperature hot-wire anemometer practice in supersonic flows. Part I: The normal wire. Experiments in Fluids **1**, 83

Smits, A. J., Muck, K. C. (1984): Constant temperature hot-wire anemometer practice in supersonic flows. Part II: The inclined wire. Experiments in Fluids **2**, 33

Townsend, A. A.: The structure of turbulent shear flow. Cambridge University Press, 1976

Zang, T. A., Hussaini, M. Y., Bushnell, D. M. (1982): Numerical computations of turbulence amplification in shock wave interactions. AIAA Paper No. 82-0293

Secondary Currents in a Straight Channel Flow and the Relation to Its Aspect Ratio

I. Nezu *, H. Nakagawa, A. Tominaga **

Department of Civil Engineering, Kyoto University, Kyoto 606, Japan

Abstract

The initiation and maintenance mechanism of multi-cellular secondary currents in a straight wide-river is quite unknown at present. This study investigated experimentally the existence of secondary currents over a flat bed by varying the aspect ratio of channel. It was suggested strongly that an initiation of cellular secondary currents may be motivated by the mutual interaction between the secondary currents and the sand bed.

Nomenclature

B	spanwise width of channel;
$\mathrm{Fr} \equiv U_m/\sqrt{g h}$	Froude number;
h	flow depth or half vertical width of duct;
l_1	distance between the bottom-vortex center and the side-wall;
l_2	diameter of the bottom-vortex;
P	mean pressure;
$R \equiv h/(1 + 2\alpha^{-1})$	hydraulic radius;
$\mathrm{Re} \equiv 4RU_m/\nu$	Reynolds number;
$\mathrm{R'e} \equiv hU_m/\nu$	Reynolds number;
U_m	mean bulk velocity;
U_{max}	maximum mainstream velocity;
U_{*1}	friction velocity evaluated directly;
U_{*2}	friction velocity evaluated from the log-law;
U, V, W	mean velocities in x, y and z directions, respectively;
u, v, w	turbulent fluctuations in x, y and z directions, respectively;
u', v', w'	turbulence intensities in x, y and z directions, respectively;
x, y, z	Cartesian coordinates in longitudinal, vertical and spanwise directions, respectively, i.e. $y = 0$ on the bed and $z = 0$ at the channel center;
$\alpha \equiv B/h$	aspect ratio of the channel;
λ	spanwise spacing of sand ribbons;
$\tau_0 \equiv \varrho U_*^2$	wall shear stress;
$\xi \equiv (\partial W/\partial y - \partial V/\partial z)$	streamwise vorticity.

Introduction

Many river engineers have hitherto suggested that there might exist cellular secondary currents in a straight wide-river. Vanoni (1946) pointed out that the spanwise distribution of sediment concentration varied periodically in wide open channel, and he inferred that this fact might have been caused by cellular secondary currents. Matthes (1947) envisaged that there might exist a strong upward-current, like a tornado, near the river bed, and also that

* Formerly (1983) Alexander von Humboldt fellow, Institut für Hydromechanik, Univ. of Karlsruhe, 7500 Karlsruhe, FRG;

** Department of Civil Eng., Gunma University, Kiryu 376, Japan it developed up to the free surface and then became a so-called "boil" which was observed as a vortex-ring-like swollen water surface.

Kinoshita (1967) found by means of aerial stereoscopic survey (Cameron effect) of flood rivers that a boil became a low-speed zone and, at both spanwise sides with the spacing of twice the flow depth h, high-speed zones were formed. He suggested then that there might exist cellular secondary currents in a straight river which had a pair of counter-rotating streamwise vortices with a vortex diameter of about h (see Fig. 1). After the flood had gone, Culbertson (1967) and Karcz (1973) found, on the other hand, that longitudinal ridges and troughs which are called "sand ribbons" were formed periodically on the river bed. Thus, they suggested an existence of cellular secondary currents in a straight river.

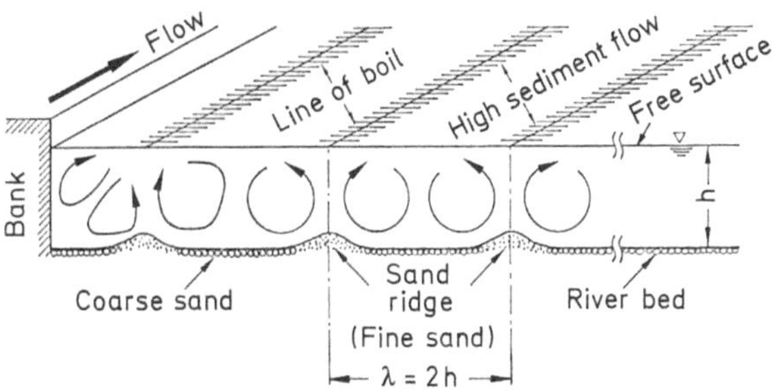

Fig. 1. Envisaged flow-pattern of multi-cellular secondary currents in a straight wide river

On the basis of these suggestions, Fig. 1 envisages a behaviour of river flow. An investigation of cellular secondary currents, i.e. three-dimensional flow patterns, is very important in hydraulic engineering, as well as in basic hydromechanics, because they might cause three-dimensional sediment transport and bed forms in river. However, the initiation and maintenance mechanisms of cellular secondary currents in a straight wide-channel are quite unknown at present.

The authors (1982) have firstly measured cellular secondary currents in straight open-channel flow (Case K: the channel width $B = 30$ cm, the flow depth $h = 4.0$ cm, the average bulk velocity $U_m = 32.0$ cm/s, the Reynolds number $R'e \equiv hU_m/\nu = 1.3 \times 10^4$ and the Froude number $\mathrm{Fr} = 0.51$), by making use of the X-type hot-films (DISA 55R61). Longitudinal ridge elements of 8 m long with a 45°-trapezoid cross section of 5 mm thick and 20 mm wide were attached onto the smooth channel bed in the spanwise spacing of $\lambda = 2h = 8$ cm. This simulated the longitudinal ridges and troughs of river bedform, i.e. sand ribbons, as shown in Fig. 1. Figure 2(a) shows the flow pattern of cellular secondary currents (V, W) obtained experimentally. A pair of cellular secondary currents is seen clearly. Speaking roughly, the upflow ($V > 0$) and downflow ($V < 0$) appear over the ridge and trough, respectively, and they behave as a circular vortex motion in which the diameter is equal to h and the position of the circular center is nearly equal to $y/h = 0.5$. The upflow is stronger rather than the downflow, and the secondary currents are larger near the bed than near the free surface. These features support well the flow mechanism envisaged by Kinoshita (1967) and Karcz (1973). It should be noticed that, although the velocity $\sqrt{V^2 + W^2}$ of the secondary currents is within only 5% of the maximum mainstream velocity U_{max}, its experimental values show comparatively large scatters. The experimental values of the turbulence quantities indicated also large scatters. We have then judged that it was fairly difficult to examine the equations of mean vorticity, mean flow energy and turbulent energy experimentally and to discuss the maintenance mechanism of cellular secondary currents even qualitatively.

Next, the authors (1984) could measure more accurately all three components of the velocity in air duct flow (Case J: the half vertical width h of the duct, which corresponds to

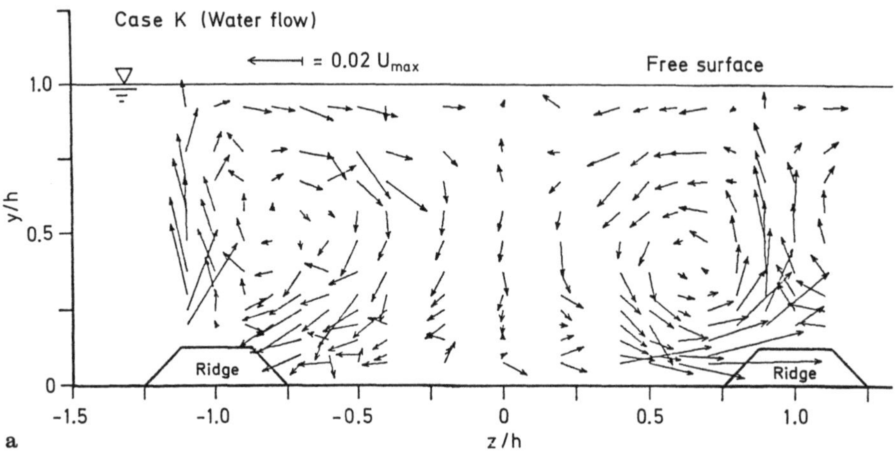

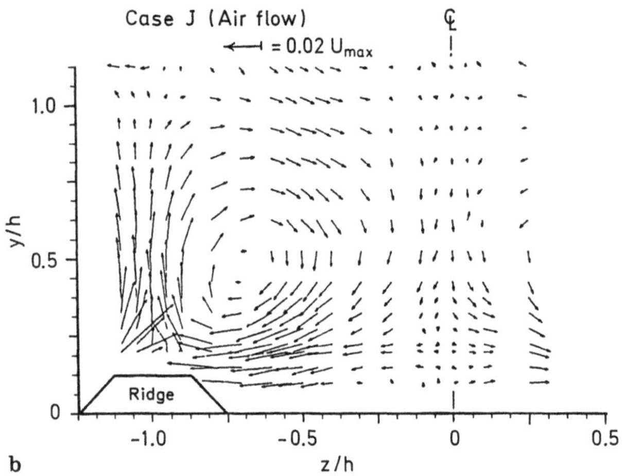

Fig. 2. Comparison of cellular secondary currents in (a) water-flow with those in (b) air-flow. Both the Reynolds numbers are nearly equal to 1.3×10^4

the flow depth, is $h = 4.0$ cm, $U_m = 5$ m/s and $R'e = 1.3 \times 10^4$), by making use of X hot-wires. In order to simulate sand ribbons on the bed, the same longitudinal ridge elements as used in water experiment of Fig. 2(a) were attached onto both the lower and upper bottoms of the duct. Figure 2(b) shows the flow pattern of secondary currents in air flow. It should be noticed that the relative magnitudes of the secondary currents normalized by U_{max} are roughly equal to each other in both Case J (air duct) and Case K (water channel) except for near the free surface, in spite of the large difference between their absolute values, i.e. $U_{max} = 540$ cm/s (Case J) and $U_{max} = 35$ cm/s (Case K). We must emphasize here that the intensity of secondary currents is surely strengthened near the free surface by its existence in comparison of Fig. 2(a) with Fig. 2(b). However, the flow patterns of cellular secondary currents are similar to each other in both cases. Hence, we can conclude that Case J makes a good reproduction of cellular secondary currents in open channel flow which is shown in Fig. 1. Then, its turbulent structure was examined through the equations of vorticity and mean flow energy. As the results, the Reynolds stress term, $(\partial^2/\partial z^2 - \partial^2/\partial y^2)\,\overline{vw}$, is nearly balanced with the production term of vorticity, $\partial^2/\partial y\,\partial z\,(\overline{v^2 - w^2})$, in most of the cell structure. The wall shear stress becomes larger on the troughs than on the ridges. A more detailed discussion is given in our recent paper (1984).

Such a spanwise non-homogeneity of flow may cause secondary currents, or the reverse may be true. However, an initiation mechanism of cellular secondary currents and sand ridges over a uniformly flat channel bed is quite unknown at present. Three possibilities of this initiation mechanism are considered, as follows:

(1) The aspect ratio $\alpha \equiv B/h$ of channel must be selected as an even number, irrespective of wall properties. It is only a kind of eigenvalue problem.
(2) An initiation of cellular secondary currents is motivated by the mutual interaction between the secondary currents and the sand bed.
(3) An existence of free surface motivates cellular secondary currents.

On the basis of our preliminary water flow experiments (1982, 1983) and the numerical calculations of Naot and Rodi (1982) who took the effect of free surface into account, we can attain a hypothesis that the existence of free surface is not an essential cause of cellular secondary currents, but it only promotes their intensity near the free surface, as seen in Fig. 2.

In the light of the above, the present study is to investigate the existence of secondary currents over a flat solid bed by varying the aspect ratio $\alpha \equiv B/h$ of channel and to examine their initiation mechanism. That is, we will examine experimentally which of (1) and (2) has a more possibility.

Apparatus and Procedure

Experiments were conducted in a straight channel 6 m long with a rectangular cross section of 8 cm × 40 cm (see Fig. 3). Air was delivered to the channel by a centrifugal fan by way of a 40 cm-square settling chamber 165 cm in length and a contraction section 100 cm in length. In order to promote the development of flow, rough sandpapers were attached onto both lower and upper bottoms at the entrance of channel, and the test section was set up at 5 m downstream from its entrance. Two partition plates of 8 cm wide and 6 m long were set into the channel as both side-walls, and thus the spanwise width B of channel could be changed arbitrary (see Fig. 3).

The half vertical width h of channel which corresponds to the flow depth of open channel, was fixed to be 4.0 cm. In this experiments, the aspect ratio $\alpha \equiv B/h$ was chosen to be equal to 10 (Case A), 6 (Case B), 4 (Case E), 3 (Case D) and 2 (Case C), as shown in Fig. 3. Of course, Case C is a square-cross channel. The maximum mainstream velocity U_{max} was fixed to be 7.1 m/s, in all cases. The Reynolds number $\mathrm{Re} \equiv 4RU_m/\nu$ (R is the hydraulic radius) became equal to $(3-5) \times 10^4$.

In order to examine in detail the turbulent structure of secondary currents, the velocity measurements were done at about 300 measuring points in the 1/4 cross section of y-z plane, as shown in Fig. 3. Two velocity pairs of (u, v) and (u, w) could be measured accurately by X-type hot-wires (DISA 55P61). The hot-wire signals were recorded in digital form by

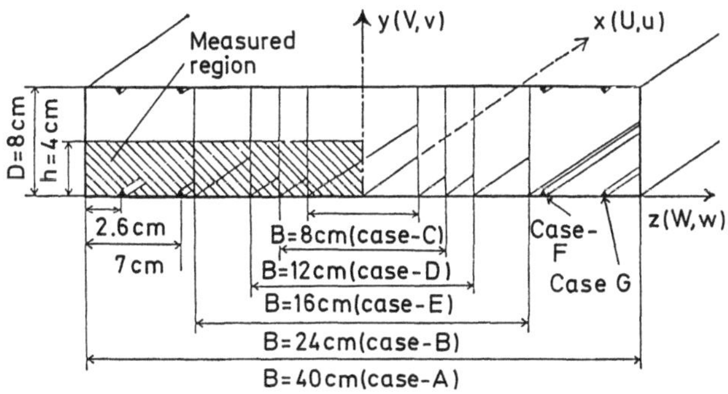

Fig. 3. Test section of the channel and its coordinate system

using an AD-converter, with the sampling size $N = 20{,}000$ and the sampling frequency $f = 2000$ Hz. Then, some statistical analyses were carried out by a large digital computer at the Data Processing Center, Kyoto University.

Experimental Results and Discussion

Mean Primary Velocity

Figure 4 shows the contours of the iso-velocity of the mean primary velocity $U(y,z)$ normalized by U_{max} in the square channel (Case C). For a comparison, the previous data of Melling and Whitelaw (1976) are shown at the right side of Fig. 4. They could measure water flows

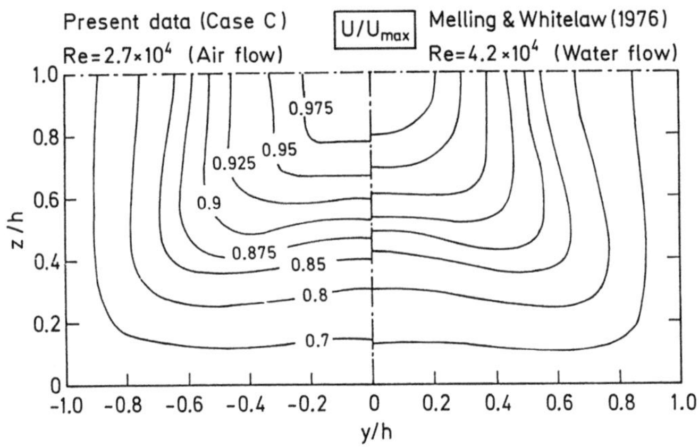

Fig. 4. Comparison of the present hot-wire data of $U(y,z)$-isovel lines with the LDA data by Melling and Whitelaw

in a square channel by making use of a Laser-Doppler anemometer (LDA). The present hot-wire data coincide fairly well with their LDA data, although the distributions of the latter are a little more peaked than those of the former. This difference may be due to the shorter channel length of Melling and Whitelaw (i.e. 147 R, compared with 250 R for the present experiment). The iso-velocity lines bulge outwards along the corner bisector. This indicates that the square channel flow is three-dimensional in all cross section.

Figure 5 shows the variations of $U(y,z)$ against the spanwise direction z. The value of $U(z)$ at the center elevation of $y/h = 1.0$ for both Cases A and B increases firstly with z/h and then attains a constant. That is to say, the central zone of $|z/h| \leqq 3$ (Case A; $B/h = 10$) and 1 (Case B; $B/h = 6$), is considered to be a two-dimensional flow. As defined later, the diameter l_2 of the bottom vortex of the secondary current is shown in Fig. 5. It should be noticed that the spanwise variation of $U(z)$ has a somewhat minimum value at the position of $(1-2)\,h$ farther from the side-wall, that is, in the bottom-vortex. As mentioned later, this minimum position corresponds to the position of an upflow of secondary current and a minimum wall shear stress. This result is one of the most important features when an initiation mechanism of cellular secondary currents is investigated.

Secondary Currents (Longitudinal Vortex)

Figures 6(a)–(e) are vector descriptions of the mean velocity V and W of secondary currents for all cases. A secondary current towards the corner of the channel is recognized clearly in all cases. The maximum value of the secondary current velocity $U_s \equiv \sqrt{V^2 + W^2}$ is nearly equal to 0.015 U_{max}, which is in a good agreement with the result of Melling and Whitelaw

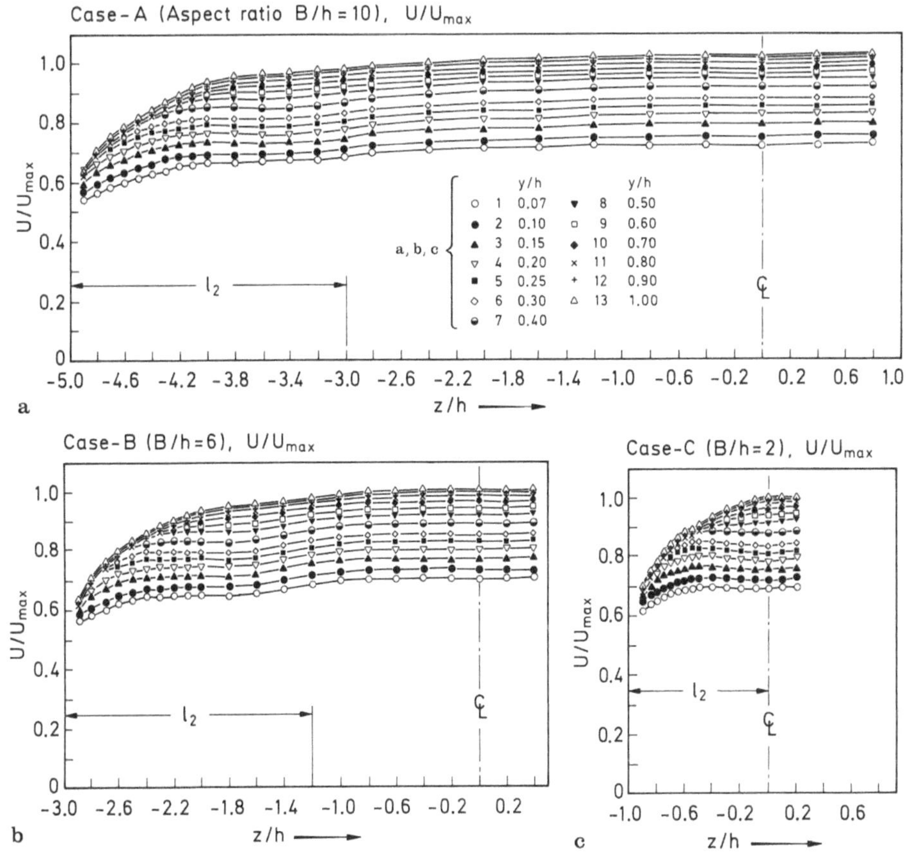

Fig. 5. Variations of mean primary velocity $U(y, z)$ against the spanwise direction z

(1976). This value is, however, smaller than the magnitude of the multi-cellular secondary currents which were shown in Fig. 2.

This well-known secondary current (we call it "corner secondary flow") flows along the corner bisector, i.e. 45°-line, irrespective of the aspect ratio B/h, and it produces only a pair of longitudinal vortices. One vortex over the corner bisector is called "side-vortex", and the other vortex under the bisector is called "bottom-vortex". Of course, in the square channel of Case C, the side- and bottom-vortices are well symmetrical with each other in regard to the bisector.

The pattern of the side-vortex is completely confined by the side wall and thus it is seldom affected by the aspect ratio $\alpha \equiv B/h$ of channel. On the other hand, the bottom-vortex is much affected by the aspect ratio α. As α increases, the size of the bottom-vortex becomes larger and then tends to attain a constant. In order to evaluate the size of the bottom-vortex, the position of the vortex center and the vortex-diameter are defined as horizontal distances l_1 and l_2, respectively, from the side-wall. l_1 and l_2 could be determined easily from Fig. 6, although the vortex-diameter l_2 was evaluated roughly. These experimental data are shown in Fig. 7, together with the corresponding Case-name. When α is smaller than 2, a side-vortex in the case of $\alpha > 2$ can be regarded as a bottom-vortex by the exchange of B for h. The vortex-diameter l_2 increases linearly with the aspect ratio α when $\alpha \leqq 3$. However, when

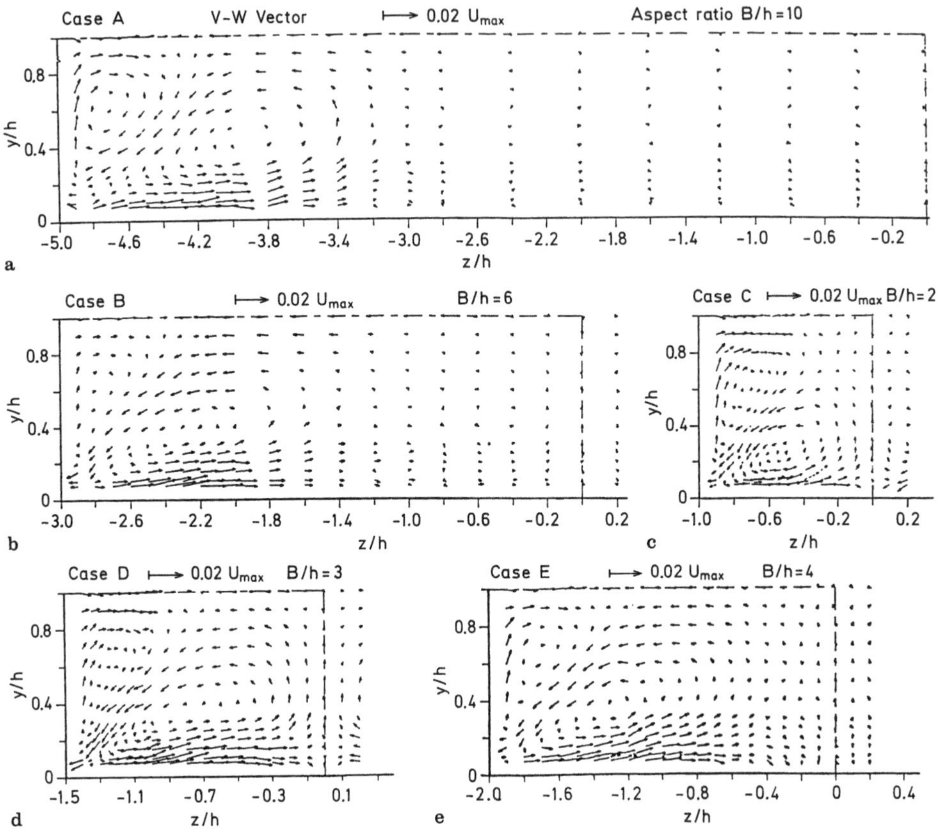

Fig. 6. Vector descriptions of the secondary currents (V, W) for all cases

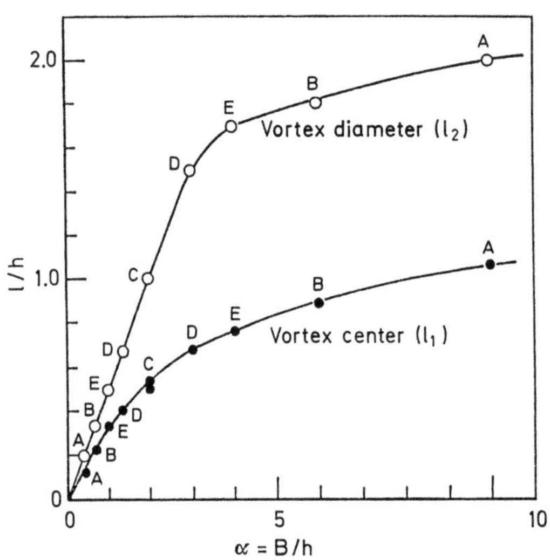

Fig. 7. Size of the bottom-vortex

$\alpha > 3$, its increase is much suppressed, and then l_2 tends to attain a constant, i.e. $l_2/h \cong 2$. The distance l_1 of the vortex center is nearly equal to a half of l_2. Also, the elevation of the vortex center attains a half depth, i.e. $y/h = 0.5$, with an increase of the aspect ratio α.

We must emphasize here that multi-cellular secondary currents observed in Figs. 1 and 2 cannot be recognized in the central zone of Cases A and B. This indicates that cellular secondary currents are produced directly neither by the corner secondary flow, nor by the variation of the aspect ratio. That is to say, the initiation mechanism (1) of cellular secondary currents mentioned in the Introduction is considered to be minor possibility.

Friction Velocity (Wall Shear Stress)

Firstly, we have examined the law of the wall, i.e. the log-law, which is described by

$$U/U_* = 5.75 \log(U_* y/\nu) + 5.5. \tag{1}$$

The friction velocity U_* can be evaluated by the least-square method so that the experimental values of U for $U_* y/\nu \leqq 300$ may coincide with Eq. (1). An example of these velocity distributions is plotted in Fig. 8 for Case A. Except for very near the side-wall (i.e. Run 25), the mean primary velocity $U(y)$ agrees very well with Eq. (1). A good agreement could be also obtained for the other cases. That is, the log-law is satisfied near the bed irrespective of the existence of corner secondary flow, although the side-wall effect appears farther from the

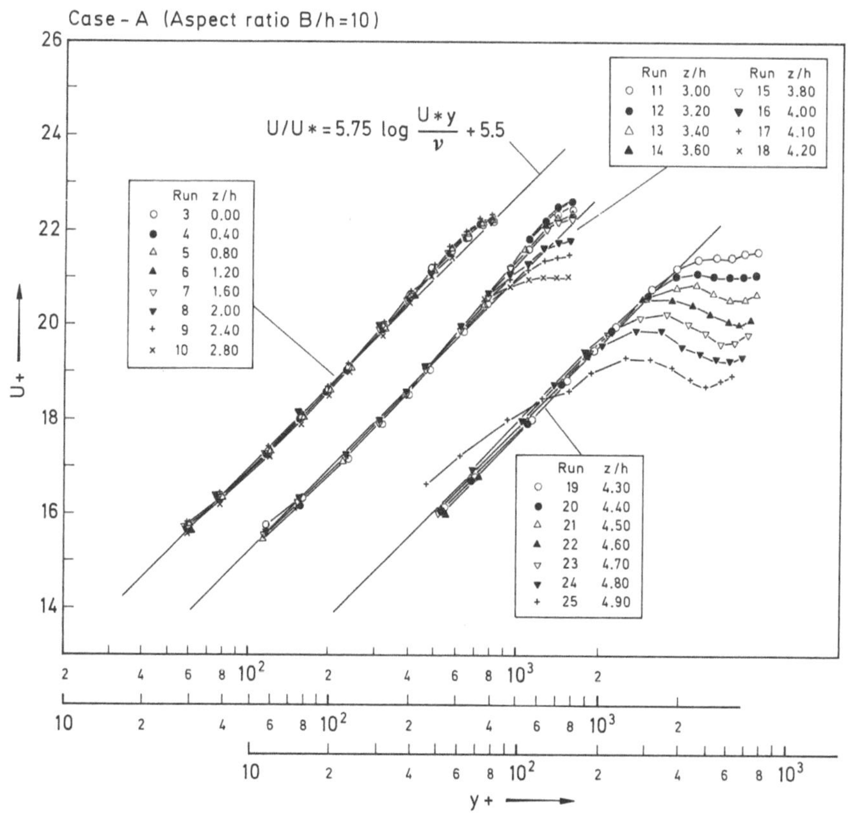

Fig. 8. Law of the wall, i.e. log-law as a parameter of spanwise direction in the channel of $B/h = 10$

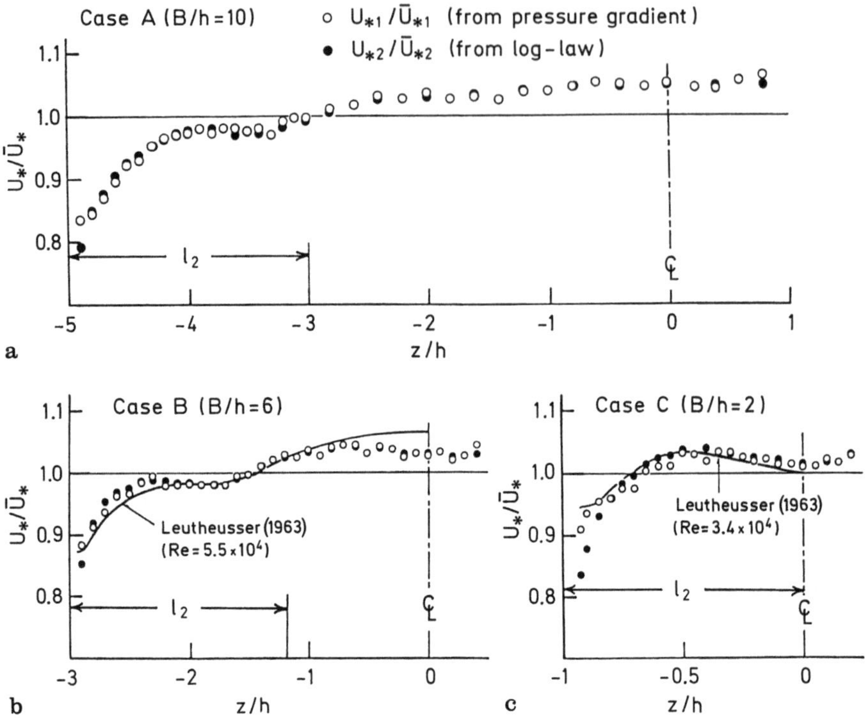

Fig. 9. Variations of friction velocity $U_*(z)$ against the spanwise direction z

bed and near the side-wall. Therefore, the friction velocity U_{*2} could be easily evaluated from the log-law, and its spanwise distributions are shown in Fig. 9.

Next, we have developed a new method by which the friction velocity can be evaluated directly. A single-sensor hot-wire was set precisely at $y = 2$ mm from the wall instead of a Preston-tube, and its output voltage E was calibrated carefully by the friction velocity U_{*1} which was obtained from the pressure gradient $\partial P/\partial x$ of the channel flow, i.e. $U_{*1} \equiv \sqrt{R \cdot \partial(-P/\varrho)/\partial x}$. Then, the calibration curve between E and U_{*1} could be described very well by

$$E^2 - E_0^2 = A \cdot U_{*1}^n \tag{2}$$

in which E_0 is the output voltage at $U_{*1} = 0$, and A and n are the calibration coefficients.

By using the calibration curve of Eq. (2), the spanwise distributions of the friction velocity U_{*1} could be obtained directly, and they are shown in Fig. 9. The experimental values of U_{*1} and U_{*2} are normalized by their spanwise average values $\bar{U}_{*1}$ and $\bar{U}_{*2}$, respectively, in Fig. 9. Then, $\bar{U}_{*1}/\bar{U}_{*2}$ was equal to 0.98, 0.98 and 0.96 for Cases A, B and C, respectively. Both the distributions of U_{*1} and U_{*2} coincide well with each other for all cases. This indicates the validity of the above-mentioned evaluation methods of the friction velocity.

For the square channel of Case C, the peak of the friction velocity appears not at the channel center $z = 0$, but at the position of $|z| = 0.5\,h$. In the same manner, the friction velocity attains a mild but important peak at the position of $1.2\,h$ for Case A and $0.8\,h$ for Case B from the side-wall. And then, it decreases somewhat and increases again in the central zone where a two-dimensional flow is established as shown in Figs. 5 and 6. Figure 9 shows also the experimental curves obtained by Leutheusser (1963). He could measure the friction velocity by making use of a Preston-tube for the same aspect ratios as Cases B and C. The present values are in a good agreement with his data.

This mild peak and valley distribution of the friction velocity in the bottom-vortex whose size is l_2, could be also observed in open-channel flow by the authors (1984), although the amplitude of peak/valley was somewhat larger in open-channel than in closed channel. The same feature of the friction velocity has been recently found by Knight et al. (1982) in both open and closed channels.

It should be noticed that the valley region of U_* at the position of $(1-2)h$ from the side-wall corresponds to the upflow ($V > 0$) region of the bottom-vortex which is produced by the corner secondary flow. In multi-cellular secondary currents shown in Fig. 2, the authors (1982, 1984) have also found the same noticeable feature that the friction velocity or wall shear stress is depressed at the place where an upflow ($V > 0$) occurs, while it is promoted at the place where the downflow ($V < 0$) occurs.

Turbulence Characteristics

We have analyzed the distributions of turbulence intensities u', v' and w', and also Reynolds stress $-\overline{uv}$ and $-\overline{uw}$ for all cases.

Turbulence Intensities. The present experimental data of u', v' and w' in the square channel have agreed well with many previous data. For example, Fig. 10 shows the contours of turbulence intensity u'/U_{max} in the square channel, and its corresponding contours of the primary mean velocity have been already indicated in Fig. 4. Surely, the present distribution of u'/U_{max} is in a good agreement with the result of Melling and Whitelaw (1976), although there are a little quantitative differences between the two. The contour lines of u' are symmetric with each other in regard to the corner bisector, and they bulge more strongly towards the corner than the contour lines of the primary mean velocity U. This suggests that the turbulence intensities are affected more strongly by the corner secondary flow than the primary mean velocity is affected, and vice versa. A close interrelation between the turbulence characteristics and the secondary currents can be explained by the vorticity equation (e.g. see our paper (1984)). On the other hand, as shown in Fig. 11, the contour lines of v' and w' indicate a good anti-symmetric distribution with each other in regard to the corner bisector for the square channel, which can be easily obtained by its boundary condition. They also bulge towards the corner, although their bulges are smaller than that of u'.

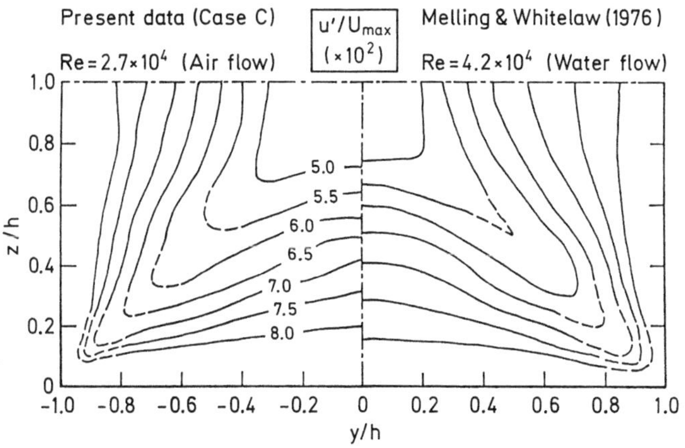

Fig. 10. Comparison of the present hot-wire data of u'-isolines with the LDA data obtained by Melling and Whitelaw (1976)

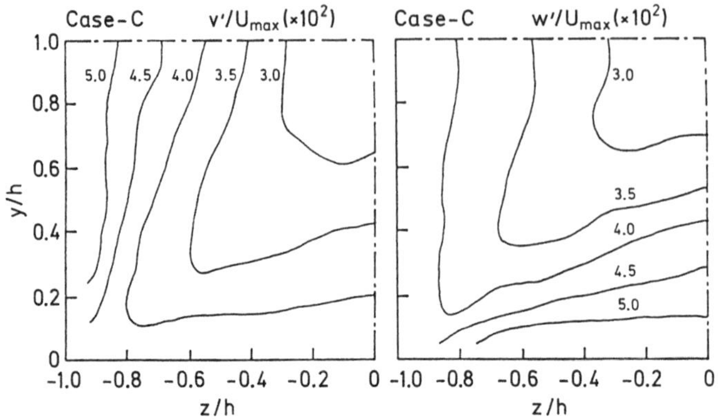

Fig. 11. Turbulence intensities v' and w' in the square channel

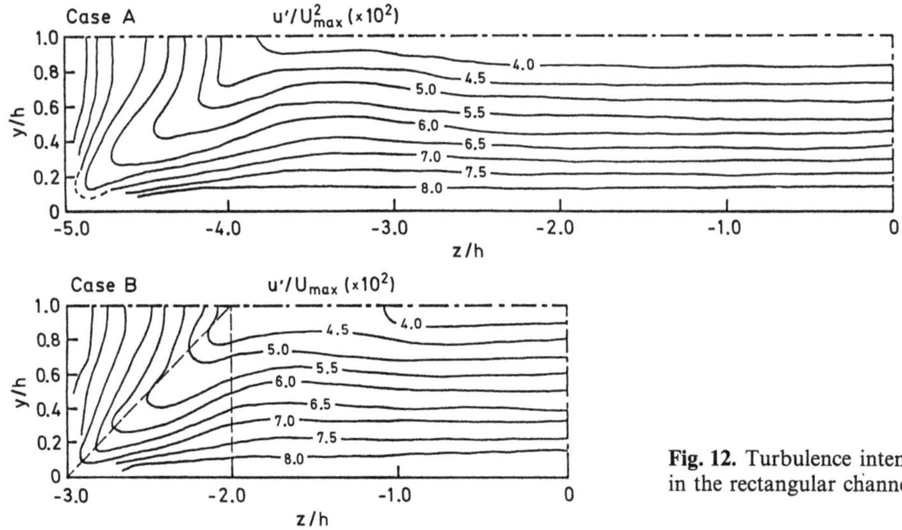

Fig. 12. Turbulence intensity u' in the rectangular channel

Even when the aspect ratio α became greater than 2, the turbulence intensities u', v' and w' showed a similar distribution to those of the square channel ($\alpha = 2$), i.e. Fig. 10, in the side-wall zone whose size is l_2, as shown in Fig. 12. In the central zone where

$$|z|/h \leqq (B/2 - l_2)/h \cong (\alpha - 4)/2, \tag{3}$$

u'/U_*, v'/U_* and w'/U_* indicated a two-dimensional distribution and they agreed well with the results of Laufer (1951) which were obtained at $z = 0$ for the aspect ratio of $\alpha = 24$.

Reynolds Stresses. Figure 13 is the vertical distributions of $-\overline{uv}/U_{max}^2$ for Case A. They indicate a linear distribution in the central zone, i.e. $|z|/h \leqq 3$ from Eq. (3), and thus a two-dimensional turbulent structure can be recognized here again. The values of $-\overline{uv}$ deviate somewhat upwards from this linear distribution in the upflow ($V > 0$) region of $|z|/h = 3.0 - 3.6$, and then they deviate downwards from it due to the side-wall effect. This deviation is, however, very small near the boundaries of $y/h = 0$ and 1, although it becomes largest near the half depth of $y/h = 0.5$. Therefore, this feature suggests an interrelation between such a deviation of $-\overline{uv}$ and a vertical mean velocity V.

Figure 14 shows the spanwise distribution of $-\overline{uw}/U_{max}^2$. Of course, $-\overline{uw}$ should be equal to zero if any secondary current does not exist. This feature is satisfied well in the central zone of $|z| \leqq h$, which is obtained from Eq. (3) for Case B. It should be noticed that the gradient of $-\overline{uw}$ against z is positive, i.e. $\partial(-\overline{uw})/\partial z > 0$, in the upflow ($V > 0$) region

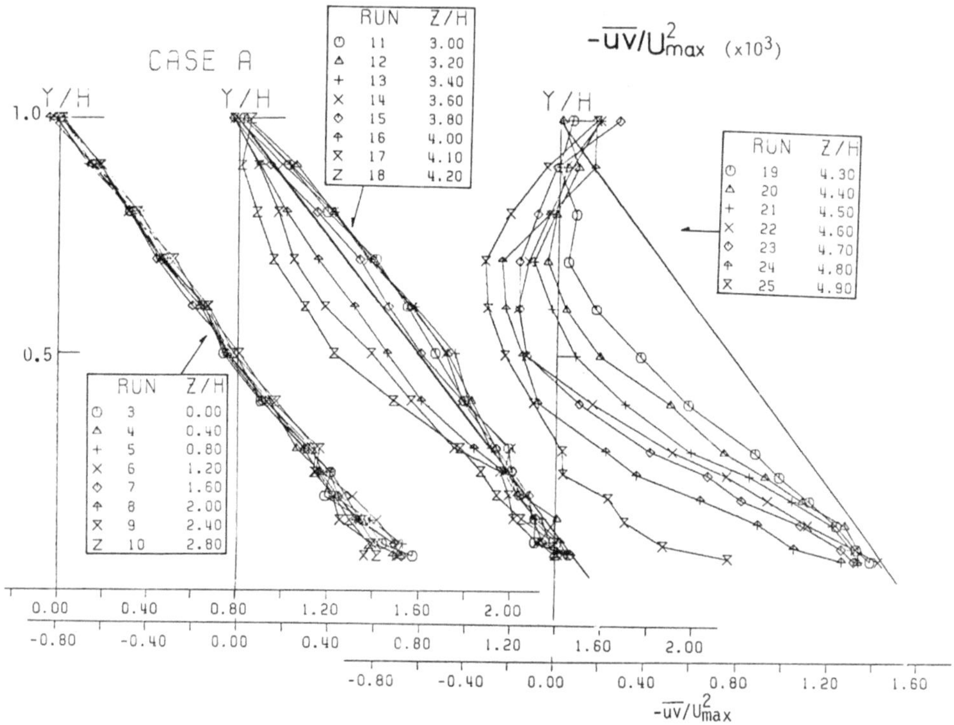

Fig. 13. Distributions of the Reynolds stress $-\overline{uv}$ against the vertical direction, y, for Case A

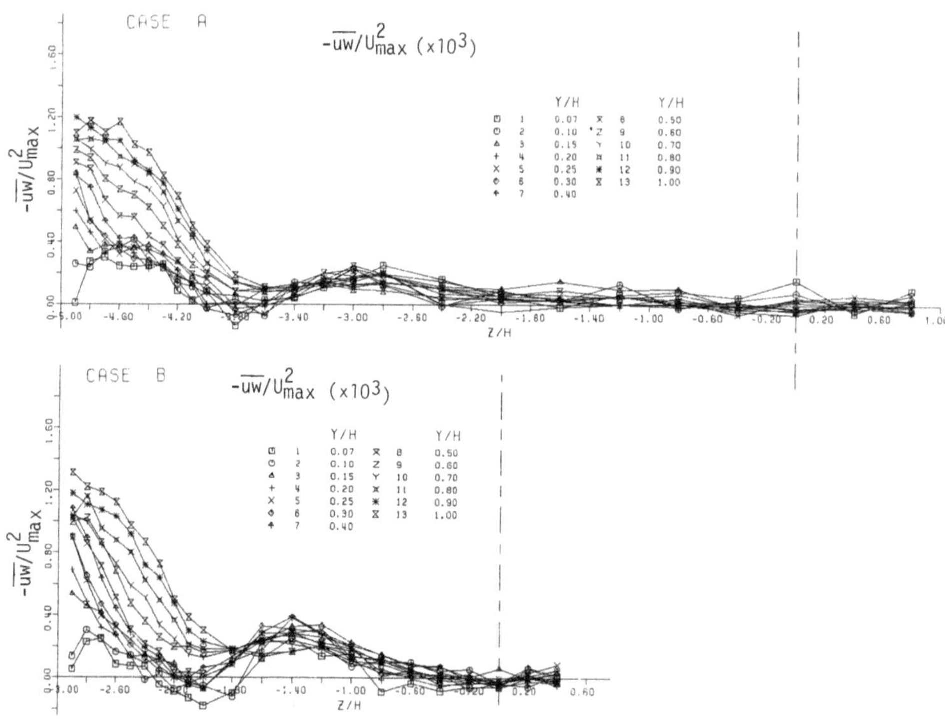

Fig. 14. Distributions of the transverse Reynolds stress $-\overline{uw}$ against the spanwise direction, z.

of $|z|/h = 1.4 - 2.0$ (see Fig. 6). This feature could be also obtained in the multi-cellular secondary currents of Fig. 2. Then, the authors (1984) verified that the loss of mean flow energy convected by the upflow was nearly balanced with the gain of energy done by the transverse Reynolds stress $-\overline{uw}$.

Vorticity. Figures 15 and 16 show the contours of the turbulence intensity difference $(\overline{w^2} - \overline{v^2})$ and the streamwise vorticity $\xi \equiv (\partial W/\partial y - \partial V/\partial z)$, respectively, for Case A. The production term $\partial^2(\overline{v^2} - \overline{w^2})/\partial y\, \partial z$ of vorticity is not at all equal to zero in the side-wall zone, and as the result it produces the vorticity ξ only in this zone, as seen in Fig. 16. As has been expected in Fig. 6, it is understood from Fig. 16 that the intensity of the bottom-vortex is stronger than that of the side-vortex. On the other hand, in the central zone of $|z|/h \leqq 3$, the vorticity is nearly equal to zero.

To sum up, cellular secondary currents are *not* produced in the central zone of solid smooth channel with the aspect ratio $\alpha = B/h > 4$.

Preliminary Experiment on Initiation Mechanism of Cellular Secondary Currents

The above-mentioned fact suggests strongly than an initiation of multi-cellular secondary currents of Fig. 1 is motivated directly neither by the existence of side-wall, nor by the existence of free surface. The former produces originally the corner secondary flow, while the latter strengthens its flow near the free surface.

Therefore, we must search the initiation mechanism of cellular secondary currents anywhere. Then, we found an important feature in the spanwise distribution of the wall shear stress $\tau_0 = \varrho U_*^2$, as mentioned previously. That is, U_* attains a mild peak in the side-wall zone of air duct, as seen in Fig. 9. This feature was also recognized in open channels by Knight et al. (1982) and the authors (1983). Thus, it is considered that this variation of U_* is created by the corner secondary flow. If the channel is composed of a movable flat-bed, the sediment transport varies spanwisely in the side-wall zone, and as the result a first pair of longitudinal sand ridge and trough may be produced. Of course, the peak zone of U_* is scoured to become a trough. Sands on the trough are swept out both downstream (by U) and at the side of the trough (by W), and then they are gathered to form a sand ridge.

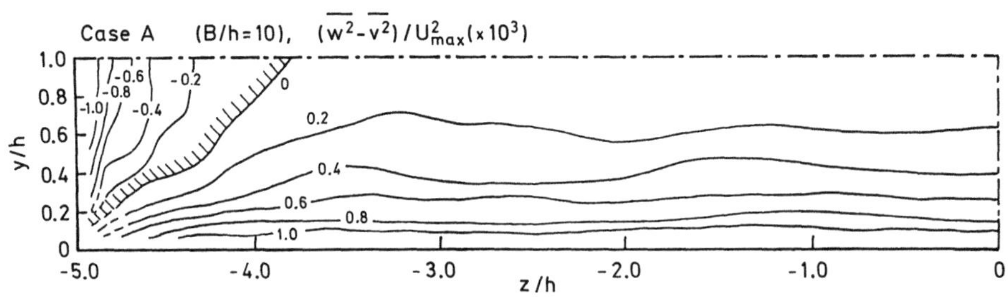

Fig. 15. Iso-lines of $(\overline{w^2} - \overline{v^2})/U_{max}^2$ in the channel of $B/h = 10$

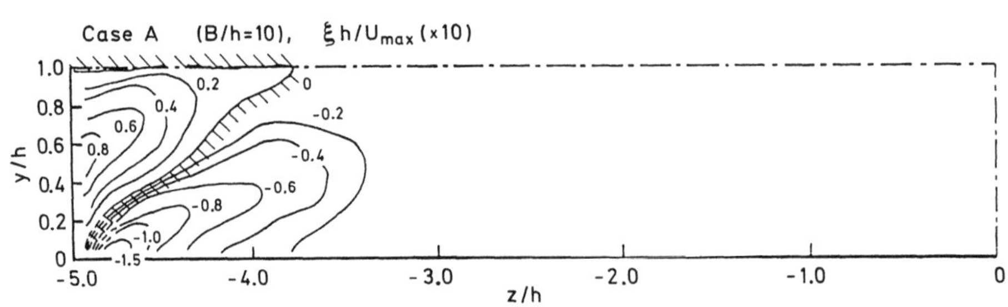

Fig. 16. Iso-lines of the streamwise vorticity ξ in the channel of $B/h = 10$

In order to simulate this situation and verify the validity of this initiation mechanism, we have conducted two experiments of Cases F and G, in which a longitudinal rod with a triangular cross section of the height $k = 0.088\,h$ was attached as a first longitudinal ridge onto the bed of the channel with the aspect ratio $B/h = 10$ (Case A), as shown in Fig. 3. In Case F, the longitudinal rod was set at the position of the minimum wall shear stress, which could be obtained from Fig. 9(a). For a comparison with Case F, its rod of Case G was set at the position of the maximum downflow ($V < 0$), which could be obtained from Fig. 6(a). The condition of Case F corresponds to the formation of a first longitudinal sand ridge.

Figure 17 shows the vector descriptions of the secondary currents (V, W) for Cases F and G. Its ridge height k is comparatively small, but a large variation of the flow patterns is recognized in comparison with Fig. 6(a). For Case F, a strong upflow ($V > 0$) is created over the ridge, and a new pair of cellular secondary currents appear on each side of the ridge. In correspondence to it, the wall shear stress varied spanwisely, and in turn it was suggested that this renewed variation of wall shear stress might produce a second pair of longitudinal ridge and trough. As seen in Case F, the flow pattern of the secondary currents over the ridge is similar to that of the cellular secondary currents of Fig. 2, although the vortex at the right side of the ridge is yet weaker than the vortex at its left side, probably because of the absence of a second longitudinal ridge. On the other hand, Case G indicates a comparatively complicated flow pattern. It should be noticed that the side-vortex of this case disappears due to the existence of the longitudinal rod, as contrasted with Cases A and F. This suggests that, if the longitudinal ridge is set on the bottom near the corner, the corner secondary flow is depressed or destroyed by its ridge.

From the above results, it is considered that an initiation of cellular secondary currents is motivated by the mutual interaction between the secondary currents and the sand bed. The wall shear stress τ_0 has a mild peak in the side-wall zone, due to the existence of the corner secondary flow which is caused by the non-homogeneity of turbulence, i.e. $\partial^2(\overline{v^2} - \overline{w^2})/\partial y\,\partial z \neq 0$ as shown in Fig. 15. This peak and valley of τ_0 produces a spanwise variation of the sediment transport and then a first longitudinal sand ridge is formed. In turn, this sand ridge creates a pair of cellular secondary currents. And then, this currents produce a second sand ridge. After this mutual interaction, a fully developed and hydrodynamic equilibrium state of the flow and the bedform will be established in the whole cross section

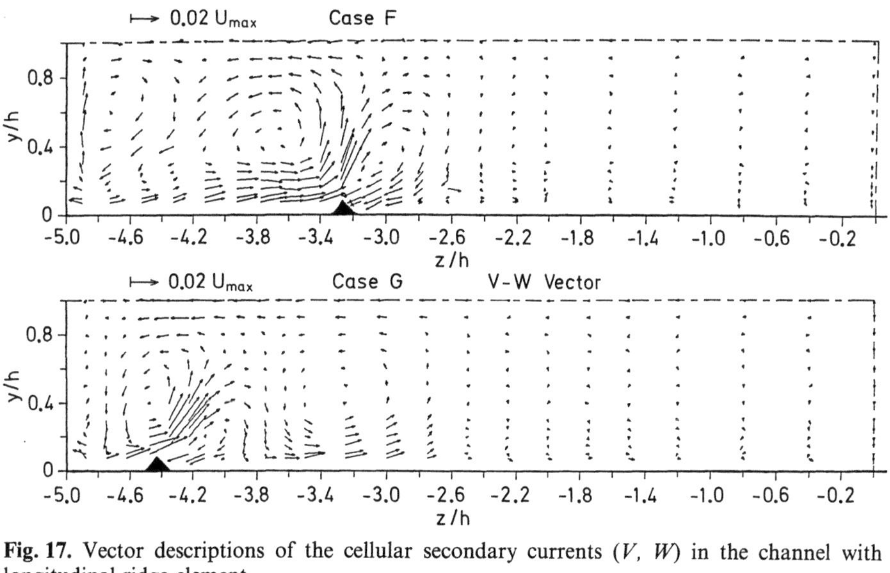

Fig. 17. Vector descriptions of the cellular secondary currents (V, W) in the channel with a first longitudinal ridge element

of channel, as shown in Figs. 1 and 2. Although the aspect ratio of channel is not an essential cause of these initiation, a final configuration of the flow and the bedform will be re-arranged self-consistently by the aspect ratio. Also, the existence of free surface may promote this interaction process, but it does not seem to produce cellular secondary currents originally.

Concluding Remarks

In this paper we have investigated the existence of secondary currents over a smooth solid bed by varying the aspect ratio of channel, and then examined their initiation mechanism. The main findings obtained from the above are as follows:

(1) The present data of the turbulent structure in square channel coincide well with the previous data obtained by Melling and Whitelaw (1976) and the others.
(2) A pair of longitudinal vortices, i.e. corner secondary flow, appear only in the side-wall zone.
(3) The size of the vortex over the corner bisector becomes constant irrespective of the aspect ratio B/h, while the size of the vortex under its bisector increases with B/h and then it approaches a constant.
(4) Cellular secondary currents *are not produced* in the central zone of solid channel with $B/h > 4$.
(5) The bed shear stress attains a mild peak in the side-wall zone.

Among the above results, (5) is the most important feature to explain an initiation mechnism of the cellular secondary currents which are observed in straight alluvial channel flow. A sand bed of alluvial channel will be scoured initially at the position where the bed shear stress attains a peak in the side-wall zone. As the result, a longitudinal sand ridge will be produced. In turn, a new cellular secondary current will be created in the neighbourhood by this sand ridge. Eventually, cellular secondary currents will be produced in all the cross section of channel, by these mutual interaction of the flow and the bedform.

Therefore, more detailed investigations will be necessary in water flow, in order to examine such a hypothesis of the initiation mechanism of cellular secondary currents and also the effects of the free surface (Froude number) and the bed roughness upon the secondary currents.

References

Culbertson, J. K. (1967): Evidence of secondary circulation in an alluvial channel. Geological Survey Research, U.S. Geol. Survey, Prof. Paper 575D, pp. D214–D216

Karcz, I. (1973): Reflections on the origin of source, small-scale longitudinal streambed scours. Fluvial Geomorphology, ed. by M. Morisawa, Proc. of 4th Annual Geomorphology Symposia Series, p. 149

Kinoshita, R. (1967): An analysis of the movement of flood waters by aerial photography, concerning characteristics of turbulence and surface flow. Photographic Surveying **6**, 1 [in Japanese]

Knight, D. W., Patel, H. S., Demetriou, J. D., Hamed, M. E. (1982): Boundary shear stress distributions in open channel and closed conduit flows. Euromech **156**, Istanbul

Laufer, J. (1951): Investigation of turbulent flow in a two-dimensional channel. NACA, TR-1053

Leutheusser, H. J. (1963): Turbulent flow in rectangular ducts. J. Hydr. Div. ASCE, HY-3 **89**, 1

Matthes, G. H. (1947): Macroturbulence in natural stream flow. Trans. Amer. Geophys. Union **28**, 255

Melling, A., Whitelaw, J. H. (1976): Turbulent flow in a rectangular duct. J. Fluid Mech. **78**, 289

Nakagawa, H., Nezu, I., Ooishi, Y. (1982): Experimental study on cellular secondary currents in open-channel flow. Annual Kansai-Branch Meeting of Japan Soc. Civil Engrs., Japan Soc. of Civil Engineers, II-40 [in Japanese]

Nakagawa, H., Nezu, I., Tominaga, A., Wakai, T. (1983): Production mechanism of cellular secondary currents in an open-channel flow. Proc. of 27th Japanese Conference of Hydraulics, pp. 591–596 [in Japanese]

Nezu, I., Nakagawa, H. (1984): Cellular secondary currents in straight conduit. J. Hydr. Eng. ASCE **110**, 173

Naot, D., Rodi, W. (1982): Calculation of secondary currents in channel flow. J. Hydr. Div. ASCE **108**, HY-8, 948

Vanoni, V. A. (1946): Transportation of suspended sediment by water. Transactions ASCE **111**, 67

A Computational Preston Tube Method

W. Nitsche, R. Thünker and C. Haberland

Institut für Luft- und Raumfahrt, Technische Universität Berlin, D-1000 Berlin 10, Federal Republic of Germany

Abstract

A computational Preston tube method to determine local wall friction forces is presented, which unlike the conventional Preston tube method offers a greater scope of application. The method is based on measuring the dynamic pressures of wall-Pitot tubes of different diameters and determing the local wall shear stress by applying a mathematical boundary-layer law to the flow velocities corresponding to the measured dynamic pressures. The method is verified experimentally in a laminar-turbulent transition flow of a flat plate, in a pipe entrance flow, in a turbulent pipe flow with sudden enlargement, in strongly non-adiabatic turbulent boundary-layer flows and in turbulent boundary-layer flows with adverse pressure gradient.

Nomenclature

A^+	Van Driest constant
a	Speed of sound
$c_f = \tau_w/\varrho \cdot u_\infty^2$	Skin friction coefficient
c_p	Specific heat
$D, D_{1,2}$	Pipe diameters
$d, d_{1,2}$	Preston tube diameters
$d^+ = u_\tau \cdot d/\nu$	Non-dimensional Preston tube diameter
K	Displacement factor
$K_{1,2\ldots i}$	Iteration parameters
$M_\tau = u_\tau/a_w$	Friction Mach number
p	Static pressure
$p^+ = (\nu/\varrho \cdot u_\tau^3)\dfrac{dp}{dx}$	Boundary-layer pressure parameter
q	Dynamic pressure, $p_{\text{total}} - p_{\text{static}}$
$q^+ = qd^2/4\varrho \cdot \nu^2$	Non-dimensional dynamic pressure
$\dot{q}$	Heat flux rate
Re	Reynolds number
T	Mean temperature
u	Mean flow velocity
$u^+ = u/u_\tau$	Non-dimensional mean velocity
$u_\tau = \sqrt{\tau_w/\varrho_w}$	Shear stress velocity
x	Longitudinal coordinate
y	Distance from wall
$y^+ = u_\tau \cdot y/\nu$	Non-dimensional distance from wall
$\beta_q = \dot{q}_w/(\varrho \cdot c_p \cdot T)_w \cdot u_\tau$	Boundary-layer heat flux parameter
κ	v. Kármán constant
$\lambda = 8 \cdot \tau_w/\varrho \cdot u_m^2$	Friction factor
λ_w	Thermal conductivity
ϱ	Density
τ_w	Wall shear stress
$\tau^+ = \tau_w \cdot d^2/4 \cdot \varrho \cdot \nu^2$	Non-dimensional wall shear stress
ν	Kinematic viscosity

Subscripts

D	Diameter
e	Boundary-layer edge
f	Fluid
m	Mean value
w	Wall condition
x	Longitudinal direction
∞	Free stream condition

Introduction and Preliminary Remarks

Local wall friction forces are one of the main quantities of interest in experimental boundary-layer investigations, as the local wall shear stress represents one of the boundary conditions of wall shear layers. Considerable effort has been made to develop dependable methods to measure these wall friction forces. A survey is given by Winter (1977).

The Preston tube method (Preston, 1954) is one of the most widely used shear stress measuring techniques due to its simple construction and practicability. It is based on the similarity law of turbulent boundary-layers

$$u^+ = f(y^+); \quad u^+ = \frac{u}{u_\tau}, \; y^+ = \frac{u_\tau y}{\nu}. \tag{1}$$

The law of the wall is used to the effect that a measured quantity proportional to the velocity (in this case the dynamic pressure of a wall-Pitot tube) can be correlated to the local shear stress at known distance of the measuring point from the wall (y_{eff}). The correlation between the dynamic pressure and the corresponding wall shear stress is usually represented by a calibration curve of the type:

$$q^+ = F(\tau^+); \quad q^+ = \frac{qd^2}{4\varrho\nu^2}, \; \tau^+ = \frac{\tau_w d^2}{4\varrho\nu^2}. \tag{2}$$

The calibration parameters q^+ and τ^+ can be obtained directly from the variables of the law of the wall Eq. (1), when half the diameter of the Preston tube is taken as the characteristic wall distance:

$$q^+ = \frac{1}{2}\left(\frac{u}{u_\tau}\frac{u_\tau d/2}{\nu}\right)^2, \quad \tau^+ = \left(\frac{u_\tau d/2}{\nu}\right)^2. \tag{3}$$

This direct relationship between the calibration parameters and the boundary-layer variables permits the calibration curve according to Eq. (2) to be calculated directly from the wall law (Nitsche, 1980), Fig. 1.

This calculation was carried out using the appropriate law of the wall of turbulent flat plate flow and fully developed turbulent pipe flow respectively according to van Driest (1956)

$$u^+ = \int_0^{y^+} \frac{2\,dy^+}{1 + [1 + 4(\kappa y^+)^2 (1 - \exp(-y^+/A^+))^2]^{0.5}}, \qquad \kappa = 0.4; \; A^+ = 26 \tag{4}$$

and the simplified relationship for the effective wall distance

$$y_{eff} = (d/2)\,K; \quad K = 1.3. \tag{5}$$

The results compare well with the empirical calibration curves according to Preston (1954), Patel (1965) and the N.P.L. (1961), and hence, this calculation illustrates some essentials of the Preston tube method:

(1) The Preston tube calibration curve is essentially the transformation of a boundary-layer similarity law into directly measurable quantities of a wall-Pitot tube. The term "calibration curve" is justified in so far as the correlation of the dynamic pressure to the effective wall distance of the corresponding velocity is an implicit part of the calibration, whereas the calculation assumes y_{eff} to be known.

(2) For boundary-layer flows in which the law of the wall deviates from the classical case, the calibration curve shown in Fig. 1 no longer holds true and unrestricted use leads to principle measuring errors.

(3) To determine wall friction forces correctly in such cases (due to the direct relationship between boundary-layer law and the calibration curve), we have to take into account

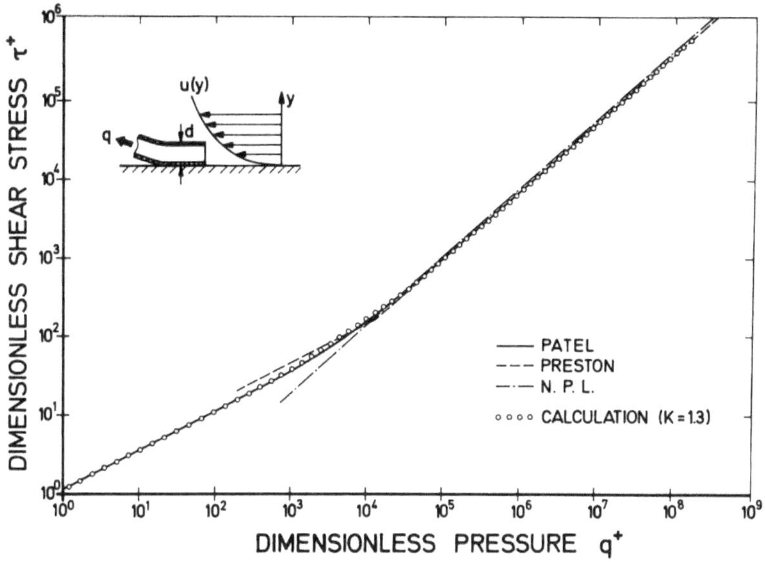

Fig. 1. Conventional Preston tube calibration curve

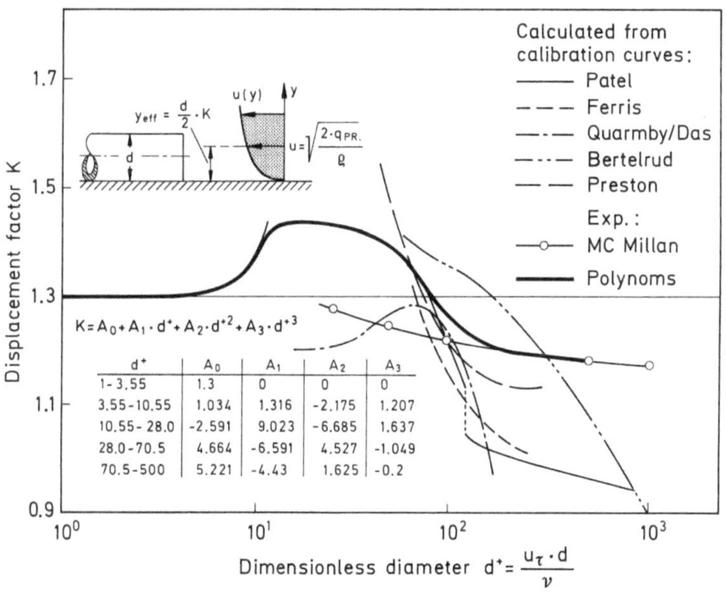

Fig. 2. Displacement factor dependent on non-dimensional diameter

calibration curves that have been modified accordingly. Investigations for some boundary-layer parameters (dimensionless pressure gradient, heat flux, friction Mach number, wall roughness) have been carried out by Nitsche (1980) and Haberland and Nitsche (1982).

(4) Calibration curves cannot be formulated for boundary-layer flows whose similarity law is unknown, and the classical Preston tube method as well as its extensions based on the use of similarity laws fail.

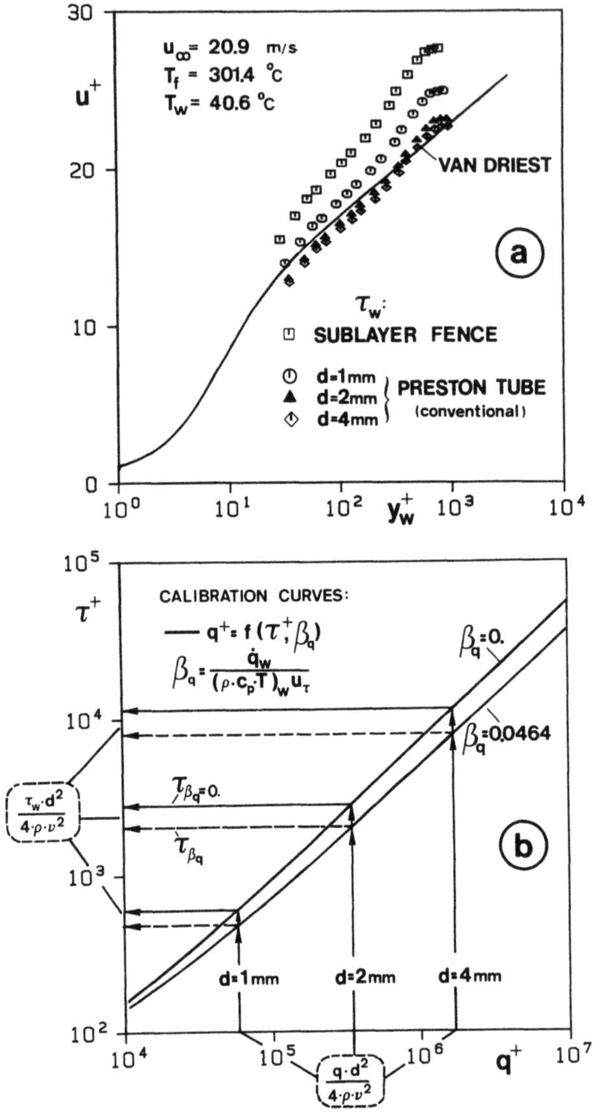

Fig. 3. Measuring errors in non-adiabatic flows; (a): Law of the wall; (b): Calibration curves

The correlation of the dynamic pressure to the effective wall distance mentioned in (1) has been simplified according to Eq. (5) in the calculation of Fig. 1. As shown in Fig. 2, a more accurate function of the displacement factor K can be obtained by means of an approximation of the calculated calibration curve to empirical formulas. For the following calculations we use a K-function close to the curve obtained from the Patel formulas but for $d^+ > 100$ fitted to the experimental results of McMillan (1956).

The principle measuring errors mentioned in (2) using the classical calibration curve in boundary-layer flows, whose law of the wall deviates significantly from Eq. (4), is exemplified in Fig. 3. To represent the measured boundary-layer profile of a strongly non-adiabatic turbulent flow (heat flux rate $\dot{q}_w \simeq 12{,}000\ \mathrm{W/m^2}$) in law of the wall coordinates, the wall shear stresses resulting from the classical calibration curve for three Preston tubes of different diameters were used. As shown in Fig. 3a, increasing probe diameter causes growing deviation in the measurement as compared to that obtained from a sublayer fence. This

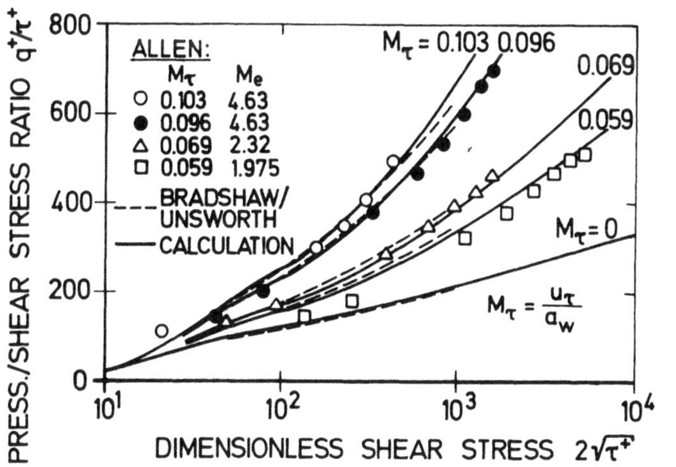

Fig. 4. Calibration curves for compressible flows

effect is clear from Fig. 3 b, which shows the calibration curve for this flow type, calculated from the law of the wall of non-adiabatic turbulent flows given by Nitsche et al. (1983), compared to the classical calibration curve. Increases in dimensionless pressure due to the larger probe diameters lead to increasing measuring errors when using the classical calibration curve and to considerable deviations when the boundary-layer is represented in law of wall coordinates, Fig. 3 a.

The need to consider modified calibration curves for changed wall laws, as mentioned in (3), is evident from Fig. 3. A further example is depicted in Fig. 4, which shows the calculated calibration curves for compressible flows given by Haberland and Nitsche (1982) as compared with experimental results according to Bradshaw and Unsworth (1973) based on these measurements.

This extended Preston tube method based on physical boundary-layer similarity laws, offers an additional range of applications. However, since we cannot assume the boundary-layer laws to be known as a matter of principle for all flow types, general validity has to be ruled out. In addition, there are hardly any promising formulations available which describe the boundary-layer for many flows by means of similarity laws. Thus, the demand for a more generally valid Preston tube method necessarily requires the separation of the method from physical boundary-layer laws in order to overcome the restrictions they always involve.

Computational Preston Tube Method

As has been mentioned in the introduction, the conventional Preston tube method as well as its extensions are based on the unambiguous correlation of a velocity, corresponding to the measured dynamic pressure, to a characteristic wall distance and the wall shear stress determination via a boundary-layer similarity law. The unambiguousness of this correlation is expressed by the fact that the diameter of the probe can be chosen freely within the range of validity of the assumed boundary-layer law, i.e. identical wall shear stresses are measured using probes of different diameters. The unambiguousness of the measured shear stress has to play naturally a very important part in extending the Preston tube method to boundary-layer flows with unknown wall law: The former method fails due to the assumption of a non-appropriate law of the wall, as can be recognized by different wall shear stresses obtained from probes of different diameters. To a certain extent, the reversal of this fact allows correct measuring values to be expected for an unknown wall law, if we succeed in formulating a boundary-layer law which yields identical wall shear stresses when applied to dynamic pressures of Preston tubes of different diameters.

Basic Method

As shown in Fig. 1, the conventional Preston tube calibration curve can be computed by numerical integration of the wall law, Eq. (4), taking into account a relationship for the effective wall distance. In practical application of this calibration curve, the paired values q^+ and τ^+ can be interpolated, however, it is more convenient to apply the numerical calculation method by an iteration process to the measured dynamic Preston tube pressure. Figure 5 illustrates this procedure: In the first step, ($i = 1$) the boundary-layer profile for a given initial wall shear stress value is calculated with the aid of the wall law up to the effective wall distance (obtained from the Preston tube diameter and the displacement function in Fig. 2).

Since the velocity u_i calculated at y_{eff} does not satisfy the velocity corresponding to the measured dynamic pressure, the wall shear stress in the boundary-layer computation is varied iteratively, until the calculation matches the measured velocity. The local wall friction

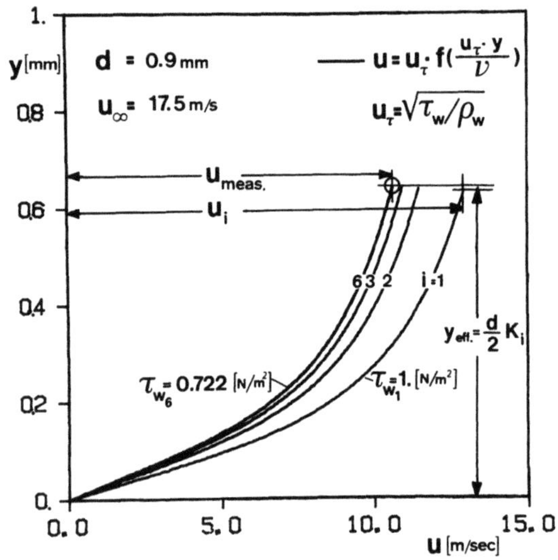

Fig. 5. Boundary-layer iteration in the basic method

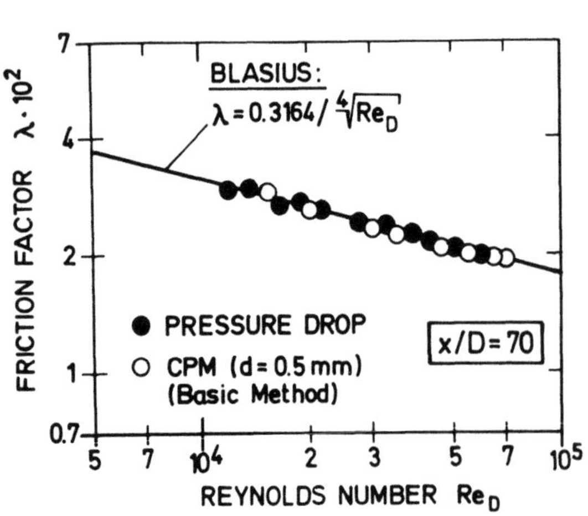

Fig. 6. Friction factors of fully developed turbulent pipe flow. CPM and pressure drop data

thus determined is equivalent to the τ_w-value obtained from the conventional calibration curve, Fig. 1, and this computer aided measuring technique can thus completely replace the classical Preston tube method. Figure 6 illustrates this with a comparison between the measured friction factors of a fully developed turbulent pipe flow, which were determined with the aid of this computational method and by means of pressure drop measurements respectively.

Extension to Boundary-Layers With Unknown Law of the Wall

As a consequence of the assumed wall law, the computational Preston tube method in this basic version is certainly subjected to the same restrictions as the conventional method. However, it may be used in boundary-layers with known parameters of influence, for example in the case of non-adiabatic turbulent flows as shown in Fig. 3, through analogous application of appropriate wall laws.

However, even this method has to fail in measuring the wall friction for unknown wall law, as the necessary unambiguous correlation of measured dynamic pressures to the wall law can no longer be formulated. The intended extension thus requires the introduction of the unambiguousness of the resulting measured value of τ_w as an additional criteria. The basic method described above can remain the starting point; it is, however, applied to the measured dynamic pressures of *two* Preston tubes of different diameters unlike the simple method based on the measured dynamic pressure of a *single* Preston tube. In case the resulting wall shear stresses do not coincide, the need for a wall law arises which – even when only mathematically formulated – will yield identical wall shear stresses for both measured dynamic pressures.

In order to achieve a great scope of application, the law of the wall used here necessarily has to permit extensive variations without yet violating fundamental principles of the boundary-layer theory, especially the boundary condition at the wall. Hence, it is convenient to base this formal law on that of turbulent boundary-layer flows, Eq. (4). By including the extension for near-separation boundary-layer flows, as proposed by Szablewski (1969), Eq. (4) can be formulated as a wall law with three free parameters

$$u^+ = \int_0^{y^+} \frac{2(1 + K_3 y^+)\, dy^+}{1 + [1 + 4(K_1 y^+)^2 (1 + K_3 y^+)(1 - \exp(- y^+ \sqrt{1 + K_3 y^+}/K_2))^2]^{0.5}} \tag{6}$$

with K_1 formally corresponding to the v. Kármán constant, K_2 to the Van Driest constant A^+ and K_3 to the dimensionless pressure parameter p^+. For $K_1 = 0.4$, $K_2 = 26$, and $K_3 = 0$, this wall law is identical to Eq. (4).

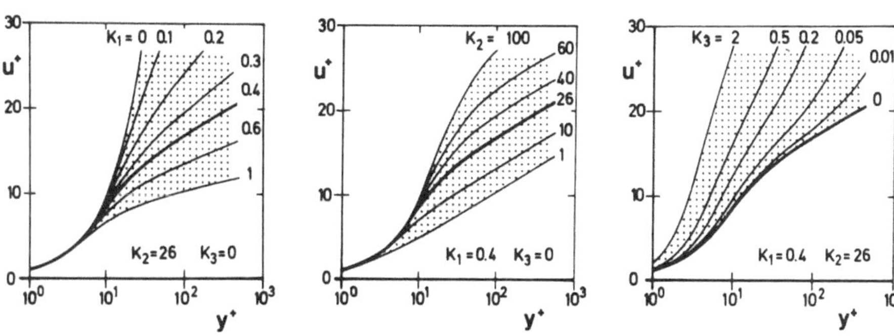

Fig. 7. Boundary-layer model used in the CPM

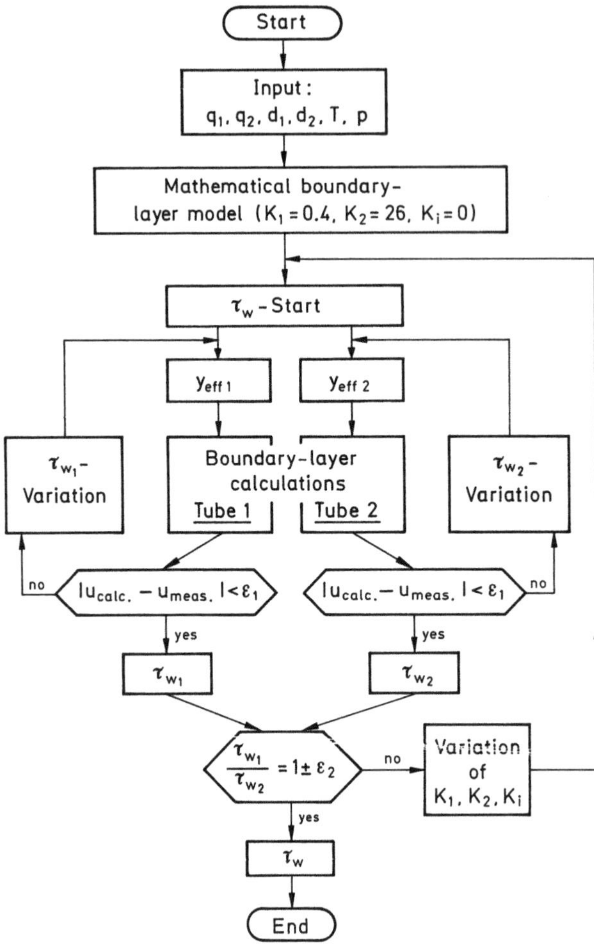

Fig. 8. CPM flow chart

Figure 7 shows a parametric representation of Eq. (6) and illustrates the influence of the free parameters as well as the range of application of this wall law.

The flow chart shown in Fig. 8 illustrates the practical application of the computational Preston tube method based on this mathematical wall law. Initially, both wall shear stresses resulting from the classical law of the wall ($K_1 = 0.4$, $K_2 = 26$, $K_3 = 0$) are calculated from the measured dynamic pressures of the two probes. This first step is identical to the basic method, Fig. 5. In case the two resulting wall shear stresses do not coincide, the parameters K_i in the mathematical wall law are gradually varied by means of an trial-and-error iteration so that the difference between τ_{w1} and τ_{w2} decreases and vanishes in the last iteration step. Figures 9 a and 9 b are examples illustrating this iteration using a K_1-variation for measuring the skin friction in a laminar-turbulent transition flow of a flat plate. The basic method ($K_1 = 0.4$) yields a lower shear stress for the smaller Preston tube than for the larger one ($\tau_{w1} = 0.238$, $\tau_{w2} = 0.287$ N/m^2), Fig. 9 b. This becomes likewise clear from the first boundary-layer calculations, Fig. 9 a, which show a larger velocity gradient at the wall for the measuring point 2.

The first step of the main iteration loop ($i = 1$, $K_1 = 0.45$) is used to determine the potential direction of the iteration procedure by comparing the resulting wall shear stresses with the basic values ($i = 0$). In the following iteration steps the value of K_1 which yields

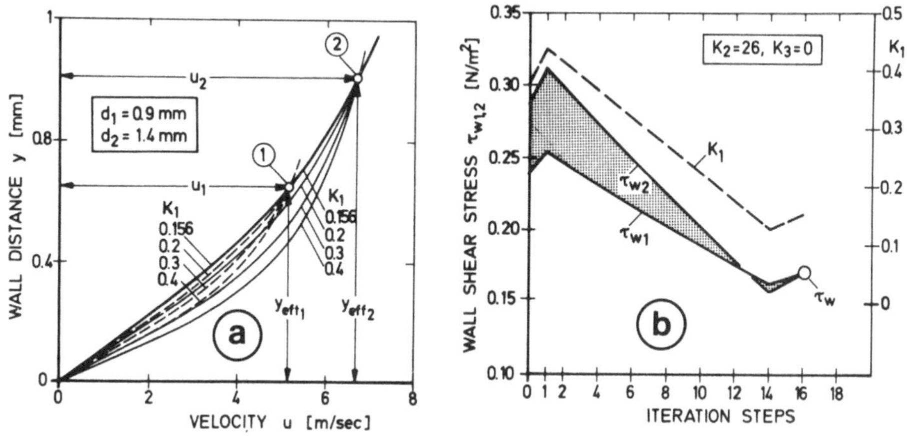

Fig. 9. Boundary-layer iteration in the CPM

identical wall shear stresses for both probes is determined iteratively ($K_1 = 0.156$, $\tau_{w1,2} = 0.171\ \mathrm{N/m^2}$), Fig. 9b. The convergence of the two wall shear stresses through the iteration process can as well be recognized from the respective boundary-layer profiles, Fig. 9a, which converge progressively and collapse at one another for $K_1 = 0.156$.

Experimental Facilities

In order to verify the computational Preston tube method, experimental set-ups were planned which could be expected to show considerable deviations in the boundary-layer laws from the classical cases of a turbulent flat plate and a fully developed turbulent pipe

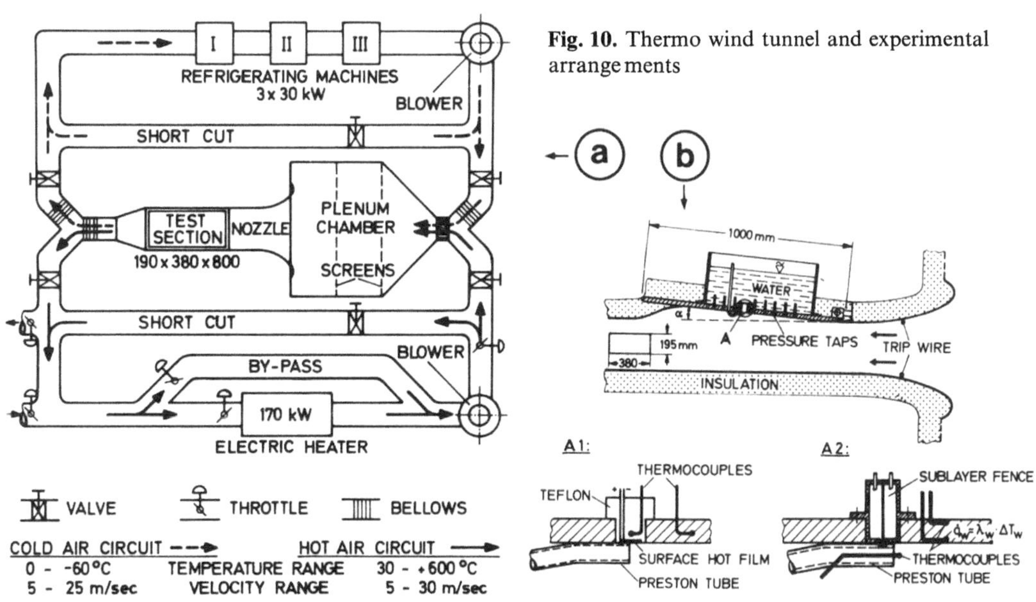

Fig. 10. Thermo wind tunnel and experimental arrangements

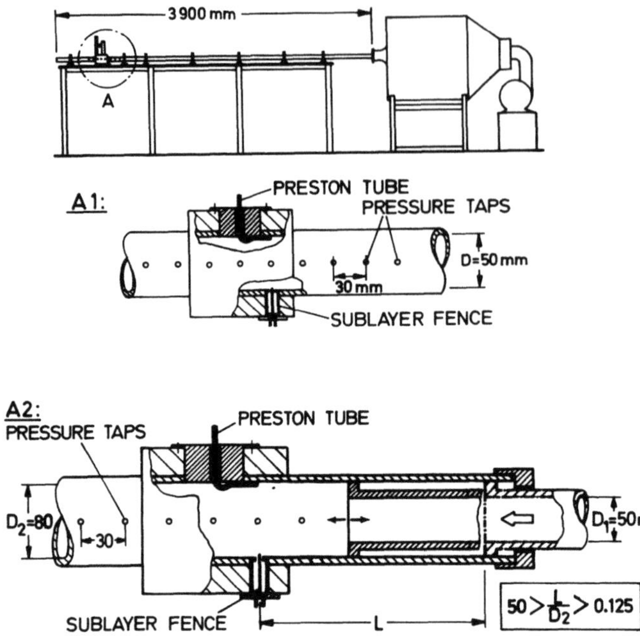

Fig. 11. Test pipe and experimental arrangements

flow. The first facility, the ILR[1]-Thermo Wind Tunnel, is shown in Fig. 10. This tunnel consists of two separate circuits for hot and cold air with infinitely variable flow velocity and temperature, Fig. 10a. The upper test section wall, Fig. 10b, could be inclined with respect to the tunnel axis to achieve a positive pressure gradient, as well as heated or cooled with water to produce a positive or negative heat flux in the boundary-layer. Furthermore, the trip wire in the inlet nozzle could be removed, to obtain a laminar-turbulent transition flow. The probe arrangements at the test section wall are shown in the details of A_1 and A_2: Skin friction measuring probes to be used as reference measuring devices were mounted in the wall close to the tip of the Preston tubes (A_1: Surface hot film probe and thermocouples to control wall and probe support temperatures in order to avoid measuring errors caused by temperature effects (Nitsche, Haberland, 1983), A_2: Sublayer fence and thermocouples in the tunnel wall to determine the heat flux rate and the reference temperature to calculate the velocity from the dynamic pressure). The static pressure distribution was measured by pressure taps along the tunnel axis, the dynamic pressure of the Preston tubes determined by the total pressure of the probe and the static pressure obtained from a pressure tap at least ten probe diameters across the probe tip.

The test pipe shown in Fig. 11 was used as the second experimental set-up, whose probe arrangement was basically the same one used in the Thermo Wind Tunnel.

The measuring station of the 50 mm test pipe, detail A_1, could be shifted towards the inlet nozzle up to $x/D = 5$ for experiments in the inlet region. In addition, a measuring station in a 80 mm pipe could be coupled at the outlet side of the 50 mm pipe, detail A_2, enabling tests in a pipe flow with sudden enlargement.

[1] Institut für Luft- und Raumfahrt (Department of Aeronautics and Astronautics) at Berlin Technical University

Results

Preliminary tests were first conducted on both the test facilities, in order to calibrate the reference measuring devices. Figure 12a shows the skin friction coefficients measured in the Thermo Wind Tunnel (adiabatic turbulent flow, constant pressure) and Fig. 12a* the measured friction factors in the 50 mm pipe at $x/D = 70$. Independent of the measuring technique and of the used Preston tube diameters, good agreement between the data and the respective theoretical friction laws can be stated.

The results of the skin friction measurements conducted for varied test conditions are also summarized in Fig. 12. The Figs. 12b and c show the results obtained in the test pipe (b: Entrance flow, c: Sudden enlargement), the Figs. 12d–f the measurements carried out in the Thermo Wind Tunnel (d: Laminar-turbulent transition, removed trip wire, e: Adverse pressure gradient, inclined plate, f: Strong heating and cooling, $-46\,°C < T$-fluid $< +300\,°C$, $+20\,°C < T$-wall $< +40\,°C$). The respective upper diagrams include the data obtained from the reference measuring devices and from the conventional Preston tube method and indicate the breakdown of this method, illustrated by different shear stresses obtained with different probe diameters. In contrast, the results of the computational Preston tube method in the respective lower diagrams are in good agreement with the sublayer fence and surface hot film data and give an impression of the wide range of application of this measuring technique.

In these investigations, the sublayer fence readings were corrected with respect to the local pressure gradient (if necessary) according to Patel (1965), and the influence of changed temperatures (Fig. 12f) on the sublayer fence calibration curve was compensated by treating the fence like a very small Preston tube, i.e. all the properties in the calibration formula were taken at wall temperature, except for the density in the dimensionless pressure, which was taken at a reference temperature at half the height of the fence.

The measurements of negative shear stresses close to the backward facing step of the pipe ($x/D_2 < 2$), Fig. 12c, became possible by calibrating the fence in the foreward and backward mode (turning of the pipe).

The CPM-results shown in Fig. 12b*–d* were obtained with a twin-probe, Fig. 13, the results in Fig. 12e* and f* with two single probes of different diameter.

Discussion and Concluding Remarks

All the results of the computational Preston tube method shown in Fig. 12 were obtained by means of an K_1-iteration without exception. Certainly the question arises about the results we would obtain using the parameters K_2 and K_3. Particularly, the unambiguousness of the measured wall shear stress had to be investigated. For that purpose, comparative computations using the K_1, K_2 and K_3 parameters have been carried out. The results of the pipe flow with sudden enlargement, Fig. 14a, clearly indicate that for this flow type, the K_1 parameter is the most useful one and renders possible shear stress determination up to the reattachement point. Closer than $x/D_2 = 2$ to the step, we obtain no results, since the mathematical wall law, Eq. (6), does not include recirculating effects. The variation parameters K_2 and K_3 are not as useful as K_1, particularly the K_2-variation yields only results for $x/D_2 > 5$. However, the main conclusion of these comparative computations is the fact that an unsuitable iteration parameter does not yield irregular wall shear stresses but *no* results at all. This is obvious from an K_1 and K_2 iteration for the measured dynamic pressures at $x/D_2 = 2.6$, shown in Fig. 15. Whereas the K_1 iteration yields coinciding wall shear stresses in the course of the computations, the K_2 iteration leads to no result, indicated by the

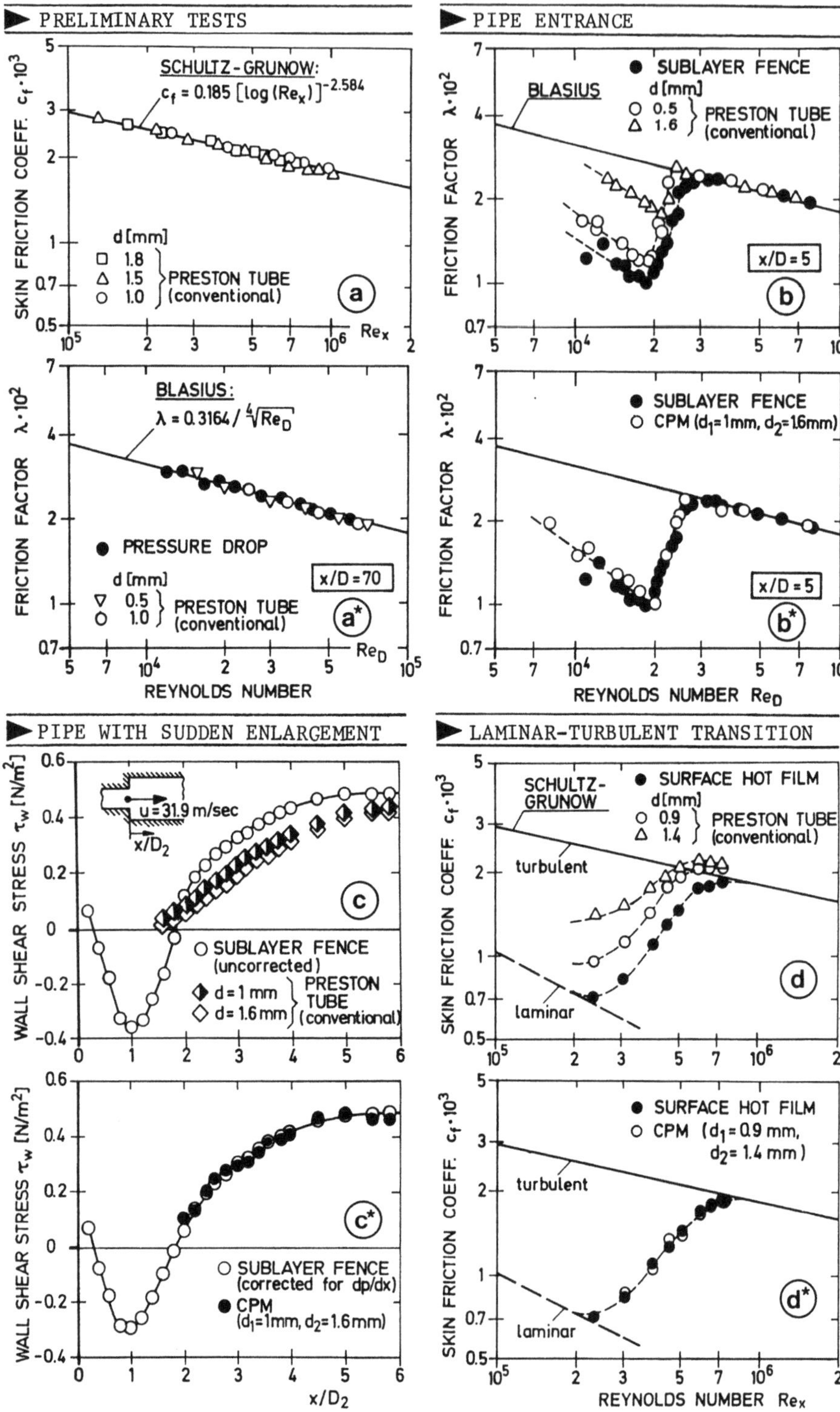
PRELIMINARY TESTS
SKIN FRICTION COEFF. c_f·10^3
SCHULTZ-GRUNOW:
c_f = 0.185 [log (Re_x)]^-2.584
d [mm]
1.8
1.5
1.0
PRESTON TUBE (conventional)
a
Re_x
PIPE ENTRANCE
FRICTION FACTOR λ·10^2
BLASIUS
SUBLAYER FENCE
d [mm]
0.5
1.6
PRESTON TUBE (conventional)
x/D = 5
b
BLASIUS:
λ = 0.3164 / ⁴√Re_D
PRESSURE DROP
d [mm]
0.5
1.0
PRESTON TUBE (conventional)
x/D = 70
a*
Re_D
REYNOLDS NUMBER
SUBLAYER FENCE
CPM (d1=1mm, d2=1.6mm)
x/D = 5
b*
REYNOLDS NUMBER Re_D
PIPE WITH SUDDEN ENLARGEMENT
WALL SHEAR STRESS τ_w [N/m^2]
u = 31.9 m/sec
x/D2
c
SUBLAYER FENCE (uncorrected)
d = 1 mm
d = 1.6 mm
PRESTON TUBE (conventional)
c*
SUBLAYER FENCE (corrected for dp/dx)
CPM (d1=1mm, d2=1.6mm)
x/D2
LAMINAR-TURBULENT TRANSITION
SKIN FRICTION COEFF. c_f·10^3
SCHULTZ-GRUNOW
SURFACE HOT FILM
d [mm]
0.9
1.4
PRESTON TUBE (conventional)
turbulent
laminar
d
SURFACE HOT FILM
CPM (d1 = 0.9 mm, d2 = 1.4 mm)
turbulent
laminar
d*
REYNOLDS NUMBER Re_x

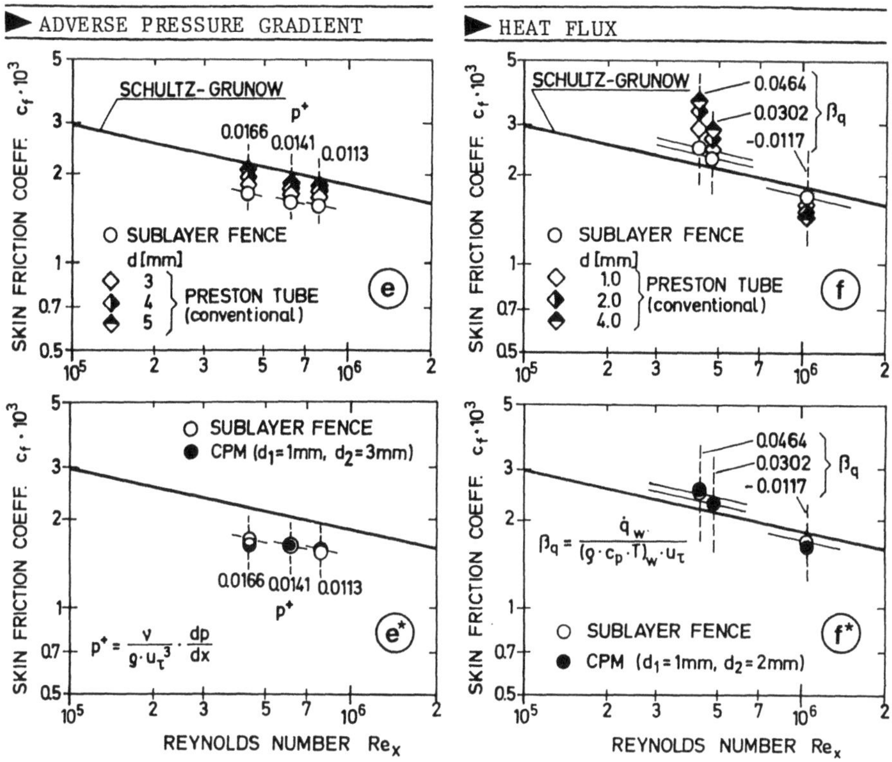

Fig. 12. Skin friction measurements at various test conditions. **a and a***: Preliminary test results for turbulent flat plate and fully developed turbulent pipe flow. **b–f**: Conventional Preston tube method compared with reference measuring techniques. **b*–f***: Computational Preston tube method compared with reference measuring techniques

convergence of the resulting wall shear stresses in the beginning of the iteration process but divergence for $K_2 < 20$ without matching in the whole course of iteration.

In case of successful iterations, the resulting shear stresses obtained with the particular iteration parameters do not agree exactly, but the differences are acceptable for wall shear stress measurements, Fig. 14a and b.

The error sensitivity of this method due to measuring errors in dynamic pressures is shown in Fig. 16. Errors of $\pm 1\%$, especially when occurring with change in sign, cause deviations in the resulting wall shear stress of 5% ($< 2\%$ when occurring in the same direction, detail of Fig. 16). To avoid these measuring errors to a large extent, the use of the twin probe shown in Fig. 13 is recommended, which allows both dynamic pressures to be measured simultaneously.

Furthermore, in case of stronger uncertainties in dynamic pressure measurements caused by special test conditions, the use of *three* different sized probes instead of the basically used *two* probes may be advantageous in regard of reducing shear stress measuring errors. For that purpose the computational procedure shown in Fig. 8 is applied to three flow velocities corresponding to the dynamic pressure data of three probes.

As an example, shear stress computations based on three-probe measurements in a strongly non-adiabatic turbulent flow ($\dot{q}_w \simeq 12{,}000$ W/m^2) are shown in Fig. 17. Since the computation yields absolutely identical shear stresses for all three probes only in exceptional cases, the truncation criteria of the main iteration loops has to be changed (two probes:

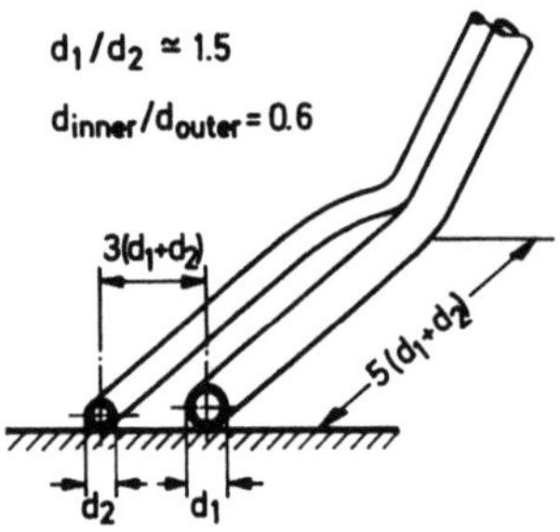

Fig. 13. CPM twin probe

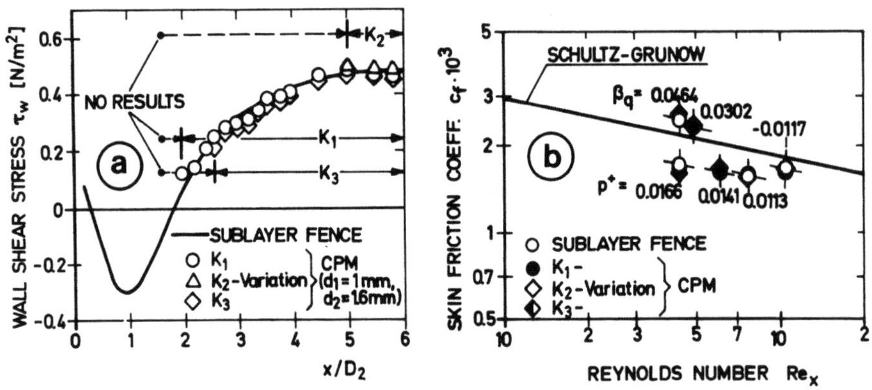

Fig. 14. CPM iteration for K_1, K_2, K_3; **a:** Pipe flow with sudden enlargement; **b:** Turbulent flat plate flow with pressure gradient and heat flux

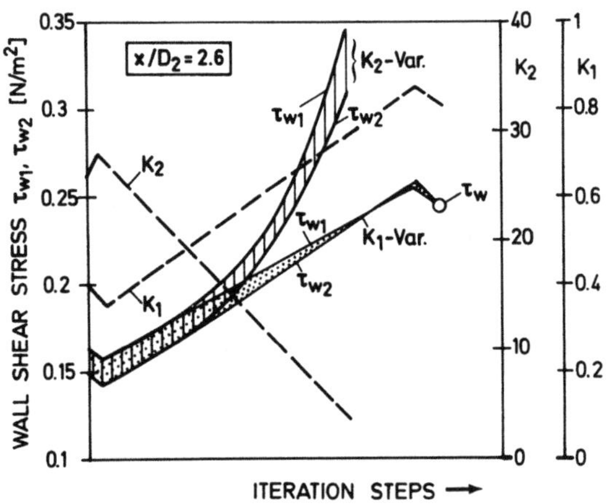

Fig. 15. CPM-iteration for K_1 and K_2

$\tau_{w1} = \tau_{w2}$, three probes: $\sum \Delta\tau_w =$ minimum), Fig. 17a. In Fig. 17b the advantage of three-probe computations is shown by comparing all the possible resulting wall shear stresses obtained from two-probe and three-probe computations based on the data of four different sized Preston tubes. It is obvious that the addition of a third probe to the computation procedure is capable of decreasing the sensitivity of the CPM due to measuring errors in dynamic pressures.

Generally, the probe diameters should be small and not thicker than 20% of the boundary-layer, as the mathematical wall laws used in the CPM are valid only in the near

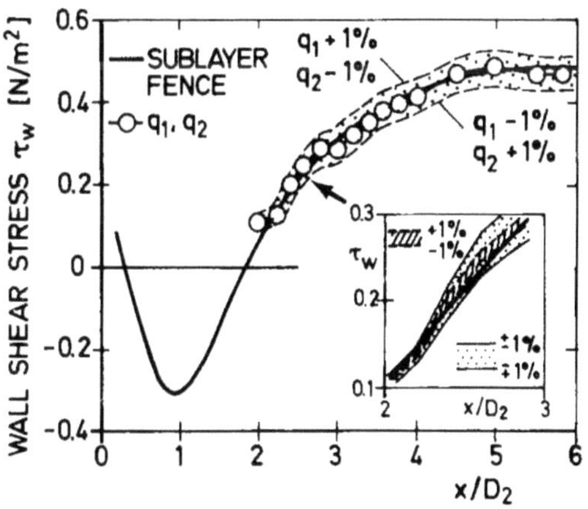

Fig. 16. Error sensitivity due to measuring errors in dynamic pressures

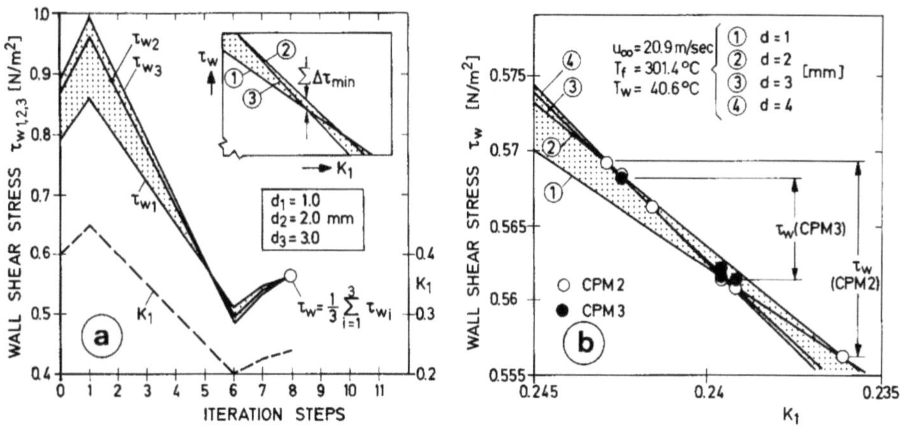

Fig. 17. Shear stress iteration (CPM 2 and CPM 3)

wall region. In addition, the probe diameter ratio should be $d_1/d_2 \simeq 1.5$, in order to get significant differences in the basic shear stresses.

Concluded, the results obtained from the computational Preston tube method are very encouraging and offer hope for skin friction measurements in arbitrary boundary-layer flows. Future efforts will concentrate on further extensions, for example the introduction of additional iteration parameters, which might be necessary for other flow types than the ones investigated here.

Acknowledgement. The authors are grateful to W. Pfingst, A. Schmiderer and N. Weiser for their assistance in carrying out the tests, to R. Suprayan and M. Prochnow for their help in preparing this paper. The research was supported by the German Science Foundation (DFG).

References

Allen, J. M. (1973): Evaluation of compressible flow Preston tube calibration. NASA TN D-7190

Bradshaw, P., Unsworth, K. (1973): A note on Preston tube calibrations in compressible flow. I.C. Aero Report 73-07

Haberland, C., Nitsche, W. (1982): Wall shear stress determination in boundary-layers with unknown law of the wall by a modified Preston tube method. ICAS Paper 82-6.4.2, Proc. of the 13th ICAS/AIAA Conf., ed. by B. Laschka, R. Staufenbiel Vol. 1, 769

McMillan, F. A. (1956): Experiments on Pitot tubes in shear flow. ARC R & M 3028
Nitsche, W. (1980): Wandschubspannungsmessung mit Prestonrohren in Grenzschichtströmungen mit zusätzlichen Einflußparametern. Z. Flugwiss. Weltraumforsch. **4**, 142. [English Transl.: Measuring skin friction with Preston tubes in boundary-layer flows with additional parameters.] Nat. Res. Counc. Canada, NRC/CNR TT-1989
Nitsche, W., Thünker, R., Haberland, C. (1983): Heat transfer and boundary-layer laws in strongly non-adiabatic turbulent flows, in *Numerical Methods in Thermal Problems*, ed. by R. W. Lewis, J. H. Johnson, W. R. Smith, Vol. 3, 707 (Pineridge, Swansea, UK)
Nitsche, W., Haberland, C. (1983): Wall shear stress measurements by means of surface hot film probes with respect to temperature effects, in Proc. of Sensor 83, Vol. 1; **1**, 42
Staff of Aerodynamics Division N.P.L. (1961): On the measurement of local surface friction on a flat plate by means of Preston tubes. ARC R & M 3185
Patel, V. C. (1965): Calibration of the Preston tube and limitations on its use in pressure gradients. J. Fluid Mech. **23**, 185
Preston, J. H. (1954): The determination of turbulent skin friction by means of Pitot tubes. J. Roy. Aeronaut. Soc. **58**, 109
Szablewski, W. (1969): Turbulente Grenzschichten in Ablösenähe. Z. Angew. Math. Mech. **49**, 215
Van Driest, E. R. (1956): On turbulent flow near a wall. J. Aerospace Sci. **23**, 1007
Winter, K. G. (1977): An outline of the techniques available for the measurement of skin friction in turbulent boundary-layers. Prog. Aerospace Sci. **18**, 1

Measurements of the Wall Shear Stress in Boundary Layer Flows

M. Acharya and M. P. Escudier

Brown Boveri Research Center, CH-5405 Baden, Switzerland

Abstract

It is argued that a direct measurement of the wall shear stress is important in all but the simplest of boundary-layer flow situations. Wall shear stress data obtained with a floating element device are presented for a fully developed turbulent boundary layer on a rough (cast) surface, and for boundary-layer transition on a smooth wall. Comparisons of the data with estimates of the wall shear stress obtained from indirect techniques in common use show considerable differences.

Introduction

An accurate measure of the wall shear stress τ_s is essential to the proper understanding of the behaviour of any boundary-layer flow. The techniques generally used to determine τ_s may be classified into three categories: momentum methods, wall similarity techniques and direct measurements. A description of these different approaches and their limitations so far as turbulent boundary layers are concerned is contained in articles by Brown and Joubert (1969) and Winter (1977).

The present paper underlines the importance of a means for the direct determination of the wall shear stress in all but the simplest of boundary-layer flow situations. Similarity techniques are all based on the existence of a region adjacent to the surface where a universal law of the wall $u/u_\tau = f(u_\tau y/\nu)$ is assumed to exist, and are practically limited to turbulent boundary layers on smooth surfaces. The limitations of momentum integral methods to obtain the wall shear stress are also well known, especially in rapidly developing flows and in the presence of pressure gradients. Even for nominally two-dimensional zero-pressure-gradient boundary layers, an extremely small change in flow cross-section suffices to introduce significant errors into the deduced skin friction.

In view of these limitations, questions are frequently raised about the validity and accuracy of skin friction data obtained using such techniques in "non-standard" boundary layers such as those undergoing transition, or flows with wall roughness, pressure gradients, or three-dimensional effects. Efforts are being made to use laser anemometry for the measurement of the mean velocity gradient close to the wall (Mazumder et al. (1981), Reynolds (1983)) which could provide a useful alternative to obtain the wall shear stress for flow over smooth surfaces. However, at present, the direct measurement of wall shear stress using a floating element device is the only viable possibility in most instances, even though the use of such devices in the presence of pressure gradients is beset with difficulties.

In the following, measurements of the wall shear stress made using a floating element device are presented for two different situations: a fully developed turbulent boundary layer on a rough (cast) surface, and a boundary layer undergoing laminar-to-turbulent transition on a smooth wall. In both cases the flow is incompressible, with a low external turbulence level and zero pressure gradient. The data are examined and discussed in the context of currently accepted ideas about such flows.

Facility and Instrumentation

Wind Tunnel

The experiments were conducted in an open-circuit low-speed boundary-layer tunnel of a design similar to that of Bradshaw (1972). An aerofoil-type centrifugal blower is followed by a straight-sided wide-angle diffuser leading to a settling chamber with nominal dimensions of 800 × 800 mm. A honeycomb and screens are installed in the settling chamber which is followed by an 8:1 two-dimensional contraction leading into the working section. This entire flow-conditioning section is mounted on a frame so as to be moveable as a unit. The 3 m-long test section has an entry cross-section of 800 × 100 mm and is mounted on an independent frame. The upper wall of the test section is the working wall on which the boundary layer of interest develops; the lower flexible wall may be adjusted to generate the desired streamwise pressure gradient. The test section and contraction may be bolted together directly or, for transition experiments, a boundary-layer bleed device may be inserted in between to obtain a laminar boundary layer with a well defined origin. Maximum airspeed for zero pressure gradient is about 45 m/s. A feedback control circuit regulates the blower speed to maintain the static pressure difference across the contraction (i.e., essentially the dynamic pressure at the contraction outlet) at a desired value. Different test plates with the desired surface characteristics may be installed. Each plate carries a series of 90 mm access ports with interchangeable plugs, so as to enable access to the boundary layer with a floating element device or probe traverse.

Instrumentation

With the exception of the floating element device, all instrumentation used is standard and will not be described here. Data acquisition is accomplished through a custom-built interface to a PDP 11/04 minicomputer. The system enables simultaneous acquisition of 8 channels of data, with a maximum data rate of 500 kHz for single-channel operation. Up to four channels of data can be pre-conditioned (biased and amplified) to make best use of the ± 10 V (10 bit resolution) dynamic range of the ADC. Boundary-layer profiles are measured under computer control and the data either processed on-line or stored for subsequent processing. Floating element data are individually recorded, with the element being checked for proper operation at each measurement.

Floating Element

The floating element, shown schematically in Fig. 1, is of the nulling type, with a range of 1000 μN, and a resolution of 1 μN. The central component, which acts as a balance, is the movement of a precision galvanometer which supports the element itself, a disc of 20 mm diameter. The movement is carried on an assembly of five aluminium slabs. Nine differential screws, each with a movement of 50 μm/turn, permit accurate positioning of the element in the surrounding baseplate. The position of the black-white interface on a target mounted on the back of the element is detected by a fibre-optic scanner with a spatial definition of 0.25 μm.

The galvanometer coil system and optical position detector operate in a feedback control loop. A force applied to the element moves it away from its null position. The resulting output of the position detector is amplified and integrated to produce a current through the coil which imposes a counteracting torque to drive the element back towards its null position. A proportional-integral-differential (PID) controller is used to ensure high accuracy and good transient behaviour. The coil current is directly proportional to the imposed

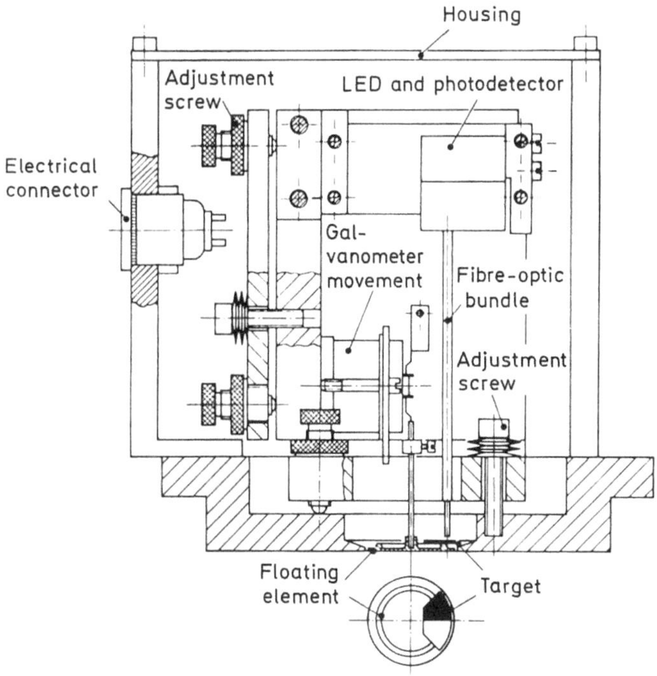

Fig. 1. Schematic drawing of floating element

force. A digital readout indicates the value of this force directly in micronewtons. The basic calibration of the instrument is against a precision (0.1 %) 100 mg weight hung from the face of the element with the coil axis and the arm supporting the element in a horizontal plane.

The entire floating-element mechanism is enclosed in a sealed housing and, during operation, the pressure difference across the element maintained close to zero.

Extensive tests were conducted to establish the limits of accuracy and reliability of the device in flows with and without pressure gradients. A detailed description of these tests, especially the behaviour of the element in the presence of pressure gradients will be presented elsewhere. For present purposes it suffices to say that in both laminar and fully developed turbulent boundary layers on smooth walls with zero pressure gradient, measurements with the floating element were in agreement with expected values to better than 5% (see Fig. 7).

The Rough Surface Boundary Layer

There have been many investigations of the influence of surface roughness on the turbulent boundary layer. A large number of these have dealt with regular roughness geometries which, while giving some insight into the problem of roughness characterisation, are not of direct practical interest.

The measurements on a cast aluminium surface described here are part of a project aimed at examining rough surfaces with microgeometries similar to those encountered in engineering practice, as produced by various manufacturing and finishing processes. The microgeometry of such a surface may be characterised by a set of statistical quantities including, for example, the arithmetic average roughness, the distribution and auto-correlation of the surface height, and the surface slope angle distribution. One of the aims is to establish which quantities are the most relevant and the minimum number which yield an acceptable description. Another aim is to determine if a correlation can be found between the skin friction developed on a rough surface and such statistical descriptors for the surface.

The zero-pressure-gradient boundary layer over the cast surface has been documented in detail, including measurements of the mean velocity, turbulence intensity and Reynolds shear stress profiles. The wall shear stress was measured using a floating element. The data were examined to investigate the feasibility of deriving the wall shear stress from a fit to the measured mean velocity distribution within the boundary layer.

In most investigations of turbulent boundary layers over rough walls that have been reported in the literature, the mean velocity profile has been represented following Perry and Joubert (1963) as a combination of the profile forms proposed by Clauser (1954) and Coles (1956):

$$\frac{u}{u_\tau} = \frac{1}{\kappa} \ln\left(\frac{y u_\tau}{\nu}\right) + B - \frac{\Delta u}{u_\tau} + \frac{\Pi}{\kappa}\,\omega[y/\delta]. \tag{1}$$

Here u is the mean velocity, y the normal distance from the aerodynamic origin (i.e., where u effectively goes to zero), ν is the kinematic viscosity of the fluid, κ and B are universal constants with values of 0.41 and 5.0 respectively, and u_τ is the friction velocity ($\equiv \sqrt{\tau_s/\varrho}$), τ_s being the wall shear stress and ϱ the fluid density). $\Delta u/u_\tau$ is a measure of the influence of the roughness on the velocity profile (zero by definition for an aerodynamically smooth surface) known as the roughness function. The last term in Eq. (1) models the wake-like behaviour of the outer region of the boundary layer. Since it is in general the case that the aerodynamic origin does not coincide with either the mean surface location or the origin chosen for the measurements, y is replaced by $y_m + \varepsilon$, where y_m is the distance measured from some arbitrary origin and ε is the incremental distance to the aerodynamic origin (again zero for a smooth surface). It has been usual in the past to use some procedure to fit measured mean velocity profiles to a representation of this kind, thereby obtaining a value for the wall shear stress together with the three other independent parameters: δ, ε, and $\Delta u/u_\tau$.

An indirect method of the kind just described to determine the wall shear stress is a direct extension of the procedure used for flow over smooth surfaces, where ε and $\Delta u/u_\tau$ are identically zero, and there are no essential difficulties in determining τ_s and δ by matching Eq. (1) to measured data. In fact, since the contribution of the wake component is usually negligible for $y/\delta < 0.5$, τ_s may easily be determined graphically using the method first suggested by Clauser (1954): the mean velocity data are plotted in semilogarithmic coordinates u/u_∞ versus $\ln(u_\infty y/\nu)$, and compared with a fan of straight lines representing the velocity variation in the fully turbulent core, with $c_f (\equiv 2\tau_s/\varrho u_\infty^2)$ as the only parameter. The line which best matches a measured velocity profile plotted on such a chart then yields a value for the corresponding skin friction coefficient c_f.

Figure 2 shows a typical Clauser chart for data taken in a zero pressure gradient fully developed *smooth* wall boundary layer. The values of c_f obtained from a Clauser chart and by fitting Eq. (1) (with ε and $\Delta u/u_\tau$ set equal to zero) to measurements according to a linear regression algorithm are generally within 2 or 3 per cent of each other, and are in good agreement with both Preston tube and direct floating element measurements of the wall shear stress.

For a rough surface, there is in principle again no difficulty in determining values for c_f and the two additional parameters ε and $\Delta u/u_\tau$ such that there is a minimum rms deviation, say, between Eq. (1) and an experimental profile. In practice, however, the process yields highly inaccurate values of c_f because of a crucial difference between the rough and smooth situations. This difference stems from the small deviations from linearity that are commonly found in the mean velocity profile in semilogarithmic coordinates (see Coles (1968)). For

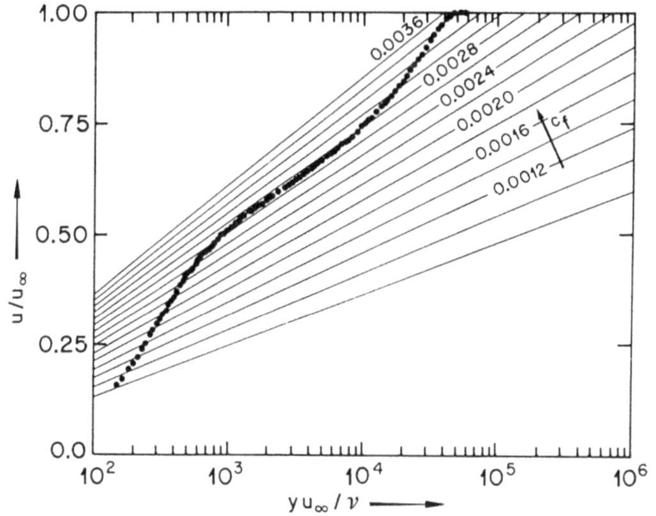

Fig. 2. Clauser chart for a mean velocity profile in a smooth-wall boundary layer

smooth surfaces these deviations are not of major significance. When a fitting procedure with only the two free parameters c_f and δ is used, these deviations from a semilogarithmic behaviour merely result in an increased rms scatter. For rough surfaces, on the other hand, the situation is quite different: the inclusion of the additional parameters ε and $\Delta u/u_\tau$ *forces* the measured profile to be as linear as possible in semilogarithmic coordinates, i.e., deviations, which may well be inherent to the data set, e.g., due to the influence of an individual roughness element, are reduced to a minimum. The additional degrees of freedom allowed by the simultaneous choice of slope (c_f) and intercept ($\Delta u/u_\tau$) generally result in a much "better" fit in a statistical sense, but can produce erroneous values for these quantities. In fact, it is possible to obtain more than one set of values for the different parameters which

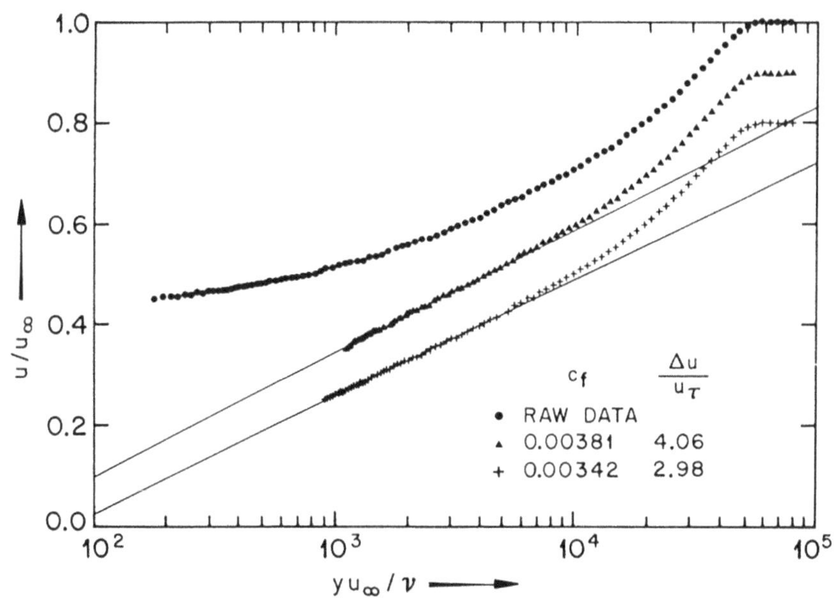

Fig. 3. Mean velocity profile in a cast surface boundary layer (the second and third curves have been shifted down relative to the first for the sake of clarity)

yield essentially the same "goodness" of fit. Figure 3 shows the result of such a fitting procedure applied to a mean velocity profile measured in the cast surface boundary layer. The first curve represents the raw data the second the results of a "best" fit. Also shown is the result obtained when the value of τ_s, independently measured using the floating element, was used as an input, thus eliminating one degree of freedom from the fitting procedure. The difference in values of c_f in this case is seen to be about 12%. Differences as large as 30% have been found in other cases. The results from a fitting procedure with all parameters to be determined were often found to be physically unrealistic: e.g. τ_s was found to increase with Reynolds number in some cases, and negative values of $\Delta u/u_\tau$ were obtained in other instances. On the other hand, when the values of τ_s, obtained from a direct floating element measurement, were used as an input to the fitting procedure, the resultant $\Delta u/u_\tau$ exhibited a consistent trend with Reynolds number. The skin friction data obtained with the floating element for the cast surface are plotted in Fig. 4 against the momentum thickness Reynolds number. The corresponding values of the roughness function $\Delta u/u_\tau$ are plotted against a roughness Reynolds number in Fig. 5.

The difficulties associated with an extended Clauser plot approach of this sort have also been noted by Perry, Schofield and Joubert (1969). The effect is illustrated most vividly if the fitting procedure for a rough surface is applied to smooth surface data, i.e. ε and $\Delta u/u_\tau$ are not set identically equal to zero. The results of such an exercise for the data of Fig. 2 are shown in Fig. 6. The upper curve shows a comparison with a "log-law" line corresponding to the value $c_f = 0.00278$ obtained from the floating element. The second curve shows the results of the fit: a value of 1.11 is obtained for $\Delta u/u_\tau$ and the value of 0.00314 obtained for c_f is 13% higher than the floating-element value. However, the match between the measured points and the logarithmic velocity distribution is excellent. About 20 smooth wall profiles

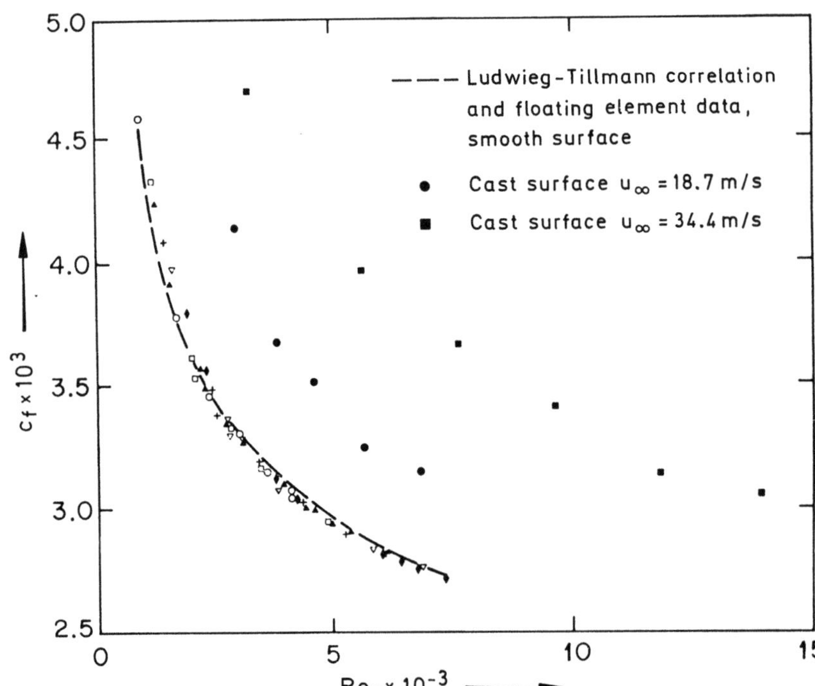

Fig. 4. Variation of skin friction coefficient with momentum thickness Reynolds number for the cast surface

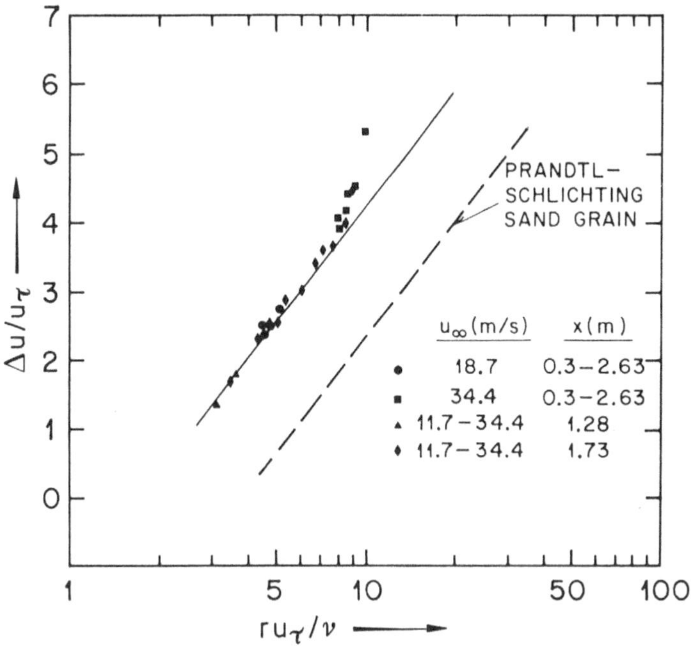

Fig. 5. Variation of roughness function with roughness Reynolds number for the cast surface. A line parallel to the Prandtl-Schlichting sand grain line has been drawn through the data to aid in comparison

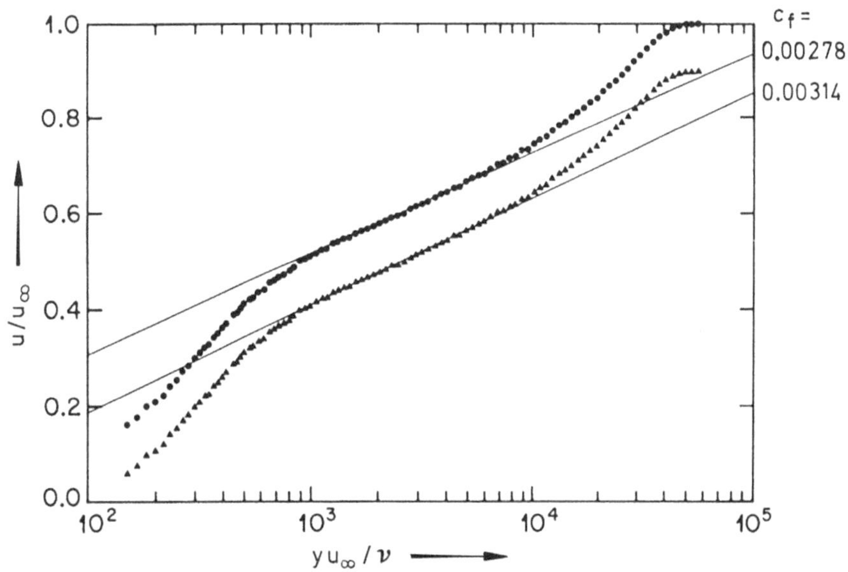

Fig. 6. Rough surface fitting procedure applied to smooth suurface data (the second curve has been shifted downward for the sake of clarity)

were processed in this fashion. The average error in c_f was 12%, with a maximum deviation of 20%, and values for $\Delta u/u_\tau$ were in the range 0.3 to 2.0.

It is thus seen that significant errors can be made in the estimation of the wall shear stress using indirect techniques which apparently lead to excellent agreement with a standard representation of the mean velocity profile.

Boundary Layer Transition

The second type of flow examined is the low free stream turbulence, zero-pressure gradient boundary layer undergoing natural transition on a smooth wall. In such a situation it is important to consider the response of the floating element, which was designed to measure forces with steady time mean values, to sudden changes in the wall shear force that occur in the transition region due to the passage of a turbulent spot. As described in an earlier section, a PID controller is used to integrate the error signal and provide the necessary restoring coil current. Since the time constant of this system to a step input is about 0.5 seconds, the instrument cannot follow rapid changes in the imposed shear force. Nevertheless, it will register a true mean value of this force, as long as the changes in the force are not so large as to drive the instrument out of the operating range of the position detector. The output of the controller was monitored during the measurements, and it was established that the element remained within this operating range and yielded consistent and repeatable values for the wall shear force, except for small values of wall intermittency ($\gamma_s < 0.1$). At the start of the transition region the element was occasionally driven out of balance, presumably by the passage of a turbulent spot, and it took some time for equilibrium to be reestablished. For the transition measurements reported here, upper and lower bounds of the force registered by the element were recorded in addition to the mean value.

Experiments were conducted in which the wall shear stress was measured with a floating element and profiles of the mean velocity and streamwise turbulent intensity through the boundary layer were determined for a number of streamwise locations covering the laminar, transition and turbulent regions, and for a number of different free stream velocities. For each case, a zero pressure gradient was maintained and the free stream turbulence level was nominally the same (about 0.4%). The mean velocity measurements upstream of transition showed excellent agreement with the Blasius profile for all cases. From Fig. 7, which shows the variation of skin friction coefficient with momentum thickness Reynolds number, Re_θ, it can be seen that the value of Re_θ at which transition begins decreases with increasing free stream velocity. This result is different from the generally accepted view that for a given pressure gradient and free stream turbulence level, the value of Re_θ at the start of transition is unique. The variation of Re_θ with Re_x for the different cases, shown plotted in Fig. 8, is consistent with the floating element data of Fig. 7.

In the transition region, assumptions regarding the behaviour of the flow near the wall, such as those for the fully developed turbulent boundary layer that lead to the law of the wall, are invalid. In order to determine the wall shear stress in the transition region, it is the usual practice either to make use of the momentum integral equation and integral quantities from measured mean velocity profiles, or to obtain the wall shear stress as the intermittency weighted sum of the corresponding laminar and turbulent boundary layer shear stress values at that Reynolds number:

$$c_f = (1 - \gamma_s)\, c_{fL} + \gamma_s c_{fT} \tag{2}$$

where γ_s is the value of the intermittency at the surface, c_{fL} is the laminar-flow value at the appropriate Reynolds number and value of the pressure-gradient parameter, and c_{fT} is obtained from any standard correlation in terms of Re_x for turbulent boundary layers with a proper choice for the origin of the turbulent boundary layer. Figure 9 shows one of the sets of skin friction data from Fig. 7, plotted against the distance from the leading edge of the plate, with bands to indicate the upper and lower bounds on the shear force as recorded by the floating element. It is seen that the excursions are largest in the transition region. Also shown are the values for c_f computed from Eq. (2) using measured values of the wall intermittency γ_s. The origin of the turbulent boundary layer was chosen as $x = 550$ mm, a

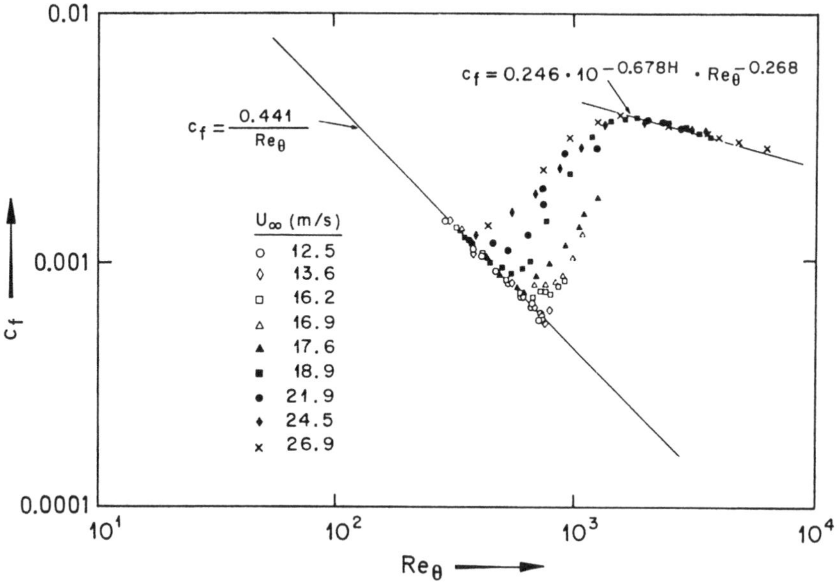

Fig. 7. Variation of skin friction coefficient with momentum thickness Reynolds number through the transition region

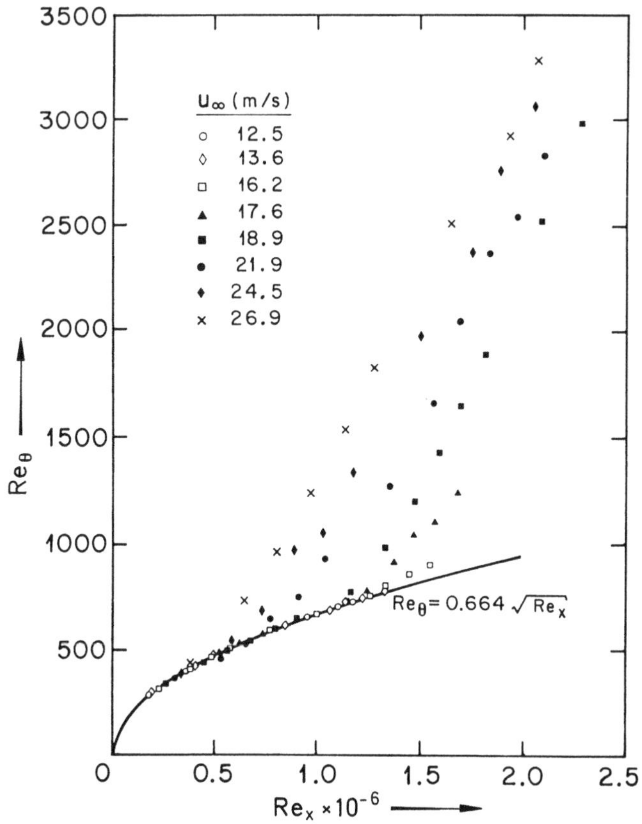

Fig. 8. Variation of momentum thickness Reynolds number with the x-Reynolds number for the data of Fig. 7

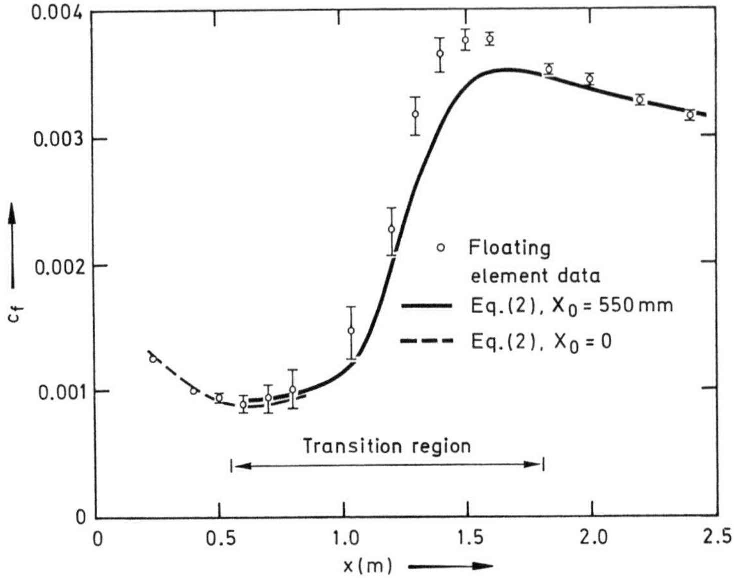

Fig. 9. Comparison of measured and computed skin friction coefficients for $u = 18.9$ m/s

value determined from measurements of the wall intermittency and extrapolation of the measured boundary layer thickness in the turbulent region. The agreement between measurements and the computed values is seen to be good both upstream and downstream of the transition region, but the values obtained in the transition region from Eq. (2) are seen to under-predict the wall shear stress even when the uncertainties in both the floating element and the intermittency data are accounted for.

Conclusions

It has been shown that indirect techniques for the determination of wall shear stress for boundary layers developing on rough surfaces and those undergoing transition lead to considerable errors. A direct measurement of the wall shear stress is seen to be essential for these and other "non-standard" situations.

References

Bradshaw, P. (1972): Two more wind tunnels driven by aerofoil-type centrifugal blowers. Imperial College Aero. Rep. 71–10

Brown, K. C., Joubert, P. N. (1969): The measurement of skin friction in turbulent boundary layers with adverse pressure gradients. J. Fluid Mech. **35**, 737

Clauser, F. H. (1954): Turbulent boundary layers in adverse pressure gradients. J. Aero. Sci. **21**, 91

Coles, D. E. (1956): The law of the wake in the turbulent boundary layer. J. Fluid Mech. **1**, 191

Coles, D. E.: The young persons guide to the data. Proc. Computation of turbulent boundary layers (AFOSR-IFP Stanford Conference, Vol. II, 1968)

Mazumder, M. K., Wanchoo, S., McLeod, P. C., Ballard, G. S., Mozumdar, S., Caraballo, N. (1981): Skin friction drag measurements by LDV. Appl. Opt. **20**, 2832

Perry, A. E., Joubert, P. N. (1963): Rough-wall boundary layers in adverse pressure gradients. J. Fluid Mech. **17**, 193

Perry, A. E., Schofield, W. H., Joubert, P. N. (1969): Rough-wall turbulent boundary layers. J. Fluid Mech. **37**, 383

Reynolds, W. C. (1983): private communication

Winter, K. G. (1977): An outline of the techniques available for the measurement of skin friction in turbulent boundary layers. Prog. Aerospace Sci. **18**, 1

Hysteresis of Turbulent Stresses in Wall Flows Subjected to Periodic Disturbances

K. Hanjalić and N. Stošić

Mašinski fakultet, Sarajevo, Yugoslavia

Abstract

The paper discusses some specific features of turbulent flows subjected to the periodic variation of external conditions. A particular emphasis is placed on the lagging response of turbulent structure and the hysteresis of turbulent stresses that were detected experimentally in a number of periodic flows. Several sets of experimental data were analysed in this light and attempts were made to obtain their prediction by means of single- and double-scale models of turbulence. Attention was focused on a case of simulated tidal flow where the above mentioned features are particularly pronounced. Some modifications of the standard stress-dissipation model of turbulence yielded a noticable improvement of the prediction of major properties of this and some other periodic flows.

Introduction

If a turbulent flow is subjected to the space- or time-variation of the external conditions, the turbulent interactions associated with different parts of the wave number spectrum will not generally respond in the same manner. Consequently, the turbulence and mean flow parameters are bound to exhibit different phase shifts and amplitude modulations. A number of recent publications on unsteady turbulent flows revealed the existence of a definite time-space lag between the response of turbulent structure and the mean flow parameters, but also a different variations of these shifts across the flow. This effect may become particularly pronounced if the external disturbances have a periodic character, the intensity of the lag depending upon the pulsation frequency and their amplitudes. Furthermore, because the response of the turbulence structure varies over the wave number spectrum, the time lags during the acceleration period are markedly different from those which occur during the deceleration phase, so that some of the turbulence parameters display a pronounced hysteresis (Anwar and Atkins, 1980). A mismatch of time scales governing the turbulence generation and its spatial and spectral transport and decay may either cause the appearance of residual turbulence or to diminish its level. Yet, most authors report that both the mean flow and turbulence parameters, normalized appropriately, yield the same quantitative behaviour as those in corresponding steady flows. These findings can hardly be justified in cases where the phase shifts of dependent variables and normalizing parameters do not remain constant across the flow.

The above mentioned features impose significant effects upon the transport processes in turbulent flows and they are, therefore, of substantial practical interest. Boundary layer development on rotating airfoils and turbomachinery blades, heat transfer and flow resistence in heat exchangers with pulsating flows such as the air intercoolers of two-stage compressors and dispersion of dissolved pollutants in coastal waters, are just some of the many such examples.

On the other hand, the inertial lag of the dynamics of turbulence structure behind the mean flow distortion in pulsating flows is a challenging feature that may serve as a severe test for the verification of modelling techniques.

It is generally recognized that the current prediction methods, verified in steady flows of varying complexity, give poorer performances when applied to the calculation of unsteady and periodic flows. This is in particular the case with eddy viscosity models, even if the transport coefficient is supplied by the differential equations, such as implied by $k - \varepsilon$ model. Shemer and Wygnanski (1981) introduced separate eddy viscosity coefficients for the steady and time dependent parts of the phase averaged field. Cousteix et al. (1981) emphasized the importance of the inclusion of the normal stress production in the kinetic energy equation (usually neglected in calculation of boundary layers), and reported that the decisive improvement had been achieved by solving the equations for all turbulent stress components in comparison with $k - \varepsilon$ model.

In fact, depending on frequency and amplitude, the pulsating flow may be expected to depart from both the spatial- and spectral-equilibrium and any model that relies upon the equilibrium assumption is bound to be more or less unsatisfactory. This evidence was recognized earlier by the present and other authors (e.g. Hanjalić, Launder and Schiestel, 1979) when analysing some steady non-equilibrium flows including simple cases such as the development of turbulence when passing through a sudden contraction of the cross-sectional area, or boundary layers subjected to a sudden deceleration (or acceleration), followed by the relaxation of the imposed pressure gradient. The latter case bears some resemblance to a half cycle period of an unsteady periodic flow. In most of the mentioned cases the standard $k - \varepsilon$ model, or even the full stress model did not yield the satisfactory reproduction of the experimental results.

The difference in the quality of predictions obtained by $k - \varepsilon$ and stress models, as found by Cousteix et al. (1981) and the present authors suggests that the structural parameter $\overline{u_1 u_2}/k$ does not remain constant as found in steady equilibrium wall boundary layers, nor do their phase shifts exhibit the same behaviour. This in fact emphasises the importance of both the spatial- and spectral-transports of individual stress components, particularly in flows with preferential straining direction. Hence an adequate account for this should be made in order to achieve a satisfactory simulation of turbulence dynamics.

Similar dissatisfaction was experienced in attempts to calculate the unsteady periodic flows discussed in the present work. However, contrary to the findings of Cousteix et al. (1981), the application of full stress model as postulated by Launder, Reece and Rodi (1975), although performing beter than simple $k - \varepsilon$ model still did not yield the acceptable agreement with the experiments. One of the reasons for a different outcome may be attributed to a different treatment of the viscosity-affected near-wall region, which in the present work is bridged over by means of wall functions. It is believed, however, that the major reason lies in a general inadequacy of the single-point methods to represent the non-equilibrium situations.

In fact, if the difference in response of various turbulence parameters can be attributed to the spectral non-equilibrium, then, in order to achieve a closer simulation of the real processes, the mathematical model should account for the spectral dynamics, in addition to the modelling of processes in the physical space. This condition can not be fully satisfied by any of the present single point closure schemes.

Yet the simplicity and a wide verification of single point models seem to justify all efforts towards their further improvement and extension of their validity.

The present work describes our efforts in exploring the suitabilities of single- and two-scale second order closure models in standard and modified forms for predicting the earlier mentioned features of wall turbulent flows subjected to periodic disturbances of external flow conditions. In the first phase of the work the testing was carried out with the $k - \varepsilon$ and the Reynolds stress-dissipation models, with several modifications, proved earlier to produce some improvement in predicting the non-equilibrium and recirculating turbulent flows. One of the modifications implies a different formulation of the source term in the dissipation equations (Hanjalić et al., 1981, Launder et al., 1981), which enshures somewhat weaker coupling of the modelled dynamics of ε with the dynamics of turbulent kinetic energy. Other modifications concern different specification of wall functions of both mean and turbulence parameters. These employ $k^{1/2}$ instead of friction velocity U_* as the characteristic velocity scale and the relaxation of the constant stress-turbulent energy constraints in the near-wall region in sense of proposals given by Chieng and Launder 1980.

In the second part of the work the use was made of the two-scale model of turbulence (Hanjalić, Launder and Schiestel, 1979), with some minor modifications concerning the division of the energy spectrum and the consequent redefinition of the empirical coefficients. This work is still in progress and the preliminary results of the application of the two-scale models are encouraging.

The models were used to calculate a number of the pulsating turbulent water and air flows having different frequencies and amplitudes of the external perturbations. Particular attention was paid to the results of experimental simulation of tidal flow of Anwar and Atkins (1980) with large period of oscillation, in which the responses of turbulent stresses during the acceleration and deceleration periods show distinctly different features and a visible hysteresis.

Considered Models and their Modifications

Single-Scale Stress-Dissipation Model

It is generally recognized that the weakest point in the single-point closure models is the transport equation governing the turbulence scale. Most of the models employ at present the turbulence energy dissipation rate ε which, in combination with the kinetic energy, yields the characteristic time or length scale of turbulence. The source term in the equation for ε is usually modelled in form:

$$S_\varepsilon = (C_{\varepsilon 1} P - C_{\varepsilon 2} \varepsilon)\, \varepsilon / k \tag{1}$$

where P is the production of turbulent kinetic energy.

Serious criticisms have been placed in the past on this formulation, the major one being that the source of ε is too closely linked with the source of kinetic energy itself. Bearing in mind that ε represents in fact the energy transfer rate across the wave number spectrum which equals the energy dissipation rate only under the conditions of spectral equilibrium, one may expect that the production of ε will depend on mean rate of strain (not necessarily in the same manner as the production of kinetic energy itself!), but also upon the mean vorticity, turbulence anisotropy (as advocated earlier by Lumley and coworkers) and other parameters that may influence the energy transfer. A functional combinations of variables at disposal, satisfying the invariance conditions, may be written in the form (retaining only the first order terms):

$$S_\varepsilon = f\left\{\overline{u_i' u_j'} \frac{\partial U_i}{\partial x_j} \frac{\varepsilon}{k},\ k\left(\frac{\partial U_i}{\partial x_j}\right)^2,\ k\left(\epsilon_{ijk} \frac{\partial U_i}{\partial x_j}\right)^2,\ A\frac{\varepsilon^2}{k},\ \frac{\varepsilon^2}{k}, \dots\right\} \tag{2}$$

where A represents the first invariant of the stress anisotropy.

Extensive testings would be required to find a usable form of the expression (2) and to prove its superiority over the current simple practice which uses only the first and the last terms. However, equally simple alternative

$$S_\varepsilon = C_{\varepsilon 1}\, C_\mu\, k \left(\frac{\partial U_i}{\partial x_j}\right)^2 - C_{\varepsilon 2}\,\frac{\varepsilon^2}{k} \tag{3}$$

which yielded noticable improvement of the calculation of both the simple non-equilibrium flows and more complex recirculating flows behind a step (Launder et al. 1981), helped also to improve the predictions of pulsating cases, considered in the present work [1].

It should be mentioned that the new term offers some potential advantages since it relaxes the linkage between the energy and ε equation as implied by standard models. The expression (3) implies also that the first term ("production") is now scaled by the time scale of the mean flow deformation $(\partial U_i/\partial x_j)^{-1}$. This seems plausible since the energy transfer rate through the spectrum is controlled primarily by the "large" eddies which are strained almost entirely by the mean rate of strain (Townsend, 1980). The second term ("destruction") remains to be scaled by turbulence scale (k/ε), implying that the attenuation of the transfer process is controlled by turbulence itself.

Other modifications, considered earlier, concern the diffusive term in the ε equation but none of the remedies proved to be equally benefitial in different flow situations and have not been pursued further in this work.

Two-Scale Model

The idea of modelling separately the dynamics of low and high wave-number parts of the energy spectrum, as implied by the two-scale model of Hanjalić, Launder and Schiestel (1979), seems particularly attractive for predicting the flows subjected to strong variation of external conditions. Indeed, the two-scale $k - \varepsilon$ model produced a more adequate response of some steady homogeneous and wall flows that were exposed to sudden changes of pressure gradient. The improvements reported in the above-mentioned paper encouraged us to employ the same model to the periodic flows, considered here. The two-scale $k - \varepsilon$ model produced, however, somewhat better predictions (e.g. Fig. 1), but they appear to be still inferior to those obtained by single-scale stress model.

As it was noted earlier, it is believed that the two-scale model failed to produce all the effects that were expected because the energy spectrum was divided at too high wave number with too much energy asigned to the low wave-number portion of the spectrum. Attempts have been made, therefore, to move the partitioning wave number into the production region as to include the mean rate-of-strain effect into the equations governing the high wave-number structure. Positive effects, achieved by considering all stress components in the single-scale model, discussed earlier, give some justification for exploring an all-stress variant of the two-scale model. This approach requires, however, an extensive testing of various steady flows to determine a number of new empirical coefficients, before the model is applied to an unsteady situation.

In the moment of writing this paper we had completed only the preliminary testing of the model in homogeneous flows. The starting set of the equations can be written in the form:

$$\frac{D\,\overline{u_i' u_j'}^{\,p}}{Dt} = P_{ij}^p - \phi_{ij}^p - \frac{2}{3}\delta_{ij}\,\varepsilon_p + \mathscr{D}\,\overline{u_i' u_j'}^{\,p} \tag{4}$$

1 If $k - \varepsilon$ model is employed, in boundary layer type of flows the Eq. (3) reduces to the standard form given by Eq. (1)

$$\frac{D\overline{u_i' u_j'}^t}{Dt} = P_{ij}^t - \phi_{ij}^t + \frac{2}{3}\delta_{ij}\varepsilon_p - \frac{2}{3}\delta_{ij}\varepsilon_t + \mathscr{D}\overline{u_i' u_j'}^t \tag{5}$$

$$\frac{D\varepsilon_p}{Dt} = C_{\varepsilon 1 p} C_\mu k_p \left(\frac{\partial U_i}{\partial x_j}\right)^2 - C_{\varepsilon 2 p}\frac{\varepsilon_p^2}{k} + \mathscr{D}\varepsilon_p \tag{6}$$

$$\frac{D\varepsilon_t}{Dt} = C'_{\varepsilon 1 t} C_\mu k_t \left(\frac{\partial U_i}{\partial x_j}\right)^2 + C_{\varepsilon 1 t}\frac{\varepsilon_p \varepsilon_t}{k_t} - C_{\varepsilon 2 t}\frac{\varepsilon_t^2}{k_t} + \mathscr{D}\varepsilon_t \tag{7}$$

where indices "p" and "t" refer to the production and transfer regions of the energy spectrum, respectively. P_{ij} is the stress production and ϕ_{ij} pressure redistribution. The latter term was modelled in the manner proposed by Launder et al. (1975).

For the beginning it was assumed that coefficients in all terms of stress equations for "p" and "t" region have the same values, equal to those in the single-scale model. The model, applied to a series of homogeneous flows, that were considered at the 1980–81 AFOSR-HTTM Stanford Conference, produced in the whole better reproduction of the experimental data than the single-scale model. This is illustrated in Fig. 1, where comparison is given between the predictions of kinetic energy distribution, obtained by two models, in an axisymmetric contraction of Tan-atichat (1980).

The solution of the full set of Eqs. (4) to (7) in a more complex flows does not seem rational at present and we turned to a simplified form which solves the contracted form of the Eq. (5) (for k_t), while the individual stress components in "t" region are calculated from algebraic expressions worked out on the basis of Eq. (5). In this way only two new differential equations are solved in addition to the single-scale stress model. This work is still in progress.

Treatment of the Near-Wall Region

A common practice to avoid the need for incorporating the viscous and near-wall effects into the modelled transport equations and solving them within the viscous sublayer is to employ wall functions by which the viscosity-affected region is bridged over. The boundary condition for the mean velocity, specified in the first point of the numerical grid, placed usually just outside the viscous layer, is defined most often by means of the logarithmic velocity law, which is assumed to prevail even under a severe variation of the external flow conditions. A number of authors presume that the standard form of the law of the wall remains valid at every time instant in an unsteady situation. This assumption relies on condition that the phase shift of the mean velocity in the logarithmic region equals that of the wall shear stress. This argument was confirmed directly for a case of periodic boundary layer by Cousteix et al. (1981) on the basis of their own experimental results and data of Patel (1977). However, one

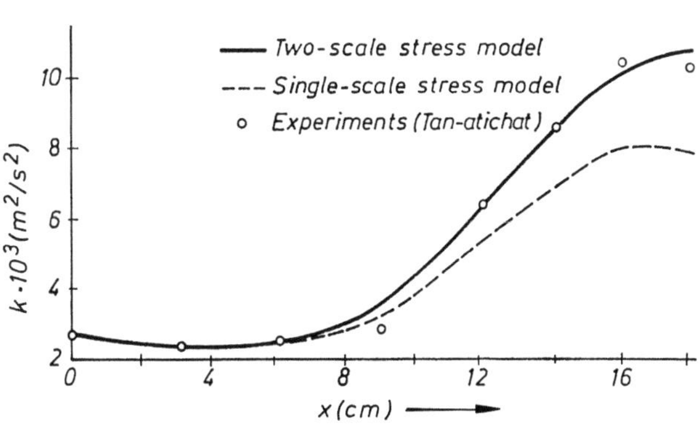

Fig. 1. Variation of kinetic energy in an axisymmetric contraction

should bear in mind that in the cases considered the amplitude of the mean velocity oscillation was not very high (in fact less than 15% of the time averaged mean velocity) and the conclusion may not hold for higher amplitudes or different frequencies. If the amplitude of the periodic velocity component increases sufficiently to bring the phase-averaged (ensemble-averaged) velocity at a certain time instant close to zero or to have negative values, logarithmic law will fail. The argument can be extended to a case of oscillating turbulent flow with zero time mean velocity. Hence the question may be raised about the limiting amplitude ratio up to which the standard law of the wall may be regarded as valid. Bearing this in mind, the use of $k^{1/2}$ as the velocity scale in the logarithmic law, as proposed by Launder and Spalding (1973) appears to be an attractive proposition for periodic flows. In this way the logarithmic law may be expected to extend its validity not only to cases of large amplitude of periodic velocity oscillations, but also to other situations in which the ratio $\overline{u_1 u_2}/k$ does not remain constant.

An attempt to confirm this assumption was made by replotting the velocity profiles of some reported cases of periodic flows, where the standard form of logarithmic law did not show a convincing universality, such as in the mentioned flow of Anwar and Atkins. However, because of the large experimental scatter of data it was not possible to draw a definite conclusion. Yet, the computations that used the wall function based on $k^{1/2}$ showed improvements and were used consistently in all cases presented here.

As far as the turbulence parameters are concerned, their modelling in the wall region requires a specific approach. Most of the numerical finite difference or finite volume schemes presume linear variation of all dependent variables between the grid nodes as computationally most economical. This practice seems to be satisfactory everywhere except in the vicinity of the rigid wall where turbulent parameters may exhibit a very steep variation. This is particularly important for the control cell closest to the wall which may encompass not only the viscous sublayer, but also a portion of fully turbulent region. Chieng and Launder (1980) split the near-wall cell into two parts. The wall-adjacent part represents the viscous layer, the edge of which is defined by a prespecified value of turbulent Re-number. By assuming a quadratic variation of kinetic energy and zero shear stress in the inner zone and linear variations of both properties in the outer part, the mean values of the energy production and dissipation over the cell were evaluated.

The idea can be extended to presume the variation of each stress component over the first numerical cell and to calculate the average values of source terms in each transport equation of a full-stress turbulence model, but the uncertainties in formulating the source terms within the viscous region discourage any attempts in this direction.

In this work we adopted a simpler route and calculated only the averaged values of ε over the cell, but related the values of normall stresses in the first grid to that of the shear stress, which was calculated from the mean momentum equation. Because the kinetic energy is used

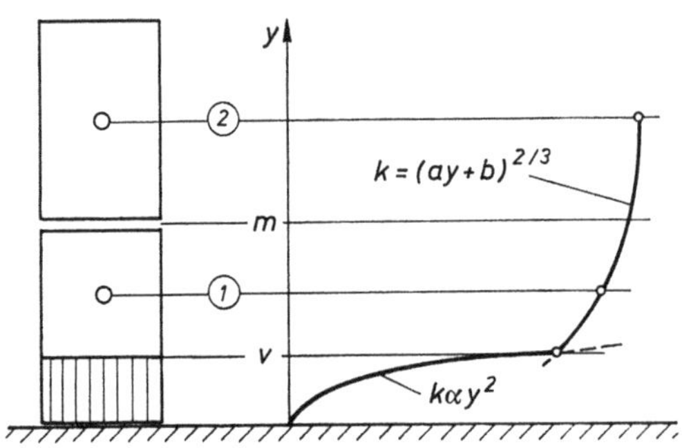

Fig. 2. Distribution of kinetic energy in the near-wall region

only to obtain the average ε, we adopted a parabolic expression $k = (a y + b)^{2/3}$ to define the variation of k in the outer part of the cell, while in the inner part we employed the quadratic distribution as Chieng and Launder (Fig. 2).

With the assumption that in the outer part the local ε may be approximated by $\varepsilon = k^{3/2}/(C_1 y)$, the adopted distribution of k yields a simple expression for the mean value of ε over the cell (notation is given in Fig. 2):

$$\bar{\varepsilon} = \frac{k_v^{3/2}}{10\, y_m} + \frac{1}{C_l\, y_m} [a(y_2 - y_1)] \tag{8}$$

where:

$$k_v^{3/2} = 20\, a \frac{v}{k_v^{1/2}} + b; \quad a = \frac{k_2^{3/2} - k_1^{3/2}}{y_2 - y_1}; \quad b = k_1^{3/2} - a\, y_1. \tag{9}$$

Presentation and Discussion of Results

Frequency and amplitude are the parameters that characterize the unsteady turbulent flows and these can vary in a wide range. In his review lecture L. W. Carr (1981) attempted to classify the available experimental data on unsteady boundary layers according to these two variables and concluded that, although low amplitude or low frequency does not necessarily mean quasisteady behaviour, there is a small range of amplitude and frequency for which no unsteady effects have been reported. He concluded further that, provided the frequency remained lower than the critical value (related to turbulent bursts), the turbulent structure remains unaffected by oscillation of the flow field. This argument holds if one considers the properties averaged over a large number of cycles in periodic situations. However, there are flows of practical relevance, such as tidal currents, where a strong influence upon the turbulence mechanism was notices due to the flow unsteadyness, although the frequency of oscillation was very low (Anwar and Atkins). The analysis of a detailed behaviour of turbulence properties and global parameters over a single cycle seems to be in such cases of a significant theoretical and practical importance.

We consider first the experiments of Anwar and Atkins (1980) that were obtained in a flume in which the water was first accelerated at a constant rate, and then decelerated for the same period of time. Three different cases were examined with acceleration/deceleration times of $T = 200$, 300 and 550 seconds. The predictions of all three cases were carried out by means of single-scale stress-dissipation turbulence model with the modifications described earlier. Also the case with $T = 200$ sec. was solved by the both, single-scale and double-scale $k - \varepsilon$ models. The solution of the equations was accomplished by use of Patankar-Spalding (1972) parabolic numerical scheme, which was modified as to be applicable to unsteady situations.

Figure 3 shows the mean velocity variation in time for $T = 200$ sec. together with the normalised turbulent shear stress close to the free surface (at the reference boundary layer edge, $y/\delta = 1$). The predicted $\overline{u_1 u_2}$ curve shows a good qualitative behaviour, with a pronounced time lag behind the mean velocity. It should be pointed out that the absolute values of $\overline{u'_1 u'_2}$ are very small and a discrepancy between the predicted and experimental data should not be regarded as serious, as could be seen in Fig. 4, where detailed distribution of $\overline{u'_1 u'_2}$ is given at several points in the cycle.

Figure 4 shows the prediction obtained by the three mentioned types of turbulence model. All three models yielded the results in acceptable agreements with experiments,

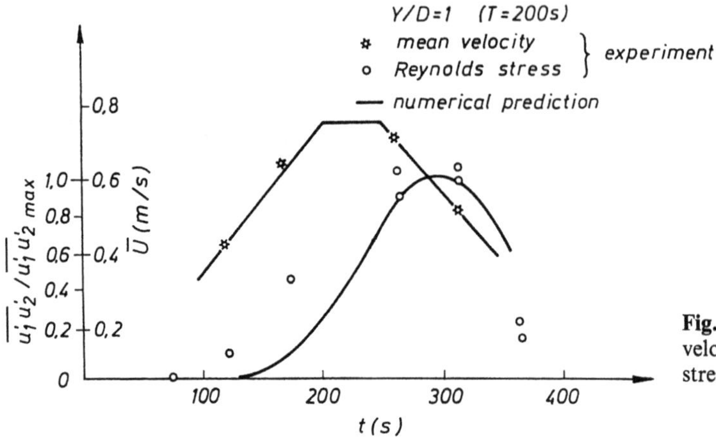

Fig. 3. Temporal variation of velocity and Reynolds shear stress

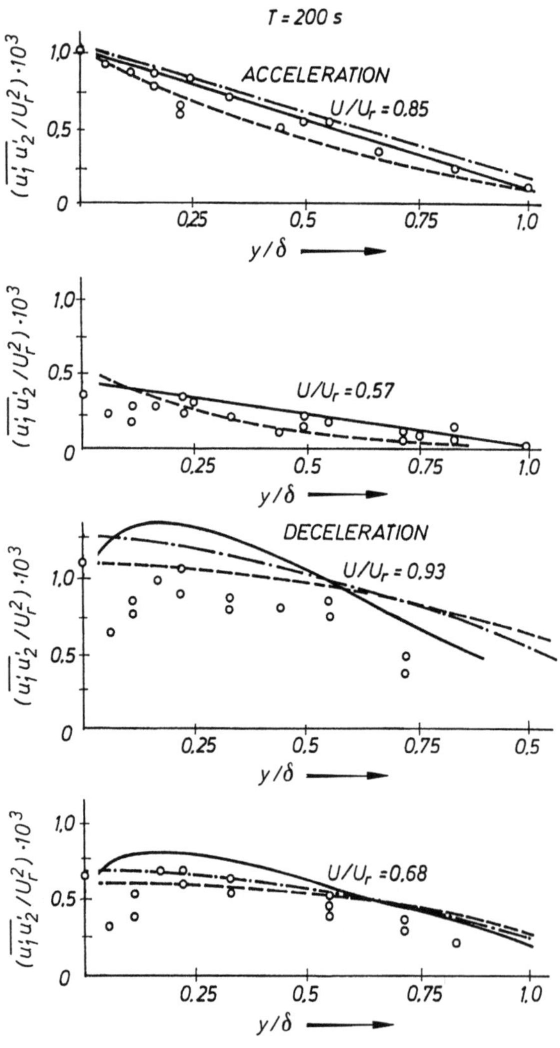

Fig. 4. Turbulent shear stress in simulated tidal flow; o— experiments, — single scale stress model, --- single scale $k-\varepsilon$ model, —·— two scale $k-\varepsilon$ model

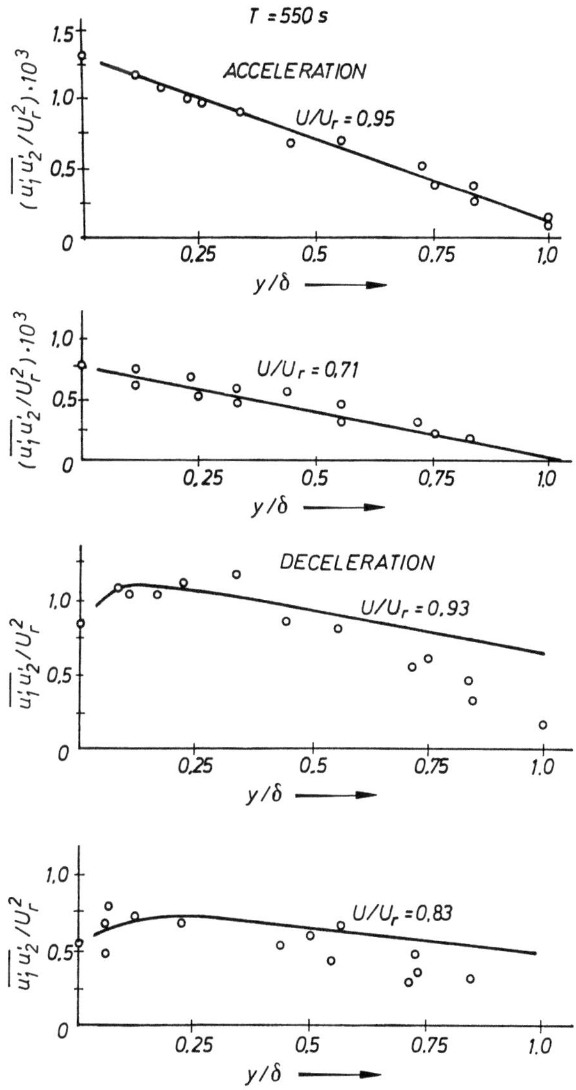

Fig. 5. Turbulent shear stress in simulated tidal flow; o— experiment, — predictions

though the single-scale stress model seems superior, in particular during the deceleration part of the cycle. Even better agreement was achieved in case with $T = 550$ as shown in Fig. 5.

Figure 6 shows the hysteresis of turbulent shear stress and kinetic energy as well as their ratio at three different flow depths. In all cases shown, the hysteresis is very pronounced what may have a significant practical consequences for example on transport of sediment or dispersion of pollutants in natural water bodies. The hysteresis of $\overline{u_1' u_2'}/k$ ratio, on the other hand, is an indication of the departure of turbulence structure from the local equilibrium and is probably the cause of discrepancy between the predictions obtained with $k - \varepsilon$ and stress models. Figure 6 indicates also that the hysteresis, though qualitatively the same, has different intensity at different depths, the larger values occuring in the outer flow region closer to the water surface.

We consider next the two flows markedly different from the previous one. These are the pulsating flow in a channel, investigated by Acharya and Reynolds (1975) with frequencies

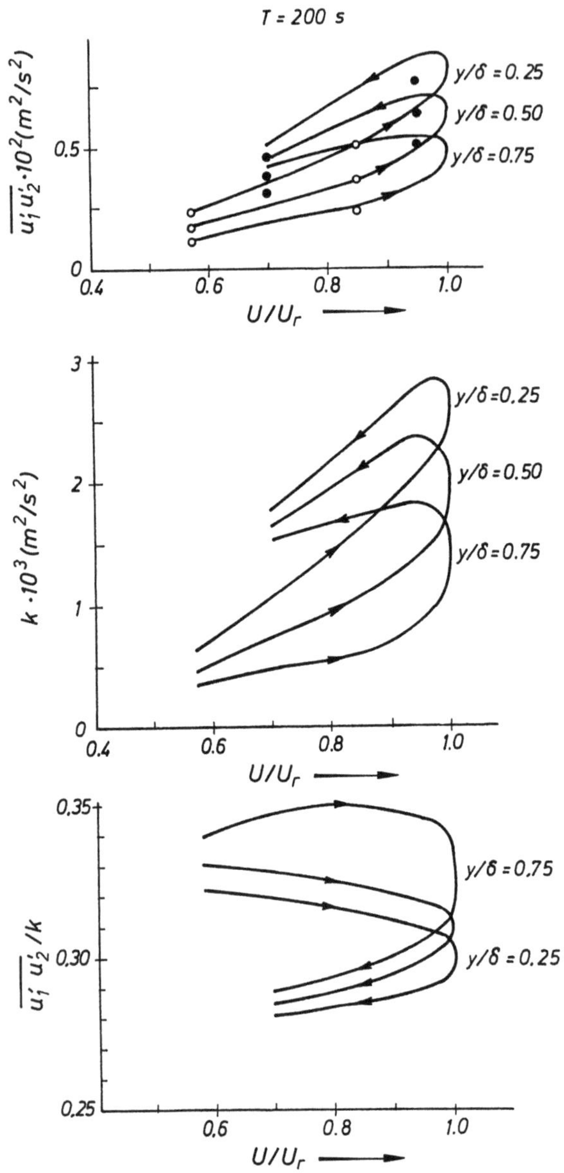

Fig. 6. Hysteresis of turbulent shear stress and kinetic energy, — predictions, experiments: o acceleration, ● deceleration (Anwar and Atkins)

of 24 and 40 Hz, but small amplitude, and the oscillating boundary layer in zero pressure gradient, explored by Cousteix et al. (1977) with large amplitude (~ 37%), and frequency of 43 Hz. A selection of predicted results, compared with experimental data is given in Figs. 7 to 12. Figure 7 presents the prediction of the phase-averaged velocity profiles at several characteristic time instants within a cycle of Cousteix et al. flow. The obtained results are in excellent agreement with measurements. The predicted phase shift to the velocity across the flow, shown in Fig. 8 may also be regarded as satisfactory. The same could be concluded for the boundary layer shape factor which is shown in Fig. 9, but in a specific way as to emphasize its hysteresis. In the same figure the maximum- and the wall values of the shear stress are presented, exhibiting also a visible hysteresis. The maximum values do not agree

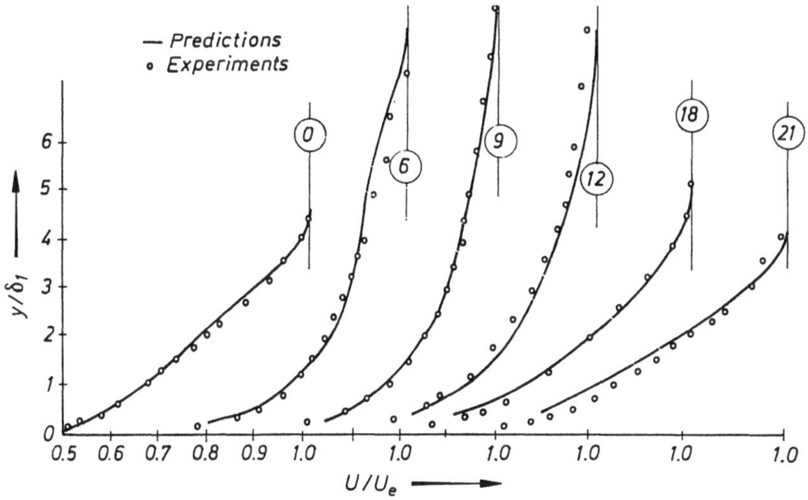

Fig. 7. Phase-averaged velocity profiles at different time instants; o— experiments (Cousteix et al. (1977)), — predictions

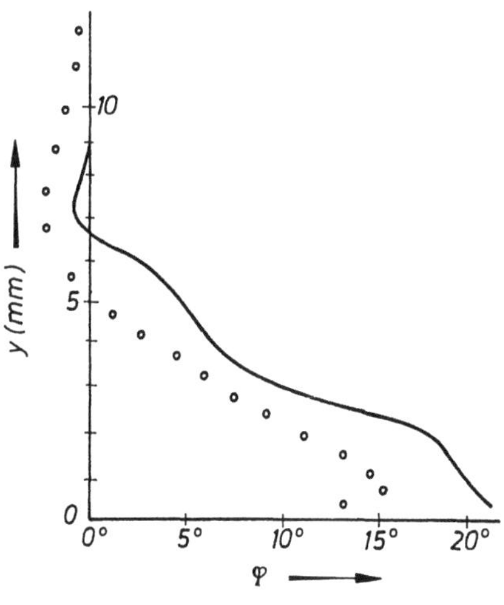

Fig. 8. Velocity phase,
— predictions, o experiments

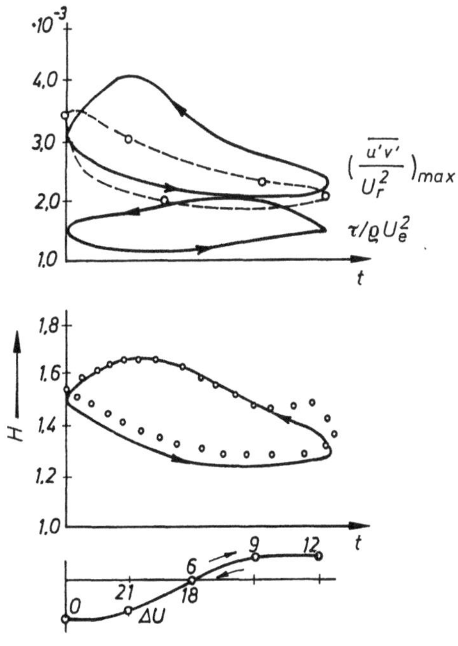

Fig. 9. Shape factor and wall and maximum shear stress,
— predictions, o experiments

well with the experiments (the authors did not quote the wall values), what is surprising, considering an excellent reproduction of the mean velocity profiles.

Similar shapes of hysteresis were exhibited by both shear and normal stresses in the experiment of Acharya and Reynolds. In this case the predictions are in much closer agreement with experiments as shown in Fig. 10 for three different positions in the flow.

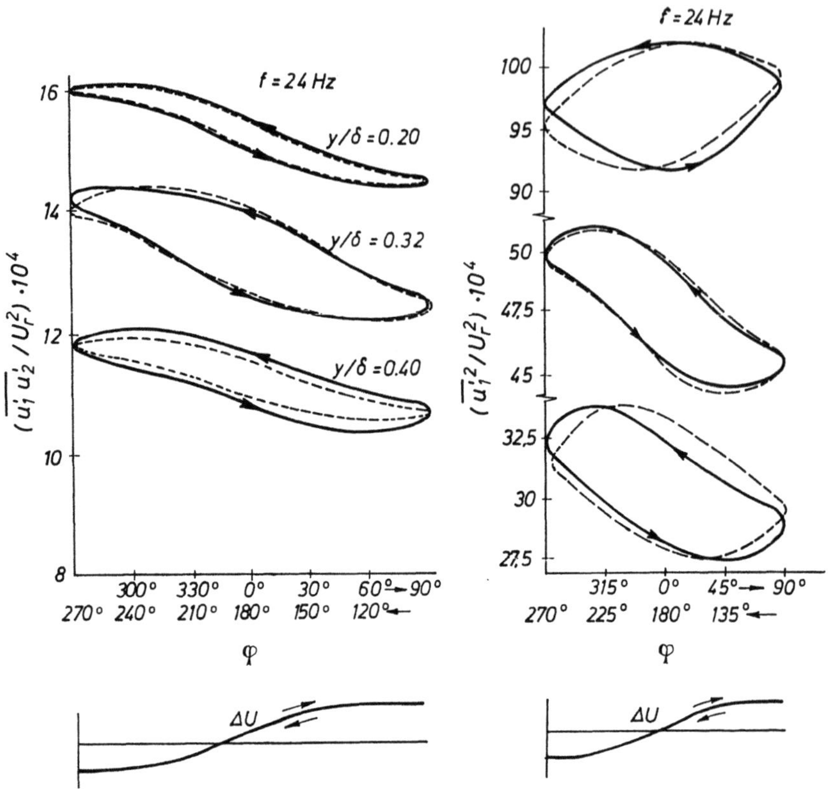

Fig. 10. Hysteresis of turbulent stresses at different flow depths in pulsating boundary layer: — predictions, --- experiments (Acharya and Reynolds)

Interesting shapes and orientation of hysteresis curves require a careful analysis, but we have not been able as yet to correlate any of the characteristic parameters.

Finally, in Fig. 11 the variations across the flow of the amplitudes and phases of the mean velocity and some components of turbulent stress are presented. The overal agreement with the experiments can be regarded as acceptable though some improvement would be desirable. At present we were not able to explain the excessive discrepancy of ΔU in the region close to the wall.

Conclusions

Three different types of periodic unsteady flows were analysed and predicted: The flow with a very low frequency and large amplitude, the second one with large frequency but small amplitude, and the third type with large values of both parameters. Attention was concentrated to the behaviour of ensemble-averaged turbulent stresses within a single cycle of the flow oscillation. All three cases showed clearly that the turbulence structure lags behind the mean flow field and that both mean flow- and turbulence-parameters show visible hysteresis. This feature may in some cases have a significant effects upon the global effects and transport processes within the flow.

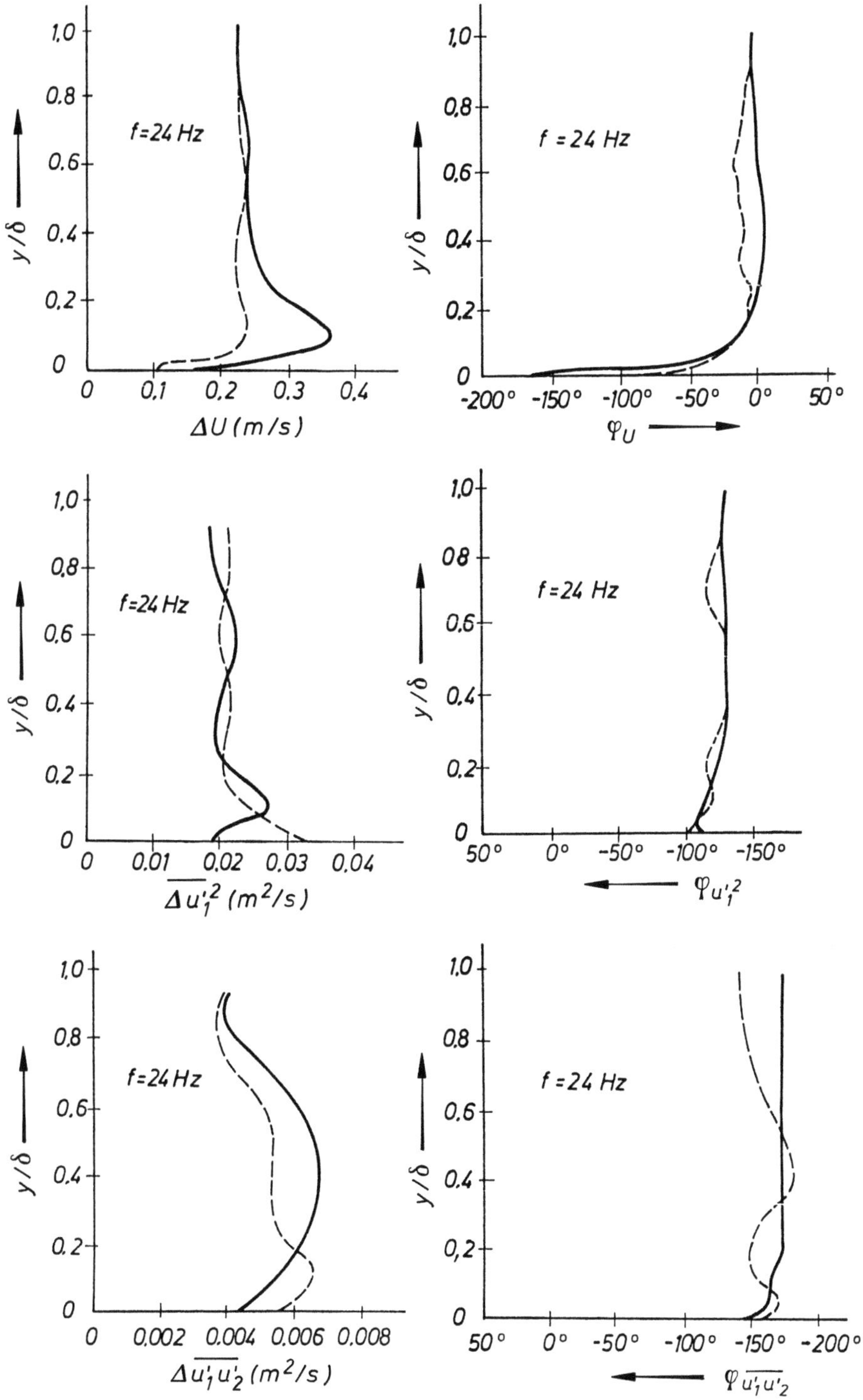

Fig. 11. Amplitudes and phase shifts of mean velocity and turbulent stresses: — predictions, --- experiments (Acharya and Reynolds, 1975)

In an attempt to predict all three considered cases, single-scale $k - \varepsilon$ and stress model, as well as two-scale $k - \varepsilon$ model were used. The closest agreement with experiments was achieved by means of the stress-dissipation model with some modifications that also helped earlier to improve the predictions of some steady non-equilibrium flows.

It is believed that still further improvements can be achieved with two-scale model that is based on separate transport equations for each stress component. This model is still in the stage of testing, but its application to a family of homogeneous flows, as shown in Fig. 1, seems very encouraging.

Acknowledgement. The authors acknowledge the financial support of Selfgoverning Community for the Interest in Science of Bosnia and Hercegovina. A large part of computation was carried out in the Computer Centre of Energoinvest. Sincere thanks are due to our colleagues R. Selimović and S. Vasić for their valuable help in carrying out the numerical computations.

References

Acharya, M., Reynolds, W. C. (1975): Rept. TF-8, Mech. Eng. Dept., Stanford University, Stanford, Calif.

Anwar, H. O., Atkins, R. (1980): J. Hydr. Div. **8**, 1273

Carr, L. W. (1981): In *Unsteady Turbulent Shear Flows*, IUTAM Symp. Toulouse, ed. by R. Michel, J. Cousteix, R. Houdeville (Springer, Berlin, Heidelberg, New York) pp. 3–34

Chieng, C. C., Launder, B. E. (1980): Numerical Heat Transfer **3**, 189

Cousteix, J., Houdeville, R., Javelle, J. (1981): In *Unsteady Turbulent Shear Flows*, IUTAM Symp. Toulouse, ed. by R. Michel, J. Cousteix, R. Houdeville (Springer, Berlin, Heidelberg, New York) pp. 120–144

Cousteix, J., Desopper, A., Houdeville, R. (1979): In *Turbulent Shear Flows 1*, ed. by F. Durst, B. E. Launder, F. W. Schmidt, J. H. Whitelaw (Springer, Berlin, Heidelberg, New York) p. 154

Hanjalić, K., Launder, B. E., Schiestel, R. (1980): In *Turbulent Shear Flows 2*, ed. by L. J. S. Bradbury, F. Durst, B. E. Launder, F. W. Schmidt, J. H. Whitelaw (Springer, Berlin, Heidelberg, New York) p. 36

Hanjalić, K., Ivanović, M., Selimović, R., Stošić, N., Vasić, S. (1982): In *The 1980–81 AFOSR-HTTM Stanford Conference on Complex Turbulent Flows*, ed. by S. J. Kline, B. J. Cantwell, G. M. Lilley (Mech. Eng. Dept., Stanford University) p. 1364

Launder, B. E., Reece, G. J., Rodi, W. (1975): J. Fluid Mech. **68**, 573

Launder, B. E., Leschziner, M. A., Sindir, M. (1982): In *The 1980–81 AFOSR-HTTM Stanford Conference on Complex Turbulent Flows*, ed. by S. J. Kline, B. J. Cantwell, G. M. Lilley (Mech. Eng. Dept., Stanford University) p. 1390

Launder, B. E., Spalding, D. B. (1973): Imperial College, Mech. Eng. Dept. Internal Report HTS/73/2

Patankar, S. V., Spalding, D. B. (1970): *Heat and Mass Transfer in Boundary Layers* (Intertext, London)

Shemer, L., Wygnansky, I. (1981): In *Turbulent Shear Flows 3*, ed. by L. J. S. Bradbury, F. Durst, B. E. Launder, F. W. Schmidt, J. H. Whitelaw (Springer, Berlin, Heidelberg, New York) p. 8.13

Townsend, A. A. (1978): J. Fluid Mech. **81**, 171

Periodic Turbulent Shear Flows

B. R. Ramaprian [1], S. W. Tu [2], and A. N. Menendez [3]

Abstract

Experimental results for a fully developed periodic turbulent pipeflow and a time-mean zero pressure gradient periodic turbulent boundary layer are presented. Both of them are characterized by large amplitude and frequency of oscillation. Their similarities and differences are discussed.

Introduction

There has been considerable interest, in recent years, in the study of unsteady turbulent shear flows. This is because of their many practical applications such as helicopter rotor blade flows, turbomachinery blade flows, aircraft flutter, gusts on structures etc. Some fundamental experimental studies of the structure of turbulent shear flows which are subjected to an externally imposed deterministic unsteadiness have been reported in the past few years. These include studies of fully developed periodic turbulent flow [e.g., Mizushina, Maruyama and Shiozaki (1973); Mizushima, Maruyama and Hirasawa (1975); Kirmse (1979); Ramaprian and Tu (1980), etc.], and studies of periodic turbulent boundary layers [e.g., Cousteix, Houdeville and Javelle (1981), Parikh, Reynolds and Jayaraman (1982), etc.]. A comprehensive review of the state of the art in this field was made by Carr (1981) and hence will not be repeated here. While these studies have provided a significant amount of information on the effect of imposed periodicity on the behavior of turbulent shear flows, there are a large number of areas in which information is either not available or is controversial. This is particularly true for flows in which the imposed periodicity is at a frequency large enough to interact with a dominant frequency of the turbulent motions – such as the "bursting" frequency, and/or has a large amplitude. There is a general feeling that imposed periodicity has no effect on the time-mean properties of the flow. However, previous work by the present authors on pipeflow (referred earlier) showed that the time-mean velocity distribution in a pipe shows a small but measurable effect of imposed periodicity. There is no general concensus on the quantitative response of wall shear stress and near-wall flow to imposed periodicity. In fact, no data on shear stress at the wall or Reynolds shear stress distribution in the unsteady shear flows are available at high frequencies and large amplitudes of imposed oscillation.

An extensive program of study of periodic turbulent shear flows has been in progress at Iowa during the last several years. Under this program a comprehensive study of fully developed periodic turbulent pipeflow has been completed. This was followed by the study

1 Professor of Mechanical Engineering, Research Engineer, Iowa Institute of Hydraulic Research, The University of Iowa, Iowa City, Iowa;
2 Engineer, Bechtel Civil and Minerals, Inc., San Francisco, California;
3 Research Scientist, INCYTH, Laboratorio de Hidraulica Aplicada, Ezeiza, Argentina

of a turbulent boundary layer in a periodically modulated freestream. This latter study is being performed in an unsteady flow water tunnel recently built at Iowa and the study has already yielded a large volume of very useful data. Both the pipeflow and boundary layer studies have been/will be described in detail elsewhere. The present paper contains some of the major results from the two studies. The experimental results are used to study the similarity and differences in the response of pipeflow and boundary layer flow to imposed external periodicity. The main features that distinguish both the present studies from most other similar studies are the combination of high frequency and large amplitude oscillation, wall shear stress measurements in the pipeflow experiments and Reynolds shear stress measurements in the boundary layer experiments. The experiments are briefly described below.

Pipeflow Experiments

Experimental Conditions

These experiments were performed in a long circular pipe of radius (R) 2.5 cm and length 10 meters through which water flow was maintained at a cross-sectional, time-average velocity ($\bar{U}_m$) of about 1 m/s, corresponding to a mean Reynolds number of about 50,000. This Reynolds number was modulated by varying the discharge through the pipe sinusiodally with time using a rotating sleeve valve at the pipe exit. The details of the apparatus and experiment are described in Ramaprian and Tu (1982). Detailed measurements including instantaneous longitudinal velocities, U (using frequency-shifted LDA) and instantaneous wall shear stress, τ_w (using a flush mounted hot film gage) were obtained.

The instantaneous velocities U were decomposed into time-mean component $\bar{U}$, periodic component U_p and random component u, by obtaining ensemble averages over 1000 cycles of oscillation. The first series of experiments included measurements of steady flow at five different Reynolds numbers ranging from 20,000 to 80,000. These experiments were used not only to test the experimental procedures but also as the basis for obtaining information on the quasi-steady flow (i.e., flow at infinitely low oscillation frequency). The quasi-steady behavior was then compared with the results of the second series of experiments, in which the flow was oscillated sinusiodally around a mean Reynolds number of 50,000. Two oscillation frequencies (f), namely 0.5 Hz and 3.6 Hz were studied. The higher frequency is of the order of the estimated mean bursting frequency in the flow. The lower frequency can be considered to be an intermediate frequency at which the flow still deviated significantly from quasi-steady behavior. In both experiments the amplitude of modulation γ_{um} of the velocity U_m was large enough (15% at 3.6 Hz and 64% at 0.5 Hz) to yield new and significant results. From the measurements made, detailed information on various aspects such as velocity, turbulence intensity and Reynolds shear stress was obtained, the last quantity being determined from the momentum equation.

Results

As already mentioned, only the major results will discussed in this paper.

Time-Mean Velocity Profile. The time mean velocity distribution across the pipe is shown for the two oscillation frequencies (Fig. 1). In each case the results are compared with those for a quasi-steady (zero-frequency) flow of the same amplitude γ_{um}. It is seen that at the higher frequency, there is a small but perceptible decrease in the velocity near the wall ($\eta = y/R = 0$) which is more clearly seen as a corresponding increase in velocity near the centerline as required by global continuity. The effect on the mean velocity at the lower frequency is,

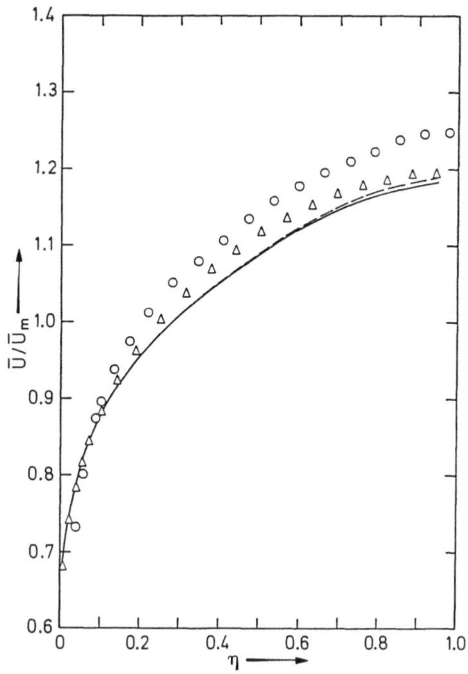

Fig. 1. Time-mean velocity profiles in unsteady and quasi-steady pipeflow at Re = 50,000. o, $f = 3.6$ Hz, $\gamma_{um} = 15\%$, Δ, $f = 0.5$ Hz, $\gamma_{um} = 64\%$; ---, quasi-steady flow, $\gamma_{um} = 15\%$; —, quasi-steady flow, $\gamma_{um} = 64\%$. The velocity profile in steady flow at the same time-mean Reynolds number is indistinguishable from the dashed line

however, much smaller. Unfortunately, the geometry of the tube did not allow measurements to be made at closer distances from the wall than shown in the figure. It was in fact observed that the time-mean velocity distribution does not follow the universal logarithmic law, especially at the higher frequency.

Amplitude and Phase Results. The phase averaged local velocity at any point y away from the wall and at a phase position θ in the oscillation cycle can be represented, to a good approximation, by

$$\langle U \rangle (y, \theta) = \bar{U}[1 + \gamma_u \cos(\theta + \phi_u)].$$

Figure 2 shows the amplitude ratio $[(\gamma_u \bar{U})/(\gamma_{um} \bar{U}_m)]$ and the phase angle ϕ_u of the phase-averaged velocity, for the two oscillation frequencies. The results, in each case, are compared with the exact solution [from Uchida (1956)] for fully developed periodic laminar flow in the same pipe at the same frequency. It is seen that the turbulent and laminar flow results are very different from each other. The turbulent flow exhibits a larger overshoot in amplitude and a very much larger "Stokes layer" within which unsteady effects are present. More striking is the result that the phase angle ϕ_u does not seem to extrapolate to a value of more than about + 10 degrees at the wall in the turbulent flow even at the higher frequency, while the laminar solution indicates the asymptotic value of + 45 degrees.

The phase lead of the wall shear stress was also found to be less than 10 degrees even at the higher frequency as shown by the filled symbols at $\eta = 0$ in Fig. 2, thereby corraborating the LDA results for the velocity.

Ensemble Averaged Turbulence Intensity. The ensemble averaged rms turbulence intensity u'_p is defined as $u'_p(y, \theta_u) = [\langle u^2 \rangle]^{1/2}$. Figures 3 and 4 show distributions of u'_p across the pipe at a few typical phase positions in the oscillation cycle, for the two oscillation frequencies. It is seen that at the intermediate frequency, the ensemble averaged turbulent intensity

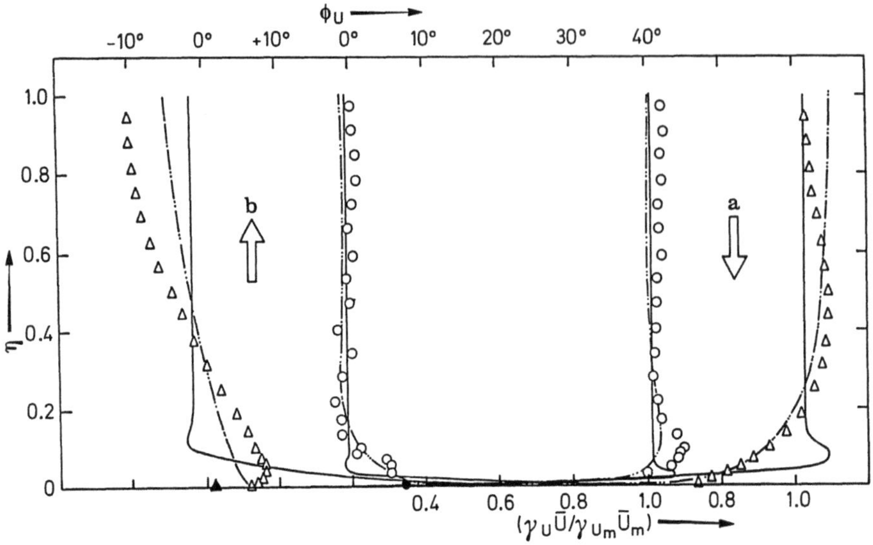

Fig. 2. Amplitude and phase of the ensemble averaged velocity in pipeflow (a) amplitude (b) phase. ○, 3.6 Hz; △, 0.5 Hz; Filled symbols represent wall shear stress data: ●, 3.6 Hz; ▲, 0.5 Hz; ——, laminar theory (Uchida 1956); —··—, numerical calculations using a typical eddy viscosity model (from Ramaprian and Tu, 1982)

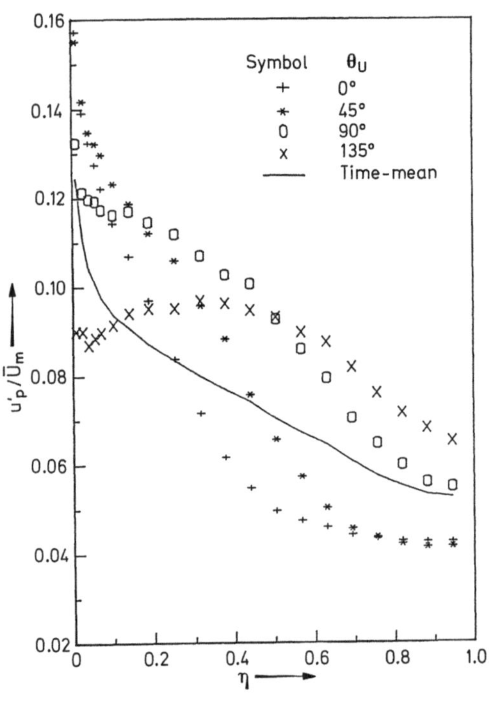

Fig. 3. Ensemble averaged longitudinal turbulent intensity in pipeflow at $f = 0.5$ Hz. Decelerating period. ——, time-mean distribution, u'

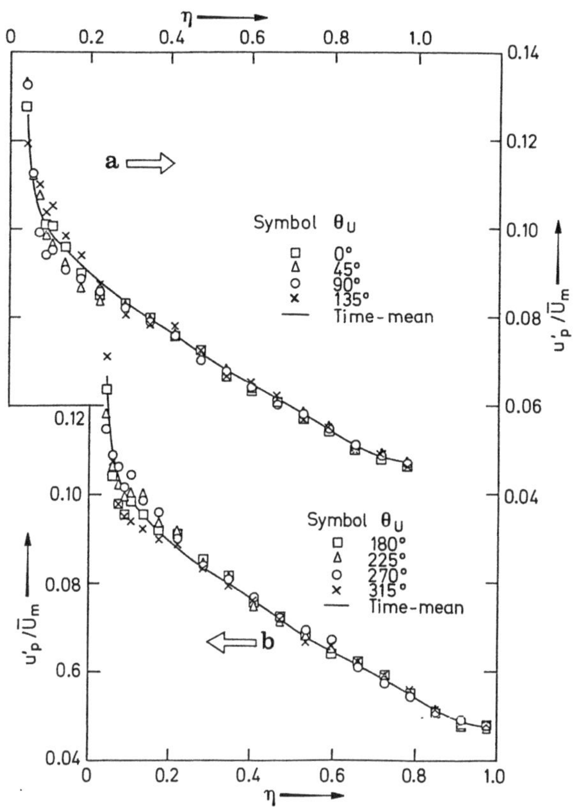

Fig. 4. Ensemble averaged longitudinal turbulent intensity in pipeflow at $f = 3.6$ Hz. (a) decelerating period (b) accelerating period. ——, time-mean distribution, u'

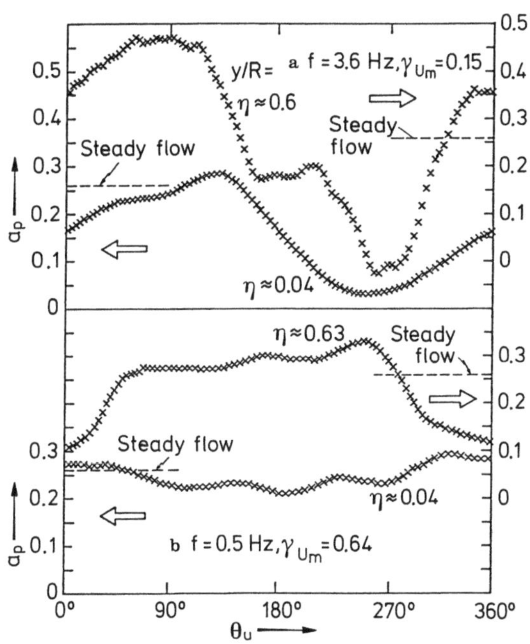

Fig. 5. Phasewise distribution of the structure parameter a_p in pipeflow

profiles are affected across the entire pipe. Crossovers and distortions can be observed in these distributions. The Reynolds shear stress distributions (not shown here) were also found to exhibit qualitatively the same trend (except in the outer region during a part of the cycle) indicating an equilibrium state between these two turbulent properties. At high frequency (Fig. 4), the ensemble averaged turbulent intensity remains frozen in the outer region. The ensemble averaged Reynolds shear stress, however, was found, in this case, to vary slightly with U_m even in the outer region.

Structural Equilibrium. It was observed that at both the intermediate and high frequency of oscillation studied, the mean flow, the turbulent kinetic energy and the Reynolds shear stress were mutually out of phase with one another. Thus, eddy viscosity type models either of the "gradient" or "bulk" type relating the turbulent transport either to the local mean flow or turbulent structure at an instant (in a quasi-steady manner) cannot accurately describe such flows. This is seen in Fig. 2, where the predictions of the unsteady flow obtained by using a well-known eddy viscosity model are shown. Details of this calculation method are discussed in Ramaprian and Tu (1982). The departure of the flow from structural equilibrium is illustrated by Fig. 5 which shows the behavior of a characteristic "structure parameter" $a_p = [-\langle uv\rangle/u_p'^2]$ during the oscillation cycle. This parameter should remain constant at a value of about 0.26 in an equilibrium flow. The variation of a_p in the periodic flow, therefore, demonstrates the breakdown of the structural equilibrium, partially during the cycle at intermediate frequency and completely at the high frequency.

Boundary Layer Experiments

The pipe flow experiments, while being very extensive, had two limitations. These are (i) the uncertainty of the measurements very near the wall (caused by the wall curvature) and (ii) the lack of direct measurements of the ensemble-averaged Reynolds shear stress $\langle uv\rangle$. These limitations have been removed in the complimentary experiments on a periodic boundary layer on a flat plate currently under progress. A two-component LDA system is being used in these experiments so that u_p', v_p' and $\langle uv\rangle$ can be measured directly. A large and unique unsteady flow water tunnel has been built for this study. The tunnel is designed such that the freestream velocity at the test section can be varied with time in a prescribed manner. Figure 6 shows a layout of the tunnel. The tunnel works under a constant head of about 8 meters. The test section is 50 cm × 22.5 cm in cross section and 2.4 m in length. A smooth brass plate 50 cm wide × 2.4 m long forming the bottom wall of the test section is used as the test surface for the boundary layer studies. The tunnel has presently been set up for a freestream velocity variation given by

$$U_\infty = \bar{U}_\infty [1 + \gamma_u \cos(2\pi f t)]$$

with $\bar{U}_\infty \approx 0.90$ m/s, $\gamma_{u\,\infty} \approx 0.4$ and $f = 2$ Hz. (The bursting frequency in the boundary layer is estimated to be about 4 Hz.) The amplitude and frequency of oscillation can be varied. It may be noted that, because of the wide separation between the oscillation frequency and the acoustic frequency in this hydraulic system, the amplitude and phase of the freestream velocity variation is constant along the test section. The description of the tunnel and a complete documentation of the flow quality in the tunnel will be available from a forthcoming report. Instantaneous velocities have been measured at 5 longitudinal locations ranging from $(2\pi f x/\bar{U}_\infty) \approx 8 - 27$, with x measured from the contraction exit. The data have been processed to yield mean, periodic and turbulent properties of the flow including Reynolds

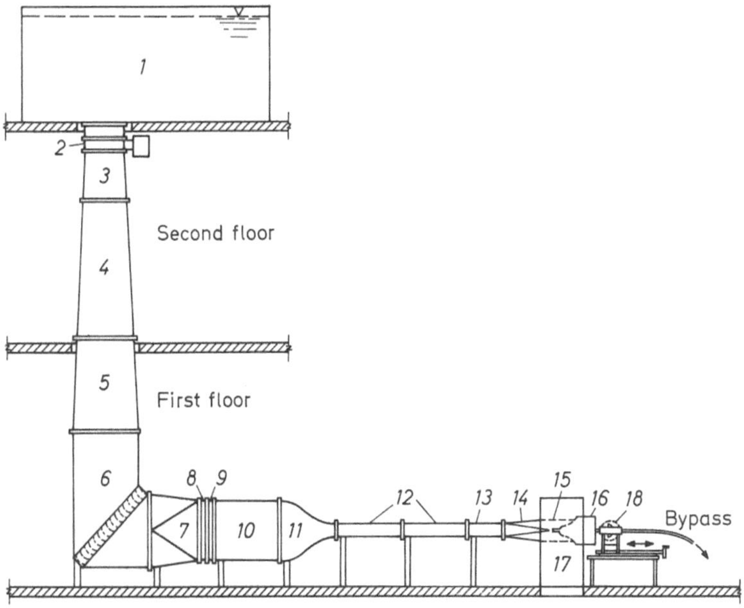

Fig. 6. Lay-out of the periodic flow water tunnel. *1* Constant head tank, *2* 24″ Butterfly valve with actuator, *3* Diffuser I, *4* Diffuser II, *5* Diffuser III, *6* Vaned elbow, *7* Circular to square transition, *8* Distance piece, *9* Aluminum honeycomb, *10* Settling chamber, *11* Three-dimensional contraction, *12* Test section, *13* Distance piece, *14* Transition piece, *15* Slotted cylinder, *16* Rotating sleeve, *17* Discharge enclosure, *18* 3 HP variable speed SCR drive for sleeve

shear stress and some triple correlations. All the raw as well as processed data are on digital tape. Experiments at a lower oscillation frequency of 0.5 Hz will soon begin.

Only, the data relevant for comparison with the "high-frequency" pipeflow results will be presented in this paper. Further, only data from the last measuring station, corresponding to $(2\pi f x/\bar{U}_\infty)$ will be used in these comparisons.

Results

Time-Mean Velocity Profile. Figure 7 shows the time-mean velocity distributions in the periodic boundary layer ($f = 2$ Hz), quasi-steady boundary layer ($f = 0$) at the same amplitude and steady boundary layer for a freestream velocity of $\bar{U}_\infty$. The quasi-steady and steady flow show negligible difference between each other suggesting that nonlinear (Reynolds number) effects are not important in this case. The unsteady flow results, however, indicate a definite, though small reduction in the time mean velocity near the wall, compared to the quasi-steady flow. This result, thus confirms the pipeflow results. The apparent *increase* in the time-mean velocity in the *outer part* of pipeflow is simply a consequence of normalizing the velocities by the time-mean cross-sectional average velocity $\bar{U}_m$, rather than by the time-mean centerline velocity $\bar{U}_{max}$.

Phase and Amplitude Results. The phase ϕ_u and amplitude (γ_u) distributions of the local ensemble averaged velocity $\langle U \rangle$ are shown in Fig. 8. The distributions are very similar, even quantitatively, to those for pipeflow. The maximum phase shift measured near the wall is still of the order of 10 degrees and the amplitude overshoot near the wall is less than 10%. The boundary layer thickness δ at this station was about 4 cm, corresponding to y/d (in Fig. 8) $= 0.4$. The nondimensional "Stokes layer" thickness (about 0.25 δ) in the boundary layer can

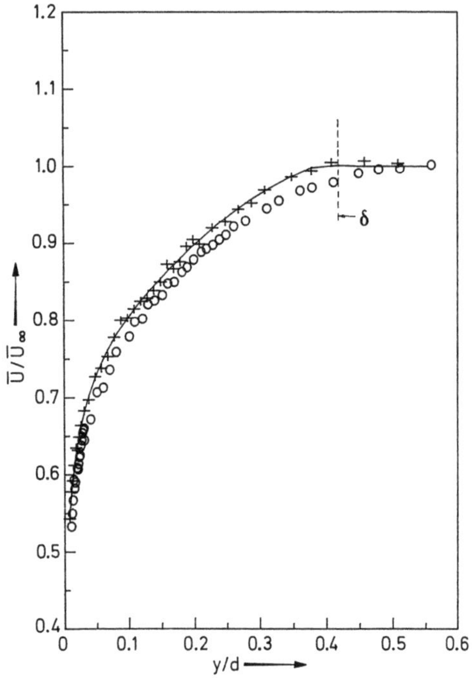

Fig. 7. Time-mean velocity profiles in unsteady and quasi-steady boundary layer flow, o, unsteady flow; ——, quasi-steady flow; +, steady flow at the time-mean freestream velocity

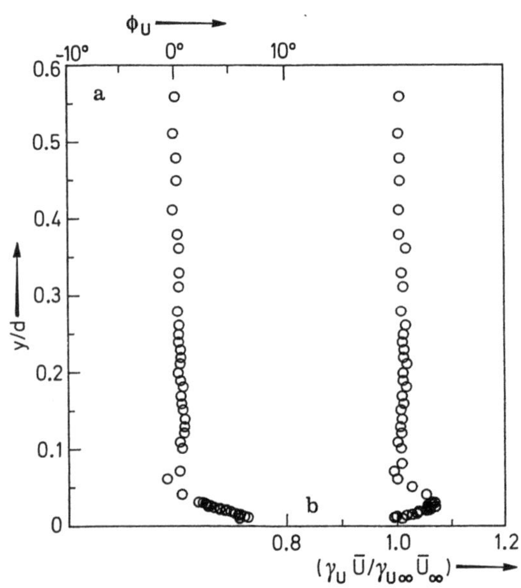

Fig. 8. Amplitude and phase of the ensemble averaged velocity in boundary layer flow. (a) amplitude (b) phase

be seen to be comparable to that in the pipeflow. The flow beyond the Stokes layer behaves slug-like in both cases.

Ensemble Averaged Turbulent Stresses u'_p *and* $\langle uv \rangle$. Figures 9 and 10 show the distributions of $\langle u'^2_p \rangle (y, \theta)$ and $\langle uv \rangle (y, \theta)$ for a few typical phase positions in the oscillation cycle. The distributions $\langle u'^2_p \rangle$ are in complete agreement with the pipeflow results at the higher oscillation frequency. The intensity remains frozen during the oscillation cycle in the outer 75 per cent of the boundary layer. A similar freeze in the outer layer is observed in the case of $\langle uv \rangle$

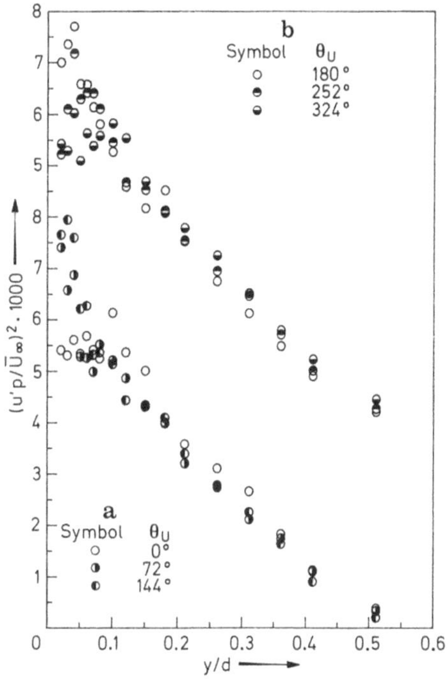

Fig. 9. Ensemble averaged longitudinal turbulent intensity in boundary layer flow. (a) decelerating period (b) accelerating period

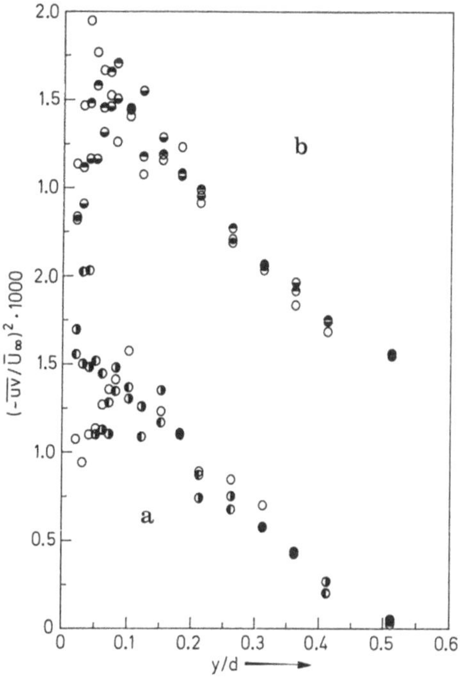

Fig. 10. Ensemble averaged Reynolds shear stress in boundary layer flow. (a) decelerating period (b) accelerating period. Symbols as in Fig. 9

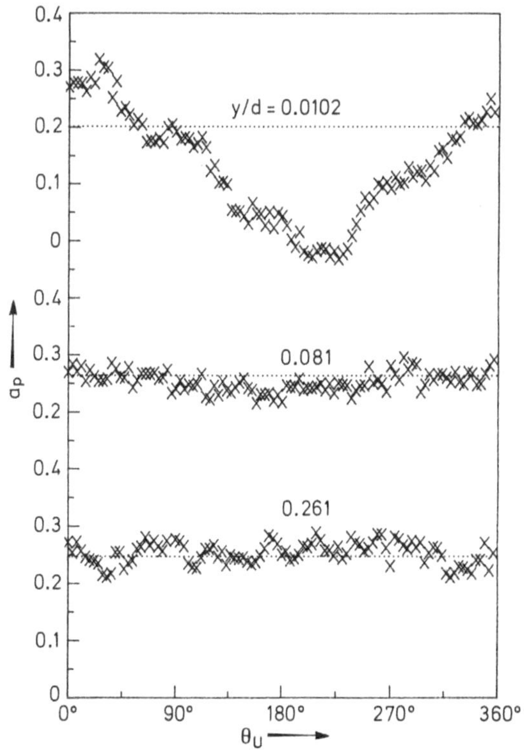

Fig. 11. Phasewise distribution of the structure parameter a_p in boundary layer flow. . . . , steady flow value

also. This last result is in some disagreement with the pipeflow result at 3.6 Hz which indicated a slight modulation in $\langle uv \rangle$ even in the outer region.

Structure parameter. Similar differences between the pipe and boundary layer results can be observed in the behavior of the structure parameter a_p shown in Fig. 11. It is seen that very near the wall the two results are very similar. The turbulent structure shows departure from equilibrium. However, beyond about $0.2\,\delta$, the turbulent structure in the boundary layer seems to be in a state of frozen equilibrium, in contrast with the pipeflow in which there is a complete breakdown of structural equilibrium in the outer flow. This difference in behavior might be ascribed to the difference between the structure of the outer flow in the boundary layer and that in fully developed pipeflow. The outer flow in the pipe is still controlled by the wall via global momentum and continuity constraints whereas the outer flow in the boundary layer is not. The boundary layer data are now being analyzed in grater detail to obtain a better understanding of the response of the different regions of the flow to imposed insteadiness.

Conclusions

(1) The experimental data on both pipeflows and boundary layers indicate that externally imposed periodicity produces a small but detectable effect on the time-mean velocity at high frequency and large amplitude of oscillation.

(2) Stokes layer thickness, as well as phase and amplitude of the velocity variation are very similar in both the flows, with the phase angle near the wall being of the order of + 10 degrees. Thus, the response of periodic turbulent shear flows is very different from that of periodic laminar shear flows.

(3) Turbulence structure very near the wall is disturbed from equilibrium in both the flows at high frequencies. The outer layer of the boundary layer flow is in a state of frozen equilibrium while that in the pipeflow appears to be still disturbed from equilibrium either partially (at moderate frequencies) or completely (at high frequencies) during the oscillation cycle.

Acknowledgement. The work reported in this paper was supported by the U.S. Army Research Office, through their Grant/Contract No. DAAG-29-79-G-0017 and DAAG-29-83-K-0004 with Dr. R. E. Singleton as Scientific Program Officer. The authors gratefully acknowledge this support.

References

Carr, L. W. (1981): A review of unsteady turbulent boundary-layer experiments, in *Unsteady Turbulent Shear Flows*, ed. by R. Michel, J. Cousteix, R. Houdeville (Springer, Berlin, Heidelberg, New York) p. 3

Cousteix, J., Houdeville, R., Javelle, J. (1981): Response of a turbulent boundary layer to a pulsation of the external flow with and without adverse pressure gradient, in *Unsteady Turbulent Shear Flows*, ed. by R. Michel, J. Cousteix, R. Houdeville (Springer, Berlin, Heidelberg, New York) p. 120

Kirmse, R. E. (1979): Investigations of pulsating turbulent pipe flow. Trans. ASME paper No. 79-WA/FE-1

Mizushima, T., Maruyama, T., Shiozoki, Y. (1973): Pulsating turbulent flow in a tube. J. Chem. Eng. Jpn. **6**, 487

Mizushima, T., Maruyama, T., Hirosawa, H. (1975): Structure of the turbulence in pulsating pipe flows. J. Chem. Eng. Jpn. **8**, 210

Parikh, P. G., Reynolds, W. C., Jayaraman, R. (1982): Behavior of an unsteady turbulent boundary layer. AIAA J. **20**, 769

Ramaprian, B. R., Tu, S. W. (1980): An experimental study of oscillatory pipe flow at transitional Reynolds numbers. J. Fluid Mech. **100**, 513

Ramaprian, B. R., Tu, S. W. (1982): Study of periodic turbulent pipe flow. IIHR Report No. 238, University of Iowa, Iowa City.

Uchida, S. (1956): The pulsating viscous flow superposed on the steady motion of incompressible fluid in a circular pipe. Z. Angew. Math. Phys. **7**, 403

Part IV

Reacting Flows

Reacting Flows – Introductory Remarks

R. W. Bilger

The University of Sydney, N.S.W. 2006, Australia

Our understanding of non-reacting turbulent flows is rudimentary enough. One might question whether research into reacting turbulent flows can make much progress when such a basic understanding of turbulence is lacking. The non-linearity inherent in all but the simplest chemical reactions compounds an already highly non-linear problem as do the often associated phenomena of heat release, strong density gradient, radiant heat transfer and two-phase transport. Such difficulties are in themselves a challenge. There are, however, strong practical incentives for endeavours in this field. Unlike non-reacting flows, scale modeling of combustion systems is highly unsatisfactory and measurements in the hostile combustion environment pose very great problems of access, probe integrity and interference with the flow. Development of combustion chambers such as for diesel engines and glass melting furnaces has progressed by trial and error in the absence of even rudimentary knowledge of the flow patterns so essential to mixing, combustion and heat transport. The advent of three-dimensional computer models of such furnaces and combustion chambers, albeit rather crude for their details at this stage, is proving to be of immense benefit in the insight that they are able to give designers and developers who have for so long been blind-folded.

The major thrust of turbulent reacting flow research is to provide a better understanding of the turbulence/chemistry interactions involved in these flows and improved approaches to their mathematical descriptions and modeling. The reviews in the text edited by Libby and Williams [1] give a good introduction to the problems and approaches involved, while Jones and Whitelaw [2] succinctly review current models and computational methods.

Scalar transport is at the heart of the problem and the papers in *Turbulent Shear Flows 3* on this topic represent a cross-section on the problems and approaches involved and the introductory remarks by Bray [3] relate these to the general literature on scalar transport as well as to flows with chemical reaction. The effects of chemical reaction and the associated heat release on turbulent transport is one major aspect of turbulence/chemistry interaction. Two papers in this collection address this question. Dibble and Schefer report the development of a joint LDV/Raman measurement technique for the measurement of velocity scalar correlations. Direct measurements of scalar turbulent fluxes $\overline{u_k' \phi'}$ and $\widetilde{u_k'' \phi''}$ (where u_k is a velocity component and a ϕ a scalar, and the second term is in density weighted or Favre-averaged form) are vital to the development of improved models of scalar transport in reacting flows. Doubts exist [4, 5] as to the accuracy of the joint LDV/Mie-scattering technique and it is hoped that the comprehensive data obtained in turbulent diffusion flames by this technique [6, 7] will be checked by the LDV/Raman technique introduced here.

In the previous volume Sreenivasan et al. [8] show that, while gradient transport works satisfactorily for cross-stream fluxes in the classical simple shear flows, it breaks down badly in flows which are only slightly more complicated such as in an asymmetrically heated wake. Lockwood and Stolakis in this volume show that the $k - \varepsilon$ gradient transport model works quite well in turbulent jet diffusion flames a result now fairly well established in the literature.

It should be noted, however, that comparison of mean mixture fraction predictions with those deduced from experimental data [5] would be a much more direct test of the adequacy of the mean scalar transport. Comparisons with mean temperature involve the further inclusion of the model for conserved scalar fluctuations g (which Drake et al. [5] found to be inadequate), the radiation model and the fast chemistry assumption used; four submodels in all beyond those used for the momentum transport. There is a great need for experimental data for such quantities as g which will allow direct testing of the results of these submodels as well as for experimental data on the scalar velocity correlations which directly test the modeling of individual terms in their balance equations. Such detailed structural validation will give more confidence in using models such as that presented by Lockwood and Stolakis beyond the range in which they have been validated. In the absence of such detailed data the data of Becker and Yamazaki [9] and Becker and Liang [10] could provide a much more extensive overall test of the modeling than that used here, in particular the effects of buoyancy on the mixing and fluctuation intensity.

The finding of Lockwood and Stolakis of intensitivity of their results to the buoyancy generation terms in the k and ε equations is surprising. Glass and Kent [11] and Tamanini [12] both found these terms to be significant. Both used non-gradient models for $\overline{\varrho' u'}$, neither of them very satisfactory. Stream-wise turbulent fluxes do not gradient model in even classical non-reacting shear flows and improved models for these terms are needed. It should be noted also that in density weighted averaging the term in the k equation is [13]

$$-\overline{u''}\frac{\partial \bar{P}}{\partial x} = \frac{\overline{\varrho' u'}}{\bar{\varrho}}\frac{\partial \bar{P}}{\partial x} \tag{1}$$

where the axial pressure gradient results from the buoyant flow field and the term is not directly proportional to that obtained in conventional averaging. Stårner and Bilger [6] found this term to be significant in non-buoyant flows with axial pressure gradients.

Buoyancy and axial pressure gradients are only a first step toward more complex flows. Weakly swirled jet flames with appreciable radial pressure gradients would be a next step in a heirarchical development of turbulence modeling under combustion conditions, and one in which the heat release and radial density gradients should provide a very severe test of modeling capabilities in a yet relatively simple parabolic flow. It is clear that gradient modeling assumptions have to be abandoned in more complex flows and modeling developments in non-reacting flows are of much interest to the combustion researcher. Champion and Libby [14] in a paper presented at the conference but to be published elsewhere use a non-gradient closure at the second order level to model combustion in a turbulent boundary in a premixed flow. The have developed further the Bray-Moss-Libby aerothermochemistry [15] model for premixed combustion, in particular the modeling of the pressure velocity fluctuation correlation terms. Second-order closure approaches are likely to remain of prime interest to combustion researchers since larger eddy simulation techniques appear to require finer grid spacings and the added complexities of the combustion situation already stretch available computing power to the limit.

The effect of the turbulence on the chemical reaction rates is the other major aspect of turbulence/chemistry interaction and the one addressed by the remaining papers in this section. In lower order closure schemes the problem is simply one of modeling the mean chemical reaction rates in terms of the mean scalar and flow variables available. Neglect of the highly non-linear nature of the reaction rate is extremely hazardous; thus in a simple case

$$\bar{w}_i = \overline{\varrho^2 Y_i Y_j A \exp(-\theta/T)} \neq A\bar{\varrho}^2 \bar{Y}_i \bar{Y}_j \exp(-\theta/\bar{T}) \tag{2}$$

where w_i is the reaction rate of species i, Y_i its mass fraction, ϱ the density, T the temperature, and A and θ parameters of the Arrhenius rate constant. It is known that under fast chemistry conditions the neglect of the correlations in this last expression can lead to errors of several orders of magnitude. Indeed, under fast chemistry conditions reaction rates are controlled by the rates of turbulent mixing. Lockwood and Stolakis used the conserved scalar/fast chemistry approach widely used in non-premixed systems. Direct calculation of the reaction rate is avoided since specifies concentrations density and temperature are calculable from the statistics of the conserved scalar. It has been shown [13] that under these conditions the mean reaction rates are given by

$$\bar{W}_i = \bar{P}_i \equiv \overline{-\varrho D (\nabla f)^2 \frac{d^2 Y_i}{d f^2}} \propto \bar{\chi}_s p(f_s) \tag{3}$$

where D is the molecular diffusivity, f is the instantaneous value of the conserved scalar (or mixture fraction) $p(f_s)$ is the probability density of f being at its stoichiometric f_s, and $\bar{\chi}_s$ is the conditional mean of its scalar dissipation $\chi (\equiv 2D(\nabla f)^2)$, conditioned for the stoichiometric value of the conserved scalar. The result of Eq. (3) is exact and analytic, involving no modeling. It states that under such fast chemistry conditions the rate of chemical reaction is equal to the rate of presentation P_i of the (reactant) species i through microscale mixing processes to the stoichiometry mixture ratio where reaction occurs. That this rate of presentation is proportional to scalar dissipation, albeit a suitably conditioned average, is intuitively gratifying. It gives analytical rigour to ideas which in many theories are justified heuristically.

This overiding significance of the scalar dissipation and its joint statistics with the scalar itself are not, unfortunately, matched by our state of knowledge. Direct measurements of scalar dissipation are very difficult and even the modeling of unconditioned mean dissipation in simple flows is a complex question [16]. The computations of Ashurst et al. in the present volume are thus welcome even if the two-dimensional asumption must limit assurance in the application of the results to three-dimensional turbulence. The finding that the dissipation and the scalar itself are correlated in inhomogeneous flow is perhaps not surprising and the appropriate form for a model of the product $\bar{\chi}_s p(f_s)$ remains an intriguing problem. Information on this quantity and its scaling could no doubt be derived from these authors calculations for the complete range of f_s.

There is much current interest in extending these ideas of the primary role of scalar dissipation to situations of finite rate chemistry, i.e. where mixing rates are of the same order as chemical reactions rates, such as in problems of flame stabilization and blow-off in nonpremixed systems. Peters [17] and the Southampton school [18] have correlated the structure and extinction behaviour of laminar hydrocarbon flames with χ_s and apply this flamelet structure to turbulent diffusion flames. Local extinction of the flame sheet is assumed to occur when χ_s exceeds a critical value and turbulent flame structure and stabilization are deduced from models of the joint statistics of f and χ. A somewhat different view [19] is that χ is proportional to the rate of presentation of reactants in mixed form P_i of Eq. (3), and if these are not consumed locally by the chemical reaction they may diffuse and be convected to regions of the flow where mixing is not so intense. A perturbation formulation is used in this latter case but it is not restricted to small perturbations and could in principle be valid through to extinction. These approaches reflect the old conflict between ideas of stretching and thinning of reaction zone flame sheets to local extinction on the one hand, and a distributed reaction zone on the other. It is becoming apparent that the former will probably pertain to systems where activation energies are high, there are large variations in fuel to air ratio and there is little backmixing of well-reacted products. The distributed reaction zone may pertain to systems with low activation energy, more uniform fuel to air ratio or with significant backmixing of well-reacted products. A gas turbine primary zone may be closer to the latter, except perhaps at the lean blow-out limit.

Progress in modeling the influence of turbulence on chemical reaction rates will accelerate when more detailed experimental data become available. The new laser techniques offer prospects of yielding a great breakthrough in this regard. In nonpremixed flames it is essential to have simultaneous space and time resolved measurements of several scalar variables, as concentrations and temperature have little meaning unless there is some measure of stoichiometry. Spontaneous Raman scattering [20] is much better than CARS in this regard and joint measurements with fluorescence for radicals such as OH [21] will give us the detail needed. The data available so far has not yet achieved this potential and it is to be hoped that data on high Reynolds number flames in which finite rate chemistry effects are appreciable will soon be available. The much lower cost approach to this problem taken by Mudford and Bilger in this present volume is already yielding extremely interesting results. The joint pdf's of reacting scalar are the first available for nonpremixed reacting flows and give great insight into the structure of the reacting flows, albeit for an isothermal system of complex flow pattern. The techniques used can be adopted for simpler flow patterns and a grid flow with half and half reactants analogous to the half heated grid [22] is of particular interest. The existing data has been used [23] to illustrate the larger errors that can be made by the improper modeling of Eq. (2) and also to show that much more accurate formulations of the mean reaction rate can be made in terms of the mean perturbation variable arising in the perturbation approach [19].

The last two papers in this section attempt a more general closure of the chemical reaction rate than the approaches considered above. A model which would handle everything from premixed to nonpremixed flows, and from fast chemistry to slow chemistry and for many simultaneous reactions would of course be ideal. The direct calculation of the joint scalar pdf [24] obviates, of course, the problem altogether, although the difficulties are then transferred to the diffusive and convective terms. The Interaction by Exchange with the Mean (I. E. M.) mode of Borghi and Pourbaix reported here has many characteristics similar to the pdf calculation approach.

A check on the formulation of such models is that they must give the same results as Eq. (3) above when the chemistry is fast in non-premixed problems. It can be shown that the formula given in Section II.3 by Borghi and Pourbaix implies that

$$\bar{\chi}_s/\bar{\chi} = \text{const.} \int_0^{u_s} (\tilde{u} - u)\, \tilde{p}(u)\, du/\tilde{p}(u_s) \tag{4}$$

where $u \equiv f/(\widetilde{f''^2})^{1/2}$ is the conserved scalar normalized by its rms fluctuation, and the normal modeling for $\bar{\chi}$ has been assumed. This form appears plausible for near-Gaussian pdf's provided τ_{ex} is taken as somewhat less than k/ε. It would be interesting to have it checked for the Ashurst et al. calculations and its plausibility checked for some experimentally measured pdf's. A similar qualification applies to the work of Byggstøyl and Magnussen and their Eq. (32) implies

$$\bar{\chi}_s/\bar{\chi} = \frac{26.4\ \mathrm{Re}_t^{-1/4}}{1 - 9.7\ \mathrm{Re}_t^{-3/4}} J_1(u_s - \tilde{u})/\tilde{p}(u_s) \tag{5}$$

where $\mathrm{Re}_t \equiv \nu \varepsilon / k^2$ and the J_1 function is that given earlier [25]:

$$J_1 (u_s - \tilde{u}) \equiv \int_0^{u_s} (u_s - u) \tilde{p}(u)\, du - (u_s - \tilde{u})\, H(u_s - \tilde{u})\,. \tag{6}$$

The Reynolds number dependence in Eq. (5) is not possible since for the complete range of f_s, $\bar{\chi}_s$ must always average out at $\bar{\chi}$. Also, the correct order of magnitude is found only for $\mathrm{Re}_t \sim 10^5$, very high Reynolds number turbulence indeed.

References

1. Libby, P. A., Williams, F. A. (eds.): *Turbulent Reacting Flows*, Topics Appl. Phys., Vol. 44 (Springer, Berlin, Heidelberg, New York 1980)
2. Jones, W. P., Whitelaw, J. H. (1982): Calculation methods for reacting turbulent flows: a Review. Combust. Flame **48,** 1
3. Bray, K. N. C.: "Introductory Remarks", in *Turbulent Shear Flows 3*, ed. by L. J. S. Bradbury, F. Durst, B. E. Launder, F. W. Schmidt, J. H. Whitelaw (Springer, Berlin, Heidelberg, New York 1982) p. 93
4. Starner, S. H., Bilger, R. W. (1983): Differential diffusion effects on measurements in turbulent diffusion flames by the Mie scattering technique. Prog. Astronaut. Aeronaut. **88**, 81
5. Drake, M., Bilger, R. W., Stårner, S. H.: Raman Measurements and Conserved Scalar Modeling in Turbulent Diffusion Flames, in Nineteenth Symposium (International) on Combustion (The combustion Institute 1982) p. 459
6. Stårner, S. H., Bilger, R. W.: Measurement of Scalar-Velocity Correlations in a Turbulent Diffusion Flame, in Eighteenth Symposium (International) on Combustion (The Combustion Institute 1981) p. 921
7. Stårner, S. H. (1983): Joint measurements of radial velocity and scalars in a turbulent diffusion flame. Combust. Sci. Tech. **30**, 145
8. Sreenivasan, K. R., Tavoularis, S., Corrsin, S.: "A Test of Gradient Transport and Its Generalizations", *Turbulent Shear Flows 3,* ed. by L. J. S. Bradbury, F. Durst, B. E. Launder, F. W. Schmidt, J. H. Whitelaw (Springer, Berlin, Heidelberg, New York 1982) p. 96
9. Becker, H. A., Yamazaki, S.(1978): Entrainment, momentum flux and temperature in vertical free turbulent diffusion flames. Combust. Flame **33**, 123
10. Becker, H. A., Liang, D. (1982): Total emission of soot and thermal radiation by free turbulent diffusion flames. Combust. Flame **44**, 305
11. Glass, M., Kent, J. H.: Predictions and Measurements in Turbulent Buoyant Diffusion Flames, in Second Australian Conference on Heat and Mass Transfer (The University of Sydney 1977) p. 445
12. Tamanini, F. (1977): Reaction rates, air entrainment and radiation in turbulent fire plumes. Combust. Flame **30**, 85
13. Bilger, R. W. (1976): Turbulent jet diffusion flames. Prog. Energy Combust. Sci. **1**, 87
14. Champion, M., Libby, P. A. (1984): Turbulent premixed combustion in a boundary layer. Combust Sci. Tech. **38**, 267
15. Bray, K. N. C., Libby, P. A., Masuya, G., Moss, J. B. (1981): Turbulence production in premixed turbulent flames. Combust. Sci. Tech. **25**, 127
16. Sirivat, A., Warhaft, Z. (1982): The mixing of passive helium and temperature fluctuations and grid turbulent. J. Fluid Mech. **120**, 475
17. Peters, N.: Laminar diffusion flamelet models in nonpremixed turbulent combustion. Progr. Energy Comb. Sci. (to appear)
18. Liew, S. K., Bray, K. N. C., Moss, J. B.: The Predicted Structure of Stretched and Unstretched Methane-Air Diffusion Flames, in *Ninth International Colloquium on Dynamics of Explosions and Reactive Systems* (Poitiers, France 1983) (to appear)
19. Bilger, R. W. (1980): Perturbation analysis of turbulent nonpremixed combustion. Combust. Sci. Techn. **22**, 251
20. Drake, M. C., Lapp, M., Penney, C. M., Warshaw, S., Gerhold, B. W.: Measurements of Temperature and Concentration Fluctuations in Turbulent Diffusion Flames Using Pulsed Raman Spectroscopy, in Eighteenth Symposium (International) on Combustion (The Combustion Institute 1981) p. 1521

21. Drake, M. C., Pitz, R. W., Lapp, M., Fenimore, C. P., Lucht, R. P., Sweeney, D. W., Larendou, N. M.: Measurements of superequilibrium hydroxyl fluorescence. Paper presented at the 1983 Fall Meeting of the Western States Section of the Combustion Institute.
22. La Rue, J. C., Libby, P. A., Seshadri, D. V. R. (1981): Further results on the thermal mixing layer downstream of a turbulence grid. Phys. Fluids **24**, 1927
23. Mudford, N. R., Bilger, R. W.: Examination of Closure Models for Mean Chemical Reaction Rate Using Experimental Results for an Isothermal Turbulent Reacting Flow, in *Twentieth Symposium (International) on Combustion* (The Combustion Institute, to appear)
24. Pope, S. B. (1981): A Monte Carlo method for the pdf equations of turbulent reactive flow. Combust. Sci. Techn. **25**, 159
25. Bilger, R. W.: *Turbulent Reacting Flows*, Topics Appl. Phys., Vol. 44 (Springer, Berlin, Heidelberg, New York 1980) p. 74

Simultaneous Measurement of Velocity and Scalars in a Turbulent Nonpremixed Flame by Combined-Laser Doppler Velocimetry and Laser Raman Scattering

R. W. Dibble and R. W. Schefer

Sandia National Laboratories, Livermore, CA 94550, USA

Abstract

This paper describes a new combined laser Doppler velocimeter-laser Raman scattering apparatus which simultaneously measures velocity and scalars. Measurements of simultaneous velocity and scalars in a ducted nonpremixed turbulent flame are presented. Measurements presented include the turbulent momentum transport and both conventional and Favre averages of variables.

Introduction

Analytical and numerical solutions of the time-dependent Navier-Stokes equations for turbulent fluid flow are not in general possible and such solutions are improbable for the forseeable future. In lieu of direct solution of the Navier-Stokes equations, the most common approach is numerical solution of the time-averaged Navier-Stokes equations. It is not possible to solve the time-averaged equations without relying on empiricism to model the new correlation terms that emerge as a consequence of time-averaging the nonlinear terms. This empirical approach has been most successful for constant-density, nonreacting flows for which the largest set of experimental observations exists. The extension of these submodels to turbulent combustion presents two additional problems: 1) how to incorporate the variable density aspect of the flow into the model and 2) how to model the chemical reactions (which have an exponential temperature dependence) when the temporal and spatial fluctuations of temperature are comparable to the mean temperature. Progress in the modeling of turbulent combustion will be accelerated by obtaining measurements which address these two problems. Measurements in turbulent reacting flows have been fewer and of greater uncertainty than measurements in nonreacting flows; this is due to the hostile environments associated with turbulent combustion which make conventional measurements difficult, if not impossible. The emergence of laser-based diagnostics in the last decade has improved the prospects of making the necessary measurements in turbulent reactive flows. This paper presents a combined laser Doppler velocimeter (LDV)-laser Raman scattering system which simultaneously measures velocity, temperature, and species concentration. The paper also presents recent measurements obtained in a turbulent nonpremixed flame. These new simultaneous measurements allow direct comparison of the aforementioned correlation terms for which an empirical model must be postulated. The data reduction scheme presented here emphasizes obtaining velocity and density measurements in an adiabatic, chemical equilibrium flow with large variation in the density. From the simultaneous measurements of density with velocity, the turbulent momentum transport as well as the density-weighted (Favre) velocity are deduced. A closing comment is made on the possibility of using the apparatus in nonadiabatic and nonequilibrium flows.

Experimental

All measurements were made in the Sandia Turbulent Diffusion Flame Facility, an induced-draft, horizontal wind tunnel with an axisymmetric fuel jet located at the upstream end of a test section. A fully windowed test section has a 30 cm-square cross section and is 200 cm long. The fuel nozzle inside diameter *D* is 0.53 cm. The bulk velocity of the fuel mixture (22 mole precent argon in hydrogen) is 154 m/s, and the velocity of the coflowing air stream is 8.5 m/s. The jet Reynolds number, based on the pipe inside diameter and bulk fuel velocity, is 24,000. The use of this fuel mixture allows direct comparison to earlier LDV-Rayleigh measurements in the same facility (Driscoll et al., 1983).

Velocity-scalar correlations are obtained from simiultaneous measurements of the velocity and the associated scalar. In this work, the simultaneous measurements are made with a combined LDV-laser Raman scattering apparatus. This appratus (Fig. 1) has recently been discussed by Dibble et al., (1983) so only the salient features will be addressed here. An LDV event triggers the pulsed dye laser. Species concentration and temperature are determined from measurements of laser Raman scattering from gases in the probe volume.

The LDV was a commercially available forward-scattering system. Velocimeter laser beams (488 nm, 4 Watt) were focused in an optical volume having a diameter of 0.5 mm and a length of 2.0 mm. Through this optical probe volume we focused light from a flashlamp-pumped dye laser (1 J/pulse, 1 pulse/s, 2 μs pulsewidth, $\lambda = 514.5$ nm, $\Delta\lambda = 0.1$ nm), which was located 20 m from the test facility in an adjacent room. Raman-scattered light was collected at right angles by a 30 cm-focal-length, *f*/2 collection lens and was relayed at X3 magnification to the entrance slit of a 3/4 m grating spectrometer. The entrance slit width determined the length of the Raman probe volume (1 mm) while the height of the probe volume was determined by the laser beam diameter (less than 1 mm). Since the slit height was 6 mm, the measured Raman signal was not sensitive to small movement of the laser beam, which could result from beam steering by the turbulent flame (Starner and Bilger, 1980). Five photomultplier tubes were positioned at the exit plane of the spectrometer to receive the Stokes-shifted vibrational Raman scattered light from N_2, O_2, H_2, H_2O, and the anti-Stokes vibrational Raman scattering from N_2. In addition, elastically scattered (Rayleigh and Mie) laser light was measured. Laser pulse energy was measured with a vacuum photodiode which received the attenuated laser light after it had passed through the test

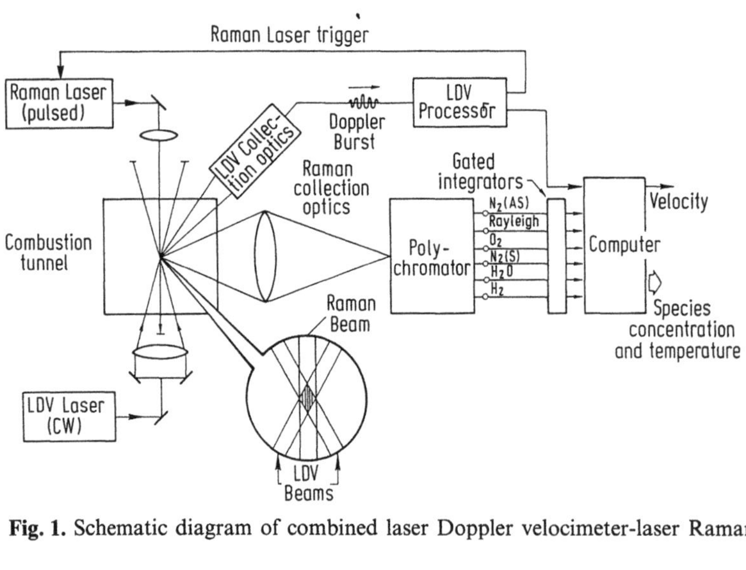

Fig. 1. Schematic diagram of combined laser Doppler velocimeter-laser Raman scattering apparatus

section. Electrical outputs from the phototubes were connected to a twelve-channel charge integrator (CAMAC), which was gated open for the duration of the laser pulse. As an indication of the overall efficiency of the Raman collection system, 3000 photoelectrons per Joule of laser light were collected from vibrational Raman scattering of N_2 in room air.

Data Reduction

The use of laser Raman scattering for thermometry and concentration measurement has been reviewed by several authors (Lapp and Penney, 1973; Ledermann, 1977; Eckbreth, 1981; and Rahn et al., 1981). In brief, the integrated charge Q_i from a given photomultiplier tube is linearly related to laser energy Q_{laser} and species concentration $[N_i]$ as follows:

$$Q_i = k_i Q_{laser} [N_i] f_i(T). \tag{1}$$

The proportionality constant k_i is dependent on the vibrational Raman cross section and optical collection efficiency. System calibration to determine k_i was accomplished using

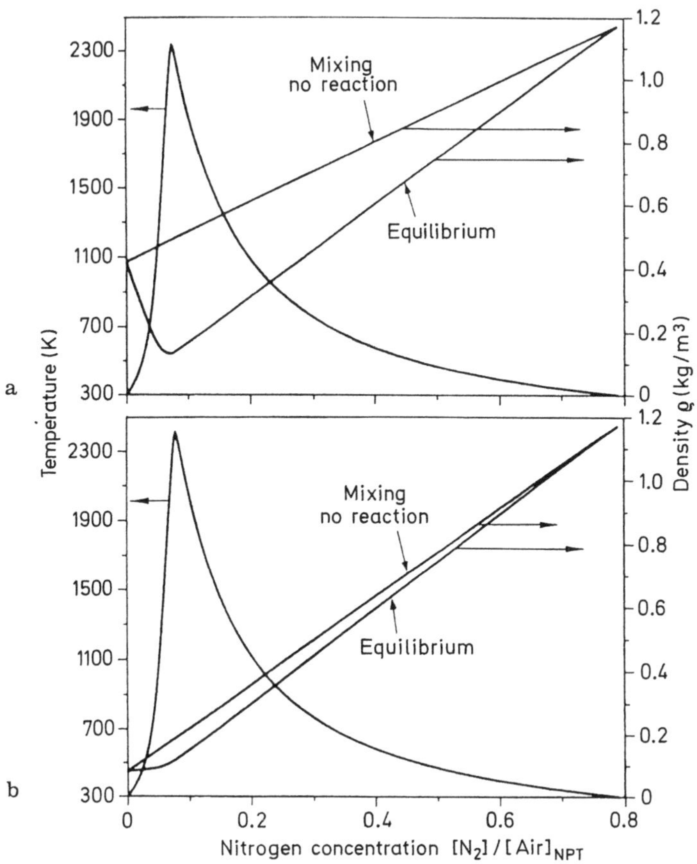

Fig. 2 (a). Relationship between density or temperature and nitrogen concentration of fuel-air mixtures. The fuel is 22 mole precent argon in hydrogen. The concentration of nitrogen has been normalized by the concentration of air at normal pressure and temperature; **(b)** Same as Fig. 2 (a) except fuel is pure hydrogen vibrational Raman scattering from gas samples at room temperature and pressure. The bandwidth factor $f_i(T)$ accounts for the temperature-dependent distribution of molecules in their allowed quantum states. Also convolved into the bandwidth factor $f_i(T)$ are the spectral location, shape, and bandwidth of the spectrometer and the bandwidth of the laser. The bandwidth factor can be calculated for N_2, O_2, and H_2; however, the calculation is less reliable for triatomic species such as H_2O (e.g. Stephenson, 1981). After calibration, a measurement of species concentration requires that laser pulse energy, Raman signal, and temperature be measured. The need for a temperature measurement can be eliminated by careful attention to the parameters that affect the bandwidth factor as was done by Eckbreth (1979) and Smith (1980). Under these circumstances, Smith (1980) showed that the bandwidth factor can be made equal to unity with an uncertainty of +/− 3% over the entire temperature range of interest here. The 3% uncertainty may be further reduced by relating the temperature to another measured quantity such as the nitrogen concentration; i.e., $f_{N_2}(T) = f_{N_2}(T([N_2]))$. Such a specification is possible when local chemical equilibrium and adiabaticity are assumed. These assumptions were previously examined by Dibble et al., (1983) and found to be acceptable for this flame. With these assumptions, there exists a unique relationship between the nitrogen concentration and the density, as illustrated in Fig. 2a for the present fuel mixture and as illustrated in Fig. 2b for the case of pure hydrogen fuel. Figure 2b will be discussed below.

Results and Discussion

A typical experimental run yields a data matrix with 160 rows (laser shots) and 8 columns: axial velocity (u), Q_{N_2}, Q_{O_2}, Q_{H_2}, Q_{H_2O}, $Q_{N_2 - \text{anti-Stokes}}$, Q_{Rayleigh}, and Q_{laser}. For the purpose of measuring temperature and density in this flow, it suffices to measure Q_{N_2} and Q_{laser} and then to infer the density and temperature through Eq. 1 and Fig. 2a.

Figure 3 shows scatter plots of velocity and density fluctuations u' and ϱ'. This data was collected in the middle of the shear layer at radial location $y/D = 4$ and at axial location $x/D = 50$ which is near the mean location of the flame tip. The fluctuations were determined by subtracting the mean velocity $\bar{u}$ and mean density $\bar{\varrho}$ from each simultaneously measured velocity-density pair $\varrho - u$. Each point represents a single LDV-Raman event. Scatter in the data is due to the fluctuations, which are fundamental to turbulence. The correlation of these points $\overline{\varrho' u'}$ is directly related to turbulent transport. (The analogous scatter plot for single molecules in a slowly moving multicomponent mixture would appear much like Fig. 3a, correspondingly, the net mass flux would be the correlation of the velocity with density.) Data presented in Figs. 3a, b, and c were collected in the same manner except that the time delay Δt between the LDV event and the Raman laser pulse was increased from $\Delta t = 40$ µs in Fig. 3a to $\Delta t = 2000$ µs in Fig. 3b to $\Delta t > 1$ s in Fig. 3c. In comparison to Fig. 3a, the data points in Fig. 3b are more randomly distributed, and, correspondingly, the correlation $\overline{\varrho' u'}$ is reduced. In Fig. 3c, the time delay is sufficiently large so that the correlation is, within statistical error, zero. By systematically varying the time delay, an integral time scale τ_ϱ is determined (Tennekes and Lumley, 1972). The results of such an experiment are presented in Fig. 4. From Fig. 4, an integral time scale τ_ϱ of ~ 1.5 ms is determined. The product $\tau_\varrho \bar{u}$ is an integral length scale $\lambda_\varrho = 2$ cm. Since the integral length scale is large compared to the largest dimension of the probe volume (1 mm), the spatial resolution of the LDV-Raman probe is adequate for the integral scale. However, the probe dimension is six times larger than the Kolmogoroff microscale (0.15 mm) calculated for the measurement location and conditions of Fig. 3 (Goulard et al., 1976). The spatial averaging is not expected to be significant since the amplitude of the microscale fluctuations is expected to be small.

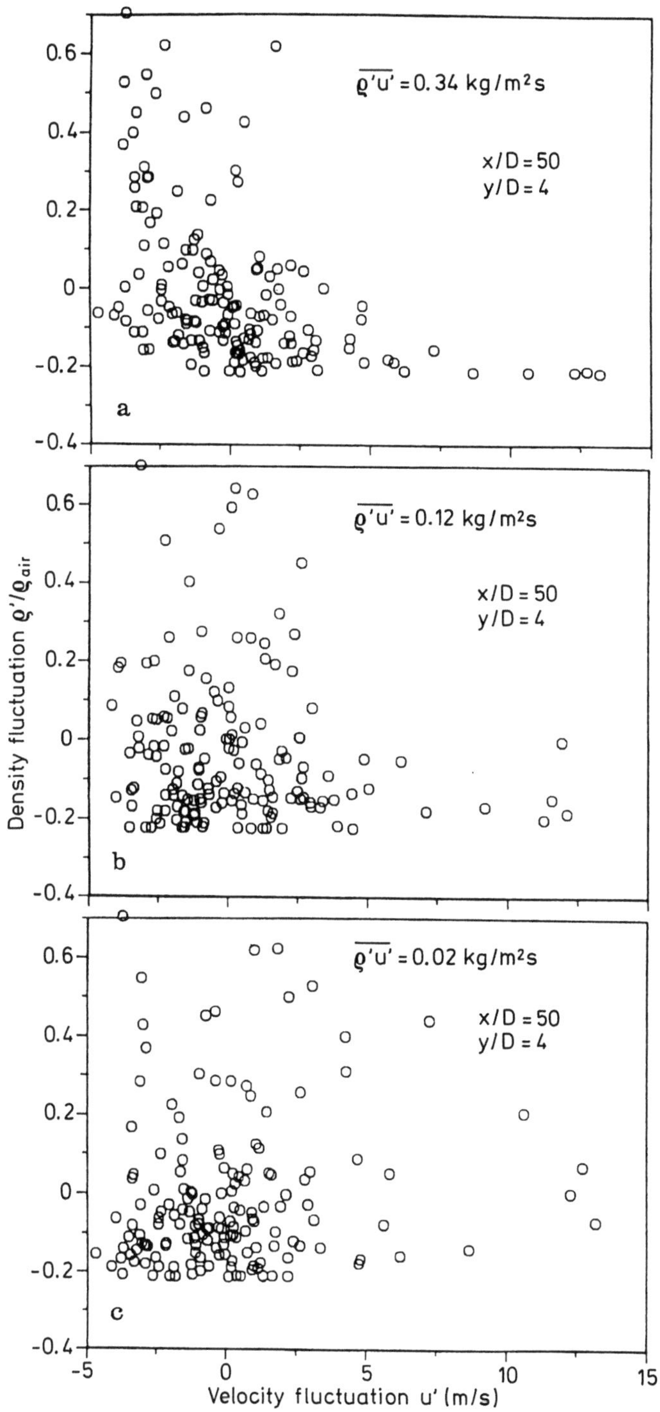

Fig. 3. Scatter plot of fluctuating density ϱ' and velocity u' pairs. In Fig. 3(a), (b), and (c), the time delay Δt between the LDV event and the laser Raman pulse has been varied as follows: (a) $\Delta t = 40\ \mu s$, (b) $\Delta t = 2000\ \mu s$, and (c) $\Delta t > 1$ s. As the time delay Δt increases, the correlation $\overline{\varrho' u'}$ decreases

Figure 4 shows the correlation approaching zero as the time delay is increased. Zero correlation means the velocity is statistically independent of the scalar, which is expected at large time delays. Interestingly enough, small correlations are obtained when the time delay is too small ($\tau_\varrho \leqq 20\ \mu s$). In this case, the laser pulse collides with the LDV particle with an attendant bright Mie scattering that competes with or overwhelms the Raman signals in spite of the large rejection, 10^{12}, of the spectrometer. When this Mie scattering event occurs, the 'Raman' signals are always large, independent of the velocity, and hence, the correlation is zero. These Mie scattering events are clearly descernable by the unusually large signal in the Rayleigh scattering channel; when this event occurs, the Raman data are rejected. As the

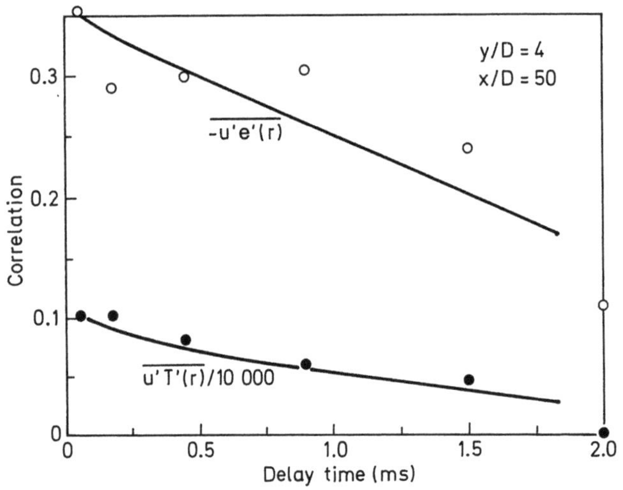

Fig. 4. The measured temporal velocity-scalar correlation for density $\overline{u'\varrho'(\tau)}$ and temperature $\overline{u'T'(\tau)}$ are plotted as a function of radius at $x/D = 50$. n.b. to facilitate plotting, the velocity-temperature correlation $\overline{u'T'(\tau)}$ has been divided by 10000

particulate number density is increased, a larger fraction of the Raman data is rejected. In the work reported here, the inlet air and fuel were filtered and subsequently LDV seed particles were sparingly added until the LDV event rate was one per second, which is commensurate with the maximum pulse rate of the dye laser. In this respect, the practice of single-pulse Raman scattering is inconvienced by particulates nearly as much is the practice of Rayleigh scattering (e.g. Arcoumanis et al., 1984). However, unlike Rayleigh scattering, Raman scattering is more tolerent of laser light scattering and reflections from test section windows and from other surfaces.

It has been suggested that notions gained from modeling constant density flows can be extended to the analogous variable density flows through the use of density-weighted (Favre) variables (Libby and Williams, 1980; Jones and Whitelaw, 1982). In addition to providing direct measurement of the turbulent momentum transport, the correlation $\overline{\varrho' u'}$ can be used to generate the Favre mean velocity $\tilde{u}$ via the identity

$$\tilde{u} = \bar{u}\left(1 + \frac{\overline{\varrho' u'}}{\bar{\varrho}\bar{u}}\right). \tag{2}$$

Radial profiles of the mean velocity and the Favre mean velocity are presented in Fig. 5a. The close agreement between the two velocities is a consequence of the small value of the correlation $\overline{\varrho' u'}$ relative to the product of the mean density and the mean velocity $\bar{\varrho}\bar{u}$. Correspondingly, the turbulent transport of axial momentum, which is equal to the density-

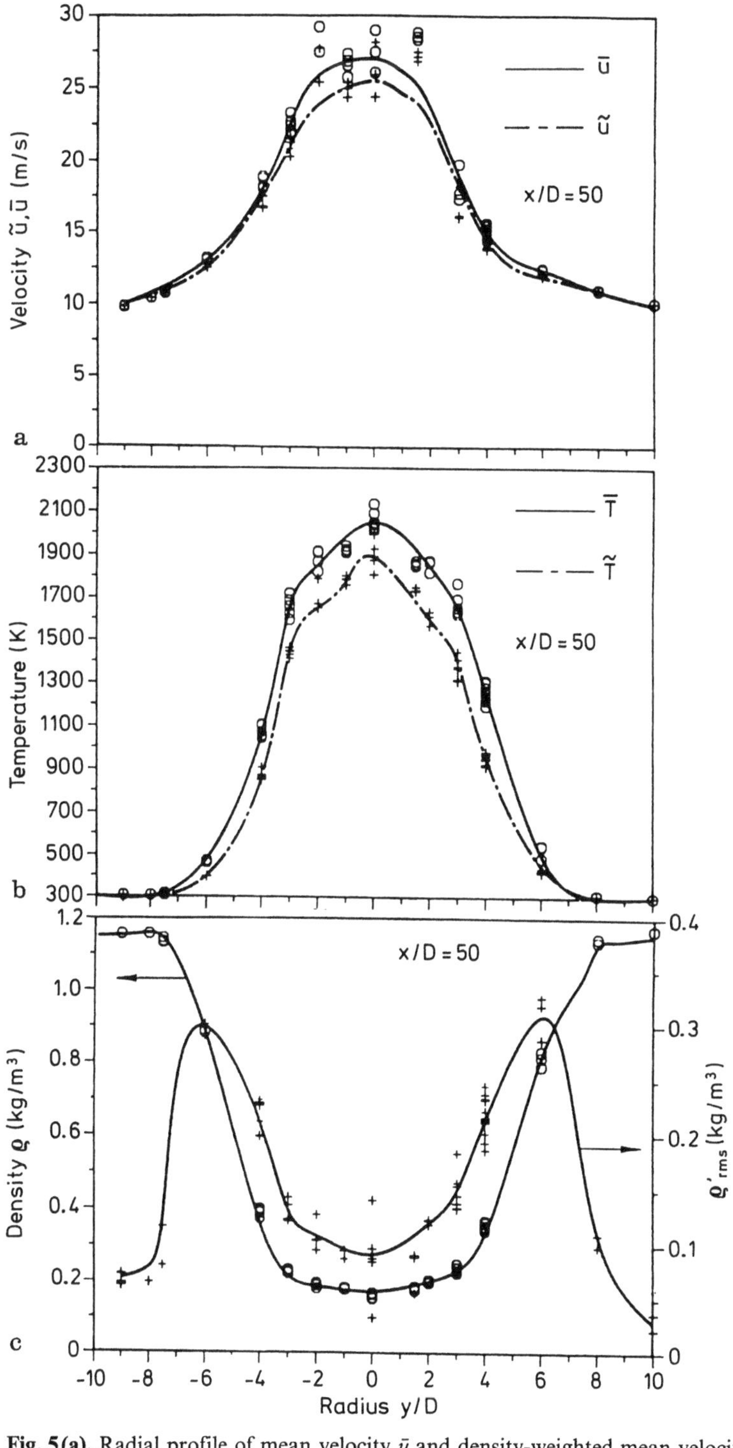

Fig. 5(a). Radial profile of mean velocity $\bar{u}$ and density-weighted mean velocity $\tilde{u}$. **(b)** Radial profiles of mean temperature $\bar{T}$ and density-weighted mean temperature $\tilde{T}$. **(c)** Radial profiles of mean density $\bar{\varrho}$ and density fluctuation rms ϱ'_{rms}

velocity correlation $\overline{\varrho' u'}$, is negligible when compared to the convective transport $\bar{\varrho}\bar{u}$. Hence, overall model predictions for this flow system are insensitive to the modeling of this term. In particular, this explains why a model which incorporates a gradient diffusion for the axial turbulence transport does not produce unreasonable agreement with observations, in spite of measurements which show that the axial turbulent transport is often counter to the gradient (Driscoll et al., 1983). The Favre average of other scalars can be generated using equations analogous to Eq. (2). The radial profiles of mean and Favre mean temperature are presented in Fig. 5 b. The difference between the two means implies a negative value for the density-temperature correlation $\overline{\varrho' T'}$, which might have been anticipated *a priori* by noting the inverse relationship between temperature and density in Fig. 2 a.

The difference between the mean and the Favre mean is relatively insensitive to the position inside the shear layer (see Fig. 5 a and b). Plotted in Fig. 5 c are the radial profiles of mean density $\bar{\varrho}$ and rms of density fluctuation ϱ'_{rms}. When density fluctuations are zero the mean and the Favre mean should be the same, as they are outside the shear layer (i.e. $y/D = 8$ or $y/D < -8$). Therefore, one might expect that since the rms of the density fluctuations go through a minimum at the centerline, the difference between the mean and Favre mean (Fig. 5 a and b) should do so as well. Such expectations overlook the corresponding minimum at the centerline of the mean density which has a compensating effect. It is this compensating effect that causes Favre averaged quantities to appear much like their constant density flow analogs.

Comments on the Potential Application to Heat Transfer Research: In both turbulent combustion and turbulent heat transfer, variable density flows are accompanied by heat transfer without mass transfer and, possibly, large departures from chemical equilibrium. In order to investigate this class of flows with the LDV-Raman apparatus (Fig. 1) it will be necessary to relax the assumptions of adiabatically and chemical equilibrium. A limiting case of chemical nonequilibrium occurs when there is mixing without chemical reaction. In this case, the relationship between density and nitrogen concentration is represented by a straight line which (see above Fig. 2 a) connects fuel density with air density. This pure-mixing line is coincident with the line which results from the assumption of any amount of chemical reaction, assuming nitrogen to be inert, but neglects chemical heat release. Between the mixing line and the adiabatic equilibrium line is the set of allowed states obtainable by the inclusion or removal of chemical energy, i.e., the relationship between density and nitrogen concentration is bounded by the pure mixing line and by the adiabatic chemical equilibrium line. Thus, there is an uncertainty in assigning a density to a given measurement of nitrogen concentration. The uncertainty is largest when the nitrogen concentration is equal to 0.08; at this concentration the density may be between 0.16 kg/m^3 and 0.51 kg/m^3. The use of hydrocarbon fuels, which have molecular weights greater than present argon-in-hydrogen fuel, would exacerbate the uncertainty problem. Since a relationship between the nitrogen concentration and the density no longer exists, it becomes necessary to simultaneously measure the temperature and the density of all of the major species. Such an undertaking is, at best, difficult, particularly if the more advanced laser diagnostics such as CARS or inverse Raman (Rahn et al., 1981) are to be used. However, an alternative does exist. One can note that the uncertainty in the relationship between density and nitrogen concentration would not exist if the fuel had zero mass. As Fig. 2 b illustrates, the ideal of a massless fuel is reasonably approximated when pure hydrogen is used as fuel. When the fuel is hydrogen, the conclusion is that the measurement of the nitrogen concentration will allow the density to be inferred with small uncertainty; in particular, the LDV-Raman apparatus presented in Fig. 1 could be used to investigate variable density flows which have both heat transfer and departures from chemical equilibrium.

Summary

This paper describes a new LDV-laser Raman system with which one can simultaneously measure velocity, density, and species concentration. Simultaneous measurements in a turbulent nonpremixed flame are reported. From the simultaneous measurements, one can compute the density-velocity correlation which is the turbulent momentum transport. In addition, this correlation and the density-temperature correlation are used to compute the density weighted (Favre) mean of the axial velocity $\tilde{u}$ and temperature $\tilde{T}$. Comparison between the unweighted means and density-weighted means show the density-weighted mean to be smaller than the unweighted mean by no more than 10%. The difference is not a strong function of position, provided one is inside the shear layer. The potential application of this LDV-laser Raman system to investigations into turbulent heat transfer is discussed.

Acknowledgements. This work was supported by the Department of Energy, Office of Basic Energy Science, Division of Chemical Science.

References

Arcoumanis, C., Green, H. G., Whitelaw, J. H. (1984): Transactions of the SAE **55**, 307
Dibble, R. W., Kollmann, W., Schefer, R. W. (1984): Combust. Flame **55**, 307
Driscoll, J. F., Schefer, R. W., Dibble, R. W. (1983): Nineteenth Symposium (International) on Combustion (The Combustion Institute) p. 447
Eckbreth, A. C., Bonczyk, P. A., Verdieck, J. F. (1979): Prog. Energy Combust. Sci. **5**, 253
Eckbreth, A. C. (1981): Eighteenth Symposium (International) on Combustion (The Combustion Institute) p. 1471
Goulard, R., Mellor, A. M., Bilger, R. W. (1976): Comb. Sci. Tech. **14**, 195
Jones, W. P., Whitelaw, J. H. (1982): Combust. Flame **48**, 1
Lapp, M, Penney, C. M.: Laser Raman gas diagnostics (Plenum, New York 1973)
Ledermann, S. (1977): Prog. Energy Combust. Sci. **3**, 1
Libby, P. A., Williams, F. A.: Turbulent reacting flows, ed. by P. A. Libby, F. A. Williams (Springer, New York 1980) p. 65
Rahn, L. A., Mattern, P. L., Farrow, R. A. (1981): Eighteenth Symposium (International) on Combustion (The Combustion Institute) p. 1533
Smith, J. R. (1980): Transactions of the SAE **89**, Paper No. 800137, 808
Stephenson, D. A. (1981): Appl. Spectrosc. **35**(6), 582
Tennekes, H., Lumley, J. L. (1974): A first course in turbulence. Third Printing (MIT Press, Cambridge, MA)

Assessment of Two Turbulence Models for Turbulent Round Diffusion Jets with Combustion

F. C. Lockwood and P. Stolakis

Imperial College of Science and Technology, London SW7, England

Abstract

The performances of the $k - \varepsilon$ model of turbulence and an algebraic stress model have been compared (for the case of an axisymmetric gaseous jet diffusion flame). The effect of radiation losses is shown to be significant. The implications of neglecting these losses – which was the general practice up to the present day – in combustion calculations are explored and discussed.

Nomenclature

a_g	acceleration of gravity
$a_{g,n}$	grey gas weighting coefficient
$b_{1,n}, b_{2,n}$	constants in grey gas expressions
c	soot concentration (kg/m^3)
c_p	specific heat
E	activation energy
f	mixture fraction
g	mean square mixture fraction fluctuations
h	enthalpy
H	calorific value of the fuel
k	turbulence energy/gas absorption coefficient
L	geometric path length
m	mass fraction
n	grey gas index for gas/soot
P	production of turbulence energy
p	partial pressure/probability density function
q'''_{rad}	radiated energy per unit volume
R	perfect gas constant
r_s	stoichiometric oxygen requirement to burn 1 kg of soot
S	source term
S_{fu}	stoichiometric air requirement to burn 1 kg of fuel
T	mean temperature
U	mean streamwise velocity
u	fluctuating part of U
V	mean cross-stream velocity
v	fluctuating part of V
w	fluctuating part of the velocity normal to the x, y-plane
x	streamwise direction
y	cross-stream direction
β	volumetric expansion coefficient
δ	width of the jet
ΔU_m	maximum velocity difference across the jet
ε	turbulence energy dissipation
ε_g	emissivity of the gas/soot mixture
μ_t	effective viscosity
$\xi = \left\lvert \dfrac{\delta}{\Delta U_m}\left(\dfrac{\partial U_{\mathrm{¢}}}{\partial x} - \left\lvert\dfrac{\partial U_{\mathrm{¢}}}{\partial x}\right\rvert\right)\right\rvert^{0.2}$	(Rodi [15])
ϱ	density
σ	Prandtl or Schmidt number, or Stefan Boltzman constant
ϕ	a scalar quantity

Subscripts

ad	adiabatic reaction
c	CO_2
co	CO
¢	centre-line
d	destruction
f	formation
fu	fuel
g	gas
j	species j in mixture
ox	oxygen
rad	radiation
s	soot
w	water

Superscripts

	fluctuating part

Introduction

There have been relatively few attempts to predict the turbulent round jet diffusion flame using modern computational procedures. In part this has been due to a scarcity of suitable data for validation purposes, although this problem has now largely been eliminated due to the perseverance of a handful of experimentalists working in the field, see for example [1 to 5]. It is also partly because there is no possibility of succeeding in the task in the absence of a good turbulence model and turbulence modellers have been hard enough pressed to devise reliable simulations of non-reacting jet flows.

The available turbulence models range from zero- to multi-equation ones, an orderly description of them is presented by Launder and Spalding [6]. The two-equation $k - \varepsilon$ model was originally applied in flow computations by Jones and Launder [7] and Launder and Spalding [8] and has since become widely used in engineering calculations. It appears to combine an optimal mix of computer economy and predictive precision. It is least satisfactory in the presence of body forces such as those due to swirl or buoyancy both of which are often significant in practical combustors. In the former case the isotropic effective viscosity implied by the $k - \varepsilon$ model is less than satisfactory, while in the latter case the striking discrepancies between horizontal and vertical buoyant shear layers are for example not well simulated by the simple $k - \varepsilon$ model [14].

Some workers, for example Launder et al. [9] and Gibson and Launder [10] have examined multi-equation Reynolds stress turbulence models with some small advantage. However, the extra computational burden is very significant and has not so far been superimposed on the already heavily charged background of combustion calculations. In an effort to enhance significantly the validity of turbulence models researchers have explored diverse avenues such as the formulation of multiple time scale equations (e.g. [11]). The reader is referred to a 'state of the art' review by Gibson et al. [12].

In sum no new turbulence model has emerged which affords more than a marginal improvement in predictive ability and this is always at the expense of greatly increased computational costs. There is no doubt that industry would pay dearly for absolutely trustworthy and precise predictions but they cannot as yet be so satisfied. Fortunately combustor predictions of useful engineering accuracy are often procurable at a correspondingly acceptable expense with the simple $k - \varepsilon$ model (e.g. [13]). To some extent this is because other processes such as the thermal radiation often exercise a dominating influence and an accurate representation of the turbulence is inessential.

The Turbulence Models

The $k - \varepsilon$ Model

The parabolic and axisymmetric forms of the equations for the turbulence energy, k, and turbulence energy dissipation, ε, are:

$$\varrho U \frac{\partial k}{\partial x} + \varrho V \frac{\partial k}{\partial y} = \frac{1}{y} \frac{\partial}{\partial y} \left(y \frac{\mu_t}{\sigma_k} \frac{\partial k}{\partial y} \right) + P - \varrho \varepsilon + \beta \varrho a_g \overline{u T'} \tag{1}$$

$$\varrho U \frac{\partial \varepsilon}{\partial x} + \varrho V \frac{\partial \varepsilon}{\partial y} = \frac{1}{y} \frac{\partial}{\partial y} \left(y \frac{\mu_t}{\sigma_\varepsilon} \frac{\partial \varepsilon}{\partial y} \right) + \frac{\varepsilon}{k} (C_{\varepsilon_1} P - C_{\varepsilon_2} \varrho \varepsilon) + C_{\varepsilon_3} \frac{\varepsilon}{k} \beta \varrho a_g \overline{u T'} \tag{2}$$

where the production term is given by:

$$P = -\varrho\left(\overline{uv}\frac{\partial U}{\partial y} + \overline{u^2}\frac{\partial U}{\partial x} + \overline{v^2}\frac{\partial V}{\partial y} + \overline{w^2}\frac{v}{y}\right).$$

From dimensional considerations the turbulence viscosity is:

$$\mu_t = \varrho C_\mu \frac{k^2}{\varepsilon}. \tag{3}$$

Presuming analogy with laminar flow the turbulence stresses may be expressed as [13]:

$$-\varrho\overline{uv} = \mu_t \frac{\partial U}{\partial y} \tag{4}$$

$$-\varrho\overline{u^2} = -\frac{2}{3}\varrho k + 2\mu_t \frac{\partial U}{\partial x} \tag{5}$$

$$-\varrho\overline{v^2} = -\frac{2}{3}\varrho k + 2\mu_t \frac{\partial V}{\partial y} \tag{6}$$

$$-\varrho\overline{w^2} = -\frac{2}{3}\varrho k + 2\mu_t \frac{V}{y}. \tag{7}$$

When the flow is reacting an additional turbulence equation for the mean square of the mixture fraction fluctuations, $g \equiv \overline{(f - \bar{f})^2}$, is required:

$$\varrho U \frac{\partial g}{\partial x} + \varrho V \frac{\partial g}{\partial y} = \frac{1}{y}\frac{\partial}{\partial y}\left(y \frac{\mu_t}{\sigma_g}\frac{\partial g}{\partial y}\right) + C_{g_1}\mu_t\left(\frac{\partial f}{\partial y}\right)^2 - C_{g_2}\varrho\frac{\varepsilon}{k}g. \tag{8}$$

The preceeding equations constitute the usual $k - \varepsilon$ model, or more properly the $k - \varepsilon - g$ model, for reacting parabolic flows. They are solved along with those for continuity, U-momentum conservation, energy conservation and mixture fraction, f, conservation. The last buoyancy production term in Eq. (1) is exact for small temperature variations and can be modelled by a simple gradient approximation. The corresponding term of Eq. (2) is a modelled one proposed by Hossain and Rodi [14]. We have found that these terms do not improve the predictions in a straightforward application of the $k - \varepsilon - g$ model and they are omitted for the predictions shown. All correlations involving density fluctuations are ignored so the system is equivalent to a Favre-averaged one. The values of the model constants are given in Table 1.

Table 1. Values of constants for $k - \varepsilon$ model

C_μ	C_{ε_1}	C_{ε_2}	C_{g_1}	C_{g_2}	σ_k	σ_ε	σ_f	σ_g
$0.09-0.04\,\xi$	1.44	$1.92-0.0667\,\xi$	2.8	2.0	1.0	1.3	0.7	0.7

The Algebraic Stress Model

The present algebraic relations for the Reynolds stresses are extracted by assuming local equilibrium of turbulence production and dissipation [16]. The relations which have been derived for non-reacting flows are [14, 16]:

$$-\overline{uv} = \frac{1 - C_2}{C_1} \frac{\overline{v^2}}{k} \left[1 + \frac{\beta a_g}{C_{1T}} \frac{k}{\varepsilon} \frac{\partial T/\partial y}{\partial U/\partial y} \right] \frac{k^2}{\varepsilon} \frac{\partial U}{\partial y} \tag{9}$$

$$\frac{\overline{u^2}}{k} = 0.94 \tag{10}$$

$$\frac{\overline{v^2}}{k} = \frac{2}{3} \left[1 - \frac{1 - C_2}{C_1} \right] = 0.53 \tag{11}$$

$$\frac{\overline{w^2}}{k} = 0.53 \tag{12}$$

$$-\overline{vT'} = \frac{1}{C_{1T}} \frac{\overline{v^2}}{k} \frac{k^2}{\varepsilon} \frac{\partial T}{\partial y} \tag{13}$$

$$\overline{uT'} = \frac{1}{C_{1T}} \frac{k}{\varepsilon} \left[\overline{uv} \frac{\partial T}{\partial y} + (1 - C_{2T}) \overline{vT'} \frac{\partial U}{\partial y} - \beta a_g (1 - C_{2T}) \overline{T'^2} \right]. \tag{14}$$

The constants in Eqs. (10) to (12) for the normal stresses are empirical and, strictly, pertain only to non-combusting flows [16]. The equation for the temperature fluctuations is solved in its full form since to do so makes no additional modelling demands. It is [12]:

$$\varrho U \frac{\partial \overline{T'^2}}{\partial x} + \varrho V \frac{\partial \overline{T'^2}}{\partial y} = \frac{1}{y} \frac{\partial}{\partial y} \left[y C'_g \varrho \frac{k}{\varepsilon} \left(\overline{v^2} \frac{\partial \overline{T'^2}}{\partial y} + \overline{uv} \frac{\partial \overline{T'^2}}{\partial x} \right) \right] - 2 \varrho \overline{vT'} \frac{\partial T}{\partial y} - \frac{2}{C'_T} \varrho \frac{\varepsilon}{k} \overline{T'^2}. \tag{15}$$

The values of additional constants needed for the algebraic stress model are given in Table 2 [14].

Table 2. Values of additional constants pertaining to algebraic stress model

C_1	C_2	C_{1T}	C_{2T}	C'_T	$C_{\varepsilon 3}$	C'_g
2.2	0.55	3.2	0.5	1.0	1.1	0.11

The Combustion Model

The formulation embodied by the 'simple chemically reacting system' concept, as originally used by Pun and Spalding [17], was adopted. This model has formed the basis of all the attempts to predict the turbulent diffusion flame by modern computational methods. The underlying assumptions are: a single step global reaction which excludes intermediate species, a chemical kinetic rate which is fast compared to the turbulence mixing rate, equality of the turbulent mass diffusion coefficients, and low Mach number.

The statistical description of non-linear functions of the mean flow parameters is assisted by the use of a probability density function (pdf) the shape of which may be calculated (e.g. [18, 19]) or assumed. Since the pdf transport equation is little explored and its computation is as yet too expensive for engineering purposes we adopt the latter course. Two assumed forms for the pdf of mixture fraction are commonly used, the clipped Gaussian function [20] and the Beta function [21]. We have explored both and found that there is little to choose between them with the latter exhibiting perhaps some marginal predictive superiority.

Mean values of all properties ϕ, which are non-linear functions of f are determined from:

$$\bar{\phi} = \int_0^1 \phi(f)\, p(f)\, df \tag{16}$$

were $p(f)$ is the assumed (Favre-averaged) pdf. In the case of the Beta function assumption:

$$p(f) = \frac{f^{a-1}(1-f)^{b-1}}{\int_0^1 f^{a-1}(1-f)^{b-1}\, df}, \qquad 0 < f < 1 \tag{17}$$

with

$$a = \bar{f}\left[\frac{\bar{f}(1-\bar{f})}{g} - 1\right], \quad b = \frac{(1-\bar{f})}{\bar{f}}\, a. \tag{18}$$

The temperature, for example, may be related to the mixture composition and enthalpy by:

$$h = \int_T m_j c_{p,j}\, dT + m_{fu} H \tag{19}$$

where $c_{p,j}$'s are determined according to Tribus [22].

Since a fast reaction is presumed it might be thought more appropriate to compute $\bar{T}$ from equilibrium chemistry. Experience reveals [23] that this is not satisfactory in the fuel-rich region where, for example, CO is significantly overpredicted. It seems likely that this fault can probably be eliminated by the imposition of flammability limits but this remedy is insufficiently explored.

Previous workers have always assumed that the laboratory scale free diffusion flame is effectively adiabatic in which case the enthalpy is, for the above specified reaction model, a linear function of f. In the present work we allow radiation losses to the surroundings. The appropriate functional relation between h and f is then not evident, and this too in an under-explored subject. We presume that $h_{rad} = h_{ad}$ at $f = 0$ and $f = 1$, $h_{rad} = h_{ad}$ – (the radiation 'source') at $f = \bar{f}$, and that the relations between these three points are linear. The prescription cannot be greatly in error.

The Radiation Model

Although it has almost always been argued that the radiation loss is probably rather small its neglect is nonetheless, as we shall see, to the detriment of model validation studies especially since the available data are mainly for temperature.

Since the experimental flames have all been situated in cool, open laboratories the transfer of radiative energy from the surroundings to the flame may be ignored. And since these laboratory flames are optically rather thin one should also, without too much error, be able to neglect self-absorption. The flame is then a simple emitter and the rate of heat emission per unit volume is:

$$\dot{q}'''_{rad} = 4 k_g \sigma T^4 \tag{20}$$

where σ is the Stefan Boltzman constant and k_g is the grey gas absorption coefficient. Strictly the emission behaviour of the emitting gases and the soot is non-grey but this complication can be circumvented by imaging the real gas and soot to be replaced by a mixture of grey gases plus one clear gas [24]. This formulation has been adopted by Truelove [25] who presumes a mixture of two grey gases, a clear gas, and soot, and provides a correlation for the total emittance of the form:

$$\varepsilon_g = \sum_n a_{g,n}(T)\,[1 - \exp(-k_{g,n}(p_w + p_c + p_{co}) - k_{s,n}c - k_{CH_4,n}\,p_{CH_4})] \tag{21}$$

where

$$a_{g,n}(T) = b_{1,n} + b_{2,n}T. \tag{22}$$

The absorption coefficient may be deduced from a pseudo-grey relation:

$$\varepsilon_g = (1 - e^{-k_g L}). \tag{23}$$

The Soot Model

The luminosity of laboratory flames is usually rather small, all the same soot is highly emitting and some account of the presence must be given if radiation transfer is not neglected. We solve a balance equation for the soot mass fraction:

$$\varrho U \frac{\partial m_s}{\partial x} + \varrho V \frac{\partial m_s}{\partial y} = \frac{1}{y}\frac{\partial}{\partial y}\left(y \frac{\mu_t}{\sigma_s}\frac{\partial m_s}{\partial y}\right) + S_s \tag{24}$$

where $S_s \equiv S_f - S_d$, $S_f \equiv$ soot formation rate and $S_d \equiv$ its destruction rate. The Schmidt member σ_s is set equal to 0.7.

The chemical kinetics of soot formation is a tremendously complex subject.

We choose a very simple expression which exhibits the right behavioural features. It has its origins in diesel engine studies and has been used with some success in steady flow combustors by Abbas et al. [26]:

$$S_f = C_f\,p_{fu}\,\phi^n \exp(-E/RT) \tag{25}$$

where p_{fu} is the fuel pressure, ϕ is the equivalence ratio $= S_{fu}(\bar{f}/1 - \bar{f})$, $S_{fu} =$ the fuel stoichiometric air requirement, the parameters E/R, n and C_f are respectively assigned the values 20140, 3, and 0.01 as a result of validation studies [26].

The soot particles are so small (typical size is of the order of 1 μm in spray flames [27] and in gaseous ones, as the present case, much smaller) that their consumption rate will be determined by laws applicable to a gas species. We adopt the simple expression of Magnussen et al. [28] who, by adopting the conventional assumption to the effect that the reaction rates are controlled by turbulence decay, link the soot burn out rate to the product of its mean concentration and the turbulence time scale ε/k:

$$S_d = \min\left[Am_s\frac{\varepsilon}{k}, \frac{Am_{ox}}{r_s}\frac{\varepsilon}{k}\frac{m_s r_s}{m_s r_s + m_{fu} S_{fu}}\right] \tag{26}$$

where A is an adjustable constant assigned [28] the value of 4. The second term restricts the soot consumption in the absence of sufficient oxygen.

The Numerical Solution Procedure

The governing differential equations were solved using a parabolic version [29] of the Imperial College TEACH code [30]. The code features a grid which is caused to expand with the flow. The expansion is accommodated by appropriate alterations of the cell face convection and diffusion coefficients. The non-orthogonality of the grid is small and the orthogonal forms of the governing differential equations are retained with little error [31]. Because substantial changes in density and temperature may occur between adjacent streamwise stations iteration is employed. For the first iteration the V-velocities at the downstream station are set equal to the corresponding upstream ones. Subsequently: (i) the downstream U-velocities are determined, (ii) new downstream V's are obtained from the continuity equation, (iii) the remaining downstream dependent variables are solved for along with the density and other ancillary information, and (iv) the sequence returns to (i) until a convergence criterion is satisfied. Note that a pressure correction equation is not solved, it is expedient and sufficiently precise to ignore the small cross-stream pressure variations.

Comparison of Predictions with Data

There now exist a good number of data sets for laboratory turbulent diffusion flames. The scope of this study is arbitrarily restricted to four experimental hydrocarbon flames. We have considered two flames of Lockwood and Moneib [1], both for methane fuel but one has the fuel stream mass diluted by 42 per cent by volume of nitrogen. We have included the German natural gas flame of Wittmer and Günther [4] and Lenz and Günther [3]. And we have considered the town gas flame of Lockwood and Odidi [5]. All of these flows had Reynolds numbers in excess of 15,000. The various flames studied are summarized in Table 3.

Table 3. Experimental flame conditions

Flame	Fuel volumetric composition	Burner diameter (mm)	Re
Methane flame [1]	93.63% CH_4 5.54% hydrocarbons 0.83% inerts	7.74	15,000
Nitrogen diluted methane flame [1]	As above but diluted with 42% by volume N_2	7.74	15,000
Methane flame of [3, 4]	81.1% CH_4 15.3% N_2 0.7% CO_2 2.9% hydrocarbons	8.00	37,000
Town gas flame [5]	55% H_2 27% CH_4 4% CO 2% C_2H_6 8% CO_2 4% N_2	5.00	24,000

Mean Velocities

Mean velocity data for the Lockwood and Moneib methane flames are reported in [2]. Velocity data are also reported by Wittmer and Günther [4]. Predictions are compared with

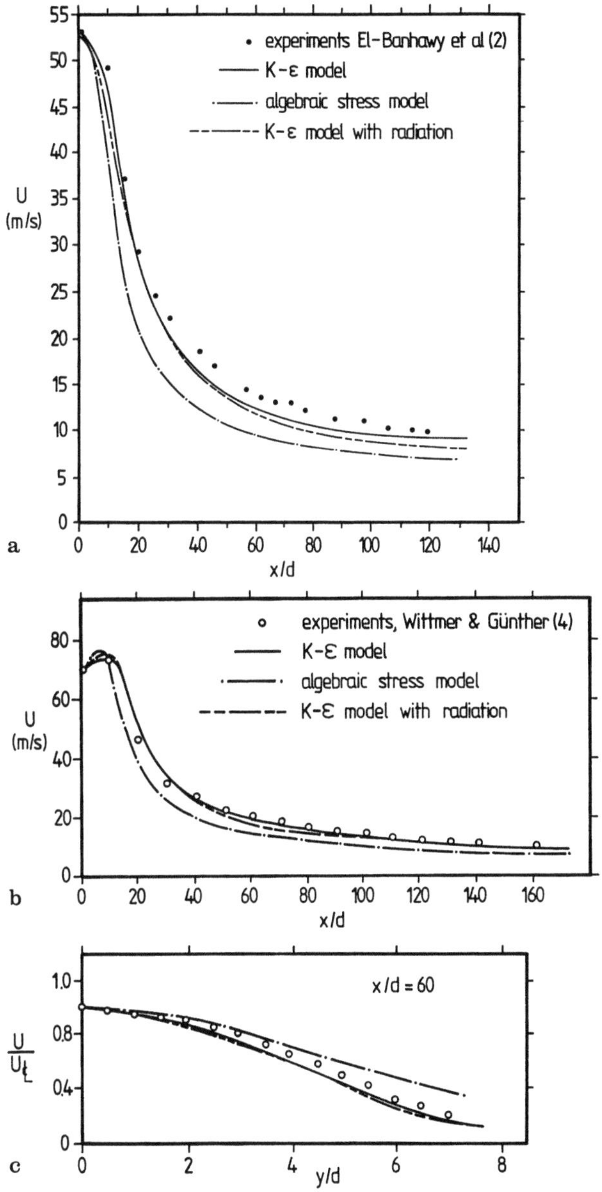

Fig. 1. Axial and radial profiles of mean velocities. (a) Methane flame [2]. (b, c) Flame of [4]

these data in panels (a) to (c) of Fig. 1. The predicted jet spread of the algebraic stress model is excessive. Tamanini [32] mentions a similar discrepancy although he shows no predictions. The $k - \varepsilon$ predictions are superior, and the effect of radiation on the velocity is evidently small.

The marked superiority of the $k - \varepsilon$ model is surprising. In common with other workers (e.g. [32]) we have found only marginal differences between the two models, which favour the algebraic stress model, used here, for inert flows. We might expect the algebraic stress formulation to exhibit more than marginal superiority for flames since the buoyancy influences should be better modelled. That it does not is presumably due to inconsistencies in the derivation of the present algebraic stress model when substantial density fluctuations occur.

We are currently rederiving the model for the variable density case. It is of interest to note that when the last (buoyancy generation) terms are dropped from Eqs. (1) and (2) the discrepancies in the algebraic stress predictions are made slightly worse.

As mentioned the last term of the k equation is an example of one for which the effect of the large density variations are incorrectly simulated since the use of β in the term implies small temperature fluctuations. The unapproximated term is $a_g \overline{u\varrho'}$. It would be of interest to retain the correct term and solve a modelled equation for the correlation $\overline{u\varrho'}$. We have so far simply expressed it as $-\varrho \overline{u\varrho'} = (\mu_t/\sigma_\varrho)\, \partial\varrho/\partial x$. The predictions are little affected by this 'improvement'. We also note that the dissipation of the $\overline{u T'}$ equation assumes a uniform fluid specific heat and thermal conductivity but in view of the preceeding observation we would expect this to be of small consequence. We conclude that the possible inconsistencies associated with the fluctuating density and the fluctuating pressure (see Bilger [33] and Jones [34]) arise in the simplifications used to procure the algebraic stress relations.

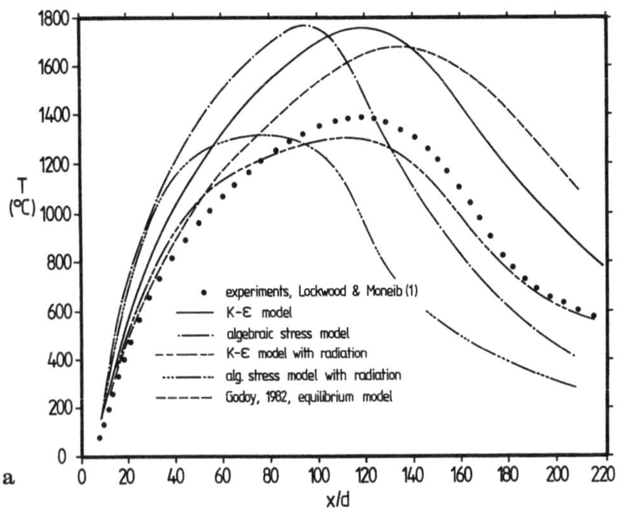

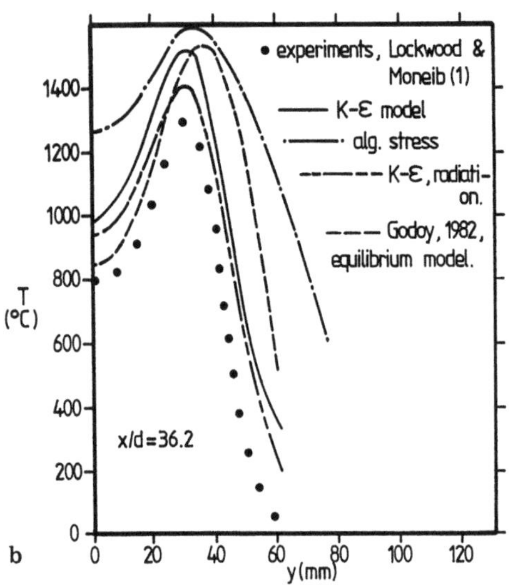

Fig. 2. Mean temperature distributions for methane flame of [1]. (a) Along symmetry axis. (b) Radial profile at $x/d = 36.2$

Mean Temperature

Mean temperature data are available for all five flames; they are compared with the predictions in the several panels of Figs. 2 to 5. The influence of the thermal radiation loss is significant, rather less so for the cleaner burning nitrogen diluted (Fig. 3) and town gas (Fig. 5) flames. The poor performance of the algebraic stress model, in particular its overprediction of the spreading rate, is seen again here.

It has mainly been assumed in previous studies that the radiation loss from these laboratory flames is small, and this caused us to believe that the present crude model would be more than sufficient. It is clear that a more elaborate radiation simulation is called for. Better simple models exist which are quite easy to incorporate [35]. We may expect the inclusion of self-absorption will raise the predicted temperatures and so enhance the already rather satisfactory performance of the $k-\varepsilon$ model.

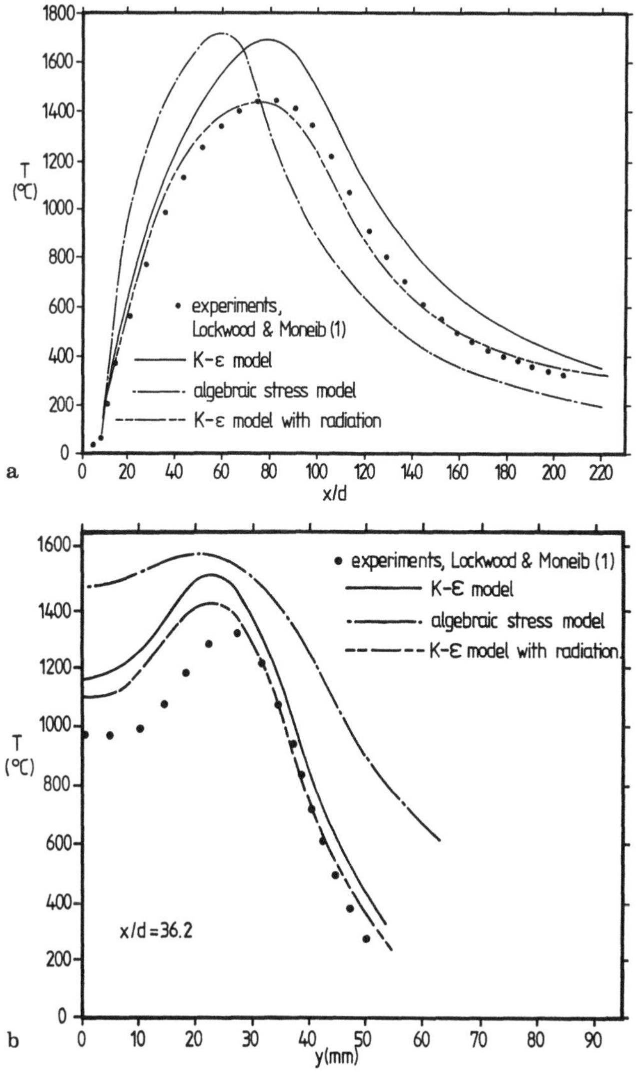

Fig. 3. Mean temperature distributions for nitrogen diluted flame of [1]. (a) Along symmetry axis. (b) Radial profile at $x/d = 36.2$

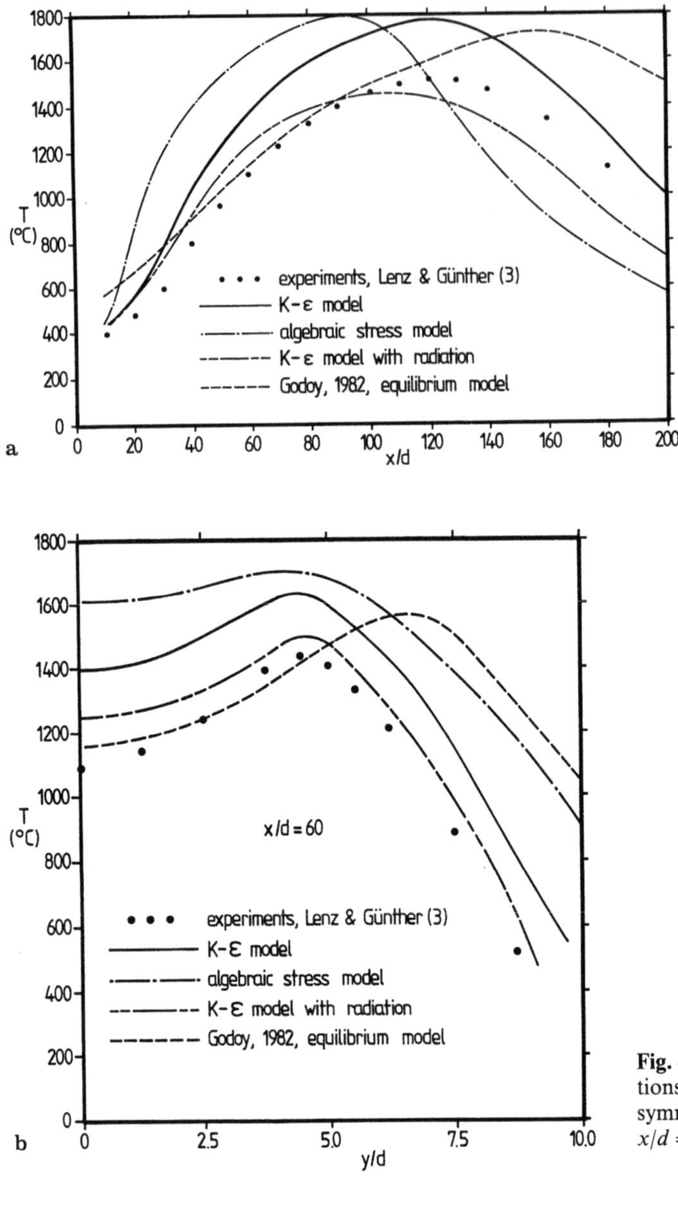

Fig. 4. Mean temperature distributions for flame of [3]. (a) Along symmetry axis. (b) Radial profile at $x/d = 60$

Godoy [36] has presented $k - \varepsilon$ predictions for the undiluted flame of [1] based on the presumption of full chemical equilibrium chemistry (Figs. 2(a), (b) and 4(a), (b)). Although radiation was neglected it is clear that the on-axis temperature peak would be somewhat too far downstream were the results to be corrected for the radiation loss. The reason is that in the fuel rich regions the chemical equilibrium supposition substantially overpredicts the CO concentration levels, delaying burn out. Combustion models based on equilibrium chemistry would appear to require the introduction of an extinction component based on the flammability limits. At this point it should be noted that Liew et al. [23] appear to obtain satisfactory concentration predictions by straightforwardly applying the concentration versus mixture fraction functions determined from laminar flame experiments.

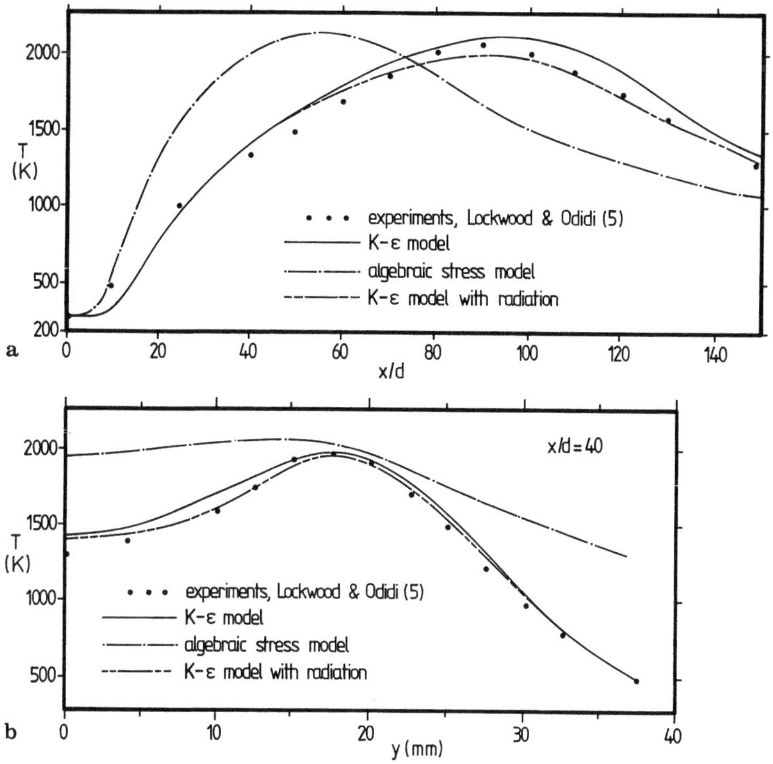

Fig. 5. Mean temperature distributions for flame of [5]. (a) Along symmetry axis. (b) Radial profile at $x/d = 40$

Fluctuating Velocity

A few data for the fluctuating velocity components exist and these are shown in Figs. 6 and 7 along with the corresponding predictions of the algebraic stress formulation. On the whole the agreement is surprisingly good. In general the level of turbulence is somewhat overpredicted, with the values of $(\overline{u^2})^{1/2}$ being least in error. The algebraic stress model overpredicts the jet spread with significant adverse consequences for the temperature predictions. The predicted $\overline{uv}$ correlation which has higher values than the measured ones, for the Wittmer and Günther flame [4] as shown in Fig. 6(b), is responsible for this discrepancy.

Some $k-\varepsilon$ predictions of $\overline{uv}$ are shown in the same Fig. 6(b) and it is obvious that they are in much better agreement with the data. Radiation losses have only a small effect on the fluctuating velocity calculations.

Fluctuating Temperature

Somewhat more fluctuating temperature data are available than for fluctuating velocities (Figs. 8, 9, 10). The on-axis r.m.s. temperature is generally underpredicted in the upstream region. The poorer performance pattern of the present algebraic stress predictions re-emerges here. Allowing for the thermal radiation both increases and decreases the magnitude of $(\overline{T'^2})^{1/2}$ but, on the whole its effect is smaller than for mean temperature and the predictions are here not generally improved by its inclusion. The peak r.m.s. temperatures of the methane flames of [1] tend to be overpredicted by both methods (Figs. 8(b) and 9(b)) and

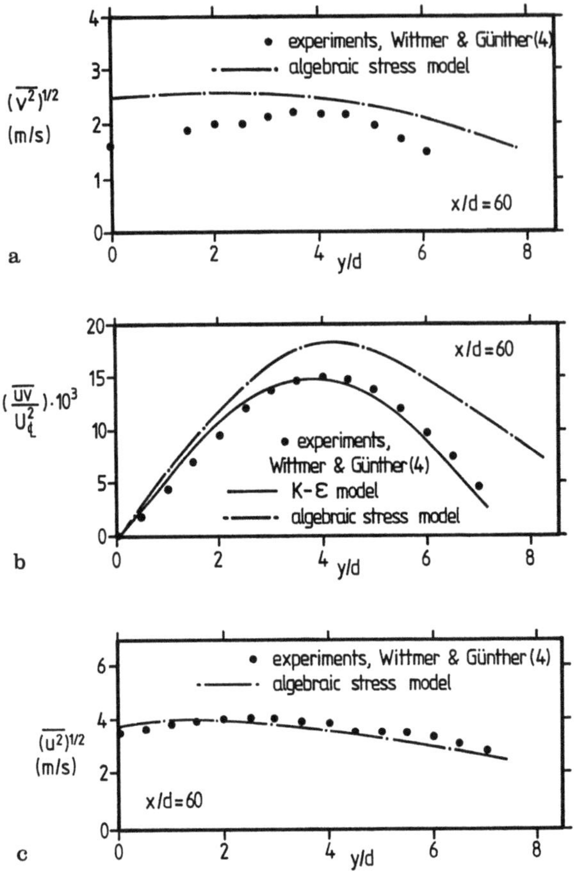

Fig. 6. Axial and radial distributions of normal and shear stresses for flame of [4]. (a) Radial distribution of $(\overline{v^2})^{1/2}$ at $x/d = 60$. (b) Radial distribution of $(\overline{uv})$ at $x/d = 60$. (c) Radial distribution of $(\overline{u^2})^{1/2}$ at $x/d = 60$

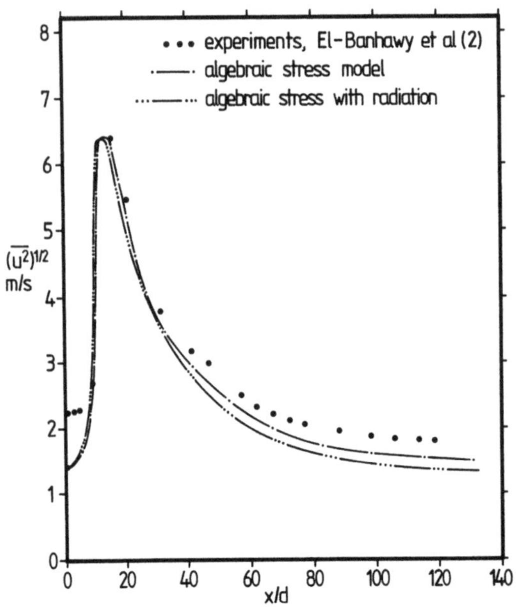

Fig. 7. Distribution of streamwise normal stress along symmetry axis for methane flame of [2]

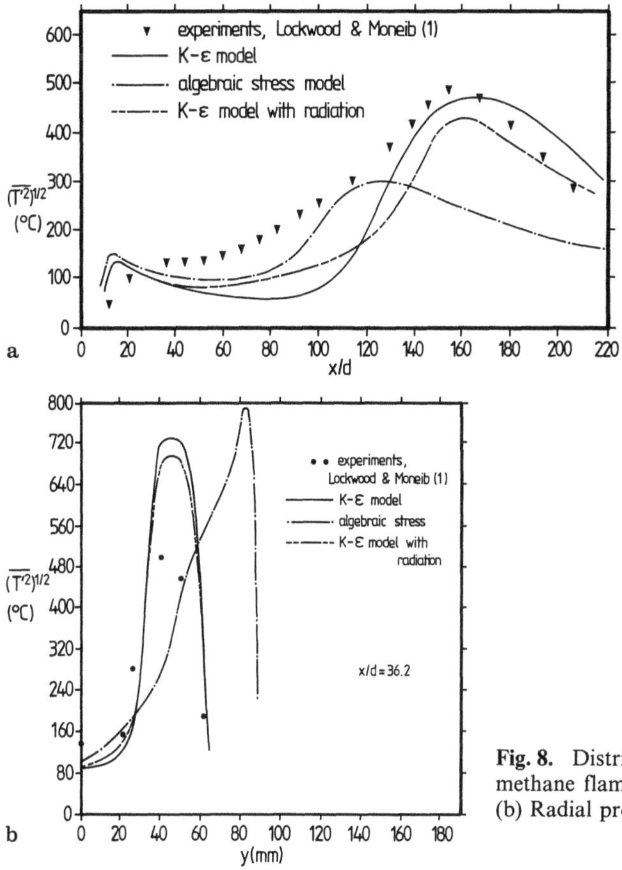

Fig. 8. Distribution of rms temperature for methane flame of [1]. (a) Along symmetry axis. (b) Radial profile at $x/d = 36.2$

this may possibly indicate insufficient compensation of the thermocouple data. The best predictions are for the nitrogen plumes. The natural gas flames of [3] may also be considered as diluted because of the high nitrogen content of the fuel.

Concluding Remarks

The most striking outcome of these straightforward validation studies has been that the role of the thermal radiation transfer is much greater than had previously been suspected. Satisfactory model validation studies can therefore not be performed for the typical laboratory turbulent jet flame if the radiation transfer is neglected.

The $k - \varepsilon$ turbulence model, combined with a standard fast reaction combustion model, is able to simulate these vertical flames fairly satisfactorily. The effect of the additional buoyancy source terms on model equations is small. These results should give heart to the mathematical modellers of combustion equipment who would prefer for reasons of economy the simple $k - \varepsilon$ model.

The present algebraic stress model performs poorly. This is not the case for inert flows of the present kind where it is only marginally superior. There are probably two reasons for this. The first is that the large density fluctuations that occur in reacting flows are not properly modelled. The second is that the pressure variation correlations that appear in the

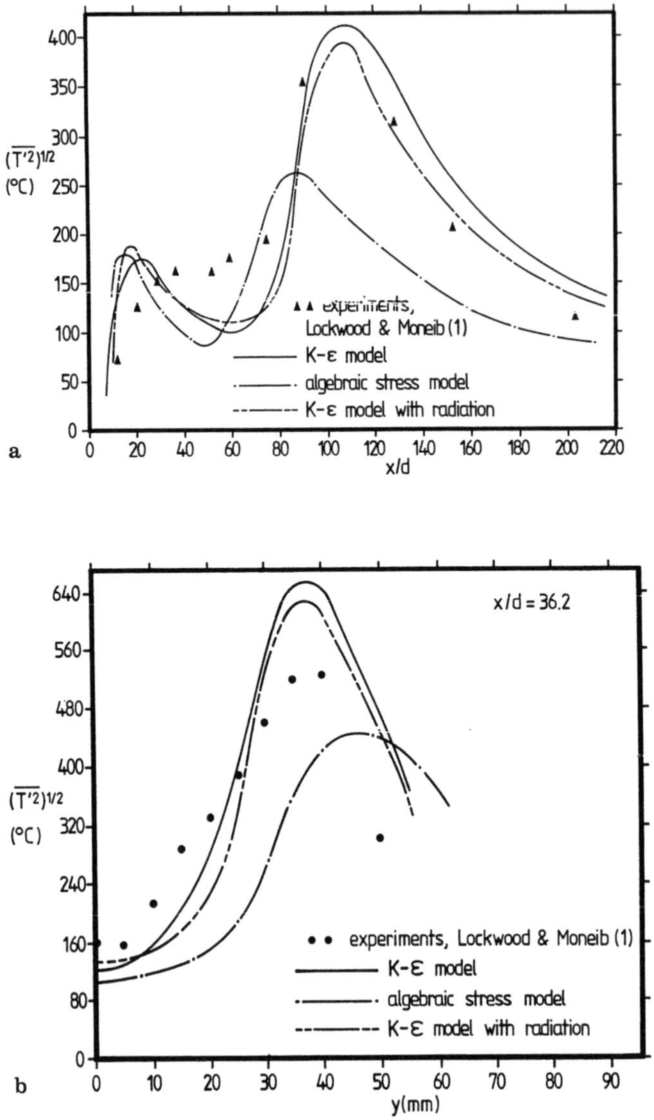

Fig. 9. Distribution of rms temperature for nitrogen diluted flame of [1]. (a) Along symmetry axis. (b) Radial profile at $x/d = 36.2$

Reynolds stress and turbulent flux equations (see [34]) may be large enough in reacting flows to influence the calculations. Bilger [33] discusses the possible effects of a similar term appearing in the turbulence energy equation. We have not yet worked out a consistent formulation of the algebraic stress expressions. We suspect that it may well prove more expedient to solve the stress equations in their full forms. Finally, we note that the time scale of the temperature fluctuations decay rate appears not to be equal to the turbulence time scale [37]. The conventional assumption in modelling the dissipation term in the $\overline{T'^2}$ equation is that the ratio of these two time scales is constant. We have assumed that this ratio (time scale of temperature fluctuations decay rate/turbulence time scale) is 0.5.

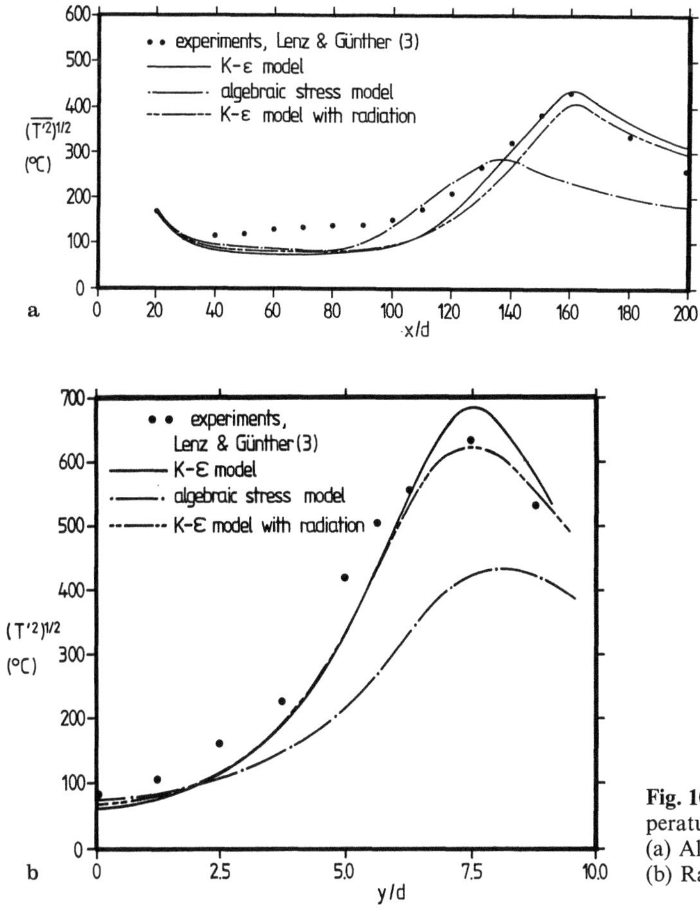

Fig. 10. Distribution of rms temperature for flame of [3].
(a) Along symmetry axis.
(b) Radial profile at $x/d = 60$

References

1. Lockwood, F. C., Moneib, H. A. (1982): Fluctuating temperature measurements in turbulent jet diffusion flames. Combust. Flame **47**, 291
2. El-Banhawy, Y., Hassan, M. A., Lockwood, F. C., Moneib, H. A.: Velocity and unburned hydrocarbon measurements in a vertical turbulent free jet diffusion flame. Combust. Flame (to appear 1983)
3. Lenz, W., Günther, R. (1980): Measurements of fluctuating temperature in a free-jet diffusion flame. Combust. Flame **37**, 63
4. Wittmer, V., Günther, R. (1980): Correlation measurements of velocity and temperature fluctuations in a free jet diffusion flame. Advisory Group for Aerospace Research and Development (AGARD) Proceedings No. 28, 10
5. Lockwood, F. C., Odidi, A. O. O. (1974): Measurement of mean and fluctuating temperature and of ion concentration in round free-jet turbulent diffusion and premixed flames. 15th Symp. Int. on Combustion (The Combustion Institute) p. 561
6. Launder, B. E., Spalding, D. B.: Mathematical models of turbulence (Academic, New York 1972)
7. Jones, W. P., Launder, B. E. (1972): The prediction of laminarization with a two-equation model of turbulence. Int. Heat Mass Transfer **15**, 301
8. Launder, B. E., Spalding, D. B. (1974): The numerical computations of turbulent flows. Comput. Meth. Appl. Mech. Eng. **3**, 269
9. Launder, B. E., Reece, G. J., Rodi, W. (1975): Progress in the development of a Reynolds-stress turbulence closure. J. Fluid Mech. **68**, 537

10. Gibson, M. M., Launder, B. E. (1978): Ground effects on pressure fluctuations in the atmospheric boundary layer. J. Fluid Mech. **86**, 491
11. Hanjalić, K., Launder, B. E., Schiestel, R.: Multiple-time-scale concepts in turbulent transport modelling. 2nd Symp. on *Turbulent Shear Flows* (Imperial College, London 1979)
12. Gibson, M. M., Jones, W. P., McGuirk, J. J., Whitelaw, J. H.: Outline lecture notes on fundamentals of turbulence and their application to calculation methods. Dept. of Mech. and Chem. Eng. (Imperial College 1980)
13. Pope, S. B., Whitelaw, J. H. (1976): The calculation of near wake flows. J. Fluid Mech. **73**, 9
14. Hossain, M. S., Rodi, W.: Influence of buoyancy on the turbulence intensities in horizontal and vertical jets, in *Turbulent Buoyant Convection*, ed. by D. B. Spalding (Hemisphere, New York 1976)
15. Rodi, W.: Turbulence models and their application in hydraulics. A state of the art review. Report SFB 80/T/127 (Universität Karlsruhe 1978)
16. Launder, B. E. (1975): On the effects of a gravitational field on the turbulent transport of heat and momentum. J. Fluid Mech. **67**, 569
17. Pun, W. M., Spalding, D. B.: A procedure for predicting the velocity and temperature distributions in a confined, steady, turbulent, gaseous diffusion flame (Proc. Int. Astronautical Federation Meeting, Belgrade 1967)
18. Janicka, J., Kolbe, W., Kollmann, W.: The solution of a pdf-transport equation for turbulent diffusion flames. Proc. of the Heat Transfer and Fluid Mechanics Institute (Stanford University Press, Stanford 1978)
19. Pope, S. B.: A Monte Carlo method for the pdf equations of turbulent flows. Dept. of Mech. Eng. (Massachussets Institute of Technology 1979)
20. Lockwood, F. C., Naguib, A. S. (1975): The prediction of the fluctuations in the properties of free round jet turbulent diffusion flames. Comb. Flame **24**, 109
21. Richardson, J. N., Howard, H. C., Smith, R. W.: The relation between sampling-tube measurements and concentration fluctuations in turbulent gas jets. Proc. 4th Int. Symp. on Combustion (The Combustion Institute 1953) p. 814
22. Tribus, M.: Thermostatics and thermodynamics (Van Nostrand, London 1961)
23. Liew, S. K., Bray, K. N. C., Moss, J. B. (1981): A flamelet model of turbulent non-premixed combustion. Comb. Sci. Tech. **27**, 69
24. Hottel, H. C., Sarofim, A. F.: Radiative transfer (McGraw-Hill, New York 1967)
25. Truelove, J. S.: A mixed grey gas model for flame radiation. United Kingdom Atomic Energy Research Establishment, Harwell, Report AERE-R8494, Dec. 1976
26. Abbas, A. S., Koussa, S. S., Lockwood, F. C.: The prediction of particle laden flows, 18th Int. Symp. on Combustion (The Combustion Institute 1980) p. 1427
27. Chigier, N. A. (1977): Instrumentation techniques for studying heterogeneous combustion. Prog. Energy Comb. Sci. **3**, 175
28. Magnussen, B. F., Hjertager, B. H.: On mathematical modelling of turbulent combustion with special emphasis on soot formation and combustion. 16th Int. Symp. on Combustion (The Combustion Institute 1976) p. 719
29. Patankar, S. V., Spalding, D. B.: Heat and mass transfer in boundary layers. 2nd ed. (Intertext Books, London 1970)
30. Gosman, A. D., Ideriah, F. J. K.: TEACH-T: a general computer program for two-dimensional, turbulent, recirculating flows (Mech. Eng. Dept. Report, Imperial College 1976)
31. Bergeles, G.: Three-dimensional discrete-hole cooling processes; an experimental and theoretical study. Ph.D. Thesis (Imperial College, University of London 1976)
32. Tamanini, F.: Reaction rates, air entrainment and radiation in turbulent fire plumes. Combust. Flame **30**, 85
33. Bilger, R. W. (1976): Turbulent jet diffusion flames. Prog. Energy Combust. Sci. **1**, 87
34. Jones, W. P.: Models for turbulent flows with variable density and combustion. Von Karman Inst. for Fluid Dynamics, Lecture Series 1979-2 Prediction Methods for Turbulent Flows (January 1979)
35. Gibb, J.: Central electricity generating board. Marchwood, England (Private Communication 1982)
36. Godoy, S.: Turbulent diffusion flames. Ph.D. Thesis (Imperial College, University of London 1982)
37. Warhaft, Z., Lumley, J. L. (1978): An experimental study of the decay of temperature fluctuations in grid generated turbulence. J. Fluid Mech. **88**, 659

Calculated Scalar Dissipation in Two-Dimensional Flows *

Wm. T. Ashurst and A. R. Kerstein

Sandia National Laboratories, Livermore, USA

E. Effelsberg and N. Peters

Institut für Allgemeine Mechanik, RWTH Aachen, Fed. Rep. of Germany

Abstract

Convection and diffusion of a conserved scalar is calculated in a two-dimensional mixing layer using the discrete vortex method. Scalar dissipation, the square of the scalar gradient, is calculated along various scalar isolines. The dissipation distribution is found to be approximately lognormal with parameters that can be described by a laminar diffusion model which has lognormally-distributed stretching of the mixing layer interface.

Introduction

For turbulent diffusion flames with fast chemistry, Bilger (1979) has formulated a conserved scalar quantity which has proven useful for predicting burning rates in jets, and which is also applicable to planar mixing layers and other geometries. These predictions involve averages over the joint distribution of the scalar and the scalar dissipation, defined as the square of the scalar gradient. The probability density function (pdf) of the scalar is readily measured, but the pdf of the dissipation and its correlation with the scalar are relatively inaccessible experimentally or theoretically. In some cases it has been assumed that these two quantities are statistically independent, an unverified assumption with significant impact on predictions.

We have directly calculated, in a stochastic fashion, the pdf of the dissipation in some model flows which have application to lift-off of turbulent jet diffusion flames. The pdf of dissipation is an input to the recent analysis by Janicka and Peters (1982) of lift-off heights. Their model, which uses assumed functional forms for the pdf's of the scalar and the dissipation in transport equations for the scalar pdf, is in good agreement with experimental lift-off data in the first ten jet diameters. When they adopt the alternative assumption of statistical independence between the scalar and its dissipation, large disagreements with experimental jet results are obtained, indicating sensitivity to assumptions concerning the correlation of the scalar and the dissipation. Since all versions of their model involve a number of assumptions, the results of our computations will aid in the validation of the assumptions concerning the pdf's of the scalar and the dissipation.

In the same analysis, Janicka and Peters found that, for the region beyond twenty diameters, the various calculations underpredict the lift-off height by almost a factor of two. Their models also do not explain the blow-off at forty diameters for a methane jet. They believe that blow-off is caused by the passage of a large coherent structure that carries the flame stabilization region downstream. Thus, they feel that blow-off predictions will require time-dependent analysis. Our computational approach can be applied to this regime, and our current calculations represent the start of such time-dependent work.

* Work partially supported by U. S. Department of Energy, Office of Basic Energy Sciences

According to Yule (1978) and Petersen (1978) the large coherent ring vortex that forms at the jet nozzle undergoes transitions to smaller three-dimensional structures within the first ten diameters of the jet (actual location depending on Reynolds number). This was determined by the loss of azimuthal correlation of velocity fluctuations. However, experimental work by Dimotakis et al. (1981) indicates that unreacted, entrained fluid may be found on the jet axis at substantial distances downstream. Their results are inconsistent with gradient diffusion models, suggesting the influence of coherent structures on flame blow-off.

At this time we have not done a full three-dimensional jet calculation. Earlier work (Ashurst, 1981) of a single vortex ring shows promise in its agreement with experiment in the number of waves formed by nonlinear instabilities. A complete jet calculation is a huge task. Therefore, we focus our efforts on a small volume of fluid and abstract from detailed calculations the statistics that can be incorporated into pdf analyses.

Method

The numerical method is the same as used by Ashurst and Barr (1981) in their conserved scalar calculations. A short description of the method will be given here. The two-dimensional velocity field is determined by a Lagrangian calculation of the motion of discrete vorticity. In the previous work the vorticity was distributed randomly in a square; in these calculations the vorticity is distributed such that it simulates a mixing layer. The discrete vortex method was calibrated with the simulation of the spatially developing mixing layer (Ashurst, 1979). In the current problem the vorticity is randomly distributed along the y axis in the middle of the square. Initially the conserved scalar Z is zero for positive x and unity for negative x. So, this small volume of fluid simulates the edge of a diffusion jet. Free boundaries are used in the x direction and periodic boundaries are used in the y direction. The mixing layer develops in time rather than in space. The total circulation is unity, which produces a velocity difference of unity across the layer. The scalar transport and diffusion are calculated with an Eulerian grid (80-by-80 mesh cells in the square) using Zalesak's flux-corrected transport scheme (1979). The numerical diffusion is ten times less than the assumed diffusion coefficient D of 0.0025 L^2/τ where L is the square edge length and τ is a reference time unit. Because of the initial random vortex locations, the time development of the nonlinear growth requires ensemble averaging. The observation time is limited by the initial conditions and by the boundary conditions. The initial conditions of a step change in the scalar value result in a rapidly changing scalar field at early times. Therefore, we must wait until transients decay before gathering statistics. The finite system size limits the large scale growth and so we must stop gathering statistics before finite-size effects predominate. The statistics thus gathered are used to calculate the scalar dissipation χ, and its pdf. We calculate χ $(= 2\,D\,(\nabla Z)^2)$ along selected scalar isolines and accumulate the quantity $l_Z/\sqrt{\chi}$ where l_Z is the arc length of the isoline in the mesh cell. This achieves the desired volumetric weighting.

Results

The averaging time period, selected according to the above criteria, is the interval between 1.3 and 1.5 τ. Figure 1 presents scalar contours at time 1.3 τ for several different initial conditions of the random number seed. The contours are drawn by using bilinear interpolation of the scalar values. The discrete vorticity (200 vortices used) is shown by its velocity vector, scaled so that the largest vortex velocity is drawn with length 0.1 L (the left-hand

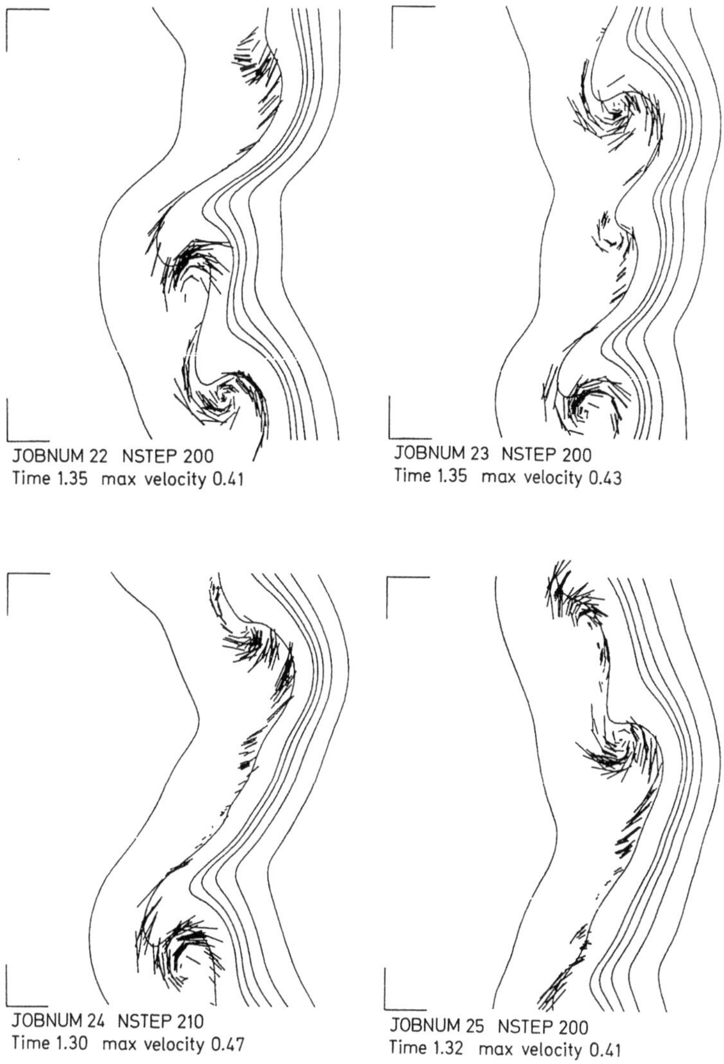

Fig. 1. Representative scalar contours and vorticity locations at time $1.3\,\tau$. The rotation is counterclockwise, so flow is down on the left side of the layer and up on the right side with a velocity difference of unity across the layer. The vorticity is shown by its velocity vectors, scaled so that the maximum has length 0.1 L. The left-hand corners of the square are drawn with lines of length 0.1 L. From left to right in each plot the scalar contour values are 0.99, 0.5, 0.2, 0.15, 0.10, 0.05 and 0.01

corners of the square are shown with lines of this length). The vorticity coalesence is apparent. From left to right, the contour lines have scalar values of 0.99, 0.5, 0.2, 0.15, 0.10, 0.05 and 0.01. Not shown in these plots are isoline values which have been calculated in increments of 0.05 between 0.2 and 0.5.

Figure 2 presents scalar dissipation pdf's for several scalar values between 0.05 and 0.50. These are ten-run averages. Notice that as the scalar value increases, *i.e.*, as one approaches the center of the mixing layer from outer edge, the distribution of χ continues to spread over a wider range. Thus, there is a significant correlation between the scalar and its gradient.

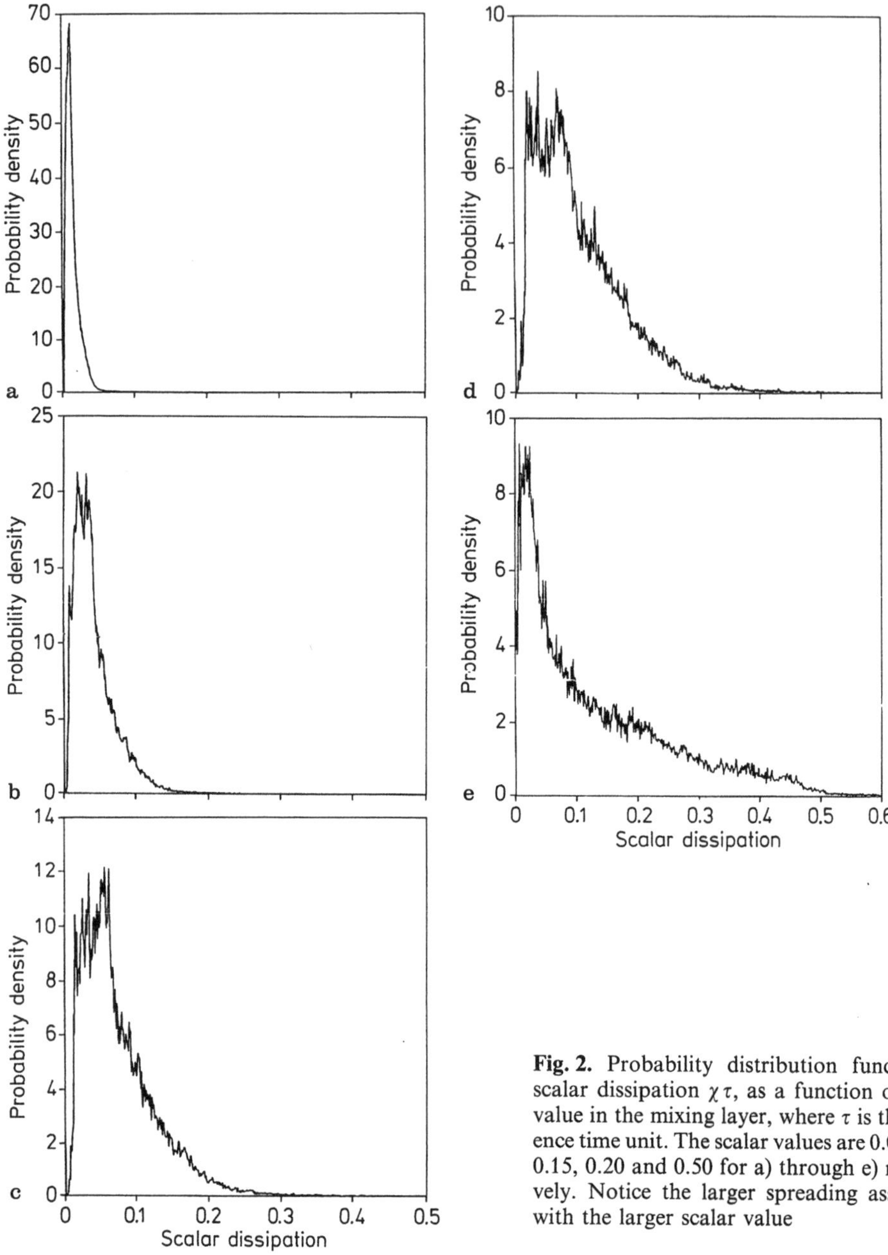

Fig. 2. Probability distribution function of scalar dissipation $\chi\tau$, as a function of scalar value in the mixing layer, where τ is the reference time unit. The scalar values are 0.05, 0.10, 0.15, 0.20 and 0.50 for a) through e) respectively. Notice the larger spreading associated with the larger scalar value

By definition, a random variable is lognormal if its logarithm is governed by the Gaussian distribution. Therefore, the cumulative distribution function of $\ln \chi$ will be a straight line one probability paper if χ is lognormal. In Figure 3, we show the data for scalar values of 0.05 and 0.5. The linearity does indicate that the dissipation is lognormal. For χ lognormal, the variance, σ^2, of the $\ln \chi$ distribution is inversely proportional to the slope of the line, and the mean value, μ, of $\ln \chi$ corresponds to the 50 percent point. Figure 3a is representative of the data for the small scalar values that we have calculated, $Z \leqq 0.3$. Figure 3b shows that the scalar of 0.5 has more departure from lognormality at large dissipation values. However,

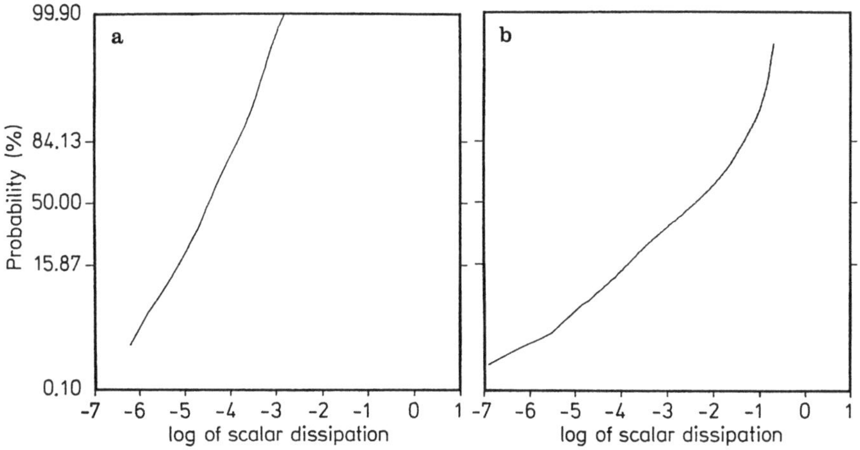

Fig. 3. Representative probability plots in which a straight line indicates a lognormal distribution, the scalar value is: a) 0.05 and b) 0.5. For low scalar values, the two ways of estimating the lognormal parameters have reasonable agreement whereas at large scalar values the estimate from the χ distribution would have a much smaller slope than the straight-line region in b)

the calculated curve is almost linear over a four *e*-fold range of χ. Table 1 presents σ^2 and μ values of the χ pdf's for each calculated scalar isoline. These lognormal parameters are determined from two estimates: 1) from the ln χ distribution, and 2) from the χ distribution using the relations $\sigma^2 = \ln(\overline{\chi^2}/\bar{\chi}^2)$ and $\mu_z = \ln \bar{\chi} - \sigma^2/2$. If the data are exactly lognormal, then the two methods would yield the same parameter values. While the mean values are approximately the same for all scalar values, the variances are quite different at large scalar value. For a scalar value of 0.5, the mean and variance values from the ln χ distribution are not very different from estimates obtained by drawing a straight line on the probability plot; 7 percent change in mean value and 24 percent in variance. Thus, we consider the data in Fig. 2 to be lognormal for low scalar values, below 0.35, but to depart from lognormality for the higher values.

Table 1. Estimated lognormal parameters for scalar dissipation

Z	From ln χ			From χ		
	σ^2	μ_Z	μ_s [a]	σ^2 [b]	μ_Z [c]	μ_s [a]
0.01	0.27	−7.30	−0.051	0.38	−7.33	−0.078
0.05	0.43	−4.51	0.039	0.35	−4.48	0.066
0.10	0.49	−3.42	0.056	0.37	−3.39	0.093
0.15	0.52	−2.85	0.062	0.38	−2.80	0.110
0.20	0.55	−2.49	0.060	0.38	−2.43	0.110
0.25	0.60	−2.25	0.041	0.38	−2.19	0.110
0.30	0.69	−2.12	−0.007	0.39	−2.03	0.082
0.35	0.86	−2.09	−0.100	0.41	−1.96	0.024
0.40	1.20	−2.19	−0.290	0.45	−2.00	−0.098
0.45	1.35	−2.35	−0.490	0.52	−2.12	−0.270
0.50	1.53	−2.57	−0.740	0.61	−2.32	−0.480

[a] $\mu_s = \mu_Z + \ln(2\pi) + 2\eta^2(Z)$, mean of ln s distribution;
[b] $\sigma^2 = \ln(\overline{\chi^2}/\bar{\chi}^2)$, variance of ln χ distribution;
[c] $\mu_Z = \ln\bar{\chi} - \sigma^2/2$, mean of ln χ distribution

In other work (Kerstein and Ashurst, 1983), we have studied the dissipation in a two-dimensional system where the vorticity is distributed in a uniform manner and the boundary conditions are periodic in both directions. Lognormality is exhibited and the calculated dissipation agrees with Kolmogorov's (1962) third hypothesis, (extended to scalar fields by Gurvich and Yaglom, 1967): σ^2 decreases as the averaging volume is increased. In those results we also find that the scalar and the dissipation are not strongly correlated.

In the inhomogeneous mixing layer, however, the scalar and the dissipation are necessarily correlated. This correlation is found to be well represented over most of the computational domain by the model presented below.

Analysis

The following analysis, which we call the lognormal stretching model, accounts for the computed results for the distribution of the scalar dissipation and for the dependence of this distribution on the scalar. We start with the one-dimensional scalar diffusion equation with no advection,

$$\frac{\partial Z}{\partial t} = D \frac{\partial^2 Z}{\partial x^2}, \tag{1}$$

and with initial conditions of constant scalar value except for a unit step discontinuity at the origin, representing our numerical initial conditions for the scalar. The time-dependent solution for the scalar is given by the complementary error function,

$$Z = \frac{1}{2} \operatorname{erfc}(\eta), \tag{2}$$

where η is the usual similarity coordinate,

$$\eta = \frac{x}{2\sqrt{D t}}. \tag{3}$$

The time-dependent solution for the scalar dissipation is

$$\chi(Z, t) = \frac{1}{2\pi t} \exp(-2\eta^2(Z)), \tag{4}$$

where the dependence of η on Z is governed by Eq. (2). With Z fixed, the scalar dissipation rate is therefore inversely proportional to time. This can be interpreted as "aging" of the mixing layer.

This solution is a highly simplified representation of transport transverse to the mixing layer interface. To generalize this one-dimensional result to the mixing layer, the gradient is modified to incorporate in a statistical manner the distortion of the interface by the flow-field. We therefore consider a mixing layer that has been distorted such that the surface at $Z = 0.5$ is given by $x = x_s(y, t)$ and we postulate that the profile can now be described by the diffusion equation in terms of the coordinate x_r relative to the distorted surface,

$x_r = x - x_s(y, t)$. Then χ is calculated as

$$\chi = \frac{1}{2\pi t} \exp(-2\eta^2) \left[1 + \left(\frac{\partial x_s}{\partial y}\right)^2\right]. \tag{5}$$

The last term in brackets describes the area increase of the mixing layer with respect to the undisturbed case. This term can be interpreted as "stretch". The notions of "aging" and "stretch" also appear in the ESCIMO model by Spalding (1979), in which they are characterized by two independent parameters. It is interesting to note (Peters, 1983) that we can combine both terms into a single stretch parameter,

$$s = \left[1 + \left(\frac{\partial x_s}{\partial y}\right)^2\right] / t. \tag{6}$$

The quantities Z, χ, η and s are now treated as random variables. Relationships among distributions involving these variables are derived based on the assumptions that η and s are statistically independent and that s is lognormally distributed:

$$P_s(s) = \frac{1}{\sqrt{2\pi}\sigma s} \exp\left[-\frac{1}{2\sigma^2}(\ln s - \mu_s)^2\right]. \tag{7}$$

(The parameters μ_s and σ^2 are the mean and variance, respectively, of $\ln s$.) The latter assumption is motivated by the fluctuating-flowfield model of Gurvich and Yaglom (1967), which predicts that scalars advected by high-Reynolds-number flows will be lognormally distributed. Finally, we assume that the functional relationship, Eq. (2), between Z and η still holds, but that χ is now given by:

$$\chi(Z) = \frac{s}{2\pi} \exp(-2\eta^2(Z)), \tag{8}$$

which follows from Eqs. (5) and (6).

Under these assumptions, we can accomplish a straightforward transformation between probability distributions based on (η, s) and those based on (χ, Z). The Jacobian of the transformation is

$$J(\eta, s) = -\frac{1}{2\pi^{3/2}} \exp(-3\eta^2). \tag{9}$$

From the definitions of marginal and conditional distributions, the conditional distribution of scalar dissipation is:

$$P_\chi(\chi \mid Z) = \frac{P_{Z\chi}(Z, \chi)}{P_Z(Z)}. \tag{10}$$

Using the transformation

$$P_{Z\chi}(Z, \chi) = \frac{P_\eta(\eta)\, P_s(s)}{|J(\eta, s)|} \tag{11}$$

and

$$P_\eta(\eta) = \frac{P_Z(Z)}{\dfrac{\partial \eta}{\partial Z}} = \frac{P_Z(Z)}{\sqrt{\pi}\exp(\eta^2)}, \tag{12}$$

we find upon substitution that the $P_Z(Z)$'s cancel and we obtain a lognormal distribution for χ conditioned on Z:

$$P_\chi(\chi|Z) = \frac{1}{\chi\sigma\sqrt{2\pi}}\exp\left[-\frac{1}{2\sigma^2}(\ln\chi - \mu_Z)^2\right], \tag{13}$$

where the Z dependence is entirely contained within the parameter

$$\mu_Z = \mu_s - \ln(2\pi) - 2\eta^2(Z). \tag{14}$$

The inverse of Eq. (2) defines the η, Z relation:

$$\eta(Z) = \mathrm{erfc}^{-1}(2Z). \tag{15}$$

The result is that the assumptions of lognormal stretch and statistical independence between the stretch and the similarity coordinate lead to a lognormal distribution of the scalar dissipation with a simple scalar dependence given by Eqs. (14) and (15). Numerical results which provide a test of the predicted scalar dependence are given in Table 1. It is found that for scalar values of 0.3 and less, the pdf's in Fig. 2 can be approximated by one lognormal distribution with $\sigma^2 = 0.38$ and $\mu_s = 0.1$, if the parameter μ_Z is calculated as given above. For scalar values greater than 0.3, μ_s is not constant. However, this is the range of scalar values for which the distribution of χ departs from lognormality as indicated, for instance, by the previously noted nonlinearity of the distribution plotted in Fig. 3b.

The change in behavior at $Z = 0.3$ is illustrated in Fig. 4, in which the four curves, from top to bottom, are plots of columns 7, 4, 6 and 3 respectively of Table 1. For $Z < 0.3$, the two methods for estimating μ_s give almost identical curves, and likewise for μ_Z. The two estimation methods will give the same results provided that the underlying distribution is lognormal, so this observation provides further evidence that χ is lognormally distributed for $Z < 0.3$. Furthermore, for Z in this range, μ_s is nearly constant, indicating that Eq. (14) correctly accounts for the Z-dependence of the conditional pdf of χ. To validate the model the constancy of σ^2 is also required. The results of Table 1 indicate that the σ^2 values are relatively constant and the two estimation methods are in relatively good agreement for $Z < 0.3$ as compared to $Z > 0.3$. The results are not as good as for μ_s, but this is not unexpected because variance estimates from data are usually less precise than estimates of the mean.

For $Z > 0.3$, Fig. 4 shows that the two estimation methods and are no longer in excellent agreement, neither for μ_s nor for μ_Z, providing further evidence of the departure from lognormality noted earlier. Furthermore, μ_s is no less Z-dependent than μ_Z for Z in this range. The model is therefore invalid in this regime. However, this limitation of the model is readily understood. The region in which $Z > 0.3$ is close to the centerline of the mixing layer. In this region, scalar isolines are highly convoluted and "zero gradient points" and "minimal gradient surfaces" of the scalar are expected to be formed (Gibson, 1968) due to vortical advection of initially uniform scalar gradients. The simple model derived above cannot be expected to explain this behavior, since the model treats the stretch as a random

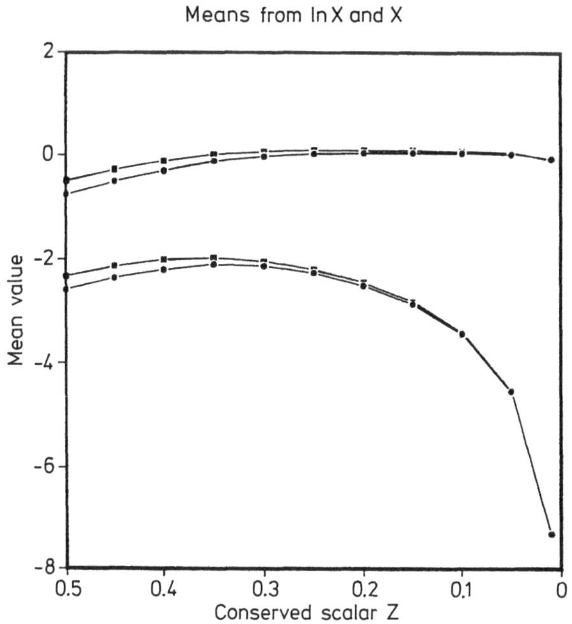

Fig. 4. Z-dependence of the estimated lognormal parameters μ_s (upper curves) and μ_Z (lower curves), each estimated by two different methods as discussed in the text

perturbation of an otherwise irrotational flowfield. However, we note that for combustion applications with hydrocarbon fuels, the range of conserved scalar values used to describe the combustion zone fall within the range of values where the lognormal stretching model is a good approximation.

There also appears to be a discrepancy in μ_s for the scalar value of 0.01. The model initial conditions assume a step discontinuity, whereas in the numerical calculation, the scalar change from zero to one was spread over two grid cells. It is possible that these initial differences are important at the leading edge of the diffusion "wave" in this transient solution. However, the more likely explanation is the lack of resolution in the calculated value. The gradient histogram is accumulated in bins whose width is independent of the scalar value. The number of nonempty bins is, for increasing scalar value: 3, 59, 179, 321, 452, 499 and the remainder greater than 500. Thus, the distribution of χ for $Z = 0.01$, discretized into only three nonempty bins, is not very precise.

Conclusions

We have found that the joint pdf of the scalar and the scalar dissipation, calculated in a two-dimensional mixing layer using the discrete vortex method, is well characterized at the edge of the layer by a simple model which we call the lognormal stretching model. The key feature of this model is the assumption that stretching transverse to a one-dimensional diffusion solution is lognormal and independent of the scalar value. For combustion applications, the scalar values of interest lie on the outer edge of the shear layer where three-dimensional effects may not be too important (depending on Reynolds number). Thus, our results can be used in the region near a jet nozzle when the diffusion flame has lifted off the nozzle. It will be interesting to perform new calculations which include chemical reaction and the resulting velocity perturbation due to chemically induced volume expansion.

References

Ashurst, W. T. (1979): Numerical Simulation of Turbulent Mixing Layers via Vortex Dynamics, in *Turbulent Shear Flows I*, ed. by F. Durst, B. E. Launder, F. W. Schmidt, J. H. Whitelaw (Springer, Berlin, Heidelberg, New York) p. 402

Ashurst, W. T. (1981): Vortex ring instability. Bull. Amer. Phys. Soc. **26**, 1267

Ashurst, W. T., Barr, P. K. (1981): Lagrangian-Eulerian Calculation of Turbulent Diffusion Flame Propagation, in *Proceedings of the Third Symposium on Turbulent Shear Flows*, University of California at Davis, pp. 3.44–3.49, Sept. 1981

Bilger, R. W. (1979): Effects of Kinetics and Mixing in Turbulent Combustion. Combustion Sci. Techn. **19**, 89; see also *Turbulent Reacting Flows*, Ed. by Libby, P. A., Williams, F. A., Topics Appl. Phys., Vol. **44** (Springer, Berlin, Heidelberg, New York 1980)

Dimotakis, P. E., Broadwell, J. E., Howard, R. D.: (1981): Experiments on chemically reacting turbulent jets. Bull. Amer. Phys. Soc. **26**, 1284

Gibson, C. H. (1968): Fine structure of scalar fields mixed by turbulence. I. Zero-gradient points and minimal gradient surfaces. Phys. Fluids **11**, 2305

Gurvich, A. S., Yaglom, A. M. (1967): Breakdown of eddies and probability distributions for small-scale turbulence. Phys. Fluids Suppl. **10**, p. S59

Janicka, J., Peters, N. (1982): Prediction of Turbulent Jet Diffusion Flame Lift-off Using a P.D.F. Transport Equation, *Nineteenth Symposium (International) on Combustion* (The Combustion Institute) p. 367

Kerstein, A. R., Ashurst, W. T. (1983): Lognormality of gradients of diffusive scalars in homogeneous, two-dimensional mixing systems. Bull. Amer. Phys. Soc. **28**, 1354

Kolmogorov, A. N. (1962): A refinement of previous hypotheses concerning the local structure of turbulence in a viscous incompressible fluid at high Reynolds number. J. Fluid Mech. **13**, 82

Peters, N. (1983): Laminar diffusion flamelet models in non-premixed turbulent combustion, to appear in *Prog. Energy Combust. Sci.*

Petersen, R. A. (1978): Influence of wave dispersion on vortex pairing in a jet. J. Fluid Mech. **89**, 469

Spalding, D. B. (1979): The Influences of Laminar Transport and Chemical Kinetics on the Time-Mean Reaction Rate in a Turbulent Flame, *Seventeenth Symposium (International) on Combustion* (The Combustion Institute) p. 432

Yule, A. J. (1978): Large-scale structure in the mixing layer of a round jet. J. Fluid Mech. **89**, 413

Zalesak, S. T. (1979): Fully multidimensional flux-corrected transport algorithms for fluids. J. Comput. Phys. **31**, 335

Nonequilibrium Chemistry in an Isothermal Turbulent Flow

N. R. Mudford and R. W. Bilger

University of Sydney, N.S.W. 2006, Australia

Abstract

The mixing of two air streams doped respectively with NO and O_3 in an isothermal turbulent flow and the resulting nonequilibrium chemical reaction have been studied using a recently developed Turbulent Smog Chamber and relatively fast chemiluminescent analysers. Joint probability density functions of reactant concentration, mixture fraction, reaction progress variable, reaction time and perturbation variable together with mean reaction rates are presented and interpreted in the light of preliminary flow characterisation results.

Introduction

Progress in the theory of turbulent reacting flow (see e.g. Libby and Williams 1980) has been hampered by the lack of experimental data with sufficient time and space resolution to elucidate the effects of turbulence and chemical kinetic time scales on the instantaneous concentrations of reactive species. While some data using laser-Raman spectroscopy (Drake et al. 1981) and fibre optic absorption probes (Batt 1977 and Sherikar and Chevray 1981) have become available the concentration requirements of these techniques are so high that the results have only been obtained where kinetic effects are small and the reaction and species concentrations are essentially mixing limited.

One chemically reacting system of particular interest is that in the atmosphere during the formation of photochemical smog. The reaction between ozone (O_3) and nitric oxide (NO) is known (Bilger 1980) to be affected by turbulent mixing and it is also probable that the rates of chemical reaction and of turbulent mixing are of the same order so that coupling between these phenomena should be evident in the instantaneous concentrations of the reacting species.

In the experiments reported here a laboratory simulation of the atmospheric reaction between ozone and nitric oxide is attempted. Large scale, slowly moving turbulence is produced in a Turbulent Smog Chamber so that the time scales of sampled species concentration fluctuations are sufficiently long (> 1 second) that adequate information can be obtained using chemiluminescent analysers with 12 Hz frequency response. Concentrations of the reactants can be varied so that a range of ratios of chemical time scale to turbulence time scales can be produced over the range of interest, i.e. where they are of the same order.

Apparatus

A detailed description of the Turbulent Smog Chamber (TSC) and associated instrumentation appears in Mudford and Bilger (1983). Briefly the component parts of the apparatus are as follows.

Turbulent Smog Chamber

The TSC consists of 0.15 mm thick polyethylene film formed into a roughly cylindrical bag of diameter 2.8 m, length 9 m and total volume 50 m^3. Two 1 m diameter, 1 $m^3 s^{-1}$ opposed air jets enter the chamber close to one end as shown in Fig. 1. The co-ordinate system used here is also shown in this figure. Turbulence and swirl in the inlet jets are minimised through the use of a perforated plate, an anti-swirl honeycomb and a woven wire screen. The flow leaves the chamber via a 0.7 m diameter butterfly valve at the far end of the chamber.

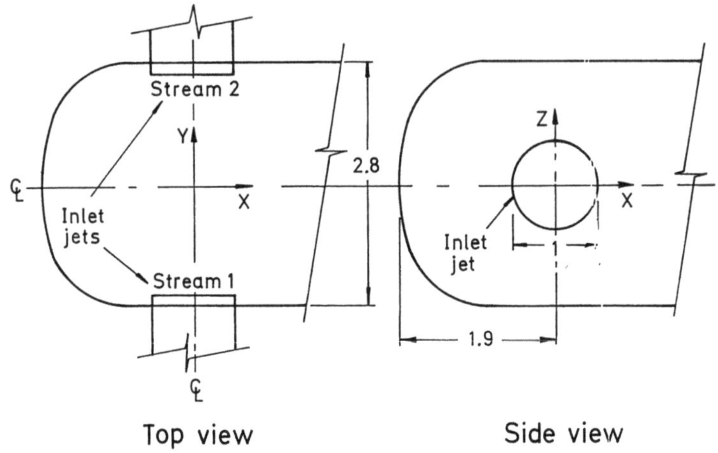

Fig. 1. Geometry and dimensions (*m*) of the inlet end of the Turbulent Smog Chamber with cartesian co-ordinate system shown. Inlet nozzle separation = 1.9 *m*

The slight reactivity of polyethylene with ozone is not considered a problem in the present experiments (Mudford and Bilger 1983). Polyethylene has the advantages of being easy to fabricate and being resistant to tearing and damage by folding.

Chemiluminecent Analysers

NO and O_3 concentrations are measured by two fast (12 Hz) high sensitivity ($\sim$ 0.4 Hz ppt^{-1}) chemiluminescent analysers which employ the chemiluminescent properties of the NO + O_3 reaction. Sample withdrawal from the TSC flow is accomplished through a 2.4 m long, 4.5 mm i.d. teflon tube with a critical metering orifice at the intake end. The presence of the sample tube in this configuration leads to negligible degradation of the frequency response which is limited by the $\sim$ 10 msec residence time in the reaction cell. Sample withdrawal rate is 220 $cm^3 s^{-1}$ at TSC conditions implying a sample volume in the TSC per cell changeover of $\sim$ 2.2 cm^3. This sample volume per cell changeover may be associated with a length scale of $\sim$ 1.3 cm which is an order of magnitude lower than the integral length scales found in the flow and discussed below. Measurements of mixture fraction obtained by heating one of the jets and measuring temperature with a resistance thermometer (spatial resolution $\sim$ 0.1 cm, frequency response $\sim$ 1 kHz) show good agreement in all respects with those obtained using the chemiluminescent analysers in the manner described below.

Swinging Hot Wire

The high turbulence intensities and the occurrence of back flow in most regions of the TSC flow preclude the use of a fixed hot wire anemometer which requires the presence of a large mean velocity for the successful interpretation of the output signals (Bradshaw 1971). In order to overcome this problem, a "swinging" hot wire has been developed. The hot wire, mounted on the end of a pendulum, intermittently sweeps through a small portion of the flow thus artificially inducing a large mean velocity over the wire. The instantaneous components of velocity being measured are then yielded by subtraction of the wire velocity from the effective cooling velocity, recorded by the anemometer, in a manner appropriate to the type of probe being used.

Twin Resistance Thermometers

Integral length scales of the turbulence, based here on the fluctuations of a conserved scalar, were estimated by heating one of the inlet streams and correlating the temperature fluctuations recorded by two resistance thermometers at various separation distances, r, along the desired direction. The integral length scale measure used is one of two given by Bradshaw (1971),

$$L \equiv \int_0^\infty R \, dr$$

where R = correlation coefficient.

Data Collection

Output from the swinging hot wire was read from the screen of a storage oscilloscope. Signals from the other instruments were recorded on a Honeywell 5600C FM tape recorder and later played back via an A/D converter into a VAX 11/80 computer for analysis.

Velocity Field

Experiments and Results

Preliminary measurements of the variation of the x component of velocity, u_x, across the horizontal midplane ($z = 0$) have been made at four downstream stations using the swinging hot wire fitted with a 'U' probe. The means and standard deviations of u_x are shown in Fig. 2 as a function of cross-stream distance, y.

For $x \leqq 2.8$ m the flow consists of a turbulent jet in a turbulent co-flowing stream. The jet width $2b$ is defined here as the distance between the points on either side of the jet where the mean velocity $\bar{u}_x = (\bar{u}_{x,p} + \bar{u}_{x,c})/2$ where $\bar{u}_{x,p}$ is the velocity at the jet peak and $\bar{u}_{x,c}$ is the mean of the minimum $\bar{u}_x$ in the two halves of the co-flowing stream. Values of the jet half width b are given in Table 1 for the first three measurement stations. The growth of the jet may be seen from Table 1 to be linear with x and to have a growth rate of $db/dx = 0.08$.

Table 1. Jet half-width b vs downstream distance, x

$x(m)$	0.8	1.7	2.8
$b(m)$	0.16	0.22	0.32

The observation of a narrow ($2b \cong 20\%$ of inlet duct separation), high turbulence intensity jet with linear growth rate and a rapidly decaying centreline velocity agrees qualitatively with the work of Witze and Dwyer (1976) on constrained radial jets and opposed

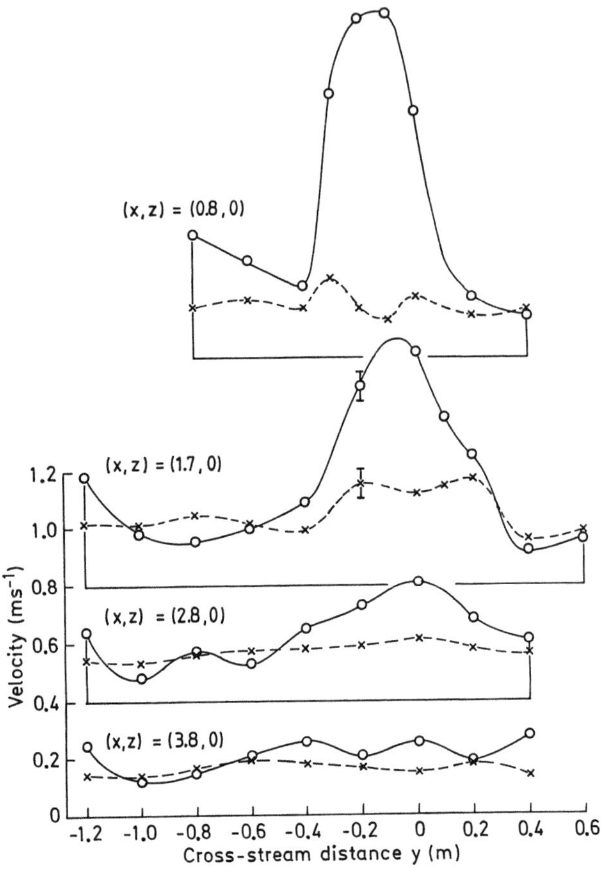

Fig. 2. Mean (0) and standard deviation (*X*) of the *x*-component of velocity, u_x, vs cross-stream distance, *y*. Lines of constant (x, z) along which measurements were taken are as shown for each pair of curves. Error bars are as shown on $(x, z) = (1.7, 0)$ curves

free axisymmetric jets. Quantitative agreement appears to be precluded by the significant differences in geometry and operating conditions between their apparatus and the present study.

By $x = 3.8$ m the jet observed upstream has been dissipated. Mean velocity is then $\bar{u}_x \cong 0.25 \text{ ms}^{-1}$ across the range of *y* surveyed. At all stations the turbulence intensity is very high varying from 0.12 to 0.5 on the jet centreline to approximately 1 elsewhere.

Integral length scales have been measured for a small number of locations and directions. The results are shown in Table 2. The scales in the cross-stream (*y*) direction show linear growth with *x*, for $x \geqq 1.7$ m and their magnitudes accord with the jet half-widths in the same region. On the basis of these velocities and length scales the Reynolds number and the turbulent Reynolds number are both $\sim 1 \times 10^4$ in the body of the flow.

Table 2. Integral length scale measurements

$x(m)$	$y(m)$	$z(m)$	Measurement direction	Integral length scale $L(m)$
1.7	− 0.06	0	+ *y*	0.14
2.45	− 0.06	0	+ *y*	0.18
3.7	− 0.06	0	+ *y*	0.26
0.8	− 0.4	0	+ *y*	0.12
1.7	− 0.7	0	+ *y*	0.14
2.45	0	0	− *z*	0.24

The flowfield in the chamber appears to be stable and repeatable. The velocity results presented in Fig. 2 were obtained over several days with the flow turned off between most runs. Deployment of an aerodynamic tuft showed that the impingement region between the jets lies within 0.2 m of the vertical centre plane ($y = 0$). This position is unaffected by start-up procedure, temporary blocking of one jet etc. A small increase in the inlet duct separation, of the order of 20%, does, however, have a marked effect on the impingement region moving it away from its central location. This is in agreement with Luna (1965) who reports that the impingement region location is indeterminate for opposed jets with separation distances between 2 and 8 jet diameters.

Extension of these preliminary flow field experiments is in progress and a more complete picture of the chamber flow should be available in the near future.

Scalar Fields

Experiments

In the scalar field experiments reported here, the inlet jet denoted STREAM 1 in Fig. 1 was doped uniformly and steadily with NO to produce $[NO]_1 \cong 1$ ppm in that jet. STREAM 2 was similarly doped with O_3 to produce $[O_3]_2 \cong 1$ ppm. The inlet jets thus formed the two streams of a two stream mixing problem. All experiments were performed in the UV dark so that the only reaction present was

$$NO + O_3 \xrightarrow{k_3} NO_2 + O_2, \tag{1}$$

with $k_3 = 0.35\ \mathrm{ppm}^{-1}\,\mathrm{s}^{-1}$ at 293 °K (Clough and Thrush 1967).

A survey was carried out in which simultaneous measurements of [NO] and $[O_3]$ were made at a large number of points in the $x > 0$ region. The sampling locations are given in Table 3 and those lying on the horizontal mid-plane are shown in Fig. 3.

Results

Species Concentrations. Insight into the mixing and chemical reaction processes occurring in the TSC can be most readily gained by consideration of the spatial variation of the joint probability density function of the raw data variables [NO] and $[O_3]$, denoted $p([NO], [O_3])$. $p([NO], [O_3])$ for sampling locations A, g, I and k, which lie along or close to the TSC

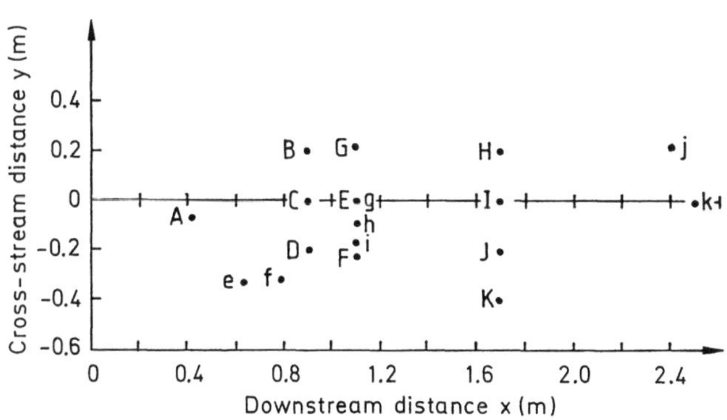

Fig. 3. Horizontal mid-plane ($z = 0$) sampling locations for the ([NO], $[O_3]$) survey. Letters refer to list of sampling locations in Table 3

Table 3. Sampling locations for the scalar field survey together with fundamental statistics of the conserved and reacting scalars at each location: mixture fraction mean ($\overline{\xi}_c$), standard deviation $\breve{\xi}_c$ and the moment coefficients of skewness and kurtosis of $p(\xi_c)$; mean ($\overline{[\;]}$) and standard deviation ($\breve{[\;]}$) of the NO and O_3 concentrations. Also shown are the inlet stream concentrations $[NO]_1$ and $[O_3]_2$ and parameters associated with the mean reaction rates

EXPT #	Sampling Location			Mixture Fraction				Reactant Concentrations				Inlet Jets			
	x (m)	y (m)	z (m)	$\overline{\xi}_c$	$\breve{\xi}_c$	Skewness	Kurtosis	$\overline{[NO]}$ (pphm)	$\breve{[NO]}$ (pphm)	$\overline{[O_3]}$ (pphm)	$\breve{[O_3]}$ (pphm)	STREAM 1 $[NO]_1$ (pphm)	STREAM 2 $[O_3]_2$ (pphm)	$\overline{[NO][O_3]}$ $(pphm)^2$	$\frac{\overline{[NO][O_3]}}{\overline{[NO]}\,\overline{[O_3]}}$
T	0.30	−0.20	−0.40	1	0	—	—	94.6	0	0	0	94.6	96.0	0	—
b	0.36	0.35	−0.45	0.957	0.081	−1.45	4.31	80.9	13.4	1.2	2.0	87.6	100.0	77.9	0.80
a	0.36	−0.35	−0.40	0.804	0.104	0.33	2.25	55.6	17.0	4.6	3.0	87.6	100.0	213.0	0.83
U	0.40	−0.28	−0.63	1	0	—	—	94.6	0	0	0	94.6	96.0	0	—
A	0.41	−0.07	0	0.333	0.366	0.68	1.95	29.8	34.7	59.5	33.8	94.6	92.0	610.7	0.34
Q	0.42	−0.31	0.45	1	0	—	—	94.6	0	0	0	94.6	96.0	0	—
c	0.44	−0.26	−0.45	0.959	0.089	−1.75	10.37	81.4	14.6	1.4	3.3	87.6	102.0	89.8	0.79
M	0.50	−0.38	−0.91	0.671	0.087	1.03	4.37	41.1	13.5	8.9	3.6	94.6	96.0	325.2	0.89
N	0.50	−0.38	−0.74	0.762	0.119	0.33	2.16	55.6	19.0	6.1	4.1	94.6	96.0	269.5	0.79
O	0.50	−0.38	−0.57	0.915	0.100	−1.21	3.23	80.3	16.6	1.7	2.7	94.6	96.0	97.2	0.71
P	0.50	−0.38	0.58	0.843	0.124	−0.38	2.70	69.5	19.5	4.7	5.0	94.6	96.0	255.1	0.78
d	0.63	−0.33	−0.40	0.780	0.069	0.59	3.43	50.8	11.6	4.8	2.5	87.6	103.0	229.3	0.94
e	0.63	−0.33	0	0.767	0.058	0.63	3.67	49.8	9.3	6.7	2.9	87.6	104.0	320.4	0.96
f	0.80	−0.32	0	0.743	0.043	0.50	4.65	44.9	6.7	6.9	2.7	87.6	106.0	301.2	0.97
D	0.90	−0.20	0	0.818	0.081	−0.63	8.53	65.9	12.9	5.0	4.4	94.6	92.0	310.8	0.94
C	0.90	0	0	0.183	0.268	1.91	5.66	12.3	25.3	70.0	26.7	94.6	92.0	287.4	0.33
S	0.90	0.08	−0.45	0.180	0.120	1.54	10.69	4.8	8.6	66.4	18.0	94.6	96.0	287.8	0.90
R	0.90	0.08	0.45	0.252	0.072	−0.40	7.18	4.4	3.3	52.2	12.4	94.6	96.0	219.4	0.96
B	0.90	0.20	0	0.266	0.094	−0.65	2.88	6.4	2.3	49.9	15.4	90.0	92.0	296.5	0.93
F	1.10	−0.22	0	0.754	0.085	−2.64	19.94	56.0	10.4	7.7	9.2	94.6	94.0	415.4	0.96
i	1.10	−0.17	0	0.761	0.080	−3.57	27.59	46.2	8.5	4.7	9.9	87.6	106.0	197.4	0.91
h	1.10	−0.08	0	0.771	0.157	−2.45	10.61	52.2	15.0	8.7	18.6	87.6	106.0	281.6	0.62
g	1.10	0	0	0.558	0.329	−0.20	1.47	40.3	30.6	38.1	34.9	87.6	106.0	582.2	0.38
E	1.10	0	0	0.478	0.304	0.13	1.47	40.4	33.5	42.3	31.0	105.0	100.0	807.2	0.47
G	1.10	0.22	0	0.356	0.069	0.76	18.16	9.8	6.9	36.5	9.9	94.6	94.0	353.1	0.99
K	1.70	−0.40	0	0.759	0.055	−3.34	27.40	55.6	6.5	5.7	5.2	94.6	92.0	309.7	0.98
J	1.70	−0.20	0	0.717	0.108	−2.09	8.24	52.9	11.4	11.5	11.1	94.6	94.0	533.6	0.88
I	1.70	0	0	0.588	0.168	−0.24	1.85	40.5	19.1	23.4	14.6	94.6	94.0	737.2	0.78
H	1.70	0.20	0	0.436	0.117	1.36	4.49	21.3	15.0	32.8	9.6	94.6	94.0	616.5	0.88
j	2.40	0.24	0	0.563	0.110	0.17	1.96	23.0	12.4	19.9	9.9	87.6	106.0	355.2	0.78
k	2.50	−0.03	0	0.619	0.103	−0.60	2.51	28.0	10.3	13.9	10.3	87.6	106.0	294.6	0.76

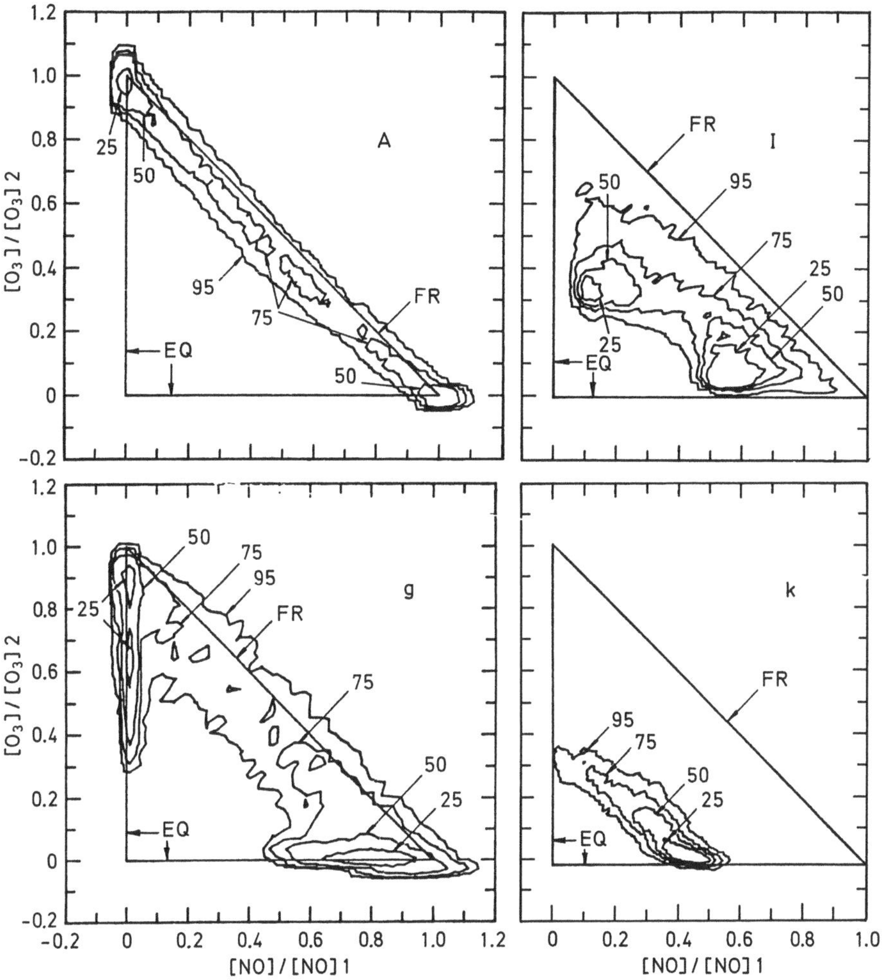

Fig. 4. Joint probability density functions of reactant species concentrations, $p([NO], [O_3])$, for sampling locations A, g, I and k on or close to the TSC centreline. Numbers associated with contours show the percentage of data enclosed by the contour summing from the highest probability density regions, e.g. contour denoted 25 encloses that 25% of data having the highest probability density. FR and EQ denote the frozen and equilibrium chemistry lines

centreline, $(y, z) = (0, 0)$, are shown in Fig. 4 in normalised form, the normalisation factors being the respective inlet jet concentrations $[NO]_1$ and $[O_3]_2$.

The lines marked FR and EQ on these plots are the frozen and equilibrium chemistry limits, respectively, for samples formed by the mixing of the inlet jet material. The majority of samples can be seen to lie far from both these limits indicating a high degree of chemical nonequilibrium. The experiment is thus well-designed with chemical kinetic and turbulent mixing time scales of the same order. The species concentration means and standard deviations are given in Table 3 for the survey sampling points.

The simultaneous progress of mixing and reaction with increasing downstream distance, x, can be seen clearly in Fig. 4. $p([NO], [O_3])$ at location A consists of two peaks at $([NO]/[NO]_1, [O_3]/[O_3]_2) = (1, 0)$ and $(0, 1)$ representing pure inlet jet material and a plateau, containing $\sim 40\%$ of the data, lying between the peaks, representing material which has mixed but has not yet reacted significantly. At location k, $p([NO], [O_3])$ is confined to the lower left hand corner of the region between the frozen and equilibrium limits indicating that both mixing and chemical reaction are well-advanced although the system is still some distance from the end-point for this experiment at $([NO], [O_3]) = (0, 0)$.

$p([NO], [O_3])$ for locations F and G, lying -0.22 m and $+0.22$ m, respectively, in the cross-stream direction (y) from location g, are shown together in Fig. 5. The well-reacted nature of the material at these locations is to be expected from the lower velocities, and thus higher mean residence times, pertaining there. Comparison of Figs. 4 and 5 shows that the two well-reacted lobes appearing in $p([NO], [O_3])$ at g arise from incursions onto the chamber centreline of material from either side of the centreline, the z-dependence of $p(\,,\,)$ in the survey region being quite weak.

The nearly frozen material at g may arise from the rapid advection by the centreline jet of frozen material from the region around sample point A or from the rapid mixing of material from the two lobes or a combination of these processes. Mixing of material from two points on the [NO], $[O_3]$ plane in the absence of chemical reaction produces samples lying on the straight line connecting the two points. Chemical reaction in the absence of further mixing moves samples along lines of constant mixture fraction (see below) towards the equilibrium chemistry lines shown in Figs. 4 and 5. It is interesting therefore to note the absence at g of samples lying on or below the line joining the low-[NO] end of the NO-rich lobe and the low-$[O_3]$ end of the O_3-rich lobe. This implies that mixing does not occur at g between the material which gives rise to these portions of the lobes at g. Considering the frozen material to be associated with the jet and the lobe material to be associated with the co-flowing stream on either side of the jet we may say that the flow is highly intermittent on the jet centreline but that the jet material forms an unbroken, although highly contorted, stream in the chamber.

Mixture Fraction. A quantitative measure of the degree of mixing in a sample can be obtained by defining a mixture fraction, ξ, as the mass fraction, in the sample, of material originating in stream 1. A measure of ξ can be formulated from the [NO] and $[O_3]$ measurements as

$$\xi_c \equiv \frac{[NO] - [O_3] + [O_3]_2}{[NO]_1 + [O_3]_2} \tag{2}$$

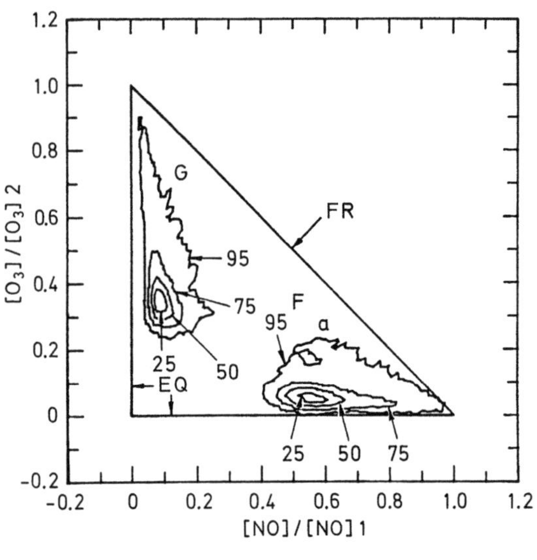

Fig. 5. Joint probability density functions of reactant species concentrations, $p([NO], [O_3])$, for off-centreline sampling locations F and G. See Fig. 4 for key to other features

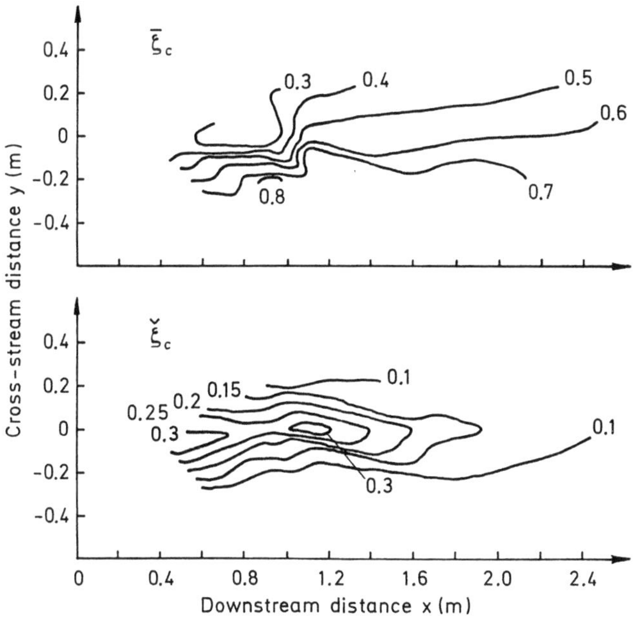

Fig. 6. Contours of mixture fraction mean ($\bar{\xi}_c$) and standard deviation ($\hat{\xi}_c$) on the horizontal mid-plane ($z = 0$)

due to the conserved nature of the scalar [NO] − $[O_3]$. Lines of ξ_c = constant are parallel straight lines of positive slope on the ([NO], $[O_3]$) plane.

The means ($\bar{\xi}_c$), standard deviations ($\hat{\xi}_c$) and the moment coefficients of skewness and kurtosis of the probability density functions $p(\xi_c)$ are given in Table 3 for each sampling location.

In Fig. 6 the contours of $\bar{\xi}_c$ and $\hat{\xi}_c$ are shown on the $z = 0$ plane. $p(\xi_c)$ for the chamber centreline sampling locations A, g, I and k and those for the off-centreline locations F and G appear in Figs. 7 and 8 respectively. The rapid progress of mixing, noted above, is seen again here with $\partial\bar{\xi}_c/\partial y$ and $\hat{\xi}_c$ decreasing markedly, with increasing x, over only two jet diameters. The decrease of $\bar{\xi}_c$ with increasing y and the $p(\xi_c)$ for F and G show that the contributions to the material on each side of the chamber originate predominantly, although not exclusively, from the jet on that side of the chamber.

An interesting feature of the development of $p(\xi_c)$ on the centreline, clearly linked to the latter phenomena, is the persistance of the bimodal nature of the distribution with the simultaneous rapid decrease in $\hat{\xi}$.

Reaction Progress and Reaction Time. In order to measure the degree of chemical reaction which has taken place in a sample we define a reaction progress variable as

$$r \equiv \frac{[O_3] - [O_3]_f}{[O_3]_e - [O_3]_f}, \qquad 0 < \xi < 1 \tag{3}$$

where subscripts f and e refer to the frozen and equilibrium concentrations at the sample mixture fraction.

The interpretation of r in terms of the 'age' of a sample is complicated by the rate of change of r at constant ξ, say, being dependent on ξ. Additionally, $r \rightarrow 1$ only as time allowed for reaction $\rightarrow \infty$. In order to provide a more direct measure of sample age, we define the parameter t_c as the time for reaction to produce the measured ([NO], $[O_3]$) from the frozen

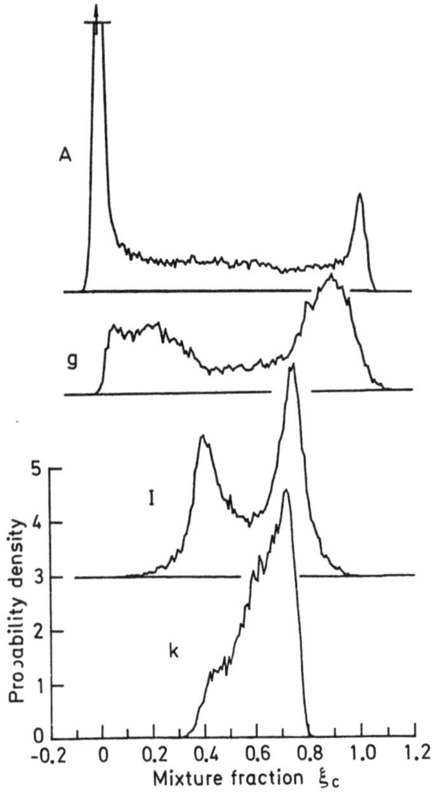

Fig. 7. Probability density function of mixture fraction $p(\xi_c)$, for sampling locations A, g, I and k on or close to the TSC centreline

reactant concentrations at the sample mixture fraction, viz

$$t_c \equiv \begin{cases} \dfrac{1}{k_3([\mathrm{NO}]_f - [\mathrm{O}_3]_f)} \ln\left(\dfrac{[\mathrm{O}_3]_f}{[\mathrm{NO}]_f}\left(1 - \dfrac{[\mathrm{O}_3]_f - [\mathrm{NO}]_f}{[\mathrm{O}_3]}\right)\right), & \xi \neq \xi_s \\ \dfrac{1}{k_3[\mathrm{O}_3]_f}\left(\dfrac{[\mathrm{O}_3]_f}{[\mathrm{O}_3]} - 1\right), & \xi = \xi_s \end{cases} \tag{4}$$

where ξ_s is the stoichiometric mixture fraction. An equivalent expression for t_c may be formulated as a function of [NO]. The significance of t_c as a flow parameter is that it is the minimum time required to produce a sample with a given ([NO], $[O_3]$) from inlet jet material via mixing and chemical reaction processes i.e. $t_c \leqq \overline{t_s} \equiv$ mean residence time of the sample constituents, where $t_s = 0$ is defined for each sample portion to be the time at which it underwent its first mixing operation.

A simple proof of this assertion may be formulated as follows. Initially, $t_c = t_s = 0$ by definition for all material entering the chamber. By definition, also, chemical reaction processes, in the absence of further mixing will maintain the inequality as $\Delta t_c = \Delta \overline{t_s}$. Now the surface t_c([NO], $[O_3]$) in ([NO], $[O_3]$, t_c) space is concave from above (i.e. from t_c large). This can be shown by plotting Eq. (4) on the ([NO], $[O_3]$) plane. Suppose that $t_c \leqq \overline{t_s}$ for two samples A and B represented by two points on this surface. Mixing these samples in proportion β then yields a mixture with

$$[\ ]_m = \beta[\ ]_A + (1 - \beta)[\ ]_B$$

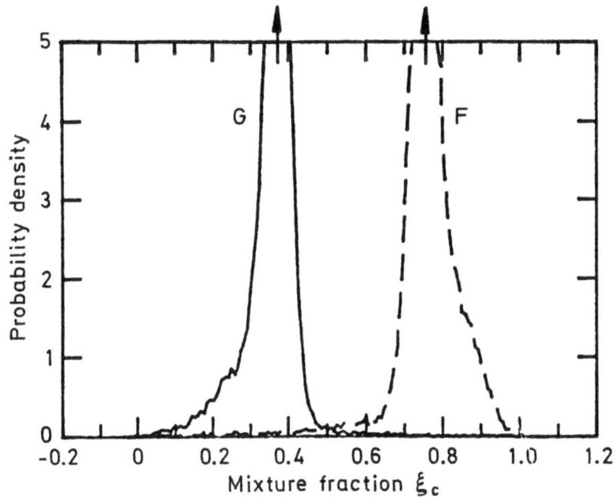

Fig. 8. Probability density functions of mixture fraction, $p(\xi_c)$, for off-centreline sampling locations *F* (----) and *G* (——)

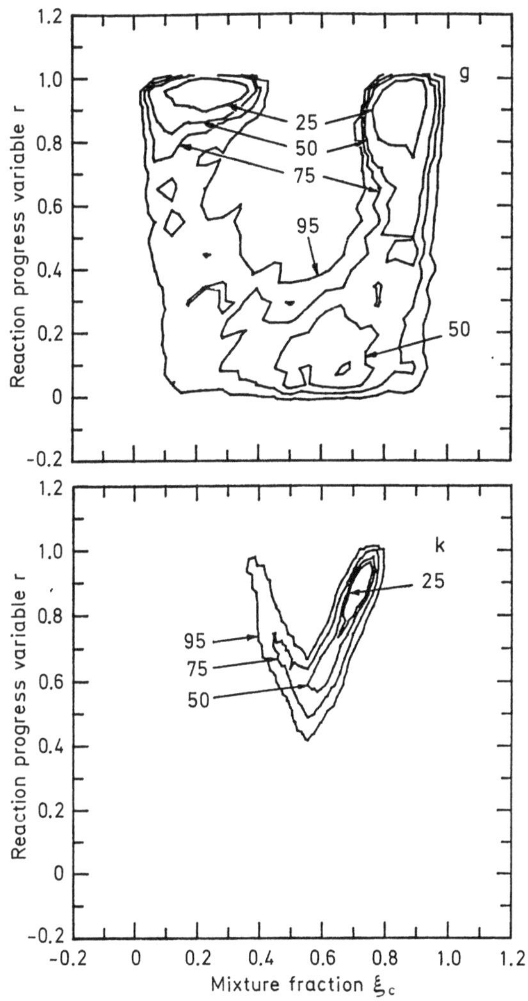

Fig. 9. Joint probability density function of mixture fraction and reaction progress variable, $p(\xi_c, r)$ for sampling locations *g* and *k* on the TSC centreline. See Fig. 4 for key to other features

and

$$\bar{t}_{s,m} = \beta \bar{t}_{s,A} + (1-\beta)\,\bar{t}_{s,B}$$

but with

$$t_{c,m} \leqq \beta t_{c,A} + (1-\beta)\,t_{c,B} \leqq \bar{t}_{s,m}.$$

Because both chemical reaction and mixing preserve the initial inequality we can conclude that $t_c \leqq \bar{t}_s$.

The joint probability density functions $p(\xi_c, r)$ and $p(\xi_c, t_c)$ are shown for sampling locations g and k in Figs. 9 and 10 respectively. These plots contain conditionally sampled data in that 35% of the data at g and 1% of that at k has been excluded on the grounds that, ξ_c was outside the range (0.05, 0.95), thereby producing indeterminate values of r, or $[O_3]$ was outside the range $([O_3]_e, [O_3]_f)$ due to instrument noise etc.

In Fig. 10, t_c is normalised by the mean residence time (25 sec) in the TSC and the mean residence times up to each sampling location, $\bar{t}_{RES}$, are also shown on the figure. Both forms of the data indicate, as expected from the foregoing, that the material having high or low ξ_c is more fully reacted and 'older' than material having intermediate ξ.

Perturbation Variable. Bilger (1980) defines a perturbation variable for species i as the difference between the mass fraction of the species in a sample and its equilibrium mass fraction at the sample mixture fraction. Here it is convenient to define the perturbation variable $y_p = y_{NO} = y_{O_3} = -\,y_{NO_2}$ in terms of species mole fractions as

$$y_P = [O_3] - [O_3]_e = [NO] - [NO]_e. \tag{5}$$

The joint pdf $p(\xi_c, y_p)$ is shown for sampling location I in Fig. 11. In this figure y_P has been normalised by its frozen value at stoichiometric viz $\xi_s[NO]_1$.

Under frozen chemistry conditions the samples lie along the frozen chemistry lines in Fig. 11 so that y_P is directly proportional to $-|\xi - \xi_s|$ with a different constant of proportionality on either side of $\xi = \xi_s$. From theories of the structure of diffusion flames it would be expected that under conditions of fast chemistry y_P would be non-zero only for mixture fractions close to stoichiometric. The present system, in which the chemistry proceeds at a finite rate, may be thought of as a reaction broadened diffusion flame. It is not surprising therefore to find that y values for the present system are intermediate between frozen and equilibrium values over a wide range of ξ with a maximum at $\xi = \xi_s$ as may be seen in Fig. 11.

Reaction Rates. By writing each reacting species concentration as the sum of a mean ($\overline{[\ ]}$) and a fluctuating component ($[\ ]'$), the mean rate of reaction (1) can be expressed as

$$\overline{\frac{d[NO]}{dt}} = -\,k_3\overline{[NO][O_3]} = -\,k_3(\overline{[NO]}\,\overline{[O_3]} + \overline{[NO]'[O_3]'}) \tag{6}$$

i.e. as the sum of a product of means and a covariance. The values of $\overline{[NO][O_3]}$ and $\overline{[NO][O_3]}/(\overline{[NO]}\,\overline{[O_3]})$ are given in Table 3 for the survey sampling locations. In some regions of the flow the latter parameter is as low as 0.33 indicating that for this flow the use of $-\,k_3\overline{[NO]}\,\overline{[O_3]}$ as a mean reaction rate could lead to significant errors.

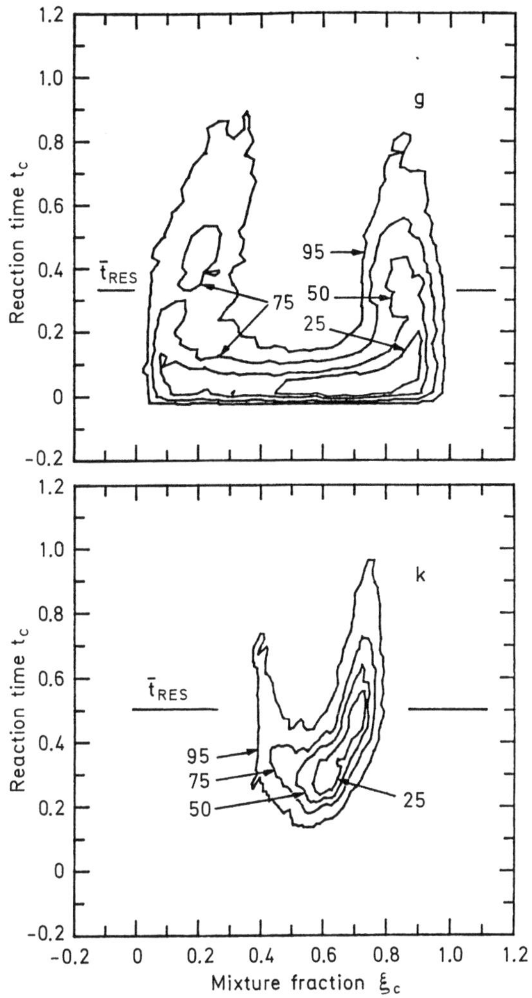

Fig. 10. Joint probability density function of mixture fraction and reaction time, $p(\xi_c, t_c)$, for sampling locations g and k on the TSC centreline. $\bar{t}_{RES}$ indicated is the mean residence time in the TSC at the sampling location. t_c and $\bar{t}_{RES}$ are normalized by the mean residence time in the TSC = 25 sec. See Fig. 4 for key to other features

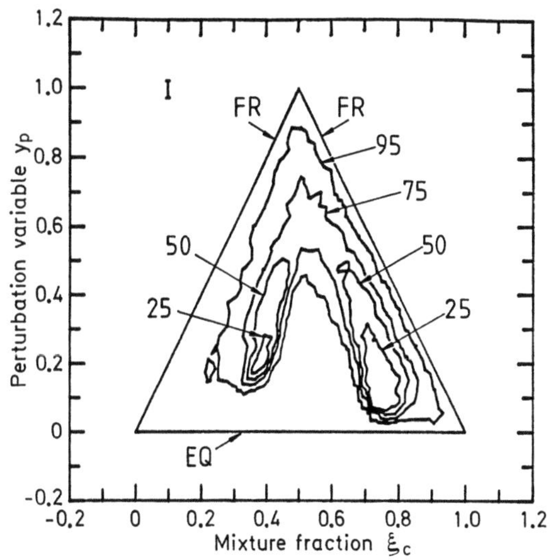

Fig. 11. Joint probability density function of mixture fraction and perturbation variable, $p(\xi_c, y_p)$, for sampling location I on the TSC centreline. y_P is shown here normalized by ξ_s[NO], the maximum y_P under frozen chemistry conditions

Conclusion

Measurements with good spatial and temporal resolution have been made of the concentrations of the two reactants, NO and O_3, in an isothermal turbulent flow in which the turbulent mixing and chemical kinetic rates are of the same order. Joint probability density functions of reactant concentrations and other parameters of interest obtained from the measurements showed in detail the development of the flow under the influence of these two competing processes.

Results of preliminary flowfield characterisation experiments showed a number of features, such as a jet on the chamber centreline, which could be simply related to the scalar field results.

Data such as that presented here should provide a good basis for tests of models of nonequilibrium chemistry in turbulent flows.

Acknowledgements. This work is supported by the Australian Research Grants Scheme. A number of the velocity and integral length scale measurements were made by Mr. R. Romanous.

References

Batt, R. G. (1977): Turbulent mixing of passive and chemically reacting species in a low-speed shear layer. J. Fluid Mech. **82**, 60

Bilger, R. W. (1980): Perturbation analysis of turbulent nonpremixed combustion. Comb. Sci. Techn. **22**, 251

Bradshaw, P.: *An Introduction to Turbulence and Its Measurement* (Pergamon, New York 1971)

Clough, P. N., Thrush, B. A. (1967): Mechanism of chemiluminescent reaction between nitric oxide and ozone. Trans. Faraday Soc. **63**, 915

Drake, M. C., Lapp, M., Penney, C. M., Warshaw, S., Gerhold, B. W.: Measurements of temperature and concentration fluctuations in turbulent diffusion flames using pulsed Raman spectroscopy. Eighteenth Symposium (International) on Combustion (The Combustion Institute 1981) p. 1521

Libby, P. A., Williams, F. A. (eds.): *Turbulent Reacting Flows*, Topics Appl. Phys., Vol. 44 (Springer, Berlin, Heidelberg, New York 1980)

Luna, R. E.: A study of impinging axisymmetric jets and their application to size classification of small particles. Ph.D. Thesis (Princeton University 1965)

Mudford, N. R., Bilger, R. W.: A facility for the study of nonequilibrium chemistry in an isothermal turbulent flow. Presented at Eighth Australasian Fluid Mechanics Conference, Newcastle, N.S.W. (Australia, Nov. 1983)

Sherikar, S. V., Chevray, R.: A chemically reacting plane mixing layer, in *Third Symposium on Turbulent Shear Flows* (Univ. of Calif., Davis 1981)

Witze, P. O., Dwyer, H. A. (1976): The turbulent radial jet. J. Fluid Mech. **75**, 401

Lagrangian Models for Turbulent Combustion

R. Borghi [1] and E. Pourbaix

Office National d'Etudes et de Recherches Aérospatiales (ONERA), BP 72 92322 Chatillon Cedex, France

Abstract

Turbulent combustion is discussed from a Lagrangian point of view. It is shown that simple models can be derived, or already known models can be very well explained. In the case of non-homogeneous turbulent flow field, a model is proposed that is able to handle multispecies, non-infinitely fast chemistry in a turbulent medium.

I. Introduction

During the five or ten past years, the prediction of turbulent flows with combustion has been classically led by numerical means, giving fairly detailed results. But these results are only approximate, because they are obtained from equations where many physical phenomena are "modelled" or even simplified, and the accuracy of each approximation is difficult to assess.

Two types of method have mainly been used: the first one is based on the calculation of the probability density function of species and temperature, (pdf) by means of a balance equation, as introduced first in the field of combustion by Dopazo and O'Brien (1974). The second one is a method where the pdf is approximately calculated from the moments, which are themselves given by the solution of their balance equations, Lockwood and Naguib (1975), Bray and Moss (1977), Borghi and Moreau (1977) and Donaldson and Varma (1976).

These two methods suffer from deficiencies of two types: the right modelling of physical phenomena is difficult, and the computation time, especially for the first method, can be prohibitive if a multidimensional pdf is needed. One way to remedy to these facts could be the use of Lagrangian models, which are able to improve the physical significance of modelling assumptions and to simplify numerical integration. The first advantage of Lagrangian models has been emphasized by Spalding (1977) and led him to propose his ESCIMO. The second one has been noted by Pope (1979), who established a correspondence between pdf balance equations and Lagrangian models (or "particle models", as they were called by Pope (1979)). Anyway, Lagrangian model have been used previously in the field of chemical engineering by Yamazaki and Ichigawa (1970) and turbulent combustion by Frost (1975).

This paper will first discuss, in detail, the perspective of Lagrangian models for turbulent combustion with respect to their ability to use more physical closure assumptions. However, for practical use (and for the moment), it will be proposed to adopt an Eulerian-Lagrangian calculation method in order to compute non-homogeneous turbulent flows with combustion; a particular example will be solved and, in addition, an Eulerian-Lagrangian version of ESCIMO will be given.

---

1 Also Lab. de Thermodynamique, Université de Rouen, 76130 Mt. St. Aignan

The Lagrangian models discussed here, are able to bring new light to the prediction of turbulent combustion influenced by the velocity fluctuations and not, on the contrary, on the velocity fluctuations modified by the combustion; this part of the general problem is to be attacked by classical Eulerian equations.

II. Lagrangian Models for Turbulent Homogeneous Reacting Flows

We consider in this section only statistically homogeneous turbulent flows.

II.1 The IEM Lagrangian Model

For many years, chemical engineers have been using a Lagrangian model called Interaction by exchange with the mean (IEM), in which the mass fraction of a fluid particle satisfies the Lagrangian balance equation

$$\frac{dY}{dt} = \frac{\tilde{Y} - Y}{\tau_{ex}} + \mathring{w}(Y).$$

The fluid particle is small enough for the reaction rate $\mathring{w}$ (which is assumed here for simplicity to depend on one variable Y only) to be given by chemical kinetics only; the mass fraction Y is randomly fluctuating, due to the turbulent exchanges between this fluid particle and the others. The term $(\tilde{Y} - Y)/\tau_{ex}$ models these exchanges, and is the core of the equation. It assumes that there is an exchange frequency $1/\tau_{ex}$, which is known, depending only on the turbulence itself, and that all the exchanges with any other particle are well represented by an exchange with a "mean particle", that has a mass fraction $\tilde{Y}$, the mean value for all the particles.

The previous equation for Y is not a random equation and there is no random term in it; but Y is random due to two things: first, the initial conditions can be random, secondly, the residence time within the homogeneous flow field is random.

In a steady homogeneous turbulent medium (an ideal configuration more or less realized in the "well stirred reactor" of chemical engineers), we can write

$$\tilde{Y} = \int_Y Y\tilde{P}(Y)\, dY$$

where, $\tilde{P}(Y)$ is the pdf of Y, as well as:

$$\bar{\varrho}\,\tilde{Y} = \int_{Y_0} \int_0^{\infty} Y(t) \cdot \varrho(t) f(t)\, P_0(Y_0)\, dt\, dY_0$$

where $f(t)$ is the residence time distribution, and $P_0(Y_0)$ the pdf at the inlet ($t = 0$); and similarly for all other moments of Y.

This Lagrangian model has been used by Aubry and Villermaux (1975), Yamazaki and Ichigawa (1970) in particular. Pope (1979), recently, demonstrated that, in the same homogeneous turbulent flow field, the pdf balance equation found by Dopazo and O'Brien (1974), namely:

$$\frac{\partial}{\partial t} P(Y) = \frac{\partial}{\partial Y} (\mathring{w}(Y)\, P(Y)) + \frac{\partial}{\partial Y} \left(\frac{6d}{\lambda_Y^2} (Y - \bar{Y})\, P(Y) \right)$$

was exactly equivalent to the IEM model with $1/\tau_{ex} = 6d/\lambda_Y^2$.

And, in general, any pdf equation, where the mixing term is $g(Y)$, is equivalent to a Lagrangian model where the exchange is modelled as: $\varphi(Y) = g(Y)/P(Y)$.

The study of Frost (1979) takes as starting point a Lagrangian model of about the same type:

$$\varrho \frac{d}{dt} = -\bar{\varrho} \frac{Y - \bar{Y}}{\tau_Y} + \varrho \mathring{w}(Y).$$

In addition, he uses a Lagrangian model for the velocity fluctuations, in the form of a Langevin equation:

$$\varrho \frac{d\vec{u}}{dt} = -\bar{\varrho} \frac{\vec{u} - \vec{\bar{u}}}{\tau_u} + (\bar{\varrho}\varrho)^{1/2} \vec{f}$$

where $\vec{f}$ is random force, of Markovian type, related to the turbulence kinetic energy. He was able to derive the joint probability density function $P(Y, \vec{u})$ as:

$$\varrho \frac{\partial P}{\partial t} + \varrho u_i \frac{\partial P}{\partial x_i} = \frac{\partial}{\partial u_i} \left\{ \bar{\varrho} \frac{u_i - u_i}{\tau_u} P \right\}$$
$$+ \bar{f}^2 \frac{\partial^2}{\partial u_i \, \partial u_i} (\bar{\varrho} P) + \frac{\partial}{\partial Y} \left\{ \varrho \frac{Y - \bar{Y}}{\tau_Y} P + \varrho \mathring{w}(Y) P \right\}$$

which is exactly similar, concerning Y, to that of Dopazo and O'Brien.

In the IEM model, τ_{ex}, as well as τ_Y and τ_u in the Frost model, or $6d/\lambda_Y^2$ in the Dopazo-O'Brien equation, has to be prescribed and related to the turbulence. Indeed, a turbulence model needs to be joint with the turbulent combustion model in order to compute the time scale τ_{ex}. A classical assumption is to simply relate τ_{ex} proportionally to τ_u, τ_u being defined as a time scale of the velocity fluctuations, for instance k/ε, if a $k - \varepsilon$ model is used.

II.2 The IEM Model Applied to Multireactive Mixtures

It is easy to generalize the IEM model to multireactive mixtures, and this will allow us to compute very simply the multidimensional pdf, knowing only the pdf for one particular species, for instance an inert tracer.

In a multireactive mixture, the IEM model reads:

$$\frac{d}{dt} Y_i = \frac{\tilde{Y}_i - Y_i}{\tau_{ex}} + \mathring{w}(Y_j,\ j = 1, \ldots, n).$$

Assuming for the moment that the $\tilde{Y}_i$ are known, the previous equations can be simultaneously integrated to give $Y_i = Y_i(t, Y_{i_0})$, $\forall i$. If we consider now the phase plane where Y_1, one inert tracer, and one Y_i are on the two axes, both relations $Y_1 = Y_1(t, Y_{10})$ and $Y_i = Y_i(t, Y_{i_0})$ represent a single trajectory beginning at (Y_{10}, Y_{i_0}) and finishing, for $t = \infty$, at the same point where $\mathring{w} = (Y_i - \tilde{Y}_i)/\tau$ and $Y_1 = \tilde{Y}_1$. In the particular case where $P_0(Y_0)$ is concentrated at the two points (Y_{i_0}, Y_{10}) and $(0, 0)$, Fig. 1a shows an example of the two trajectories obtained.

Of course, we do not know for the moment of either $\tilde{Y}_1$, nor $\tilde{Y}_i$, but it is easy to see that, if we know $P_0(Y_{10}, Y_{i_0})$ and $f(t)$, or equivalently $\tilde{P}_1(Y_1)$, it is possible to find $\tilde{P}_i(Y_i)$ and

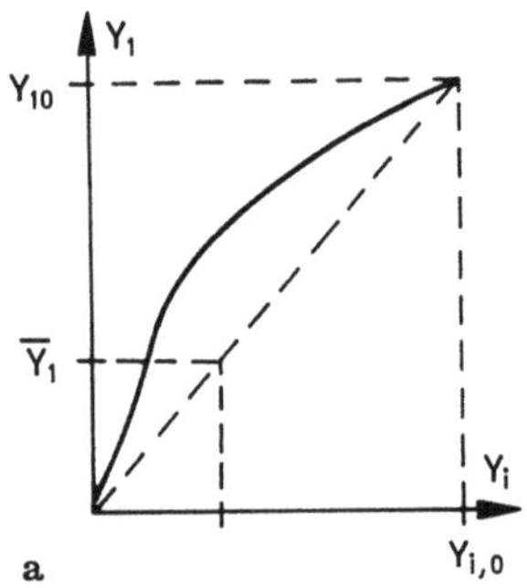

a

Fig. 1 a.

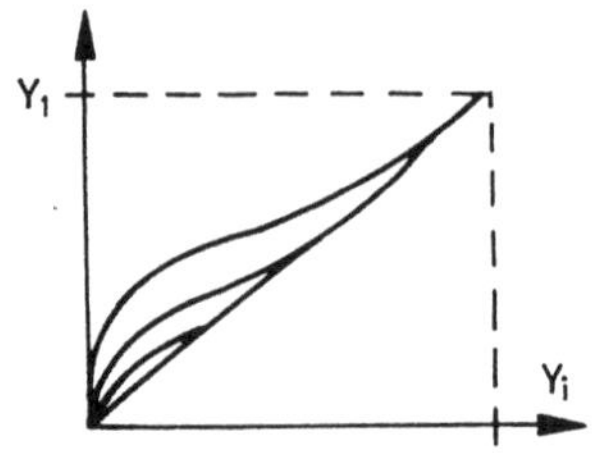

b

Fig. 1 b. Each trajectory corresponds to a value of $1/\tau$

consequently $\tilde{Y}_i$; indeed

$$\tilde{P}_i(Y_i) = \int\int P_0 \cdot \tilde{P}_1(Y_1(Y_i, Y_{10}, Y_{i_0})) \frac{dY_1}{dY_i} \, dY_{10} \, dY_{i_0}$$

or simply, in the case of Fig. 1:

$$\tilde{P}_i(Y_i) = \tilde{P}_1(Y_1^*(Y_i)) \, Y_1^{*\prime}$$

where $Y_1^*(Y_i)$ is the equation of the two trajectories and $Y_1^{*\prime} \equiv dY_1^*/dY_i$. Whatever $\tilde{P}_1(Y_1)$ can be, the IEM model leads to the result that the multidimensional pdf occupies, in the multidimensional phase plane, only the trajectories determined by the Lagrangian balance equation and the known inlet conditions. In the simple case where the homogeneous flow field is fed by two different jets of known composition, *the multidimensional pdf is one dimensional*: it occupies only a segment of a curved line. More details on these trajectories have been reported by Borghi and Pourbaix (1981).

This very simple conclusion will allow us to attempt the computation of multireactive turbulent flows. Previously, more than two-dimensional pdf was clearly improper to numerical computation, either with the pdf equation or with the moment equations (may be the Monte Carlo method of Flagan and Appleton (1974) or Pope (1981) remained to only hope). In addition, we can see that simplifications like that of Janicka and Kollman (1978), which assumes that a reactive species Y_i and an inert one Y_1 are fluctuating independently:

$$\tilde{P}(Y_i, Y_1) = \tilde{P}_i(Y_i) \cdot \tilde{P}_1(Y_1)$$

are clearly in contradiction with the simple IEM model: here Y_i and Y_1 are so strongly related that $Y_1 = Y_1^*(Y_i)$.

II.3. Possible Improvements of the IEM Model

Of course, the IEM model is just a model, and we have for the moment no direct proof that it is realistic. In fact, we already know that the Dopazo-O'Brien closure assumption is in some aspect unrealistic, and it follows that the IEM model possesses the same discrepancy. The Curl model, or the one proposed by Janicka et al. (1978), seem more satisfactory in that they have been proposed as closure assumption for the pdf equation, and can be incorporated into a Lagrangian model. The differential equation would be more complicated that the one with the IEM model, and the mixing term would involve $\tilde{P}_1(Y_1)$ itself and not only Y_1.

It is easy to see that the important property found in Sect. II.2, that is $\tilde{P}(Y_1, Y_i)$ occupies only known lines in the (Y_1, Y_i) plane, holds also with these new models. In fact any mixing term of the type $1/\tau_{ex} \cdot g(P_i(Y_i))$ will give the same result. The crucial point is the knowledge of the exchange frequency: if τ_{ex} is a non-random quantity, the previous property holds.

A better model would probably involve a random exchange frequency. Indeed k/ε represents only a mean turbulent time scale, and a full spectrum of τ_{ex} exists, ranging from a few times k/ε to τ_k, the Kolmogorov time scale ($\tau_k = (\varepsilon/\nu)^{-1/2}$). In order to incorporate this in the IEM model, or the others extracted from the Curl or Janicka assumptions, we need only to consider the same differential equation. but with $1/\tau$, random parameter, instead of $1/\tau_{ex}$.

The joint pdf $\tilde{P}(Y_1, Y_i)$ in the (Y_1, Y_i) plane is then thickened as shown by the trajectories of Fig. 1 b.

Of course, the pdf of $1/\tau$ must be given, or calculated, in order to get the complete $\tilde{P}(Y_1, Y_i)$. In an homogeneous turbulent medium, this new pdf is to be related to the autocorrelation curve for the velocity fluctuations.

It is to be noticed that, if Y_i was an inert species like Y_1, with the same molecular diffusion, the joint pdf of $\tilde{P}(Y_1, Y_i)$ would not be thickened by the fluctuations of τ. Indeed, if we consider the simple case of Fig. 1, the joint pdf would always be the straight line $Y_1 = Y_i \cdot Y_{10}/Y_{i_0}$ following from the similarity of the exact balance equation for Y_1 and Y_i (see Borghi (1979)). Then, the effect of the random nature of τ is amplified by the reaction, and because combustion is a non-linear reaction (that is a reaction with a strongly varying time scale) the mean line of the joint pdf of Fig. 1 b is not a curve corresponding to the mean $1/\tau$; it corresponds to another trajectory given by a Lagrangian model where $1/\tau$ is not constant but depending on the $\dot{w}$. This "mean Lagrangian model" corresponds to a differential equation of the same form:

$$\frac{dY_i}{dt} = \frac{\tilde{Y}_i - Y_i}{\tau_{ex}} + \dot{w}_i(Y_j);$$

if it is assumed that the pdf of $1/\tau$ is known and depends only on two parameters, τ_u and τ_k, and that the chemistry depends only on one parameter, τ_c, it follows from dimensional analysis that τ_{ex} can be written:

$$\frac{1}{\tau_{ex}} = \frac{1}{\tau_u} f\left(Y_j, \frac{\tau_c}{\tau_u}, \frac{\tau_c}{\tau_k}\right) \quad \text{with} \quad \tau_u = \frac{k}{\varepsilon}.$$

The relevance of the adimensional parameter τ_c/τ_k, in addition to the more classical one τ_c/τ_u for reacting turbulent flows has been experimentally pointed out by Bilger (1976).

In the limiting case where the combustion is very fast with respect to the mixing (and the Reynolds and Peclet numbers are very large), the previous trajectories are easier to find. Let us consider first the case of a premixed combustion in a ideally homogeneous burner feeded by fresh mixture and perfectly burnt gases (for stabilization). If the combustion is very fast the pdf of the fuel, for instance, is composed only of two peaks, corresponding to the fresh mixture or to perfectly burnt gases. On the other hand the pdf for the tracer involves intermediate states. We obtain therefore Fig. 2 a.

Let us consider now the case of a diffusion flame when the chemical reactions are very fast. Figure 2 b shows the trajectories obtained, and related $\tilde{P}_F(Y_F)$.

The heavy lines, on the Y_1, Y_F plane correspond to the case where a single reaction occurs between fuel and oxidizer, for instance $F + r0 \rightarrow P$, and where Y_0 and Y_F have the same diffusion coefficient; in this case the very fast reaction assumption implies that either Y_F or

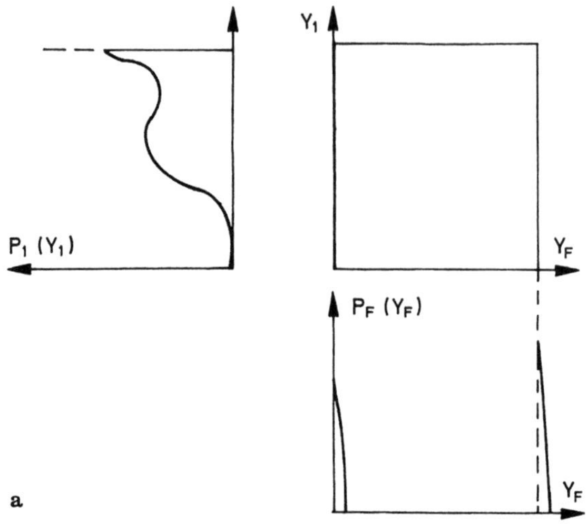

a

Fig. 2a. Joint pdf for Y_1, Y_F in a fuel lean premixed flame with very fast combustion (Y_F is the fuel mass fraction)

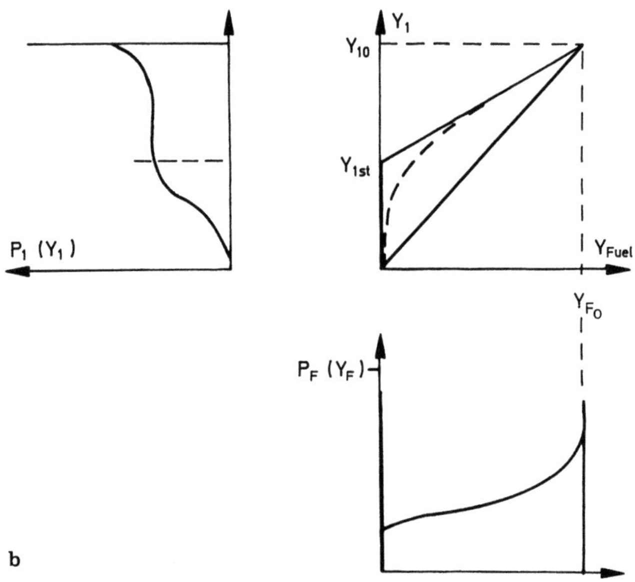

b

Fig. 2b. Joint pdf Y_1, Y_F for a diffusion flame with very fast reaction

Y_0 zero, and that $\phi = r\,Y_F - Y_0$ is a conserved species. The trajectories are then composed of $Y_F = 0$ when $Y_1 < Y_{1\text{st}} = Y_{00}/(r\,Y_{F_0} + Y_{00})$ and a straight segment joining $(Y_{1\text{st}}, 0)$ to (Y_{10}, Y_{F_0}) when $Y_1 > Y_{1\text{st}}$. In the case where the combustion cannot be represented by a single reaction, and if the Lewis numbers of all involved species are unity, the two straight lines become one curve, corresponding to the chemical equilibrium to the mixture, which is shown as the broken heavy line on the diagram.

All Lagrangian models must exhibit such a behaviour in these limiting cases. The IEM model with infinitely small chemical time gives the right answer in both cases.

The differential equation

$$dY_i/dt = (\tilde{Y}_i - Y_i)/\tau_{ex} + \mathring{w}_i(Y_j)$$

with very large $\mathring{w}$ leads in a very short time to the quasi-steady state for reactive species, where $\mathring{w}_i = (Y_i - \tilde{Y}_i)/\tau_{ex}$ while the inert tracer behaves gently.

In the case of a premixed lean flame, the trajectory issued from Y_{10}, Y_{F_0} is a straight line parallel to the Y_F-axis, followed by a line very close to the Y_1-axis, corresponding to the quasi-steady state, which is, for Y_F, $Y_F \simeq 0$. The quasi-steady state leads to

$$\mathring{w}_F(Y_F) = -\frac{\tilde{Y}_F}{\tau_{ex}}.$$

If we integrate this relation in order to obtain the mean reaction rate $\tilde{\mathring{w}}_F$, we find that

$$\tilde{\mathring{w}}_F = \delta(Y_F = 0) \cdot (-\tilde{Y}_F/\tau_{ex}),$$

where $\delta(Y_F = 0)$ is the area of the peak situated in $Y_F = 0$ on the Y_F pdf; but

$$\delta(Y_F = 0) = 1 - \delta(Y_F = Y_{F_0}) = 1 - \tilde{Y}_F/Y_{F_0},$$

which gives

$$\tilde{\mathring{w}}(\tilde{Y}_F) = -\frac{1}{\tau_{ex}}\,\tilde{Y}_F\left(1 - \frac{\tilde{Y}_F}{Y_{F_0}}\right).$$

This formula is exactly that derived from the Bray and Moss model (1977), and very similar to the Eddy break up model of Spalding.

In the case of the diffusion flame, the trajectory issued from (Y_{10}, Y_{F_0}) attains instantaneously the quasi-steady state but it corresponds here to $Y_0 \simeq 0$ and Y_F varies according to $\phi + Y_{00}/r\, Y_{F_0} + Y_{00} = Y_1$ with $\phi = r\, Y_F$ until ϕ becomes zero (and $Y_1 = Y_{1\mathrm{st}}$). At this point we cross the stoichiometric line and, after that, Y_F remains close to zero. The instantaneous reaction rate $\mathring{w}_F$ can also be deduced from the IEM model; when $Y_1 > Y_{1\mathrm{st}}$, $\mathring{w}_0 = r\,\mathring{w}_F = -\tilde{Y}_0/\tau_{ex}$ and when $Y_1 < Y_{1\mathrm{st}}$, $\mathring{w}_F = -\tilde{Y}_F/\tau_{ex}$, we obtain

$$\tilde{\mathring{w}}_F = -\frac{1}{\tau_{ex}}\left[\frac{\tilde{Y}_0}{r}\int_{Y_{1\mathrm{st}}}^{Y_{10}} \tilde{P}_1(Y_1)\,dY_1 + \tilde{Y}_F\int_0^{Y_{1\mathrm{st}}} \tilde{P}_1(Y_1)\,dY_1\right].$$

This formula is the diffusional version of the Eddy break up model. Of course it is fully possible to avoid the use of such a formula in order to compute $\tilde{Y}_F$, since:

$$\tilde{Y}_F = \int_{Y_{1\mathrm{st}}}^{Y_{10}} Y_F^*(Y_1)\,\tilde{P}_1(Y_1)\,dY_1$$

where $Y_F^*(Y_1)$ is the equation of the straight segment joining $(Y_{1\mathrm{st}}, 0)$ to (Y_{10}, Y_{F_0}). This later formula is evidently the same as that derived following the ideas originated by Toor (see for instance Lin and O'Brien (1974)).

The real Lagrangian equation is, in any case:

$$\varrho\frac{d}{dt}\,Y_i = \frac{\partial}{\partial x_\alpha}\left(\varrho\, d_i \frac{\partial Y_i}{\partial x_\alpha}\right) + \varrho\,\mathring{w}_i.$$

The role of Lagrangian models is to approximate the first term on the right hand side, because it is impossible to compute for turbulent flows.

When we suspect that the turbulent motion acts only to randomly displace flame fronts or diffusion fronts (that is any length scale and time scale of the turbulence are large enough with respect to the ones related to the laminar flame), we can replace this equation by its known solution (by experimental or theoretical means) for a purely laminar flame. Then the trajectories in any plane Y_1, Y_i are given by the deterministic relation $Y_i(Y_1)$ existing in the corresponding laminar flame and the model is linked to the "flamelet model" used recently by Bray and Moss (1981) to compute CO production in a turbulent diffusion flame.

When the time scales of the turbulence are small enough to interact with these of the laminar flame, these laminar flamelets are stretched by the turbulence. There are many trajectories, $Y_i(Y_1, \gamma)$, each one corresponding to a laminar flame with a particular stretching rate γ. If we can know, theoretically or experimentally, the relation $Y_i(Y_1, \gamma)$, we can use it jointly with the probability density function of Y_1 and γ (which, perhaps, may be separated in two independent pdf's), to compute all Y_i etc. We join here the recent discussion of Peters (1983), and similarly the ideas of Spalding for ESCIMO (1977), and of Broadwell and Marble for their "Coherent flame model" (1977). But when the length scales of the turbulence are small enough compared to the laminar flame length scales, the flamelets do not exist; instead of flamelets, the turbulent reacting medium is occupied by interactions of flamelets, and the solution of the Lagrangian equation cannot be performed for each individual flamelet. Physical modelling is then needed in this case and, in particular, the model IEM addresses these phenomena.

III. Eulerian-Lagrangian Methods for Non-Homogeneous Turbulent Reacting Flows

III.1. Extension to Non-Homogeneous Turbulent Reacting Flows

We consider now a non-homogeneous turbulent flow, and, more precisely, the case of a turbulent flame within the mixing zone of two parallel streams A and B, as schematized Fig. 3a.

The problem that we intend now to discuss is the use of Lagrangian models for non-homogeneous flows. The model of Frost (1975) is a way to solve this problem; but Frost, in fact, did not solve directly the Lagrangian equations but rather the pdf equation to what the model led.

In the case of very fast reactions and when the Lewis numbers of all involved species are unity, the generlization to non-homogeneous flows is straightforward. Neither for premixed flames, nor for diffusion flames, the trajectories of Fig. 2 and 2b have to be influenced by the non-homogeneity. It suffices, then, to get a correct calculation of $\tilde{P}_1(Y_1, \vec{x})$ at each point of the flow field to solve the problem.

We shall now propose two Lagrangian non-homogeneous models, which will lead to Eulerian-Lagrangian calculation methods. One is the extension of the IEM model, the other one appears almost equivalent to the ESCIMO model of Spalding (1977).

III.2. The Eulerian-Lagrangian Extension of IEM Model

A Lagrangian model as discussed previously cannot be applied directly, because neither $\tilde{Y}_i$ or $\tilde{P}_i(Y_i)$, nor τ_{ex} are constants along the physical trajectories between A, or B, the inlet streams, and the point P. But we can continue to use the homogeneous model, in the vicinity of any particular point P, provided that we consider that the initial points Y_{i_0} of the differential equations do not correspond to the streams A and B, but constitute a set of parameters, to be chosen conveniently. With this principle in mind, we can propose the following Eulerian-Lagrangian method.

i) We know that the turbulent flow field can be calculated with Eulerian balance equations for mean velocity, turbulent kinetic energy k, dissipation rate ε. With equations for the mean concentration $\tilde{Y}_1$ of an inert species (assumed injected with B only) and the variance $\tilde{Y}_1'^2$, we can compute the pdf of Y_1 at each point of the flow assuming a particular shape (for instance a β-function); it is possible also to compute $\tilde{P}_1(Y_1)$ more exactly but less easily, with a balance equation conveniently closed.

ii) Then, considering all the reactive species Y_i, we can compute the two trajectories issued from A and B, with the homogeneous Lagrangian model applied to each point of the flow, and with the balance equation for $\tilde{Y}_i$:

$$\frac{\partial}{\partial x_\alpha}(\bar{\varrho}\tilde{u}_\alpha \tilde{Y}_i) = \frac{\partial}{\partial x_\alpha}\left(\varrho D_t \frac{\partial \tilde{Y}_i}{\partial x_\alpha}\right) + \bar{\varrho}\int_{Y_i}^{0} \dot{w}_i(Y_j(Y_1))\, \tilde{P}_1(Y_1)\, dY_1 .$$

We need to relate D_t and τ_{ex} to k and ε, and we use classical relations:

$$\tau_{ex} = k/\varepsilon, \quad D_t \alpha k^2/\varepsilon.$$

The chemical source term is closed since we know the trajectories in the phase space and $\tilde{P}_1(Y_1)$.

iii) We can also compute all the $\tilde{P}_i(Y_i)$, and recalculate $\tilde{Y}_i$ by

$$\tilde{Y}_i = \int_{Y_i} Y_i \tilde{P}(Y_i)\, dY_i \quad (= \tilde{Y}_i^L).$$

This new $\tilde{Y}_i^L$ will not coincide with the one computed with the Eulerian balance equation because the pdf's are wrong: only two trajectories have been considered, as in the case of homogeneous flows. But we can correct all the $P_i(\tilde{Y}_i)$ with other trajectories, one for each Y_i, in order to get coincidence in the two values of Y_i. This implies only a linear set of equations to be solved if the initial points Y_{i_0} are chosen. If these initial points are chosen on the limits of the domain allowed for $\tilde{P}(Y_1, Y_i)$, as shown in Fig. 3b, the linear set of equation will have always a physical solution.

iv) It is possible again to refine the shape of $\tilde{P}(Y_1, Y_i)$ adding new trajectories, in order to verify the Eulerian balance equations for the $\tilde{Y}_i'^2$; after that, we can consider the $Y_1' Y_i'$ and so on.

If the heat release is low, the steps ii), iii) and iv) would not modify the step i) but this is generally not the case for combustion, and iterations have to be performed. There is no

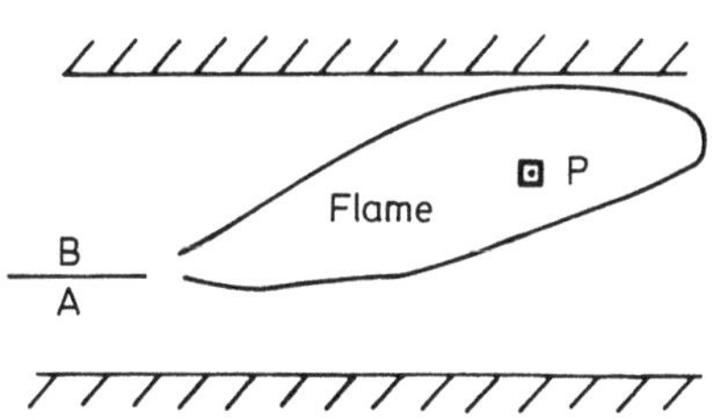

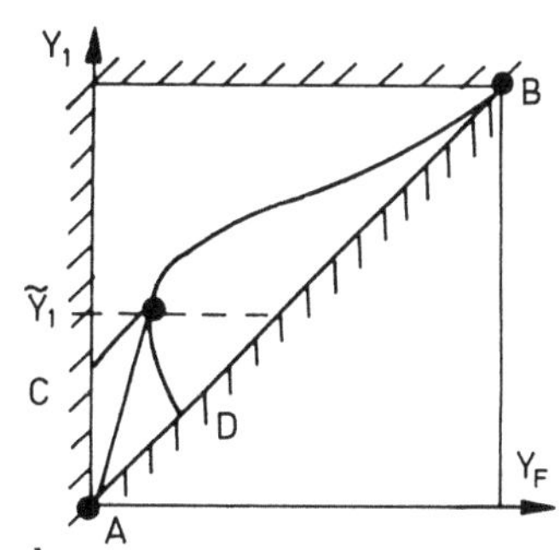

Fig. 3. Allowed domain and trajectories for $P(Y_1, Y_F)$ where Y_F in the fuel consumed by reaction, at the point P of the flame

guarantee that they will converge, but our experience has shown that they do. The undertaken turbulence by this method is fully classical but, of course, we can improve that when necessary, following the results of new researches on the influence of combustion on turbulence.

The Eulerian-Lagrangian method explained previously has been successfully applied to the case of a premixed turbulent flame in a parabolic mixing layer, where the combustion has been modelled as a single reaction, and using the simple IEM model. The steps i), ii), iii) were operated with satisfactory convergence in Borghi and Pourbaix (1983). A first attempt to compute more complicated chemistry has also been made, but only with the step i) and ii) (see Borghi and Pourbaix (1981)).

III.3. The Eulerian-Lagrangian Version of ESCIMO

ESCIMO is also a Lagrangian model but the fluid particle that is considered is not as small as the previous one. There is a structure inside, which consists in a laminar flame (or quasi-laminar) propagating and the turbulence acts at the scale of the fluid particle (called here a "fold"), stretching it without any exchanges of mass. These particular types of fluid particles have to be produced in the mixing and reaction zone and, consequently, in addition a birth rate and a $P_b(Y_b)$ of these folds have to be prescribed.

This is, in summary, the essence of ESCIMO.

One can represent, in the plane (Y_1, Y_F), the $\tilde{P}(Y_1, Y_F)$ that is implied by the model. Here Y_1 and Y_F are the mass fraction of one inert and the fuel, respectively, within the fold: they are mean values at the scales smaller than the fold itself. Figure 4 shows the trajectories obtained.

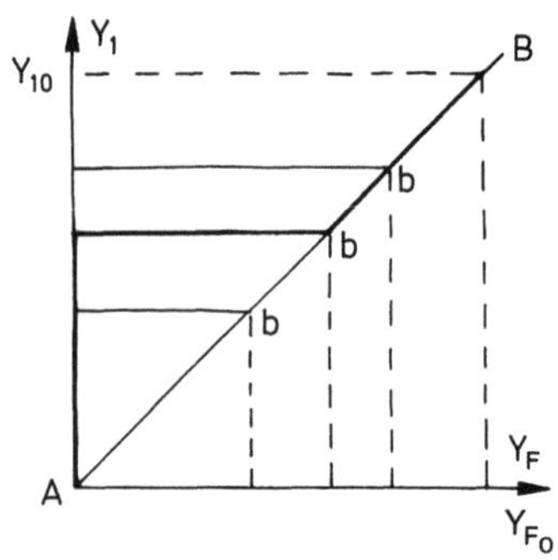

Fig. 4. Trajectories in the case of a model of the type of ESCIMO, with very sudden combustion

Before the birth points, nothing can occur except, perhaps, small scale mixing; then all birth points are on the straight line AB. After each fold, evolution is such that the inert matter inside is constant and the fuel is decreasing. Of course, this stops when there is no more fuel.

The ESCIMO model appears as a Lagrangian model with very sudden combustion, but not very fast combustion. In the first calculation, Spalding chose only one birth point which can, at first sight, be justified by the fact that the "folds" are not fluid particles, but "coherent structures", and their pdf are consequently thinner than $\tilde{P}_1(Y_1)$ itself. As a first approximation, we can assume that $P_b(Y_1) = \delta(Y_1 - Y_b)$, and, in an homogeneous medium, it follows that $Y_b = \tilde{Y}_1$.

The Lagrangian equations of ESCIMO are:

$$\frac{dY_F}{dt} = + \mathring{w}_F = -\frac{u_F}{\lambda},$$

u_F is the laminar flame speed (inside the fold), λ is the fold thickness and $d\lambda/dt = -\lambda/\tau$; by elimination of t and λ, it comes:

$$\mathring{w}_F = dY_F/dt = -(\tilde{Y}_1 - Y_F)/\tau - u_F/\lambda_0$$

where τ is the turbulent stretching time (again k/ε), and λ_0 the size of the folds at their birth, that is a turbulence length scale for the inert species Y_1. By averaging $\mathring{w}_F$, we simply obtain:

$$\tilde{\mathring{w}}_F = -\frac{\tilde{Y}_1 - \tilde{Y}_F}{\tau} - \frac{u_F}{\lambda_0}$$

which constitutes the mean reaction rate of ESCIMO, where we are interested only in Y_F and not in other species, and neglecting the influence of the stretching on u_F. It avoids any computation of residence time distributions, in contrast to the original ESCIMO, which was of purely Lagrangian type. Instead, a simple calculation of $\tilde{Y}_1$, an inert species diffusing like the fuel, is needed.

The physical meaning of τ is such that it can be identified with k/ε, but the one of λ_0 is not too precise; in particular, is λ_0 to be proportional to the integral scale $l = k^{3/2}/\varepsilon$ or the Taylor microscale λ_1 for the species Y_1? The question is of importance: in the first case $\tilde{\mathring{w}}_F$ would be influenced by the molecular diffusivity and a chemical time, both appearing through u_F; in the second one the molecular diffusivity would disappears, as $u_F \propto d^{1/2}$ as well as λ_1^2.

Taking into account only one birth point ($Y_b = \tilde{Y}_1$) is probably a shortcomming of the ESCIMO model. It is possible to avoid that, taking advantage of $\tilde{P}_1(Y_1)$.

We will now identify the folds with fluid particles, and the birth points will be defined as points where a turbulence time is equal to a given chemical time. Suppose we know an ignition delay, as a function of temperature and concentrations, for a given reaction; non-burning fluid particles corresponding to a point on the diagonal in the plane (Y_1, Y_F) have, each one, an ignition delay $\tau_i(Y_1)$; if, for the same particle, there is a range of turbulence times around the mean value k/ε, the probability that the particle will be ignited is the probability that $\tau > \tau_i$.

We propose then a model where $P_b(Y_b)$ is a curve depending on two parameters: the first one, the mean value $\tilde{Y}_b$, is to be related to the ignition chemistry and k/ε; the second, characterizing $\tilde{Y}_b'^2$, is to be related, in addition, to $\tilde{Y}_F$. In terms of $P_b(Y_b)$, referring to Fig. 4, one can easily show that $\tilde{Y}_F^L$ (as defined in Sect. III.2) is given by:

$$\tilde{Y}_F^L = \int_0^{Y_{10}} \frac{Y_{F_0}}{Y_{10}} Y_1 X_b(Y_1) \tilde{P}_1(Y_1) \, dY_1$$

where

$$X_b(Y_1) = \int_0^{Y_1} Y_b P_b(Y_b) \, dY_b$$

one obtain also:

$$\tilde{\mathring{w}}_F = -\tilde{Y}_F \int_0^{Y_{10}} \tilde{P}(Y_1) \int_{Y_1}^{Y_{10}} \frac{P_b(Y_b)}{\tau(Y_b)} \, dY_b \, dY_1 .$$

$\tilde{Y}_b$ can be prescribed such that $k/\varepsilon = \tau_i(\tilde{Y}_b)$, τ_i being an appropriate chemical model for the ignition delay. After that $\tilde{Y}_b'^2$ has to be adjusted to identify $\tilde{Y}_F^L$ and $\tilde{Y}_F$ calculated from the Eulerian balance equation with the previous $\mathring{w}_F$. Simple shapes can be used for P_b, depending on three parameters, as the clipped Gaussian, or one with peaks and rectangles.

IV. Concluding Remarks

The use of a Lagrangian view is usefull to study turbulent combustion. The main results of the discussions presented here can be summarized as follows.

i) A Lagrangian view in the "phases plane" for a reactive (consumed) species and the inert one give us simple, or less simple, turbulent combustion models.
ii) The probelm of multispecies, non-infinitely fast turbulent combustion can be attacked also with some chance of success, contrary to the previous pdf approaches.
iii) A coupling between Lagrangian calculations in the phase planes, and Eulerian calculations in the physical plane is suitable for non-homogeneous flow fields. First practical calculations have been successfully performed by Borghi and Pourbaix (1981).

References

Aubry, C., Villermaux, J. (1975): Representation du mélange imparfait de deux courants de réactifs dans un réacteur agité continu. J. Chem. Eng. Sci. **30**, 457

Bilger, R. W. (1976): Turbulent jet diffusion flames. Prog. Energy Combust. Sci. **1**, 87

Borghi, R., Moreau, P. (1977): Turbulent combustion in a premixed flow. Acta Astronautica **3–4**, 321

Borghi, R., Pourbaix, E. (1981): On the coupling of complex chemistry with a turbulent combustion model. Physico Chemical Hydrodyn. **2**, 65

Borghi, R., Pourbaix, E. (1983): Une modélisation eulérienne-lagrangienne pour la combustion turbulente. La Recherche Aérospatiale. 1983-4

Borghi, R. (1979): Reactions chimiques en milieu turbulent. ONERA Pub. 1979/2

Borghi, R., Pourbaix, E. (1981): Numerical investigation of a model of turbulent combustion of hydrocarbons. AIAA 19th Aerospace Sciences Meeting St. Louis 12–15 January, 1981

Bray, K. N. C., Moss, B. (1977): A unified statistical model of premixed turbulent flame. Acta Astronautica **4**, 291

Donaldson, C. D., Varma, A. K. (1976): Remarks on the construction of as second order closure description of turbulent reacting flows. Comb. Sci Tech. **13**, 1

Dopazo, C., O'Brien, E. E. (1974): An approach to the autoignition of a turbulent mixture. Acta Astronautica **1**, 1239

Flagan, R. C., Appleton, J. P. (1974): A stochastic model of turbulent mixing with chemical reaction. Combust. Flame **23**, 249

Frost, V. A. (1975): Model of a turbulent, diffusion-controlled flame jet. Fluid Mechan. **4**, 124

Janicka, J., Kollmann, W. (1978): A two variable formalism for the treatment of chemical reactions in turbulent H_2-air flames. 17th Symp. (Int.) on Combustion, p. 421

Janicka, J., Kolbe, W., Kollmann, W. (1978): The solution of a pdf transport equation for turbulent diffusion flames. Proc. Heat Transfer and Fluid Mech. Inst. (Stanford University Press, Stanford) p. 296

Liew, S. K., Bray, K. N. C., Moss, J. B. (1981): A flamelet model of turbulent nonpremixed combustion. Comb. Sci Tech. **27**, 69

Lin, O'Brien, E. E. (1974): Turbulent shear flow mixing and rapid reactions: an analogy. J. Fluid Mech. **64**, 195

Lockwood, F. C., Naguib, A. S. (1975): The prediction of the fluctuations in the properties of free, round, jet, turbulent, diffusion flames. Combust. Flame **24**, 109

Marble, F. E., Broadwell, J. E. (1977): The coherent flame model for turbulent chemical reactions. Project Squid Tech. Rep. TRW-9-Pu

Peters,N. (1982): Flamelets models for nonpremixed turbulent combustion. Conference INRIA, Paris Déc. 1982

Pope, S. V. (1979): The relationship models for reaction in homogeneous turbulence. Combust. Flame **35**, 41

Pope, S. V. (1981): A Monte Carlo method for the pdf equations of turbulent reactive flow. Comb. Sci. Tech. **25**, 159

Spalding, D. B. (1977): A general theory of turbulent combustion. J. Energy **2**, 16

Yamazaki, H., Ichigawa, A. (1970): Int. Chem. Eng. **10**, 471

A Model for Flame Extinction in Turbulent Flow

S. Byggstøyl

Researcher, SINTEF Avd. 15, Division of Energy- and Fluid Dynamics, N-7034 Trondheim-NTH, Norway

B. F. Magnussen

Professor, Engineering Thermodynamics, The Norwegian Institute of Technology, Division of Thermodynamics, N-7034 Trondheim-NTH, Norway

Abstract

In this paper a model for local extinction in turbulent flow is developed. The model is based on the Eddy Dissipation Concept (EDC) for chemical reaction in turbulent flow developed by Magnussen (1981) [1]. A review of this concept is included in the paper. The (EDC) is a reactor concept which is unique in that it takes into account the intermittent behaviour of the smallscale structures of turbulence, as well as the effect of the fine structures on the chemical reactions. Both fast and slow chemical reactions can be treated simultaneously and consequently the extinction process. In this paper the extinction is assumed to take place in the fine structure. The lift-off and blow-out conditions for turbulent diffusion flames are calculated and compared with experimental data. The turbulent flame propagation velocity is also discussed with regards to the same concept.

Nomenclature

c_i	concentration (kg/m^3)
c_p	specific heat
D	nozzle diameter
F	flatness factor
F	characteristic fine structure flatness factor
ΔH_R	reaction enthalpy difference
k	turbulence kinetic energy
L^*	characteristic length scale of fine structures
L'	characteristic turbulence length scale, mixing length
L'', L^n	characteristic turbulence length at different structure level
$\dot{m}$	exchange rate of mass with fine structures
Re_λ	turbulence Reynolds number
r	stoichiometric oxygen requirement to burn 1 kg fuel
T	temperature (K)
ΔT	excess temperature of reacting fine structures
U_B	turbulent flame velocity
$\bar{U}$	mean flow velocity
u^*	characteristic velocity of fine structures
u'	turbulence velocity
u'', u^n	characteristic turbulence velocity at different structure level
x	axial coordinate
Y	fractional conversion parameter
y	lateral coordinate
ϱ	density
ε	rate of dissipation of turbulence kinetic energy
μ_t	effective turbulent viscosity
ν	kinematic viscosity
γ	intermittency factor
γ^*	mass fraction occupied by fine structures
γ_λ	mass fraction occupied by fine structures regions
τ^*	time scale for the fine structure
τ_{ch}	chemical time scale
τ_λ	time scale for the fine structure regions
τ_M	bulk mixing timescale

Introduction

Local extinction in a practical turbulent combustion situation can influence important combustion characteristics like stability, noise generation, flame propagation, fuel consumption and heat transfer. These characteristics are of main importance for the performance and operation of combustion engines, combustion chambers, flare burners and fire fighting equipment.

From a fundamental point of view the extinction characteristics of chemical reactions in turbulent flow can give some new information about the geometry and the dynamics of the small scale turbulent structure and the interaction between the structure and chemical reactions which is some of the main problems in turbulence. The classical way to treat extinction phenomena is to apply Karlovitz-Kovasnays criterium recently used in a turbulent combustion situation by Chomiak (1982) [2]. Turbulent structure models have been developed and used by Lockwood and Megahed (1978) [3] and Tabaczynski (1981) [4]. Common for these models is that the chemical kinetical properties of the fluid is expressed through the laminar flame speed.

Recently Peters (1982) [5] treated the extinction process in turbulent flow by considering the quenching of diffusive flamelets where both the intermittant behaviour of turbulence and chemical kinetics were taken into account.

Review of EDC

Chemical reactions take place when reactants are mixed at molecular scale at sufficiently high temperature [1]. In turbulent flow the reactant consumption is strongly dependent on the molecular mixing. It is known that the microscale processes which are decisive for the molecular mixing as well as dissipation of turbulence energy into heat are severely intermittent i.e. concentrated in isolated regions whose entire volume is a small fraction of the volume of the fluid.

These regions are occupied by fine structures whose dimensions are small in one or two directions, however not in the third. These fine structures are believed to be vortex tubes, sheets or slabs whose characteristic dimensions are of the same magnitude as the Kolmogorov microscale [6–10].

The fine structures are responsible for the dissipation of turbulence into heat. Within these structures one can therefore assume that reactants will be mixed at molecular scale. These structures thus create the reaction space for non-uniformly distributed reactants.

In a modelling context one can assume that the reactants are homogeneously mixed within the fine structures. Thus, in order to be able to treat the reactions within this space, it is necessary to know the reaction volume and the mass transfer rate between the structures and the surrounding fluid.

The following describes a concept for treating chemical reactions in turbulent flow which include basic features of the preceeding.

Turbulence energy dissipation

In turbulent flow energy from the mean flow is transferred through the bigger eddies to the fine structures where mechanical energy is dissipated into heat. This process is schematically described in Fig. 1.

In general, high Reynolds number turbulent flow will consist of a spectrum of eddies of different sizes. Mechanical energy is mainly transferred between neighbouring eddy structures as indicated in Fig. 1. For the same reason the main production of turbulence kinetic energy will be performed by the interactions between bigger eddies and the mean flow.

The dissipation of kinetic energy into heat, which is due to work done by molecular forces on the turbulence eddies, on the other hand mainly takes place in the smallest eddies.

Important turbulent flow characteristics can for nearly isotropic turbulence be related to a turbulence velocity, u', and a turbulent length, L'. These quantities are linked to each other

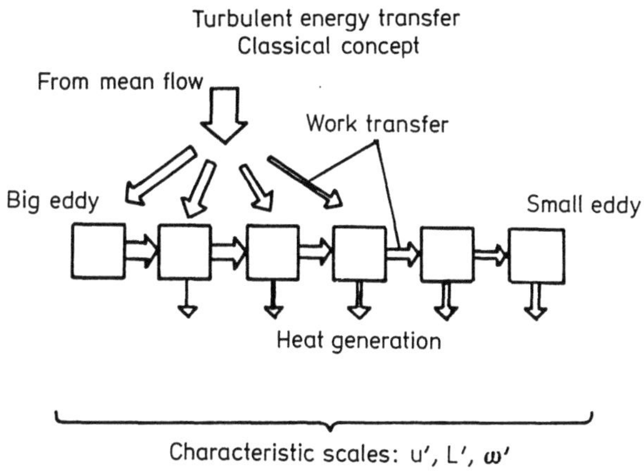

Fig. 1. Turbulent energy transfer

through the turbulent eddy velocity:

$$v_t = u' \cdot L' . \tag{1}$$

Modelling interstructural energy transfer

Figure 2 schematically illustrates a model for the transfer of mechanical energy from bigger to smaller turbulent structures [11].

The first structure level represents the whole spectrum of turbulence which in an ordinary way is characterized by a turbulence velocity, u', a length scale, L', and vorticity, or characteristic strain rate

$$\omega' = u'/L' . \tag{2}$$

The rate of dissipation can for this level be expressed by

$$\varepsilon' = \zeta^2 \left(12 \frac{u'}{L'} \cdot u''^2 + 15 \cdot v \left(\frac{u'}{L'} \right)^2 \right) \tag{3}$$

where ζ is a numerical constant.

The next structure level represents part of the turbulence spectrum characterized by a vorticity

$$\omega'' = 2\,\omega' \tag{4}$$

velocity, u'', and length scale, L''. The transfer of energy from the first level to the second level is expressed by

$$w' = \zeta^2\, 12 \frac{u'}{L'} u''^2 . \tag{5}$$

Similarly the transfer of energy from the second to the third level where

$$\omega''' = \omega'' \tag{6}$$

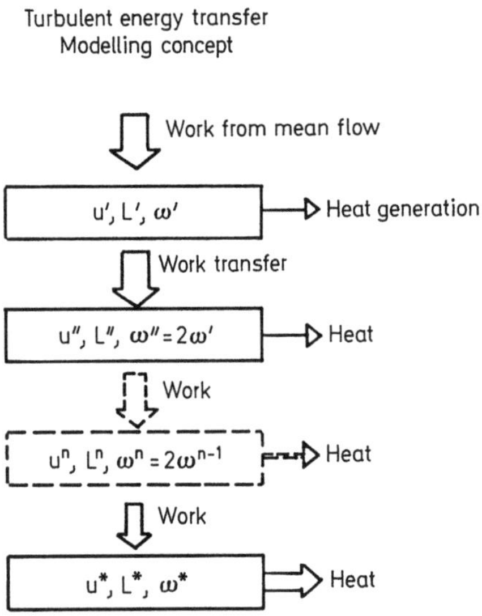

Fig. 2. A modelling concept for transfer of energy from bigger to smaller turbulent structures

is expressed

$$w'' = \zeta^2\, 12 \frac{u''}{L''} \cdot u'''^2 . \tag{7}$$

The part which is directly dissipated into heat is expressed

$$q'' = \zeta^2 \cdot 15\, \nu \left(\frac{u''}{L''}\right)^2 . \tag{8}$$

The turbulence energy ballance for the second structure level is consequently given by

$$\zeta^2\, 12 \frac{u''}{L''} u''^2 = \zeta^2 \left(12 \frac{u''}{L''} \cdot u'''^2 + 15 \cdot \nu \left(\frac{u''}{L''}\right)^2\right). \tag{9}$$

This sequence of turbulence structure levels can be continued down to a level where all the produced turbulence kinetic energy is dissipated into heat. This is the fine structure level characterized by, u^*, L^*, and ω^*.

The turbulence energy transferred to the fine structure is expressed by

$$w^* = \zeta^2 \cdot 6 \frac{u^*}{L^*} \cdot u^{*2} \tag{10}$$

and the dissipation by

$$q^* = \zeta^2 \cdot 15\, \nu \left(\frac{u^*}{L^*}\right)^2 . \tag{11}$$

According to this model nearly no dissipation of energy into heat takes place at the highest structure level. Similarly it can be shown that 3/4 of the dissipation takes place at the fine structure level.

Taking this into account and by introducing $\zeta = 0.18$ the following three equations are obtained for the dissipation of turbulence kinetic energy for nearly isotropic turbulence:

$$\varepsilon = 0.2 \frac{u'^3}{L'} \tag{12}$$

$$\varepsilon = 0.267 \frac{u^{*3}}{L^*} \tag{13}$$

$$\varepsilon = 0.67\, \nu \left(\frac{u^*}{L^*}\right)^2 . \tag{14}$$

Introducing the Taylor microscale a fourth equation is obtained

$$\varepsilon = 15\, \nu \left(\frac{u'}{\lambda}\right)^2 . \tag{15}$$

By combination of Eqs. (13) and (14) the following characteristics (scales) for the fine structures are obtained

$$u^* = 1.74\, (\varepsilon \cdot \nu)^{1/4} \tag{16}$$

and

$$L^* = 1.43\, \nu^{3/4}/\varepsilon^{1/4} \tag{17}$$

where u^* is the mass average fine structure velocity. The scales are closely related to the Kolmogorov scales.

The fine structures

The tendency towards strong dissipation intermittency in high Reynolds number turbulence was discovered by Batchelor and Townsend [11], and then studied from two points of view; different statistical models for the cascade of energy starting from a hypothesis of local invariance or selfsimilarities between motions of different scales, and then by consideration of hydrodynamic vorticity production due to stretching of vortex lines.

It can be concluded that the smallscale structures who are responsible for the main part of the dissipation are generated in a very localized fashion. It is assumed that these structures

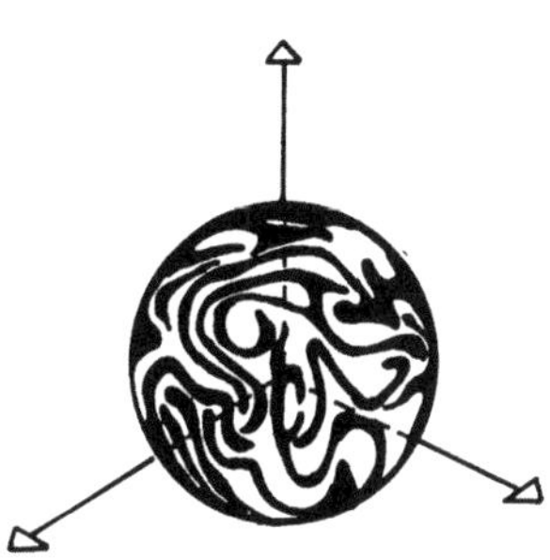

Fig. 3. Schematic illustration of fine structures developed on a constant energy surface

consist typically of large thin vortex sheets, ribbons of vorticity of vortex tubes of random extension folded or tangeled throughout the flow (Fig. 3).

The fine structures are localized in certain fine structure regions whose linear dimensions are considerably larger than the fine structures therein [10]. These regions appear in the highly stained regions between the bigger eddies.

Modelling characteristics of the fine structures

It is assumed that the mass fraction occupied by the fine structures, on the basis of consideration of the energy transfer to these structures (Eqs. 12 and 13) can be expressed by

$$\gamma^* = \left(\frac{u^*}{u'}\right)^3 . \tag{18}$$

If it is assumed that the fine structures are localized in nearly constant energy regions, then the mass fraction occupied by the fine structure regions can be expressed by

$$\gamma^* = \gamma_\lambda \cdot \left(\frac{u^*}{u'}\right)^2 \tag{19}$$

giving the following expression

$$\gamma_\lambda = \frac{u^*}{u'} . \tag{20}$$

Assuming nearly isotropic turbulence and introducing the turbulence kinetic energy and its rate of dissipation the following expressions are obtained:

$$\gamma^* = 9.7 \cdot \left(\frac{\nu \cdot \varepsilon}{k^2}\right)^{1/3} \tag{21}$$

and

$$\gamma_\lambda = 2.13 \cdot \left(\frac{\nu \cdot \varepsilon}{k^2}\right)^{1/4} . \tag{22}$$

Similarly by introducing the turbulence Reynolds number

$$\gamma^* = 40.2 \cdot \mathrm{Re}_\lambda^{-3/2} \tag{23}$$

and

$$\gamma_\lambda = 3.42 \cdot \mathrm{Re}_\lambda^{-1/2} . \tag{24}$$

Kuo and Corrsin [8] have given some results for the flatness factor of $\partial u/\partial t$ as a function of Re_λ (Fig. 5). In order to compare the above results with these results an empirical expression has been developed for the relationship between the flatness factor and the intermittency factor:

$$F_M = 1.5 \cdot (1 + 1/\gamma) . \tag{25}$$

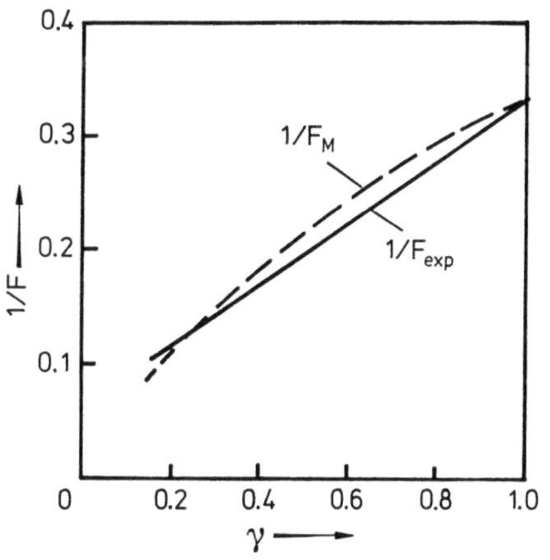

Fig. 4. Comparison between experimental results and empirical expression for the flatness factor as a function of the intermittency factor

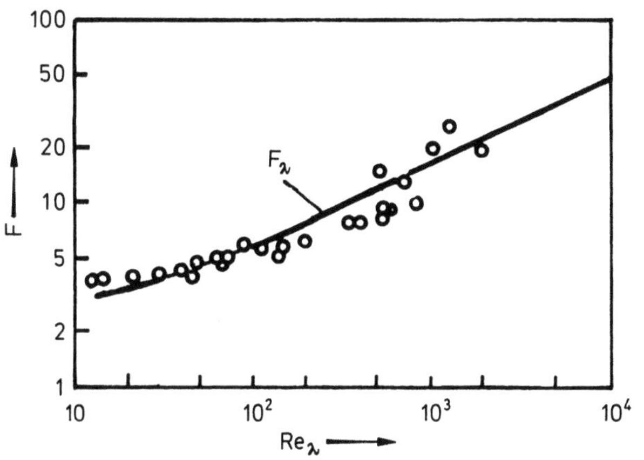

Fig. 5. Flatness factor of $\partial u/\partial t$ as a function of Re_λ compared with theoretical flatness factor for the fine structures.

Figure 4 shows a comparison between some experimental results [10] and the given empirical expression.

When Eq. (25) is applied the following expression for the flatness factor for the fine structures is obtained.

$$F_\lambda = 1.5 + 0.44 \cdot \mathrm{Re}_\lambda^{0.5}. \tag{26}$$

Equation (26) is compared with the experimental results of Kuo and Corrsin [9] in Fig. 5.

From this comparison it can be concluded that the overall features of the physics seem to have been taken care of by suggested expressions.

On the basis of simple geometrical considerations the transfer of mass per unit of fluid and unit of time between the fine structures and the surrounding can be expressed as follows

$$\dot{m} = 2 \cdot \frac{u^*}{L^*} \cdot \gamma^* \quad (1/s). \tag{27}$$

Expressed by k and ε for nearly isotropic turbulence eq. (27) turns into

$$\dot{m} = 23.6 \cdot \left(\frac{\nu \cdot \varepsilon}{k^2}\right)^{1/4} \cdot \frac{\varepsilon}{k} \quad (1/s) \tag{28}$$

Molecular mixing and reaction processes

The rate of molecular mixing is determined by the rate of mixing between the fine structures and the surrounding fluid.

The mean mass transfer rate between a certain fraction, χ, of the fine structures and the rest of the fluid, R_i, can for a certain species, i, be expressed as follows:

$$R_i = \dot{m} \cdot \chi \cdot \left(\frac{c_i^0}{\varrho^0} - \frac{c_i^*}{\varrho^*}\right) \tag{29}$$

where * and ° refer to conditions in the fine structures and the surrounding.

The mass transfer rate can also be expressed per unit volume in the fine structure fraction, χ, as

$$R_i^* = \frac{\dot{m} \cdot \varrho^*}{\gamma^*} \cdot \left(\frac{c_i^0}{\varrho^0} - \frac{c_i^*}{\varrho^*}\right). \tag{30}$$

Finally, the concentration of a species, i, in the fraction, χ, of the fine structures and in the surrounding is related to the mean concentration by:

$$\frac{\bar{c}_i}{\bar{\varrho}} = \frac{c_i^*}{\varrho^*} \cdot \gamma^* \cdot \chi + \frac{c_i^0}{\varrho^0} \cdot (1 - \gamma^* \cdot \chi). \tag{31}$$

It is now possible to put up balance equations for reacting fine structures and the surrounding fluid including chemical kinetic rate expressions.

Combustion rates

If the rate of reaction between fuel and oxygen is considered infinitely fast, the rate of reaction will be limited by the mass transfer between the bulk and the fine structures. In this case the concentration of fuel or oxygen will be very small within the reacting fine structures. If the reaction took place in all the fine structures, the rate of combustion would be expressed by:

$$R_{fu} = \dot{m} \cdot \frac{\bar{c}_{\min}}{1 - \gamma^*} \tag{32}$$

where $\bar{c}_{\min}$ is the smallest of $\bar{c}_{fu}$ and $\bar{c}_{o_2}/r_{fu}$, where $\bar{c}_{fu}$ and $\bar{c}_{o_2}$ are local mean concentrations of fuel and oxygen, and r_{fu} the stoichiometric oxygen requirement.

Not all the fine structures will be sufficiently heated to react. This is obviously the case in combustion of premixed gases where both fuel and oxygen are present in the fine structures.

The fraction of the fine structures which reacts can be assumed proportional to the ratio between the local concentration of reacted fuel and the total fuel concentration:

$$\chi = \frac{\bar{c}_{pr}/(1 + r_{fu}) \cdot \gamma_\lambda}{\bar{c}_{pr}/(2 + r_{fu}) + \bar{c}_{fu}} \tag{33}$$

where $\bar{c}_{pr}$ is the local mean concentration of reaction products. Equation (33) implies the assumption that the reaction products are kept within the fine structure region until a concentration is reached which yields χ equal to unity.

By combination of Eqs. (32) and (33) the following general equation is obtained for the rate of combustion at infinite reaction rate between fuel and oxygen:

$$R_{fu} = \dot{m} \cdot \frac{\chi}{1 - \gamma^* \chi} \cdot \bar{c}_{\min} \,. \tag{24}$$

The reacting fine structures under these conditions will have a temperature, ΔT, in excess of the local mean temperature:

$$\Delta T = \frac{\Delta H_R \cdot \bar{c}_{\min}}{\bar{\varrho} \cdot c_p} \tag{35}$$

where ΔH_R is the heat of reaction and c_p the local specific heat capacity. The temperature, T^*, of the reacting fine structures is consequently:

$$T^* = \bar{T} + \Delta T \tag{36}$$

and the surrounding temperature:

$$T^* = \bar{T} - \Delta T \cdot \frac{\gamma^* \cdot \chi}{1 - \gamma^* \cdot \chi} \tag{37}$$

where $\bar{T}$ is the local time mean temperature.

Chemically controlled reaction rate

On the basis of the previous there can be defined characteristic mixing time scales:
The bulk mixing time scale

$$\tau_M = 1/\dot{m} \,. \tag{38}$$

The fine structure time scales

$$\tau^* = \gamma^*/\dot{m} \tag{39}$$

$$\tau_\lambda = \gamma_\lambda/\dot{m} \,. \tag{40}$$

These time scales can be compared with chemical kinetic time scales in order to establish whether the reaction is mixing or chemically controlled, and even establish criteria for extinction of flames.

Modelling Flame Extinction and Turbulent Flame Propagation According to the EDC

Extinction

According to the previous concept heat and mass balance including chemical kinetics can easily be set up for the fine structures and the surrounding fluid. Consequently, knowing the chemical kinetics, a criterium for the extinction of the fine structures can be obtained. When a typical timescale for the combustion reaction is small compared to τ^* the fuel consumption is independent of chemical kinetics and is a purely hydrodynamic problem. However when the residence time becomes smaller than a typical chemical time scale the reaction will not be completed in the fine structure and the fuel consumption is strongly influenced by chemical kinetics. As the residence time decreases extinction will finally occur.

The chemical time scale, however, is not a uniquely defined quantity. Many different expressions can be found in the literature describing the various steps in the reaction process. In addition it is very difficult to find relevant data for different fuels.

When studying turbulent extinction a common substitute for the lacking chemical kinetic data has been time scales based on the laminar flame propagation behaviour of the mixture [3, 4]. These time scales are, however, more dependent on the diffusive properties of the fluid than the chemical kinetics. Another time scale that has been used is the chemical induction time or ignition delay time. This time scale is usually measured in shock tube experiments where the reaction starts from the very low radical concentration level of the unheated gases. This is not the case in flames and other reacting systems where there exists a source of free radicals somewhere in the system. Here the induction time is much shorter. In a typical turbulent combustion situation it is believed that radicals are already present in the regions where the fine structures are located. When it is further assumed that the internal mixing inside the fine structures are fast, then the fine structures can be considered as a well stirred reactor.

On the basis of the previous the fractional conversion parameter for the reactor is defined as (Fig. 6)

$$Y = \frac{c_{fu}^0 - c_{fu}^* \dfrac{\varrho^0}{\varrho^*}}{c_{fu}^0} . \tag{41}$$

For a one step irreversible reaction the mass and heat balance for the fine structure can be expressed as

$$Y = \frac{R_{fu}^* \cdot \tau^*}{c_{fu}^0} \frac{\varrho^0}{\varrho^*} \tag{42}$$

$$Y = \frac{c_p (T^* - T^0)\, \varrho^0}{\Delta H_R \, c_{fu}^0} \tag{43}$$

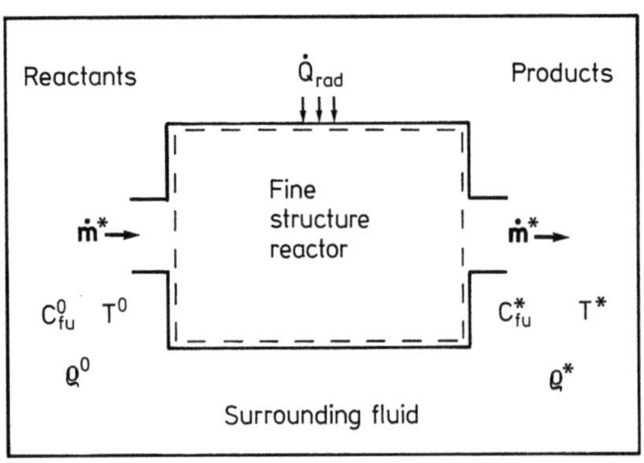

Fig. 6. Schematic illustration of a reacting fine structure

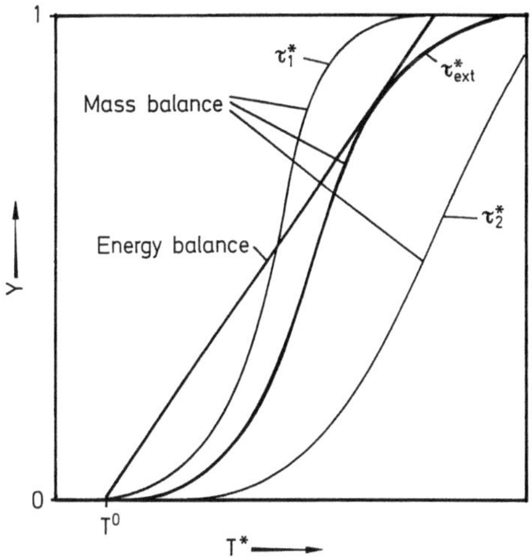

Fig. 7. Schematic illustration of Extinction in the fine structure $\tau_1^* > \tau_{ext}^* > \tau_2^*$

where the fine structure is considered to be adiabatic and the specific heat of the fluid is assumed to be constant. The extinction time scale can then be found from Eqs. (42) and (43) according to Fig. 7, where the extinction time scale τ_{ext}^* is indicated.

Turbulent flame propagation

The propagation of a plane flame in turbulent flow can be expressed by

$$U_B \frac{\partial c_{fu}}{\partial x} = \frac{v_T}{\sigma} \frac{\partial^2 c_{fu}}{\partial \chi^2} + R_{fu} \tag{44}$$

where U_B is the turbulent flame propagation velocity, v_T is the turbulent viscosity and σ is the turbulent Schmidt number. With R_{fu} expressed according to Eq. (34), the only physical meaningful solution to Eq. (44) reads

$$U_B = 2\left(\frac{\dot{m}\, v_t}{\gamma_\lambda \sigma}\right)^{1/2}. \tag{45}$$

By introducing the turbulent viscosity from the $k - \varepsilon$-model this transforms into

$$U_B = 2\sqrt{\frac{k}{\sigma}}. \tag{46}$$

In a typical shear flow situation $k \cong u'^2$ and $\sigma \cong 1$. This gives a turbulent flame propagation velocity

$$U_B \cong 2u'. \tag{47}$$

This is generally in relatively close agreement with the experimental data collected by Bradley et al. (1981) [14]. Some high Reynolds number data is shown in Fig. 8

When the fine-structure time-scale τ^* is reduced to the same magnitude as the chemical time scale, the turbulent flame speed is reduced due to local extinction in the fine structures.

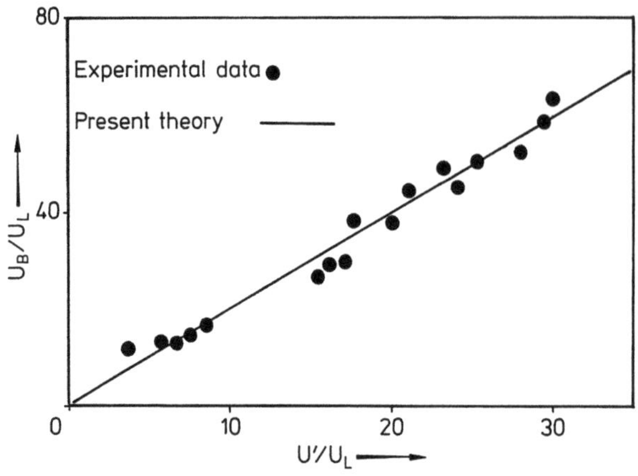

Fig. 8. Comparison between experimental and predicted turbulent flame propagation velocity

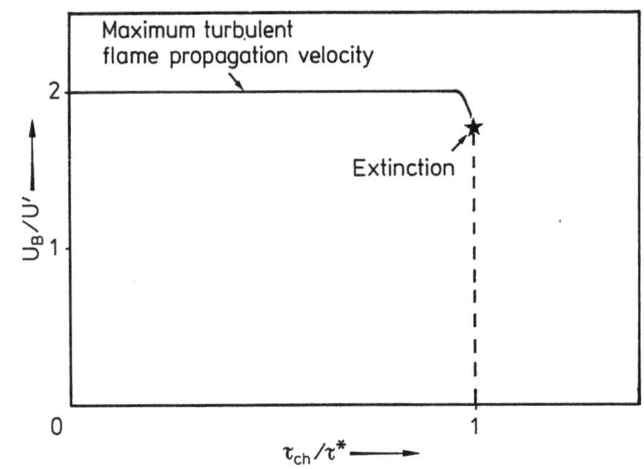

Fig. 9. Schematic illustration of the turbulent flame propagation velocity variation

At a critical value of the fine-structure time-scale the flame will be quenched by the turbulence. This leads to the following general qualitative relation for the turbulent flame speed:

$$\frac{U_B}{u'} = f\,(\tau_{ch}^*/\tau^*). \tag{48}$$

Due to the strong interaction between T^* and τ_{ch}^*/τ^*, the flame quenching appears like a catastrophy as indicated in Fig. 9.

The abrupt extinction has been shown experimentally by Chomiak et al. (1982) [3]. In these experiments extinction for a methane mixture occured at a fine structure time scale of approximately $\tau^* \cong 10^{-4}$ s. This is in close agreement with the chemical time scale deduced from the global reaction rate used by Bradley et. al. (1976) [15, 16] and is also in reasonable agreement with the multistep kinetic calculations for methane-air mixture reported by Martenay (1970) [17].

Lift-off and blow-off of turbulent diffusion flames

According to the extinction model and the result for the turbulent flame speed a lifted diffusion flame will be stabilized near the position on the stoichiometric contour where the fine structure time scale equals the chemical time scale. The reason why the stabilization

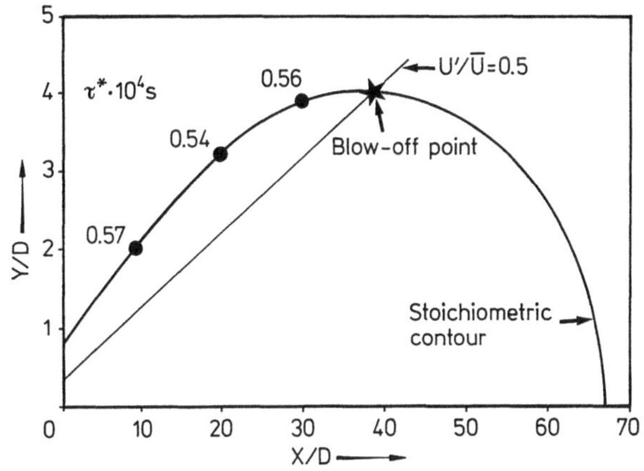

Fig. 10. Localisation of lift-off and blow-off points for a round free jet. Experimental data from Horch (18) for $D = 0.01$ m

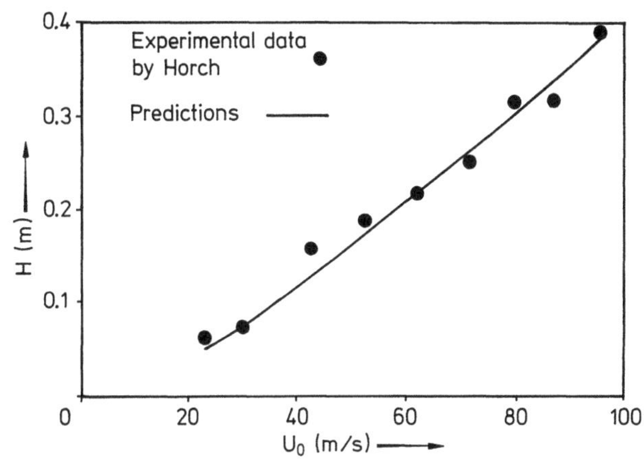

Fig. 11. Comparison between experimental and predicted lift-off heights

point is located at or near the stoichiometric contour is that the reaction rate has its maximum near this contour. This is shown in Fig. 10 where also some calculated fine structure time scales at the stabilization points are indicated. By gradually increasing the exit velocity the stabilization point moves further downstreams. This continues until the stabilization point is at a position where the mean flow velocity is nearly equal to the maximum flame propagation velocity, then the flame is blown-off. This is shown in Fig. 10 where also some calculated fine structure timescales at the stabilization points are indicated.

The extinction model has been tested by performing numerical calculations of the flow based on the $k - \varepsilon$-model of turbulence and then extracting the quenching timescales by comparison with the experimental data of Horch (1978) [18].

Figure 11 shows a comparison between experimental and predicted lift-off heights. The predictions have been based on a constant fine structure timescale, $\tau^* = 0.56 \cdot 10^{-4}$ s.

For a constant nozzle diameter the fine structure timescale at the stabilization points was nearly constant as shown both in Figs. 10 and 11, indicating that the fine structures are of main importance in the extinction process. The time scale did, however, decrease with increasing diameter, as indicated in Table 1.

Table 1. Mean structure time scale at the stabilisation point for different nozzle diameters

D (mm)	4	6	8	10
$\tau^* \cdot 10^4$ (s)	1.18	0.80	0.67	0.56

We assume that this diameter effect is due to radiative heat transfer between the fine structures at the flame stabilization point and its surroundings. This effect will tend to increase the fine structure temperatures and consequently reduce the extinction time scales for increased nozzle diameter. As such this effect can readily be handled by the described concept. At the blow-off point the ratio of a typical turbulence velocity to the mean flow velocity was nearly equal to $u'/\bar{U} \cong 0.5$. This is shown in Fig. 10 and Table 2.

Table 2. Turbulence intensity at blow-off point for different nozzle diameters

D (mm)	4	6	8	10
$u'/\bar{U}$	0.51	0.47	0.49	0.48

According to the previous, the mean velocity is nearly equal to the maximum turbulent flame propagation velocity at the blow-off point.

Conclusions

The previous have shown that extinction processes in turbulent combustion can be treated by the eddy dissipation concept (EDC) of Magnussen. Predictions deduced from the proposed extinction model compares well with experimental data.

The flame stabilization point for lifted diffusion flames is located at the stoichiometric contour and at positions where the fine structure mixing time scale τ^* is approximately equal to the chemical time scale τ^* is approximately equal to the chemical time scale τ^*_{ch}.

An expression for the turbulent flame propagation velocity is deduced from the EDC. This expression gives results which are in close agreement with experimental data. It is shown that blow-off for turbulent diffusion flames occurs at a point where the maximum turbulent flame propagation velocity is equal to the mean flow velocity.

Acknowledgements. This work has been financially supported by STATOIL. The authors would like to thank Mrs. I. Eidissen and Miss P. Magnussen for assistance in typing the paper.

References

1. Magnussen, B. F. (1981): "On the Structure of Turbulence and a Generalized Eddy Dissipation Concept for Chemical Reactions in Turbulent Flow", 19th AIAA Sc. Meeting, St. Louis, USA
2. Chomiak, J., Jarosinski, J. (1982): Flame quenching by turbulence. Comb. Flame **48,** 241
3. Lockwood, F. C., Megahed, I. E. A. (1978): "Extinction in turbulent Reacting Flows", Comb. Sci. Tech. **19,** 77
4. Tabaczynski, R. J. (1982): Premixed turbulent flame blowoff velocity correlations based on coherent structures in turbulent flows. Comb. Flame **42,** 19
5. Peters, N. (1982): Local quenching due to flame stretch and non-premixed turbulent combustion. Comb. Sci. Tech. **30,** 1
6. Kolmogorov, A. N. (1962): A refinement of previous hypotheses concerning the local structure of turbulence in a viscous incompressible fluid at high Reynolds number. J. Fluid Mech. **13,** 82
7. Corrsin, S. (1962): Turbulent dissipation correlations. Phys. Fluids **5,** 1301
8. Tennekes, H. (1968): Simple model for small-scale structure of turbulence. Phys. Fluids **11,** 3
9. Kuo, A. Y., Corrsin, S. (1971): Experiments on internal intermittancy and fine-structure distribution function in fully turbulent fluid. J. Fluid Mech. **50,** 285
10. Kuo, A. Y., Corrsin, S. (1972): Experiments on the geometry of the fine-structure regions in fully turbulent fluid. J. Fluid Mech. **56,** 477
11. Magnussen, B. F. (1975): Some features of the structure of a mathematical model of turbulence. Report, NTH

12. Batchelor, G. K., Townsend, A. A. (1949): The nature of turbulent motion at large wave-numbers. Proc. Roy. Soc. **A 199,** 238
13. Wygnanski, I., Fiedler, H. J. (1969): Some measurements in the self-preserving jet. J. Fluid Mech. **38,** 577
14. Bradley, D., Abdel-Gayed, R. G. (1981): A two eddy theory of premixed turbulent flame propagation. Proc. Roy. Soc. **A 1457,** 1
15. Bradley, D., Chin, S. B., Draper, M. S., Hankinson, G. (1976): *16 Int. Symp. on Combustion*, 1571
16. Byggstøyl, S., Magnussen, B. F. (1982): "Flame Extinction in Turbulent Structure", Task Leaders Conf. IEA, Italy
17. Martenay, P. (1970): Analytical study of the formation of nitrogen oxides in hydrogen air combustion. Comb. Sci. Tech. **1,** 461
18. Horch, K. (1978): "Zur Stabilität von Freistrahl-Diffusionsflammen", Ph.D. Thesis, Universität Karlsruhe, FRG

Index of Contributors

Springer Series in Computational Physics

Editors: **H. Cabannes, M. Holt, H. B. Keller, J. Killeen, S. A. Orszag**

R. Peyret, T. D. Taylor

Computational Methods for Fluid Flow

1983. 125 figures. X, 358 pages
ISBN 3-540-11147-9

Contents: Numerical Approaches: Introduction and General Equations. Finite-Difference Methods. Integral and Spectral Methods. Relationship Between Numerical Approaches. Specialized Methods. – Incompressible Flows: Finite-Difference Solutions of the Navier-Stokes Equations. Finite-Element Methods Applied to Incompressible Flows. Spectral Method Solutions for Incompressible Flows. Turbulent-Flow Models and Calculations. – Compressible Flows: Inviscid Compressible Flows. Viscous Compressible Flows. – Concluding Remarks. – Appendix A: Stability. – Appendix B: Multiple-Grid Method. – Appendix C: Conjugate-Gradient Method. – Index.

Y. I. Shokin

The Method of Differential Approximation

Translated from the Russian by K. G. Roesner
1983. 75 figures, 12 tables. XIII, 296 pages
ISBN 3-540-12225-7

Contents: Stability Analysis of Difference Schemes by the Method of Differential Approximation. – Investigation of the Artificial Viscosity of Difference Schemes. – Invariant Difference Schemes. – Appendix. – References. – Subject Index.

D. P. Telionis

Unsteady Viscous Flows

1981. 132 figures. XXIII, 408 pages
ISBN 3-540-10481-X

Contents: Introduction. – Basic Concepts. – Numerical Analysis. – Impulsive Motion. – Oscillations with Zero Mean. – Oscillating Flows with Non-Vanishing Mean. – Unsteady Turbulent Flows. – Unsteady Separation. – Index.

F. Thomasset

Implementation of Finite Element Methods for Navier-Stokes Equations

1981. 86 figures. VII, 161 pages
ISBN 3-540-10771-1

Contents: Introduction. – Notations. – Elliptic Equations of Order 2: Some Standard Finite Element Methods. – Upwind Finite Element Schemes. – Numerical Solution of Stokes Equations. – Navier-Stokes Equations: Accuracy Assessments and Numerical Results. – Computational Problems and Bookkeeping. – Appendix 1: The Patch Test of the *P*1 Nonconforming Triangle: Sketchy Proof of Convergence. – Appendix 2: Numerical Illustration. – Appendix 3: The Zero Divergence Basis for 2-D *P*1 Nonconforming Elements. – References. – Index.

Finite-Difference Techniques for Vectorized Fluid Dynamics Calculations

Editor: **D. L. Book**
1981. 60 figures. VIII, 226 pages
ISBN 3-540-10482-8

Contents: Introduction. – *D. L. Book, J. P. Boris:* Computational Techniques for Solution of Convective Equations. – *D. L. Book, J. P. Boris, S. T. Zalesak:* Flux-Corrected Transport. – *R. V. Madala:* Efficient Time Integration Schemes for Atmosphere and Ocean Models. – *J. P. Boris:* A One-Dimensional Lagrangian Code for Nearly Incompressible Flow. – *M. J. Fritts:* Two-Dimensional Lagrangian Fluid Dynamics Using Triangular Grids. – *R. V. Madala, B. E. McDonald:* Solution of Elliptic Equations. – *N. K. Winsor:* Vectorization of Fluid Codes. – Appendices A–E. – References. – Index.

F. Bauer, O. Betancourt, P. Garabedian

A Computational Method in Plasma Physics

1978. 22 figures. VIII, 144 pages
ISBN 3-540-08833-4

Contents: Introduction. – The Variational Principle. – The Discrete Equations. – Description of the Computer Code. – Applications. – References. – Listing of the Code with Comment Cards Index.

M. Holt

Numerical Methods in Fluid Dynamics

2nd revised edition. 1984. 114 figures. XI, 273 pages
ISBN 3-540-12799-2

Contents: General Introduction. – The Godunov Schemes. – The BVLR Method. – The Method of Characteristics for Three-Dimensional Problems in Gas Dynamics. – The Method of Integral Relations. – Telenin's Method and the Method of Lines. – Subject Index.

Springer-Verlag
Berlin
Heidelberg
New York
Tokyo

MIX
Papier aus verantwortungsvollen Quellen
Paper from responsible sources
FSC® C105338

If you have any concerns about our products,
you can contact us on
ProductSafety@springernature.com

In case Publisher is established outside the EU,
the EU authorized representative is:
Springer Nature Customer Service Center GmbH
Europaplatz 3, 69115 Heidelberg, Germany

Printed by Libri Plureos GmbH
in Hamburg, Germany